Historical Studies
in the
Physical Sciences
7

Notice to Contributors

Historical Studies in the Physical Sciences, an annual publication issued by Princeton University Press, is devoted to articles on the history of the physical sciences from the eighteenth century to the present. The modern period has been selected since it holds especially challenging and timely problems, problems that so far have been little explored. An effort is made to bring together articles that expose new directions and methods of research in the history of the modern physical sciences. Consideration is given to the professional communities of physical scientists, to the internal developments and interrelationships of the physical sciences, to the relations of the physical to the biological and social sciences, and to the institutional settings and the cultural and social contexts of the physical sciences. Historiographic articles, essay reviews, and survey articles on the current state of scholarship are welcome in addition to the more customary types of articles.

All manuscripts should be accompanied by an additional carbon- or photocopy. Manuscripts should be typewritten and double-spaced on 8½" X 11" bond paper; wide margins should be allowed. No limit has been set on the length of manuscripts. Articles may include illustrations; these may be either glossy prints or directly reproducible line drawings. Articles may be submitted in foreign languages; if accepted, they will be published in English translation. Footnotes are to be double-spaced, numbered sequentially, and collected at the end of the manuscript. Contributors are referred to the *MLA Style Sheet* for detailed instructions on documentation and other stylistic matters. (*Historical Studies* departs from the MLA rules in setting book and journal volume numbers in italicized Arabic rather than Roman numerals.) All correspondence concerning editorial matters should be addressed to Russell McCormmach, Department of History of Science, Johns Hopkins University, Baltimore, Md. 21218.

Fifty free reprints accompany each article.

Historical Studies in the Physical Sciences incorporates *Chymia,* the history of chemistry annual.

Albert Einstein *circa* mid-1920s. Courtesy of the Estate of Albert Einstein

Historical Studies in the Physical Sciences

RUSSELL McCORMMACH, *Editor*

Seventh Annual Volume

PRINCETON UNIVERSITY PRESS
PRINCETON, NEW JERSEY

Published by Princeton University Press, Princeton and London

ISBN: 0-691-08169-7
LCC: 77-75220

This book has been composed in IBM Selectric Aldine Roman

Printed in the United States of America
by Princeton University Press, Princeton, New Jersey

Contents

Editor's Foreword

In the lead article in this seventh volume of *Historical Studies in the Physical Sciences,* Tetu Hirosige discusses the genesis of Einstein's theory of relativity against the background of the mechanized world picture. He again treats a subject—the development of physical thought around the turn of this century—to which he devoted a large share of his historical researches, tragically cut short by his death earlier this year. See pp. 488–489 of this volume for Sigeko Nisio's moving appreciation of Hirosige and his work.

Hirosige's article in this volume belongs to a growing number of major studies on recent physical thought; through them we are gaining a fuller understanding of the events that closed the period—the two centuries following Newton—during which, with major qualifications, the mechanized world picture provided physical scientists with their aim.[1]

Dijksterhuis in the epilogue to his authoritative *Mechanization of the World Picture* looks beyond his period to the passage from "classical" to "modern," or twentieth century, science. Critical of the view that the passage was one primarily from mechanization to mathematization, he observes that the concepts of classical mechanics were mathematical and that the world picture associated with quantum and relativistic mechanics is still in a sense mechanized.[2] His observation suggests that it might be fruitful to study jointly world pictures and mathematization in the post-Newtonian period and, especially, from the turn of this century when the concepts of classical mechanics and their application in other branches of physical science were severely challenged from different quarters.

We might expect that the challenge to the mechanical foundations of physics—to the assertion, in principle, of the unrestricted validity within mechanics and applicability throughout physics of the mathematical concepts of classical mechanics—would be accompanied by an explicit concern with the basic mathematical concepts in all areas of physics. In this editor's foreword I will explore certain preliminary

[1] E. J. Dijksterhuis, *The Mechanization of the World Picture,* trans. C. Dikshoorn (London, 1969), p. 495. For a discussion of some of the qualifications, see P. M. Heimann, "'Nature Is a Perpetual Worker': Newton's Ether and Eighteenth-Century Philosophy," *Ambix, 20* (1973), 1–25.

[2] *Ibid.*, p. 500.

problems connected with this expectation. I will only briefly point to the problem of the challenge to the mechanical world picture, since it has received considerable historical attention recently. It is the problem of the physicists' explicit concern with mathematics that I will focus on here. To motivate my discussion of this problem I will draw on Einstein and his Central European context for historical examples. In particular I will draw on Einstein for his explicit sensitivity to the disparate mathematical expressions of fundamental concepts of the different branches of physics.

Heinrich Hertz's treatise on mechanics appeared in 1894, the year before Einstein began his advanced study of physics; it opened with the famous observation that today every physicist agrees that the problem of physics is to completely reduce all natural phenomena to mechanical laws.[3] In that same year, in his inaugural lecture to the Berlin Academy, Max Planck described the problem in its fuller complexity. He reaffirmed the goal of a mechanically founded physics, but he pointed to the difficulties of practicing theoretical physics at a time when that goal was in question.[4] The questioning mood was reflected in the published lectures by the German-speaking masters of theoretical physics—Gustav Kirchhoff, Hermann von Helmholtz, and Ludwig Boltzmann—which Einstein is said to have read as a student.[5] Kirchhoff explained in his heat lectures that the reduction of all of physics to mechanics is a goal that is in the "fullest measure worth striving for," but that, e.g., if we view matter as having a continuous structure we cannot reduce the concept of temperature to the concepts of mechanics.[6] In his dynamical lectures Helmholtz indicated an extension of Hamilton's least action principle—which he regarded as having "universal validity"—from mechanics to other branches of physics, thermodynamics and electrodynamics, whose phenomena he did not think were necessarily

[3]Heinrich Hertz, *The Principles of Mechanics,* trans. D. E. Jones and J. T. Walley (New York, 1956). "Author's Preface."

[4]Max Planck, "Antrittsrede," *Sitzungsber. Berlin,* Pt. 2 (1894), pp. 641–644; reprinted in *Max Planck in seinen Akademie-Ansprachen* (Berlin, 1948), pp. 1–5.

[5]Philipp Frank, *Einstein. His Life and Times,* trans. G. Rosen (New York, 1947), p. 20.

[6]Gustav Kirchhoff, *Vorlesungen über mathematische Physik.* Vol. 4: *Theorie der Wärme,* ed. M. Planck (Leipzig, 1894), pp. 1–4, 51, 134–135.

reducible to the motions of ponderable masses.[7] In his mechanical lectures Boltzmann acknowledged the challenge of the energetic and phenomenological world pictures, but he allowed no more than the possibility, not the certainty, that one day another "world picture will drive out the mechanical."[8] These references point to the ferment in the foundations of physics at the time. I will turn now to my main subject: Einstein and the mathematical relations of physics.

Throughout his life Einstein held strong attitudes toward mathematics, and like so many of his attitudes these changed in marked ways over the years. Early in his career he had a low opinion of the value of the subtler parts of mathematics for the work of the theoretical physicist. His attitude was not owing to any mathematical difficulties of his own; indeed mathematics was a subject toward which he was especially drawn and in which he received early encouragement. Near the end of his life he still remembered the rapt wonder he felt at age twelve when a text on Euclidean geometry came into his hands.[9] He learned the calculus between twelve and sixteen.[10] In the Gymnasium he attended in Munich he was ahead of his classmates in mathematics and apparently in nothing else.[11] Quitting the Gymnasium before earning a diploma, he obtained a written statement from his mathematics teacher to the effect that his knowledge of mathematics was uncommon and qualified him for advanced study in the subject.[12]

Einstein went to the Zurich Polytechnic with the intention—one most likely encouraged by his family—of preparing for an engineering career. In the Polytechnic entrance examination, he scored ahead of the other candidates in mathematics, and he did well in physics, too, while failing in other subjects.[13] His mathematical performance

[7] Hermann von Helmholtz, *Vorlesungen über theoretische Physik.* Vol. 1, Pt. 2: *Vorlesungen über die Dynamik discreter Massenpunkte,* ed. O. Krigar-Menzel (Leipzig, 1898), pp. 368–369, 373.

[8] Ludwig Boltzmann, *Vorlesungen über die Principe der Mechanik,* Pt. 1 (Leipzig, 1897), pp. 4–5.

[9] Albert Einstein, "Autobiographical Notes," in *Albert Einstein: Philosopher-Scientist,* ed. P. A. Schilpp (Evanston, Ill., 1949), pp. 1–95, on pp. 9, 11.

[10] Carl Seelig, *Albert Einstein. Eine dokumentarische Biographie* (Zurich, 1954), p. 14.

[11] Frank, *op. cit.* (note 5), p. 16.

[12] *Ibid.*

[13] *Ibid.,* p. 18.

was so strong that the director of the Polytechnic urged him to attend a cantonal school, earn a diploma there, and come back. Einstein spent a preparatory year in Aarau, followed by four more at the Zurich Polytechnic, where he enrolled in the division for teachers of mathematics and physical science and not in an engineering division. Although in his final diploma examination at the Polytechnic he did well in physical subjects, he again scored highest in mathematics.[14]

In the latter half of the curriculum for teachers at the Polytechnic, students were expected to take seminars in advanced mathematics. Einstein apparently often stayed away. "No one," Einstein's biographer A. Reiser wrote, "could stir him to visit the mathematical seminars."[15] According to Hermann Minkowski, one of his mathematics teachers, Einstein "never bothered about mathematics at all," which made his later success in theoretical physics mystifying to Minkowski.[16]

Although Einstein eventually regretted that he had passed up the opportunity of gaining a fine mathematical education at the Zurich Polytechnic from teachers like Minkowski and Adolf Hurwitz,[17] for years after graduation he was evidently not conscious of a mathematical handicap. Shunning mathematical virtuosity that complicated rather than clarified physical ideas,[18] he told his students at the University of Zurich around 1910 that "in fact with mathematics one can prove anything" and that all that mattered was the content.[19] In 1910 he disagreed with a Swiss friend, who believed reasonably enough that the total length of several matches laid end to end is the sum of the lengths of the individual matches; Einstein gave as his reason: "*Car moi, je ne crois pas à la mathematique.*"[20]

[14]Seelig, *op. cit.* (note 10), p. 54.

[15]Quoted in Gerald Holton, "Mach, Einstein, and the Search for Reality," *Daedalus*, 97 (1967), 636–673, on 638.

[16]Seelig, *op. cit.* (note 10), p. 33.

It should be remarked that Zurich Polytechnic generally and Minkowski in particular did not attract many mathematics students. In Einstein's second year at the Polytechnic, Minkowski observed that there was only one student with more than three semesters' mathematics, that the colloquium was sustained chiefly by assistants, and that in each of his three classes he had only about eight students. (H. Minkowski to D. Hilbert, 23 November 1897, Göttingen University Library, Hilbert Nachlass 258/65.)

[17]Einstein, *op. cit.* (note 9), p. 15.

[18]Frank, *op. cit.* (note 5), p. 206.

[19]Seelig, *op. cit.* (note 10), p. 122.

[20]*Ibid.*, p. 127.

Einstein came to the Zurich Polytechnic knowledgeable in both mathematics and physical science. During his stay at the Polytechnic, he showed little interest in formally furthering his mathematical knowledge and much in furthering his physical. Immediately after leaving the Polytechnic he began publishing on fundamental physics and not on mathematics. It was evidently during his Zurich years that his career inclinations became decidedly physical.

Since we have Einstein's own testimony that he preferred to spend time doing experimental work rather than attending mathematics sessions, we might look at the Polytechnic physical institute and also at the technical orientation that its director, Heinrich Friedrich Weber, gave to it. Weber had several characteristics worth noting. For one he was an active, but not prolific, researcher who was still publishing at the time Einstein came to the Polytechnic. His concern with research was related to the strong research tradition at the Polytechnic and to his earlier promising academic apprenticeship in Gustav Wiedemann's physical laboratory in Karlsruhe and as Helmholtz' assistant in physics in Berlin. Unlike many academic physicists elsewhere, Weber did research that was useful from the engineers' standpoint. His career was linked to two leading representatives of German industrial physics, Ernst Abbe in applied optics and Werner von Siemens in applied electricity. The first, Abbe, was Weber's teacher, whom Weber admired as a "striking example" of the cooperation of physics and technology.[21] Siemens he approached as a fellow technologist. In 1878 Weber began working on problems of alternate current technology, publishing in 1884 on Siemens' mercury unit. He attended congresses on electric units from 1883 on[22] and, according to his Polytechnic successor Pierre Weiss, was one of the group that included Weber's acquaintance William Thomson who mediated between the measuring techniques of the physical laboratories and the needs of electrotechnology.[23] Weber wanted a physical institute for the Polytechnic that was equipped not only for basic physical research but as well for the coming demands of electrotechnology. His plan for the institute met opposition, and he only had his way with the help of Siemens during a visit to Zurich.[24] Weber's institute was approved in 1886, built between

[21]Pierre Weiss, "Prof. Dr. Heinrich Friedr. Weber, 1843–1912," *Schweizerische Naturf. Ges. Verh.*, *95* (1912), 44–53, on 44.

[22]*Ibid.*, p. 49.

[23]*Ibid.*, pp. 48–49.

[24]*Ibid.*, p. 47.

1887 and 1890, and outfitted for physical and electrotechnical research. For some years it was unique.[25]

After graduation Einstein believed that he deserved an assistant's post with Weber, or if not with him then with someone else, and that Weber had stopped him from getting one. Weber passed over Einstein to take on two mechanical engineers in his laboratory that year.[26] Einstein was unhappy with Weber's obstruction, and he regretted that Weber's theoretical physics lectures did not include Maxwell's theory.[27] All the same, Weber was an established physicist, and Einstein evidently heard Weber's lectures on theoretical physics as he did not some mathematical lectures; moreover, we have reason to think that Weber's lectures, while not completely up to date, were well received.[28] Most important, Einstein worked gladly in Weber's laboratory, and for sound scientific reasons. The experimental facilities there were extraordinary. On visiting the laboratory in 1895, five years after the building was completed and a year before Einstein entered the Polytechnic, the American physicist Henry Crew recorded his impressions:

> H. F. Weber and Dr. Pernet are at the head of the dept. of Physics in the Polytechnicum. They not only have the most complete instrumental outfit I have ever seen but also the largest building I have ever seen used for a physical laboratory. . . . Tier on tier of storage cells, dozens and dozens of the most expensive tangent & high resistance galvanometers. Reading telescopes of the largest & most expensive form by dozens, 2 or 3 in each room. . . . The apparatus in this building cost 400000. francs ($80000.)—the building—the Phys. Laby. alone cost one million francs.[29]

In addition to applying to Weber, Einstein applied to Wilhelm

[25] *Ibid.*

[26] Seelig, *op. cit.* (note 10), pp. 35, 57, 62.

[27] Recollections of Einstein's Polytechnic classmate Louis Kollros, referred to in Gerald Holton, "Influences on Einstein's Early Work in Relativity Theory," *American Scholar, 37* (1967–1968), 59–79, on 64.

[28] Recollection of Kollros (Holton, *ibid.*) and of Einstein's cantonal classmate Adolf Fisch (Seelig, *op. cit.* [note 10], pp. 34–35).

[29] Entry for 20, 21, and 22 July 1895 in Henry Crew's "Notes of Travel, Europe, 1895," American Institute of Physics. I wish to thank Paul Forman for alerting me to Crew's impressions of the Zurich physical laboratory, and William Henry Crew for permitting me to quote from his father's European notebook.

Ostwald to work as a "mathematical physicist" in his Leipzig laboratory and to H. Kamerlingh-Onnes to work in his Leiden laboratory. As it turned out, Einstein's first regular post was not in an academic laboratory at all, but in the Swiss federal patent office in Berne. Beginning in 1902 he worked as patent examiner for seven years and, in his spare hours, working as a purely theoretical physicist, he helped lay down the lines of much of twentieth century physical research. Einstein was unusual, his job was not; the rapid industrialization and associated technology of Central Europe at the turn of the century made patent examining one of the standard technical jobs for scientists. Einstein did not consider the nonacademic, technical job beneath him;[30] he later remembered his years at the patent office with nostalgia, and his only complaint at the time was that the eight hour day was demanding and encroached on his own research.[31] His favorable attitude toward applied physics and technology was at least compatible with attitudes at his engineering school and, in particular, with those of Weber, his physics professor.

At the Zurich Polytechnic Einstein would have been exposed to certain mathematical attitudes, and I want to discuss these here. Engineers at the time debated the role of mathematics in their training; in part the debate was rooted in the social antagonism between men of practice and theory and the related institutional antagonism between the Technischen Hochschulen where engineers taught and studied and the universities where their mathematical teachers came from. Engineers at the time often complained that their mathematics teachers refused to recognize that their disciplines were empirical and their problems real. They charged that mathematicians did not know how to solve a differential equation, let alone carry a calculation through to a numerical solution, and that mathematicians with their narrow interest in number theory and other abstract branches of mathematics were unable to teach the mathematics that applied to reality; e.g., approximation methods, theory of errors, and descriptive geometry.[32]

[30]Martin J. Klein, "Einstein and Some Civilized Discontents," *Physics Today, 18* (1965), 38-44.

[31]Martin J. Klein, "Thermodynamics in Einstein's Thought," *Science, 157* (1967), 509-516, on 5 of reprint.

[32]Armin Hermann, "Sommerfeld und die Technik," *Technikgeschichte, 34* (1967), 311-322, on 312.

Around the turn of the century there was a broad movement to reorient mathematics toward application. In the genesis of this movement the Zurich Polytechnic played a special role. It was there that the practice originated of bringing mathematicians to the higher engineering schools to teach their specialties.[33] Gustav Zeuner moved from the Zurich Polytechnic to the polytechnic in Dresden where, from 1875, he organized its mathematical instruction on the Zurich model and prepared the ground for the "mighty opposition of the engineers starting in 1890."[34] A leader of the opposition, the German engineer Alois Riedler, deemphasized the importance of mathematics relative to the empirical and the practical, arguing that "technical problems must be dealt with otherwise than mathematically."[35] Riedler's efforts were important and "largely successful," as the Polytechnic mathematician C. F. Geiser had to acknowledge.[36] Many members of the Society of German Engineers and professors of applied scientific subjects at the technical schools shared Riedler's view that "scholarly, unfruitful theory flies out of sight of the real world above the clouds to Abel and Riemann, to where theta functions vanish and the special concept of dimension is replaced by the general concept of manifold and to where one can do gymnastics in a world of four and more dimensions."[37] Riedler's "engineer's movement" of the late 1890's was directed against mathematicians in the teacher and general divisions of Technischen Hochschulen who adhered to the "principle of the formal educational value of mathematics," rather than of the practical.[38]

In addition to the engineers, a group of mathematicians, mostly teachers in Technischen Hochschulen, under the leadership of the Göttingen mathematician Felix Klein were sharply critical of academic mathematicians' disdain for applications and their tendency to regard their posts in Technischen Hochschulen as waiting rooms

[33]Felix Klein, "The Development of Mathematics at the German Universities," in *Lectures on Mathematics* (New York, 1894), pp. 107–108; Wilhelm Lorey, *Das Studium der Mathematik an den deutschen Universitäten seit Anfang des 19. Jahrhunderts* (Leipzig, 1916), p. 142.

[34]Lorey, *ibid.*, pp. 142, 147, and 149.

[35]Karl-Heinz Manegold, *Universität, Technische Hochschule und Industrie* (Berlin, 1970), p. 154.

[36]C. F. Geiser, "Zur Erinnerung an Theodor Reye," *Vierteljahrsschr. d. Naturf. Ges. Zürich, 66* (1921), 158–180, on 167.

[37]*Ibid.*, p. 168.

[38]*Ibid.*, p. 167; Lorey, *op. cit.* (note 33), p. 149.

for calls to universities.[39] They sought to change the direction of mathematical research and teaching on the grounds that the real force in contemporary society was industry and its scientific-technical base,[40] and that the rigorist impulse had ceased to be fruitful in mathematics. Typical of their programmatic rhetoric was Karl Heun's warning in 1900 that it would be "unwise for mathematicians on their side to simply ignore the actual situation, to worsen it by continuous one-sided emphasis on hypercritical speculations which in part have lost all touch with the healthy ground of reality and practical applicability."[41] In contrast to the engineer Riedler, Klein and the mathematicians who supported him laid stress on the fundamental importance of mathematics for all of the physical and technical disciplines.

Having remarked generally on mathematics in German-language engineering schools, I want to stress that because of the quality of its mathematical staff the Zurich Polytechnic was not typical of higher engineering schools and resembled a major university in this respect. Many of the greatest mathematicians of the second half of the nineteenth century—R. Dedekind, E. B. Christoffel, G. Frobenius, Hurwitz, and Minkowski, among others[42]—taught there. Moreover, as one might expect, at the Zurich Polytechnic one could find engineers with a strong, vocal conviction of the importance of the exact mathematical treatment of engineering problems. The eminent Polytechnic engineer A. Stodola sought to convince colleagues of the need for substantial mathematical training in engineering education. In 1897 he proposed as a basis for mathematical instruction in Technischen Hochschulen the maxim that "mathematics is a fundamental science for the technologist."[43]

Although antimathematical attitudes were not institutionalized at the Zurich Polytechnic as they evidently were in some other technical schools, the Polytechnic nonetheless experienced the ferment over mathematical instruction that was typical of engineering schools at the time. Writing on Wilhelm Fiedler, one of the mathe-

[39] Hermann, *op. cit.* (note 32), p. 313; Manegold, *op. cit.* (note 35), p. 86.

[40] Manegold, *ibid.*, p. 105.

[41] Karl Heun, "Die kinetischen Probleme der wissenschaftlichen Technik," *Jahresber. d. Deutsch. Math.-Vereinigung, 9* (1900), 1–120, on 118.

[42] Lorey, *op. cit.* (note 33), pp. 143–144.

[43] A. Stodola, "Die Beziehungen der Technik zur Mathematik," *Zeitschr. d. VDI, 41* (1897), 1257–1260, on 1258.

maticians at the Zurich Polytechnic at the time Einstein studied there, the Klein supporter A. Voss recalled the "passionate struggles" between mathematicians and engineers "over the organization of mathematical instruction in the German-language Technischen Hochschulen toward the end of the last century."[44] According to Marcel Grossmann, the mathematician and classmate of Einstein, the "well-known attacks on the position of mathematics at the [Zurich] Technischen Hochschule" led to the demand that mathematical instruction adapt to the "*special needs of the technologist.*"[45] In 1878, Ferdinand Rudio, who subsequently held a mathematics chair at the Zurich Polytechnic while Einstein was a student there, defended the thesis that the "value of a mathematical discipline cannot be measured by its applicability to empirical sciences."[46] His mathematical colleagues Fiedler and Geiser, who was also at the Polytechnic while Einstein studied there, were criticized by Zurich engineers for holding similar views.[47] Fiedler rejected this criticism, explaining that "specifically technical examples and applications are impossible, because they are not of generally valid significance for science."[48] Geiser's biographer recalled that Geiser's overly abstract mathematics lectures at the Polytechnic resulted in student demonstrations during periods when engineers were forced to take his courses.[49]

In general in the education of physicists and engineers at the turn of the century, experimental as opposed to mathematical training was stressed.[50] The experience that physicists shared with engineers at the Zurich Polytechnic may have gone a step beyond this. Weber's physical institute linked the concerns of physicists and engineers; working there Einstein had plenty of opportunity to learn engineers' attitudes first hand. He had opportunity, too, to learn

[44] A. Voss, "Wilhelm Fiedler," *Jahresber. d. Deutsch. Math.-Vereinigung, 22* (1913), 97–113, on 103.

[45] Marcel Grossmann, *Der mathematische Unterricht an der Eidgenössischen Technischen Hochschule* (Basle and Geneva, 1911), p. 13.

[46] C. Schröter and R. Fueter, "Ferdinand Rudio. Zum 70. Geburtstag," *Vierteljahrsschr. d. Naturf. Ges. Zürich, 71* (1926), 115–135, on 117.

[47] Geiser, *op. cit.* (note 36), p. 165; F. R. Scherrer *et al.*, "Carl Friedrich Geiser (1843–1934; Mitglied der Gesellschaft seit 1883)," *Vierteljahrsschr. d. Naturf. Ges. Zürich, 79* (1934), 371–376, on 372.

[48] Geiser, *loc. cit.*

[49] Scherrer, *loc. cit.*

[50] H. E. Timerding, "Die Verbreitung mathematischen Wissens und mathematischer Auffassung," in *Kultur der Gegenwart,* Ser. 3, Vol. 1, Pt. 2: *Die mathematischen Wissenschaften,* ed. F. Klein (Berlin, 1914), p. A156.

their attitudes in the classes he attended; in the beginning semesters the classes for teachers corresponded closely to those for engineers, and certain subjects such as mathematics and mechanics were taught to students of different divisions together.[51] Einstein's relative neglect of mathematics during his Zurich years and his later slighting remarks on advanced mathematics in physics might reveal a sympathy for the extreme engineers' position. But to suggest this would be to go beyond the evidence and to ignore the fact that Einstein did not intend to become an engineer. My point in mentioning the engineers' movement is that at the Zurich Polytechnic as at other German language engineering schools in the late 1890's, there were forcefully stated opinions on mathematical education.

To the extent that he was aware of them, Einstein undoubtedly found physicists' judgments on mathematics interesting. Theoretical physicists often remarked that physical concerns were primary and mathematical ones secondary.[52] Indeed the role of mathematics in physics could be a sensitive disciplinary matter, as is clear from the theoretical physicist Paul Volkmann's observation to Arnold Sommerfeld in 1899: "Theoretical physics is an independent discipline, which has been enormously served by mathematics, but which will tolerate no mathematical baby halter."[53]

Einstein certainly would have been aware of the mathematical views of some recent and contemporary masters of theoretical physics. In publicized debates in the 1890's on the foundations and methodologies of theoretical physics, there was much reference to the relation of mathematics to physical pictures. The debates were reflected in certain of the lectures that Einstein reportedly read while a student. Boltzmann, e.g., in his gas theory lectures in 1895 argued that the mathematical description by differential equations

[51] H. W. Tyler, "The Federal Polytechnic at Zürich from an Administrative Standpoint," reprinted in 1906 from volume 8 of *Technology Review*, on pp. 13 and 28.

[52] Helmholtz, e.g., pointed out in his lectures on theoretical physics that in contrast to the pure mathematician who turns to physics for examples of difficult mathematical problems, he was going to do "physics not mathematics." (Hermann von Helmholtz, *Einleitung zu den Vorlesungen über theoretische Physik* [Leipzig, 1903], p. 25.)

[53] Paul Volkmann to Arnold Sommerfeld, 3 Oct. 1899, Archive for History of Quantum Physics, American Philosophical Society. I wish to thank Ernst Sommerfeld for permission to quote from his father's correspondence.

of the inner motions of bodies leads compellingly to the concept of heat as the motion of the smallest particles.[54] Two years later, in his mechanics lectures, he went further in associating the concepts of physical and mathematical atomicity, and in countering the phenomenologists' reputed claim for the "picture"-free nature of bare differential equations. In reference to finite time elements he suggested that the differential equations of physics may represent only average values constructed from elements that are not themselves rigorously differentiable.[55]

Einstein's reasons for choosing a career in theoretical physics rather than in mathematics had to do in part with his attraction to central, unifying problems. As background to his career decision, I will report a few observations by mathematicians and physicists on unifying trends in their disciplines in the twentieth century. I will begin with the mathematicians.

David Hilbert, the Göttingen mathematician and close friend of Minkowski and Hurwitz, held a firm conviction of the methodological unity of mathematics; he believed that the "question is forced upon us whether mathematics is once to face what other sciences have long ago experienced, namely to fall apart into subdivisions whose representatives are hardly able to understand each other and whose connections for this reason will become even looser. I neither believe nor wish this to happen; the science of mathematics as I see it is an indivisible whole, an organism whose ability to survive rests on the connection between its parts."[56] Minkowski, who moved from the Zurich Polytechnic to Göttingen, shared Hilbert's conviction, writing in 1905 that in the half century since the death of Dirichlet the development of all parts of mathematics showed Dirichlet's spirit; namely, the desire for "fraternization of the mathematical disciplines, for the unity of our science."[57] Hermann Weyl, who

[54] Ludwig Boltzmann, *Lectures on Gas Theory,* trans. S. G. Brush (Berkeley, 1964), p. 27.

[55] Boltzmann, *op. cit.* (note 8), pp. 3–4, 26–27.

[56] Hermann Weyl, "Obituary: David Hilbert 1862–1943," *Obituary Notices of Fellows of the Royal Society, 4* (1944), 547–553; in Weyl, *Gesammelte Abhandlungen,* ed. K. Chandrasekharan (Berlin, 1968), *4,* 121–129, on 123.

[57] Hermann Minkowski, "Peter Gustav Lejeune Dirichlet und seine Bedeutung für die heutige Mathematik" (1905); in Minkowski, *Gesammelte Abhandlungen* (Leipzig, 1911), pp. 447–461, on p. 450.

moved to the Zurich Polytechnic as professor of mathematics and successor to Geiser in 1913, compared physics and mathematics in their contrasting tendencies toward internal specialization and focus: "Whereas physics in its development since the turn of the century resembles a mighty stream rushing on in one direction, mathematics is more like the Nile delta, its waters fanning out in all directions."[58] He recognized as well that the "tendency of several branches of mathematics to coalesce is another conspicuous feature in the modern development of our science."[59]

Unlike Hilbert, Minkowski, and Weyl, Einstein felt that he did not have a sense for the central problems of mathematics. In his "Autobiographical Notes," Einstein recalled how the choice of a mathematics career had looked to him at the time. He decided against mathematics in part because he was unsure of his way around in it, believing that he lacked an intuitive feeling for what was important as opposed to mere erudition. He felt like Buridan's ass, unable to decide toward which specialty within mathematics he should move. He was certain only that each specialty could exhaust a lifetime.[60]

Like mathematicians, theoretical physicists around the turn of the century were sensitive to specializing trends. In 1894, in his inaugural lecture at the Berlin Academy, Max Planck, who had recently been made ordinary professor of theoretical physics at the University of Berlin, observed that the theoretical physicist's "summarizing activity is needed as a complement to increasing specialization."[61] He spoke repeatedly on the goal of a unified physical world picture,[62] and in this connection he thought closely about the differences between mathematics and physics. He believed that in mathematics one could not speak of a contradiction between independently developed theories like algebra and geometry. By contrast, relatively independently developed physical theories like mechanics and electrodynamics might conflict with one another and had to be modified so that they did not. Planck saw that "in this

[58]Hermann Weyl, "A Half-Century of Mathematics," *Amer. Math. Monthly, 58* (1951), 523–553; in Weyl, *op. cit.* (note 56), pp. 464–494, on p. 464.

[59]*Ibid.*, p. 465.

[60]Einstein, *op. cit.* (note 9), p. 15.

[61]Planck, *op. cit.* (note 4), p. 4.

[62]See, e.g., Max Planck, "The Unity of the Physical Universe," in Planck, *A Survey of Physical Theory*, trans. R. Jones and D. H. Williams (New York, 1960), pp. 1–26.

harmonizing of the various theories [in physics] lay the main source of their fruitfulness and progress toward a higher unity."[63]

Physics was uncommon among the basic sciences in that it included within itself a well developed, semiautonomous specialty devoted to purely theoretical concerns.[64] Einstein was drawn to that specialty, theoretical physics, a career that bridged his physical and mathematical interests and that dealt with problems whose central significance he was reasonably assured of. He explained in his "Autobiographical Notes" that he early recognized that physics, like mathematics, was internally divided into specialties, but that he soon acquired an intuitive understanding of what was important in physics as he had not in mathematics.[65] He responded to the unifying goals of theoretical physics; after sending his first paper for publication to the *Annalen der Physik* in 1901, he wrote to his former Polytechnic classmate Marcel Grossmann that "it is a magnificent feeling to recognize the unity of a complex of phenomena which appear to be things quite apart from the direct visible truth."[66] Toward the end of his career, in discussing the characteristics of good theories, Einstein restricted his remarks to "such theories whose object is the *totality* of all physical appearances,"[67] and in a certain sense in characterizing his own lifework nothing less than total theories spanning all of the specialties of physics mattered.

Einstein's famous 1905 light quantum paper[68] was as much about the problem of the unity of physics as about the theory of light, which is evident from the way Einstein introduced it; there is a "profound formal distinction," he wrote, between the concepts of the atomic theory of matter and those of Maxwell's electromagnetic

[63]Max Planck, "Verhältnis der Theorien zueinander," in *Physik,* ed. E. Warburg (Berlin, 1915), pp. 732–737, on p. 732.

[64]Wilhelm Wien, "Ziele und Methoden der theoretischen Physik," *Jahrbuch der Radioaktivität und Elektronik, 12* (1915), 241–259, on 241.

[65]Einstein, *op. cit.* (note 9), p. 15.

[66]Seelig, *op. cit.* (note 10), p. 62; quoted in Klein, *op. cit.* (note 31), p. 2 of reprint.

[67]Einstein, *op. cit.* (note 9), p. 23.

[68]Albert Einstein, "Über einen die Erzeugung und Verwandlung des Lichtes betreffenden heuristischen Gesichtspunkt," trans. A. B. Arons and M. B. Peppard as "Einstein's Proposal of the Photon Concept—a Translation of the *Annalen der Physik* paper of 1905," *Amer. Journ. Phys., 33* (1965), 367–374. Einstein's paper is closely analyzed in Martin J. Klein, "Einstein's First Paper on Quanta," *Natural Philosopher, 2* (1963), 59–86.

theory of light. The emphasis is on "formal," for he went on to describe the differences in the formalisms by which physicists conceived of these two branches of physics. He pointed out that in the atomic theory the energy of a body is given by a discrete "sum" of the energies of a finite number of individual atoms and electrons, whereas in Maxwell's theory the energy is given by a "continuous spatial function." Observing that the continuous functions of Maxwell's theory refer only to time-average values of observations, he suggested that the "theory of light which operates with continuous spatial functions may lead to contradictions with experience"[69] in describing the emission and transformation of light. He believed that physicists would better understand such phenomena if they thought of light as behaving like a "finite number of [independent] energy quanta which are localized at points in space, which move without dividing, and which can only be produced and absorbed as complete units."[70] In this way he accompanied his "very revolutionary"[71] hypothesis of light quanta with the suggestion that the concepts of matter and light may turn out to be not as formally different as physicists had believed up to now. In later writings Einstein formally characterized the separateness of atomic and field theories by the mathematics used in each: total, or ordinary, differential equations in the former and partial differential equations in the latter.[72] I want to suggest that in calling attention to the different formalisms in the two theories he had a similar kind of distinction in mind in 1905.

Mathematics is the mathematical physicist's "special language," indeed the "only language he can speak."[73] That is how Henri Poincaré characterized the role of mathematics in physics at the turn of this century. He went on to say that the mathematical physicist uses mathematics not only for calculation but "above all, to reveal

[69] Einstein, *ibid.*, p. 368.

[70] *Ibid.*

[71] "Very revolutionary" is Einstein's own judgment at the time on his light quantum hypothesis (Albert Einstein to Konrad Habicht, 1905, undated; quoted in Seelig, *op. cit.* [note 10], p. 89).

[72] See, e.g., Albert Einstein, "Maxwells Einfluss auf die Entwicklung der Auffassung des Physikalisch-Realen," in Einstein, *Mein Weltbild* (Amsterdam, 1934), pp. 208-215.

[73] Henri Poincaré, "Analysis and Physics," in Poincaré, *The Value of Science*, trans. from the original 1905 edition by G. B. Halsted (New York, 1958), pp. 75-83, on p. 76.

to him the hidden harmony of things,"[74] and he believed that the mathematical formalisms of physics cannot be separated from the physical conceptions. If physicists found, as he did, the conceptual dualism of atomic and electromagnetic theories unacceptable, then they must recast the foundations of physics in a nondualistic formalism.

Einstein's critique of the foundations of physics in 1905 involved not only his belief in the inseparability of physical concepts from their mathematical expression but also his belief in the relative mathematical simplicity of nature. The voluminous, mathematically elaborate papers on the electron theory by Sommerfeld, Woldemar Voigt, Max Abraham, and others at the time contrast strikingly with Einstein's brief, mathematically uncomplicated paper on light quanta. The standard electron theory problems entailed intricate calculations of cases: surface and volume electron charge distributions, rigid and deformable electron structures, slowly and rapidly accelerated electrons, electron velocities over and under the velocity of light, and so on. Einstein showed little interest in these problems and in the associated mathematical investigations that held interest for some theoretical and mathematical physicists. Like the atomic and electromagnetic theories taken together—the theories Einstein compared in his light quantum paper—the electron theories were usually built on the dual concepts of discrete particle and continuous field and on their respective formalisms of total and partial differential equations. By a mathematically elementary analysis, Einstein argued that Maxwell's equations had to be revised before the conceptual and mathematical dualism in the foundations of physics could be removed. In his subsequent efforts to build a unified theory comprehending both the continuous and the discrete aspects of the field, charge, and light, he followed a different direction than that of the Continental electron theorists in their early efforts to solve the same basic problem of relating the continuous field to the discrete particle: Einstein came to seek field equations that yielded point solutions representing discrete electrons and light quanta.[75]

Einstein's recent biographer Jeremy Bernstein has remarked on the "strikingly nonmathematical character" of Einstein's early, pre-general relativity papers, noting that the papers "contain relatively

[74] *Ibid.*, p. 79.

[75] Russell McCormmach, "Einstein, Lorentz, and the Electron Theory," *Hist. Stud. Phys. Sci.*, *2* (1970), 41–87.

few equations—and often the equations they do contain are not numbered sequentially . . . if only as an aid in following and referring back to the arguments."[76] (Planck, e.g., numbered his equations at this time.) It seems that Einstein around 1905 did not believe that basic changes in physical theory demanded advanced mathematics, and he was right in a sense. Given the problems of physical theory at the time, he was able to carry through a profound critique of the foundations of physics using elementary algebra, differential equations, and probability concepts, and with that light mathematical equipment he was able to formulate the kinematics of special relativity and other bodies of conceptually new physical knowledge.

After meeting Einstein at the 1911 Solvay Congress, the physicist F. A. Lindemann reported that Einstein "says he knows very little mathematics," but added that "he seems to have had a great success with them."[77] Einstein made the same self-deprecating observation in connection with the first mathematical text on relativity by Max Laue in 1911, complaining in a half joking way that he could "hardly understand Laue's book."[78] By 1911 he had begun his sustained work on the general theory of relativity, in the course of which he became aware of a mathematical complexity in nature that he had not imagined before and at the same time of his own crippling mathematical ignorance.

In 1911 Einstein was professor of theoretical physics in Prague where, according to his Prague successor Philipp Frank, one of his closest colleagues was the mathematician Georg Pick; during their long walks together, Einstein told Pick about the mathematical difficulties he encountered in generalizing his relativity theory. Pick suggested to him that the appropriate mathematical language was the absolute differential calculus of the Italian mathematicians G. Ricci and T. Levi-Civitá.[79] Soon after, in 1912, Einstein returned to the school he admired, the Zurich Polytechnic, this time as professor of theoretical physics (he was brought there by Pierre Weiss, Weber's

[76] Jeremy Bernstein, "The Secrets of the Old One," *New Yorker* (10 and 17 March 1973), pp. 44–101 and 49–91, respectively; on p. 62 (17 March 1973).

[77] F. W. F. Smith, Earl of Birkenhead, *The Professor and the Prime Minister* (Boston, 1962), p. 43; quoted in Martin J. Klein, "Einstein, Specific Heats, and the Quantum Theory," reprinted from *Science, 148* (1965), 173–180, on p. 7 of reprint.

[78] Frank, *op. cit.* (note 5), p. 206.

[79] *Ibid.*, p. 82.

successor, to replace a professor of mathematics).[80] His former classmate Grossmann was there, too, as professor of mathematics and Fiedler's successor, and together they constructed the first formulation of Einstein's general relativity and gravitational theories in terms of the absolute differential calculus.[81] From Zurich Einstein wrote to Arnold Sommerfeld in 1912 that "I am now occupying myself exclusively with the problem of gravitation and believe that, with the aid of a local mathematician who is a friend of mine [Grossmann] I'll now be able to master all difficulties. But one thing is certain that in all my life I have never struggled as hard and that I have been infused with great respect for mathematics the subtler parts of which, in my simple-mindedness, I had considered pure luxury up to now!"[82] Einstein's earlier slight esteem for mathematical complexity in physical theory had been replaced by respect.

From that time on Einstein never doubted the physical relevance of certain of the subtler parts of mathematics; his gravitational theory, which he completed in 1915, and his subsequent unified field theory, which occupied him for more than thirty years, presented continually new and demanding mathematical problems. His correspondence with his valued confidant, the Dutch theoretical physicist H. A. Lorentz, sometimes dealt with arcane mathematical matters that were now at the center of the physical argument; e.g., writing to Lorentz in 1913 Einstein remarked that mathematicians had not developed group theory sufficiently for his needs.[83] His practice now of numbering equations in his papers reflected the prominence of mathematical considerations in the field-theoretic unity he sought for physics. He had departed so far from his early attitudes that in his Herbert Spencer Lecture at Oxford in 1933 he suggested that in theoretical physics the "creative principle resides in mathematics."[84] Abandoning his former Polytechnic practice of

[80]Lorey, *op. cit.* (note 33), p. 144.

[81]Albert Einstein and Marcel Grossmann, "Entwurf einer verallgemeinerten Relativitätstheorie und einer Theorie der Gravitation," *Zeitschr. f. Math. u. Phys., 62* (1913), 225–261.

[82]Albert Einstein to Arnold Sommerfeld, 29 Oct. 1912, in *Albert Einstein/ Arnold Sommerfeld Briefwechsel,* ed. A. Hermann (Basle, 1968), pp. 26–27, on p. 26.

[83]Albert Einstein to H. A. Lorentz, 14 Aug. 1913, Einstein Archive, Princeton.

[84]Quoted in Holton, *op. cit.* (note 15), p. 650.

maintaining his distance from advanced mathematics, he selected his closest collaborators in part for their mathematical expertise.

In his "Autobiographical Notes" Einstein took pains to clarify his later understanding of the relation of mathematics and theoretical physics. There he explained that he had abandoned his early attempts to discover by trial and error the physically unifying field equations accounting for light quanta and electrons. He realized that even if he had hit upon them, he would not have gone deeply into the matter; the equations would remain arbitrary, at best a happy guess. He recognized that the development of physics depended on finding principles applicable to all of physics. The principles took the form of statements about physically permissible equations,[85] so that the development of physics might be seen as the progressive reduction of the ad hoc or empirical element in establishing its mathematical foundations. Einstein used his special and general relativity principles as heuristic guides in seeking one of the goals of theoretical physics: the "simplest" equations for describing the total physical world.[86] The lesson Einstein learned from his general relativistic and gravitational theories was that even the simplest equations are so complex that they "can be found only through the discovery of a logically simple mathematical condition which determines the equations completely or [at least] almost completely."[87] So mathematically demanding was his later way of constructing physical theories that he felt constantly inadequate. He likened his mathematical efforts to "a man struggling to climb a mountain without being able to reach its peak."[88] On his deathbed Einstein is said to have lamented to his son: "If only I had more mathematics!"[89]

Einstein was by no means alone among physicists in coming to acknowledge an increasing mathematical complexity in physics. In a general address at the 1905 Naturforscherversammlung, Wilhelm Wien observed that the problems of the electron theory demanded everything that mathematical analysis could yield.[90] He addressed the German Mathematical Society the next year, 1906, on partial

[85] Einstein, *op. cit.* (note 9), pp. 37 and 53.
[86] *Ibid.*, p. 69.
[87] *Ibid.*, p. 89.
[88] Peter Michelmore, *Einstein. Profile of the Man* (New York, 1962), p. 197.
[89] *Ibid.*, p. 261.
[90] Wilhelm Wien, *Über Elektronen* (Leipzig, 1909), p. 5.

differential equations in physics, remarking that today theoretical physicists needed the "comprehensive cooperation" of mathematicians.[91] From the side of mathematics, there was programmatic anticipation of a mathematics oriented toward applications. J. Rosanes (from whom Max Born learned of the matrix calculus, which he later used in formulating the new quantum mechanics) observed in 1903 that the estrangement of mathematics and physics of the last several decades was past and that now an "epoch of closer union" had opened.[92] The prospects of closer union greatly attracted Minkowski; in 1905 he projected that even the most abstract branch of mathematics, number theory, would soon have its "triumph in physics and chemistry."[93] During World War I, Lorey recalled the prediction of the Berlin mathematicians K. Weierstrass and L. Kronecker that the present era of abstract mathematics would be superseded by one of application and that the problems of theoretical physics, astronomy, and technology were already ripe for the application of mathematics; Lorey believed that in theoretical physics their prediction was close to realization.[94] In 1914 Klein heard the "cry of distress of modern physics which in its stormy, even revolutionary development calls for the help of mathematicians and threatens to devour a large part of our working energy!"[95]

Minkowski was overly optimistic about the applicability of number theory, but the theories of invariants, groups, vectors, tensors, matrices, and other relatively unfamiliar mathematical entities figured increasingly in physical theory before World War I. The widespread acceptance of vectors by physicists at the end of the nineteenth century was a most important mathematical development for conceptual change in physics. Michael Crowe has analyzed the consolidation of modern vector theory as a branch of mathematics in the 1890's and 1900's, concluding that the "most influential force in producing acceptance stemmed from the association of vectorial analysis with electrical theory" and that the "vast majority" of

[91] Wilhelm Wien, "Über die partiellen Differentialgleichungen der Physik," *Jahresber. d. Deutsch. Math.-Vereinigung, 15* (1906), 42–51, on 42.

[92] J. Rosanes, "Charakteristische Züge in d. Entwicklung d. Mathematik des 19. Jahrhunderts," *Jahresber. d. Deutsch. Math.-Vereinigung, 13* (1904), 17–30, on 23.

[93] Minkowski, *op. cit.* (note 57), pp. 451–452.

[94] Lorey, *op. cit.* (note 33), p. 263.

[95] *Ibid.,* p. 307.

authors of vector texts were physicists.[96] It was above all the change in the physical view of nature following Maxwell's and Hertz's work on electromagnetism that prompted an extension of the standard mathematics of physics to include vectors.[97]

It was largely as a consequence of their development of vector analysis that theoretical and mathematical physicists recognized a widening applicability of other mathematical objects and calculuses. They tried vectors of more than three dimensions, matrices, tensors, and the absolute differential calculus as mathematical aids in developing an appropriate vector analysis for formulating new electromagnetic and gravitational field theories. Poincaré in 1905, in connection with his development of Lorentz' electron theory, made an important application of group theory and four-dimensional vectors to physics.[98] In exhibiting the Lorentz covariance of the fundamental electrodynamic equations, Minkowski in 1908 made another application of four-dimensional vectors to physics and as well of Cayley's matrix calculus, explaining at length the elementary properties of matrices and remarking that matrices were preferable for his purposes to Hamilton's quaternion calculus.[99] Arnold Sommerfeld published a long, two-part paper in 1910 devoted wholly to the new algebraic and analytic methods appropriate to the Minkowski "absolute world" and its characteristic four-dimensional vector of space and time.[100] Gustav Mie exploited an elaborate ap-

[96]Michael Crowe, *A History of Vector Analysis* (Notre Dame, 1967), pp. 225 and 242.

[97]Vectors did not enter the physics literature only in direct connection with electromagnetism. In Germany E. Budde, e.g., gave an early, full treatment of vectors in his mechanical text, *Allgemeine Mechanik der Punkte und starren Systeme,* 2 vols. (Berlin, 1890–1891).

[98]Henri Poincaré, "Sur la dynamique de l'électron," *Comptes rendus, 140* (1905), 1504–1508, and *Palermo, rend. Circ. mat., 21* (1906), 129–176. Poincaré's introduction of four vectors and group theory into Lorentz' theory is discussed in Camillo Cuvaj, "Henri Poincaré's Mathematical Contributions to Relativity and the Poincaré Stresses," *Amer. Journ. Phys., 36* (1968), 1102–1113, especially 1109–1111 and 1113.

[99]Hermann Minkowski, "Die Grundgleichungen für die elektromagnetischen Vorgänge in bewegten Körpern," *Gött. Nachr.* (1908), pp. 53–111, on pp. 78–98. For an important analysis of Minkowski's "crucial innovation" of the four-dimensional space-time element, see Gerald Holton, "The Metaphor of Space-Time Events in Science," *Eranos Jahrbuch, 34* (1965), 33–78, on 68.

[100]Arnold Sommerfeld, "Zur Relativitätstheorie. I. Vierdimensionale Vektoralgebra," *Ann. d. Phys., 32* (1910), 749–776, and ". . . II. Vierdimensionale Vektoranalysis," *ibid., 33* (1910), 649–689.

paratus of matrices and higher dimensional vectors in his development of an electromagnetic theory of matter from 1912,[101] as did Max Born[102] and others who investigated Mie's theory.

A radical possibility for a mathematical redirection in basic physical theory arose from a juncture of the electromagnetic and atomic theories. With the help of the statistical interpretation of the second law of thermodynamics, Planck introduced into radiation theory a new constant that became an increasingly pervasive and unsettling part of physics. Both the inescapability of Planck's constant and its apparent incompatibility with classical physics were sufficiently clear to Poincaré in 1912 for him to announce that the greatest revolution in physics since Newton may be in progress. He characterized this second revolution by the way it upset physicists' assumptions about the branch of mathematics appropriate for expressing fundamental physical laws. He explained that ever since Newton the laws of theoretical physics were thought to be inseparable from differential equations, but that the discontinuities associated with Planck's theory now seemed to demand another mathematics.[103]

From the turn of the century, tensors were an increasingly central tool for theoretical physicists. Voigt, who coined the word "tensor" for mathematical objects that had already found limited use in physical science, and who argued for the importance of tensors for crystal physics and for other branches of physics,[104] included tensors along with vectors in a revision of his standard text on theoretical

[101]Gustav Mie, "Grundlagen einer Theorie der Materie," *Ann. d. Phys., 37* (1912), 511–534; *39* (1912), 1–40; and *40* (1913), 1–66. Max Jammer remarks on Mie's use of matrices in *The Conceptual Development of Quantum Mechanics* (New York, 1966), p. 206.

[102]Max Born, "Der Impuls-Energie-Satz in der Electrodynamik von Gustav Mie," *Gött. Nachr.* (1914), pp. 23–36. In a footnote Born had to explain that in matrix multiplication one multiplies row by column, which points up the physicists' unfamiliarity with the properties of matrices as late as 1914 (*ibid.*, p. 33).

[103]Henri Poincaré, "Sur la théorie des quanta," *Journal de physique théorique et appliquée,* Ser. 5, *2* (1912), 5–34, on 5; Russell McCormmach, "Henri Poincaré and the Quantum Theory," *Isis, 58* (1967), 37–55, on 43, 49, and 55.

[104]Woldemar Voigt, *Die fundamentalen physikalischen Eigenschaften der Krystalle* (Leipzig, 1898), pp. 20ff; Salomon Bochner, "The Significance of Some Basic Mathematical Conceptions for Physics," *Isis, 54* (1963), 179–205, on 193.

physics in 1901.[105] Tensors proved especially useful in electrodynamics and field physics for expressing energy, momentum, stress, and even mass; Abraham stressed the tensor character of the electron mass in his efforts at placing physics on electromagnetic foundations beginning in 1902.[106] Einstein above all made fundamental use of tensors in his development of general relativity. So extensive was his reliance on tensors that in the first tensor formulation of his general relativity and gravitational theories he included a full, elementary exposition of the tensor calculus. However, instead of drawing up the mathematical exposition himself, he collaborated with the mathematician Grossmann who wrote the "Mathematical" Part II which, in agreement with Einstein's understanding of the relation of physics and mathematics, followed Einstein's own "Physical" Part I.[107] Non-euclidean and four-dimensional geometries[108] entered physics before World War I, just as did the "multiple algebras" of complex numbers,[109] dyadics, quaternions, vectors, tensors, matrices, and groups. Einstein's application of Riemannian geometry to gravitation in his general relativistic extension of his and Minkowski's conceptions of space and time entailed what may be the most radical revision of the physical world picture before the invention of quantum mechanics in the interwar years.

In closing I will make some observations on possible directions for historical research into the physical sciences that articles in this and

[105] Woldemar Voigt, *Elementare Mechanik als Einleitung in das Studium der theoretischen Physik,* 2nd rev. ed. (Leipzig, 1901), pp. 10–26.

[106] Max Abraham, "Dynamik des Electrons," *Gött. Nachr.* (1902), pp. 20–41, on p. 28.

[107] Einstein and Grossmann, *op. cit.* (note 81).

On the relations of mathematics to theoretical physics in the context of general relativity theory, see Lewis Pyenson, *The Göttingen Reception of Einstein's General Theory of Relativity* (diss., Johns Hopkins University, 1973). See also his "La réception de la relativité généralisée: disciplinarité et institutionalisation en physique," *Revue d'histoire des sciences, 28* (1975), 61–73.

[108] E. Wölffing, "Die vierte Dimension," *Umschau, 1* (1897), 309–314.

[109] Although complex numbers entered classical physics and especially relativity theory, "basic conceptualizations and basic formulations continued to be presented and expressed in real variables only." It was not until quantum mechanics that the "very basic equations" of the theory displayed the symbol *i* "openly and directly." (Bochner, *op. cit.* [note 104], p. 196.)

recent volumes of *Historical Studies in the Physical Sciences* have suggested and that I have explored further in this introduction. Interactions between sciences—involving shared methods, problem areas, research programs, and facilities and institutions, among other things—sometimes seem ephemeral, if not accidental, and of limited significance; at other times they reflect deep, ongoing interdependencies. The historical relations between physics and mathematics are rich in interactions of the latter kind; the methods of physics depend strongly on the language of mathematics. A study of Einstein's career suggests that we need to understand historically the precise nature of the physicist's dependence on mathematics and his attitude toward that dependence and its implications for his conception of the physical world. We need to know which branches of mathematics the theoretical physicist was expected to command and why, and how they were learned and by whom they were taught.[110] We need to know the part mathematics played in the theorist's assumptions about physical nature and their relation to the kinds of mathematics he learned and used in constructing theories. We need to know the ways mathematics has figured historically in physical, epistemological, and methodological modes of argument in physics, and the ways in which the physical applications of different branches of mathematics may have helped define new experimental and theoretical problems for physical research. We need to know, too, the ways theoretical physicists and mathematicians have interacted.

In the editor's foreword in volume three of this series, I discussed aspects of the historical problem of understanding the development of a *single* physical science—physics. Here I will discuss briefly the historical problem of understanding the parallel development of a *family* of physical sciences. To begin I will recall Planck's observation in 1910 that changes in the fundamentals of physics affect all

[110]Helmholtz, e.g., from his own experience had formed decided views on the proper mathematical education for physicists. Writing to his son Robert, the future physicist, who at the time was entering his university studies, Helmholtz advised him to study mathematics before studying theoretical physics. Speaking of himself, he wrote: "As far as mathematics is concerned, my interest developed only through its applications, especially those of mathematical physics, and I studied everything I know of mathematics only occasionally for purposes of application. But that is a method that takes a great deal of time, and in which one reaches complete knowledge only very late." (*Anna von Helmholtz. Ein Lebensbild in Briefen,* ed. Ellen von Siemens-Helmholtz [Berlin, 1929], *1,* 249.)

of the natural sciences. It was clear to him that the current challenges to the mechanical world picture were felt by the "allied sciences" of chemistry and astronomy, even by epistemology.[111] The challenges he was speaking about originated above all from Einstein's special theory of relativity. Others of Einstein's theories illustrate Planck's point as well. E.g., the bearing of Einstein's theory of Brownian motion on colloid chemistry, of his light quantum hypothesis and his quantum theory of specific heats on physical chemistry, of his general theory of relativity on astronomy is, of course, well known.

If we wish to look beyond the individual researcher like Einstein and the examples his work suggests, we may anticipate more general studies that compare developments in physics, mathematics, chemistry, and other allied sciences. The studies might illuminate relations between intellectual traditions, origins of problems and applications, sources of material support, and the recruitment, training, and employment of the members of the sciences. They might compare methods, concepts, and instruments. They might compare the growth of specialist societies, academic posts, research facilities, and other institutional factors. They might look at the boundaries of the knowledge domains of the sciences, at how they are defined and how they shift in time and why, at what is unique to each and what is shared. In short, they might offer us a basis for a comprehensive social and intellectual mapping over time and place of the family of the physical and mathematical sciences. Ultimately, we may approach one of the general problems to which *Historical Studies in the Physical Sciences* is committed, that of understanding the development of the modern physical sciences individually, collectively, and within their larger historical context.

This work was supported in part by the National Science Foundation.

[111]Max Planck, "The Place of Modern Physics in the Mechanical View of Nature," in Planck, *op. cit.* (note 62), pp. 27–44, on pp. 27–28.

Historical Studies
in the
Physical Sciences
7

The Ether Problem, the Mechanistic Worldview, and the Origins of the Theory of Relativity

BY TETU HIROSIGE*

1. INTRODUCTION

Since the first systematic account by Max von Laue,[1] it has been, and still is, the common practice to introduce the theory of relativity with a survey of the nineteenth century ether problem. By "ether problem" I mean the theoretical and experimental investigations of possible influences of the earth's motion relative to the ether on optical and electromagnetic phenomena. I shall cite a few arbitrarily chosen examples from recent textbooks. Christian Møller begins his book with "a short historical survey of the numerous optical experiments which have been performed in an attempt to detect effects depending on the motion of the apparatus with respect to an absolute space."[2] He says that for Maxwell and his contemporaries "the ether was supposed to represent the absolute system of reference, thus giving a substantial physical meaning to Newton's notion of 'absolute space'."[3] But "the fruitless attempts to find out any influence of the motion of the earth on mechanical, optical, and electromagnetic phenomena gave rise to the conviction among physicists that the principle of relativity was valid for all physical phenomena."[4] W. G. V. Rosser, who aims at filling the gap between advanced textbooks and semi-popular books, gives a detailed account of the aberration of light from the stars and of experiments by Fizeau, Hoek, Airy, Michelson and Morley, and others "to illustrate how the theory of special relativity arose out of classical electromagnetism."[5]

Requests for reprints of this article should be addressed to Dr. Sigeko Nisio, Department of Physics, College of Science and Engineering, Nihon University, Kanda-Surugadai, Chiyoda-Ku, Tokyo, Japan.

[1] M. von Laue, *Das Relativitätsprinzip* (Braunschweig, 1911), pp. 8–18.

[2] C. Møller, *The Theory of Relativity,* 2nd ed. (London, 1972), p. 5.

[3] *Ibid.,* p. 6.

[4] *Ibid.,* p. 30.

[5] W. G. V. Rosser, *An Introduction to the Theory of Relativity* (London, 1964), p. xiii.

Such a tradition has produced the common understanding that the theory of relativity was formulated as an answer to the ether problem. Scholars holding this view overlook that the ether problem had already received an answer before Einstein's theory in the work of H. A. Lorentz and Henri Poincaré, and that logic would therefore require them to admit Edmund Whittaker's much disputed view that the theory of relativity was formulated by Lorentz and Poincaré.[6] Since Whittaker's book appeared in 1953, many authors, beginning with Max Born,[7] have debated Whittaker's view. Heinrich Lange[8] and G. H. Keswani,[9] agreeing with Whittaker, have asserted that the main results of the theory of relativity were obtained by Poincaré. Charles Scribner Jr.,[10] although considering Whittaker's view too extreme, considers Poincaré's work a valuable contribution. Their opponents T. Kahan,[11] Gerald Holton,[12] Stanley Goldberg,[13] Kenneth Schaffner,[14] M. A. Tonnelat,[15] and Arthur I. Miller[16] have insisted on the difference between the Lorentz-Poincaré and Einstein's relativity theory and reject Whittaker's view. O. A. Starosel'skaya-Nikitina,[17] although without referring to Whittaker, has discussed the limitation of Poincaré's scientific thought which

[6]E. Whittaker, *A History of the Theories of Aether and Electricity. The Modern Theories 1900–1962* (London, 1953), Chap. 2.

[7]M. Born, "Physics and Relativity," a lecture given at the International Relativity Conference in Berne on 16 July 1955, in Max Born, *Physics in My Generation* (London, 1956), pp. 189–206.

[8]Heinrich Lange, *Geschichte der Grundlagen der Physik, Vol. 1: Die formalen Grundlagen—Zeit, Raum, Kausalität* (Freiburg/München, 1954), Chap. 10.

[9]G. H. Keswani, "Origin and Concept of Relativity," *British Journal for the Philosophy of Science, 15* (1965), 286–306; *16* (1965), 19–32, 273–294.

[10]C. Scribner, "Henri Poincaré and the Principle of Relativity," *Amer. Journ. Phys., 32* (1964), 672–678.

[11]T. Kahan, "Sur les origines de la théorie de la relativité restreinte," *Revue d'Hist. Sci., 12* (1959), 159–165.

[12]Gerald Holton, [a] "On the Origins of the Special Theory of Relativity," *Amer. Journ. Phys., 28* (1960), 627–636; Gerald Holton, *Thematic Origins of Scientific Thought. Kepler to Einstein* (Cambridge, 1973), pp. 165–195. [b] "On the Thematic Analysis of Science: The Case of Poincaré and Relativity," *Mélange Alexandre Koyré. II. L'aventure de la Science* (Paris, 1964), pp. 257–268; abridged under the title "Poincaré and Relativity," in *Thematic Origins,* pp. 185–195.

[13]Stanley Goldberg, [a] "Henri Poincaré and Einstein's Theory of Relativity," *Amer. Journ. Phys., 36* (1967), 934–944. [b] "The Lorentz Theory of Electrons and Einstein's Theory of Relativity," *ibid., 37* (1969), 982–994.

prevented him from reaching the theory of relativity. I, too, have briefly discussed the problem.[18]

These discussions have conclusively shown that Lorentz' and Poincaré's theory was *not* equivalent to the theory of relativity as properly understood, and that Lorentz and Poincaré did not accept the latter theory. But there still remains the question of the origin of the difference between the two theories, that is, the question of the root of Einstein's innovation. In this respect Gerald Holton has contributed the first step forward.[19] He has effectively criticized the traditional view that Einstein put forward his theory chiefly to surmount the difficulty caused by the negative result of the Michelson-Morley ether drift experiment. After a careful investigation he reached the conclusion that "the role of the Michelson experiment in the genesis of Einstein's theory appears to have been so small and indirect that one may speculate that it would have made no difference to Einstein's work if the experiment had never been made at all."[20] In contrast to Einstein, Lorentz, Poincaré, and most other contemporary physicists saw the Michelson-Morley experiment as one of the most urgent problems requiring their theoretical efforts. This difference of attitude toward the experiment between Einstein and the others stems from the difference between the problems which then preoccupied them. The problem that Einstein viewed as fundamental for physics at that time was different

---

[c] "Poincaré's Silence and Einstein's Theory of Relativity," *Brit. Journ. Hist. Sci., 5* (1970), 73–84.

[14] Kenneth F. Schaffner, "The Lorentz Electron Theory of Relativity," *Amer. Journ. Phys., 37* (1969), 498–513.

[15] M. A. Tonnelat, *Histoire du principe de relativité* (Paris, 1971), Chap. V.

[16] Arthur I. Miller, "A Study of Henri Poincaré's 'Sur la Dynamique de l'Electron'," *Arch. Hist. Exact Sciences, 10* (1973), 207–328.

[17] O. A. Starosel'skaja-Nikitina, "Rol' Anri Puankare v sozdanii teorii otnositel'nosti," *Voprosy Istorii. Estestvoznanija i Tekhniki, 5* (1957), 39–49.

[18] Tetu Hirosige, [a] "A Consideration Concerning the Origins of the Theory of Relativity," *Japanese Studies in the History of Science,* No. 4 (1965), pp. 117–123. [b] "Electrodynamics Before the Theory of Relativity, 1890–1905," *Jap. Stud. Hist. Sci.,* No. 5 (1966), pp. 1–49, esp. pp. 44–45. [c] "Sôtairon wa doko kara umareta ka?" ("Where did the Theory of Relativity Emerge from?"), *Butsuri* (*Proceedings,* Physical Society of Japan), *26* (1971), 380–388.

[19] Gerald Holton, "Einstein, Michelson, and the "Crucial" Experiment," *Isis, 60* (1969), 133–197; Gerald Holton, *Thematic Origins,* pp. 261–352.

[20] *Ibid.,* p. 195; p. 327.

from the central issue of the ether problem which had been discussed by Lorentz, Poincaré, and other contemporary physicists. He saw the situation with a perspective that was quite distinctive. We therefore may say that a fundamental change in the aspect from which problems of physics were viewed was essential for the conception of the theory of relativity. What was the nature of that change? What were the factors that brought about the needed transformation of perspective? The origins of the theory of relativity must be sought in the answers to these questions.

To elucidate the origins of the theory of relativity it is necessary first to consider the actual nature and scope of the ether problem in the nineteenth century. The first part of the present paper is devoted to such a consideration. I do not, however, pretend to give a comprehensive history of the ether problem,[21] but intend only to clarify the nature of the problem with which Lorentz, Poincaré, and others wrestled at the turn of the century. The discussion of the ether problem will help to establish the novelty of Einstein's theory as compared with Lorentz' and Poincaré's theory. The consideration of the novelty of Einstein's approach, especially of his conceptual attitude towards physical problems, requires a reevaluation of the great influence of Mach on Einstein's thought. Differing from the common view, I find Mach's main influence upon Einstein, as far as the genesis of the theory of relativity is concerned, in his devastating criticism of the mechanistic world view. In the last part of the present paper I try to show, by discussing some aspects of Lorentz' and Poincaré's thoughts as well as some remarkable developments in the process by which Einstein's theory became accepted, that a complete emancipation from the mechanistic worldview was of crucial importance for the formation of the theory of relativity.

2. ABERRATION OF LIGHT AND THE VALIDITY OF THE WAVE THEORY OF LIGHT

Histories of the ether problem usually begin with the attempt to explain the aberration of light from the stars and the experiment by

[21] For a historical survey of optical problems in moving bodies, cf. U. I. Frankfurt and A. M. Frenik, "Ocherki razvitsiya optiki dvizhushchikhsya tel," *Trudy Instituta Istorii Estestvoznaniya i Tekhniki, 43* (1961), 3–49.

François Arago that showed that the refraction of light from the stars was not affected by the motion of the earth. Speaking of these investigations we are unconsciously inclined to believe that they were conducted in an attempt to prove the existence of the ether as an absolute reference system. Such a belief, however, is only a projection into the past of the prejudice of those who have already encountered the theory of relativity. In the early stages of the development of the ether problem, what absorbed the physicists' interest was not the issue of an absolute reference system, but rather the implications of the ether problem for the controversy over the nature of light.

When Arago, in 1810, performed his famous experiment,[22] he designed it on the basis of the emission theory of light, which was then predominant in France. In the wave theory the velocity of propagation of waves is determined exclusively by the properties of the medium, and consequently it has a constant value with respect to the medium irrespective of the motion of the source of light. On the other hand, in the emission theory the velocity of light, in general, depends on the initial velocity with which the light particles are emitted. It therefore must depend on the nature and state of the source of light. Astronomical determinations of light velocity by Ole Rømer and James Bradley had shown that it was constant irrespective of the distance over which the light was propagated. "Some astronomers, however," argued Arago, "doubted that stars having different sizes might emit rays with different velocities . . . ."[23] To test this conjecture it was necessary to determine the velocity of light from various stars with great precision. Arago proposed that this be done by observing the refraction of the light from stars. Some authors had also pointed out that such an experiment would give them a means of investigating the motions of the planets and the sun. Arago thought that "the result [of the experiment] must offer certain data concerning the true nature of light."[24] He used the translatory motion of the earth, "because the motion of our system [the earth] combined with the former [the motion of the whole solar system] would give rise to a sufficiently large inequality [of the light velocities]."[25] It was not his intention to determine

[22] F. Arago, "Mémoire sur la vitesse de la lumière" [1810], *Comptes rendus, 36* (1853), 38–49.

[23] *Ibid.*, p. 40. [24] *Ibid.*, p. 43. [25] *Ibid.*, p. 43.

the velocity of the earth with respect to an absolute reference system.

Attaching a prism to the front of a meridian circle, Arago observed the deflection by the prism of bundles of light from stars moving toward and receding from the earth. The conclusion which he drew from his observation was not that it was impossible to detect the earth's motion. His conclusions were, first, that light rays are emitted by stars with different velocities, and, second, that, of rays emitted by bodies, only those having velocities within certain limits can be perceived by the human eye.[26] A scientist will generally draw conclusions from a scientific investigation according to the purpose for which he designed the investigation. Arago's conclusions clearly indicate that he intended to solve the currently debated problem of the properties of light particles.

Arago shortly turned to the wave theory of light. In 1815 he acknowledged that Augustin Fresnel's first attempt to explain diffraction by a wave theory was very promising, and he began to encourage and help Fresnel. Three years later, in 1818, he encouraged Fresnel to examine if the wave theory could be compatible with the result of his experiment as well as the aberration of light. In response to Arago's suggestion, Fresnel attempted to explain these experimental facts by means of hypotheses of a stationary ether and a drag coefficient of ether within refracting bodies.[27] In the same year, 1818, he finished his Academy Prize paper in which the theory of diffraction was developed to its full extent. His theory of transverse waves was put forward three years later, in 1821. Thus, in the year 1818, the emission theory still reigned and the wave theory was a heresy or, at most, an inferior competitor. Prominent members of the Paris Academy, when they chose the theory of diffraction as the subject of the Academy Prize, expected that a paper on the subject would provide a vindication of the emission theory.[28] Under these unfavorable circumstances Fresnel set out to show the superiority of the wave theory over the emission theory in the explanation of the

[26] *Ibid.*, p. 46.

[27] A. J. Fresnel, "Sur l'influence du mouvement terrestre dans quelques phénomènes d'optique," *Annales de chimie et de physique, 9* (1818); *Oeuvres complètes d'Augustin Fresnel* (Paris, 1866), *2*, 627–636.

[28] É. Verdet, "Introduction aux Oeuvres d'Augustin Fresnel," *Oeuvres complètes d'Augustin Fresnel, 1,* ix–xcix, esp. xxxvi.

aberration of light and of Arago's experiment. Referring to Arago's conclusions mentioned above, he stated that the necessity of the hypotheses of the diversity of velocities of light and the limited visibility of light "is not one of the smallest difficulties of the emission theory."[29]

Since the measurement of the velocity of light in a transparent body, which is often cited as the *experimentum crucis* for the wave theory,[30] was not performed until 1850, the wave theory of light still had not succeeded even in the 1840's. The theory of the aberration of light and Arago's experiment continued to be debated in relation to the legitimacy of the wave theory. When in 1842 Christian Doppler theoretically predicted the effect named after him by discussing the mechanism of propagation of longitudinal waves, he rejected the transverse wave theory of light and stated that, as the difficulties in explaining aberration showed, the assumptions of the transverse wave theory seemed "to contain great *inherent improbability*."[31] It is, to be sure, upon the hypothesis that light is a transverse wave and not upon the wave theory in general that Doppler cast doubt here. But no one can fail to recognize his strong distrust of Fresnel's theory. His criticism of Fresnel's theory was fully developed in a paper published in the following year, 1843.[32]

In his 1843 paper Doppler classified existing theories of the aberration of light into four groups, each of which, he asserted, had a difficulty of its own. To our eyes all four kinds of explanation seem to be based on the same kinematical principle, but Doppler considered them different from each other because of differences in the physical nature of the motion to which the principle was applied. The four kinds of explanation are: first, the analogy with the phenomenon that rain appears to fall obliquely when we see it from aboard a moving vehicle; second, the consideration of the path of

[29] Fresnel, *op. cit.* (note 27), p. 628: "La nécessité de cette hypothèse n'est pas une des moindres difficultés du système de l'émission."

[30] E. Whittaker, *A History of the Theories of Aether and Electricity, the Classical Theories* (London, 1951), pp. 126–127.

[31] C. Doppler, "Ueber das farbige Licht der Doppelsterne und einiger anderer Gestirne des Himmels," *Abhandlungen kön. Böhm. Ges. Wiss., 2* (1841–1842), 465–482, esp. 468.

[32] C. Doppler, "Ueber die bisherigen Erklärungsversuche des Aberrationsphänomens," *Abh. kön. Böhm. Ges. Wiss., 3* (1843–1844), 747–765.

light in the interior of the tube of a telescope, which requires us to tilt the telescope; third, explanation by the combination of the velocities of the earth and light; fourth, William Herschel's explanation that the eyeballs must be rotated forward in the direction of the motion of the earth for the light from a star to reach the center of the retina. Doppler argued that to make the second explanation acceptable one must assume that the ether does not change its position with respect to the solar system. To fill this requirement, however, one must assume that the earth traverses the ether freely without resistance, a hypothesis that is hardly tenable, particularly since it is incompatible with the opaqueness to light of the earth and of other terrestrial bodies. After pointing out the difficulties inherent in the other modes of explanation—immaterial for our present discussion—Doppler concluded that the transverse wave theory, however many facts it might be able to account for, could not be right simply because it clearly contradicted so simple a phenomenon as the aberration of light.[33]

George Gabriel Stokes, too, when he propounded his theory of aberration,[34] directed his criticism to the absurdity of the fundamental hypothesis of the Fresnel theory that the ether moved freely through the earth. But Stokes, contrary to Doppler, believed in the transverse theory of light. His theory of aberration was a part of his efforts to save the transverse wave theory from objections such as Doppler's. Stokes' theory is based on two assumptions: that the ether around the earth moves without the earth having any relative velocity at its surface, and that the motion of the ether is irrotational, that is, that it has a velocity potential. He thus approached the problem hydrodynamically, an approach which came naturally to Stokes who had begun his scientific career as a theoretical hydrodynamicist. On 14 April 1845, four weeks before his theory of aberration was presented at the Cambridge Philosophical Society, Stokes presented to the same society a long memoir on equations of

[33] C. Doppler, *ibid.*, p. 765: "Wenn eine Hypothese selbst eine so grosse Anzahl der complicirtesten Naturerscheinungen, für die sie gelten soll, ganz genügend erklärt, mit einer einzigen Erscheinung derselben Art aber in einem offenbaren Widerspruche steht, oder zum wenigsten sie überhaupt nicht erklärt: so ist dies ein ganz unleugbares Kennzeichen davon, dass diese Hypothese im Ganzen genommen nicht die wahre und richtige sein könne."

[34] G. G. Stokes, "On the Aberration of Light," *Phil. Mag.*, *27* (1845), 9–15; *Mathematical and Physical Papers, 1*, 134–140.

motion for viscous fluids and elastic bodies.[35] Toward the end of this memoir he asserted, on the supposition that solid bodies having small shear elasticity and large plasticity would vanishingly differ from fluids having large viscosity, that the ether, even if it is a fluid, would be able to transmit transverse waves of light.[36] He based his conclusion on the inference that the displacements of the ether particles are small because the wavelength of light is extremely short. It is clear that this argument is intended to solve the then urgent problem of the wave theory of light, that is, the contradiction that whereas the ether, as the medium of transverse light waves, must possess the elasticities of a solid body, it nevertheless exerts no resistance to the motion of the earth. Stokes' attention had been drawn to viscous fluids when, in 1842 or 1843, he learned of James South's experiment that suggested that the air around the plumb of a swinging pendulum moves with it. South's experiment occasioned Stokes to think that fluids might take part in the motion of solid bodies and would naturally have suggested to him the assumption of the "ether drag" in his theory of aberration.[37] If a moving body is to impart motion to the ether, however, there must be tangential stress acting across boundaries within the ether. Such stress would give rise to a shear elasticity and make possible the propagation of transverse waves. Thus, Stokes expected the model of the ether that provided a reasonable explanation of aberration to be, at the same time, the solution to a grave difficulty for the wave theory of light, namely, the enigma of how the fluid ether can transmit transverse waves.[38]

In later years he several times discussed the question of what physical properties should be ascribed to an ether that could satisfy the requirements deriving from his theory of aberration.[39] These discussions ultimately also threw light on the above enigma. He took

[35]G. G. Stokes, "On the Theories of the Internal Friction of Fluids in Motion, and of the Equilibrium and Motion of Elastic Solids," *Trans. Cambridge Phil. Soc., 8* (1849), 287-319; *Mathematical and Physical Papers, 1,* 75-129.

[36]*Ibid., Mathematical and Physical Papers, 1,* 126-127.

[37]David B. Wilson, "George Gabriel Stokes on Stellar Aberration and the Luminiferous Ether," *British Journal for the History of Science, 6* (1972), 57-72, esp. 61-62, 71.

[38]Cf. David B. Wilson, *ibid.,* p. 70: "Among other things, therefore, Stokes's theory of stellar aberration constituted a defence for the concept of an elastic-solid ether."

[39]G. G. Stokes, "On the Constitution of the Luminiferous Ether, Viewed with Reference to the Phenomenon of the Aberration of Light," *Phil. Mag., 29*

up the problem in response to James Challis' criticism[40] of his theory of aberration. Refuting Challis' criticism he argued as follows. No stable motion of incompressible fluids has a velocity potential. A fluid that has internal friction, however, can satisfy the requirement of a velocity potential. As he had shown in his earlier paper, a fluid ether having internal friction would behave as an ordinary fluid for the translatory motions of a gross material body and as a solid elastic body for extremely small vibrations. Hence "the astronomical phenomena of the aberration of light should afford an argument in support of the theory of transverse vibrations."[41]

To return to our theme, in the first half of the century all those who dealt with the ether problem approached it with the intention of establishing a legitimate theory of light. Arago's experiment and the aberration of light were considered the touchstone of such a theory. The theory of aberration later came to be discussed also with the expectation that it might furnish a model of the ether which could solve what seemed the most serious difficulty of the transverse wave theory of light. The ether problem in this period, therefore, should be viewed against the background of controversy over the validity of the theories of light. Apparently no one attempted to associate the ether with any privileged reference system for motion.

3. PROPAGATION OF LIGHT WAVES—THE ASTRONOMER'S PROBLEM

The attached diagram shows the number of papers, grouped according to publication dates in five year periods, that are listed under the headings "aberration of light" and "light propagation in moving media" in the *Catalogue of Scientific Papers 1800–1900* edited by the Royal Society.[42] Of course this diagram cannot be taken too seriously. It only gives a rough idea of the general trend, but it shows clearly that there was little discussion of the ether

---

(1846), 6-10; *Mathematical and Physical Papers, 1,* 153-156. "On the Constitution of the Luminiferous Ether," *Phil. Mag., 32* (1848), 343-349; *Mathematical and Physical Papers, 2,* 8-13.

[40] J. Challis, "On the Aberration of Light," *Phil. Mag., 27* (1845), 321-327. On the controversy between Challis and Stokes, see Wilson, *op. cit.* (note 37).

[41] G. G. Stokes, *op. cit.* (note 39), *Mathematical and Physical Papers, 2,* 11.

[42] Royal Society of London, *Catalogue of Scientific Papers 1800–1900.* Subject Index, Vol. 3: *Physics,* Parts I and II (Cambridge, 1912, 1914).

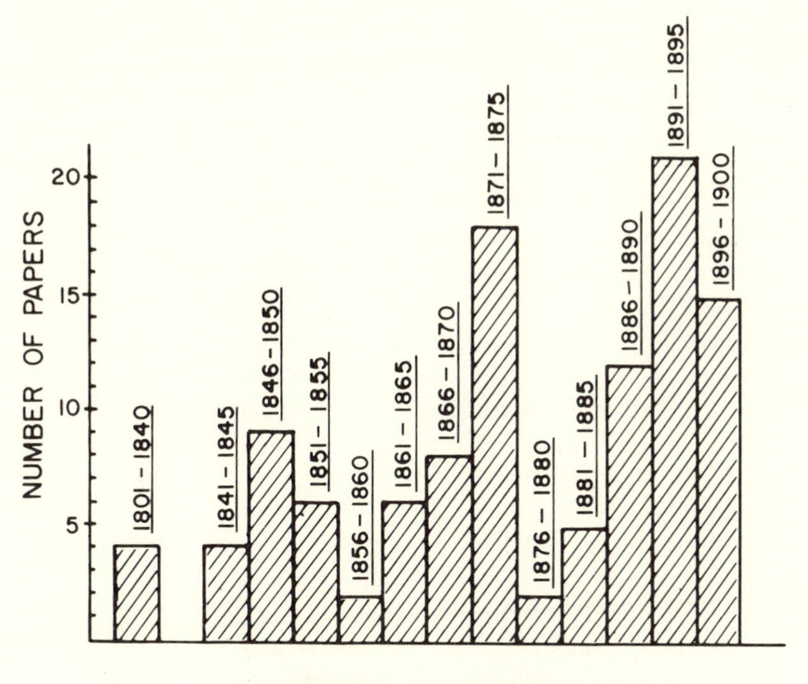

NUMBER OF PAPERS
CONCERNING THE ETHER PROBLEM

problem in the 1850's and 1860's and increasing interest in the years around 1870.

Of course, even in the years of stagnation one finds a few significant investigations. Especially the confirmation of the Fresnel drag coefficient by Hippolyte Fizeau in 1851[43] furnished a cornerstone for discussion of the ether problem in subsequent years. In 1859 Fizeau made an experiment to see if differences in the direction of light rays with respect to the earth's motion alter the change of the azimuth of the plane of polarization of polarized light produced by refraction in a layer of glass.[44] Fizeau designed his experiment to

[43] A. H. L. Fizeau, "Sur les hypothèses relatives à l'éther lumineux, et sur une expérience qui paraît démontrer que le mouvement des corps change la vitesse avec laquelle la lumière se propage dans leur intérieur," *Comptes rendus, 33* (1851), 349-355.

[44] A. H. L. Fizeau, "Sur une méthode propre à rechercher si l'azimut de polarisation du rayon réfracté est influencé par le mouvement du corps ré-

confirm the Fresnel coefficient for solid bodies. Since the change of the azimuth of the plane of polarization depends on the refractive index of the glass, Fizeau expected to find from the change of the azimuth the change in the refractive index and, consequently, in the velocity of light in the transparent body. Fizeau alleged that the expected change had been found; his conclusion was later often doubted. In 1862 Jean Babinet predicted that the influence of motion would change the position of diffraction fringes.[45] He discussed this effect in connection with an astronomical problem. He imagined that if the velocity of the translatory motion of the solar system as a whole was determined, the distances to fixed stars could be measured with precision by triangulation, with the displacement of the solar system as the base. Probably because of his astronomical illustration, Babinet's paper was published in the *Comptes rendus* under the heading "astronomy." The many discussions of the ether problem around 1870 were also motivated by astronomical interests.

One of the achievements of nineteenth century astronomy was the publication of a number of voluminous catalogues of stars.[46] The first was Friedrich Wilhelm Bessel's compilation of the results of James Bradley's observations published in 1818. The results of Bessel's own observations during the years 1821 to 1833 were published after his death in 1846. The most famous, Friedrich Wilhelm August Argelander's *Bonn Durchmusterung,* was published in 1859–1862. Its extension to the stars of the southern skies by Eduard Schönfeld was published in 1875–1885. In 1885 John Macon Thome began to publish the *Cordoba Durchmusterung* covering the skies from twenty-two degrees south to the south pole. To compile catalogues of stars it was necessary to correct the observational data. The effect of atmospheric refraction had been investigated since Newton's time. Since it was also necessary to correct the effect of

fringent," *Comptes rendus, 49* (1859), 717–723; *Ann. de chim., 58* (1860), 129–163.

[45] J. Babinet, "De l'influence du mouvement de la Terre dans les phênoménes optiques," *Comptes rendus, 55* (1862), 561–564.

[46] I owe the general knowledge of astronomy in the mid-nineteenth century to the following two books: Robert Grant, *History of Physical Astronomy* (New York and London, 1966; originally London, 1852), Chaps. 14, 16, 19, and 21; Arthur Berry, *A Short History of Astronomy* (New York, 1961; originally 1898), Chaps. 12 and 13.

aberration, precise determination of the aberration constant became an important task in astronomy. Aberration also attracted attention in connection with the solar parallax which was determined by the diameter of the orbit of the earth, one of the fundamental constants of positional astronomy. Fizeau's experiment in 1849 proved it possible to measure by terrestrial means the velocity of light which had thus far been determined only by astronomical methods. It followed that, if, by combining the velocity of light measured by terrestrial means with the aberration constant, one obtained the orbital velocity of the earth, one could readily derive the diameter of the orbit.

The new correction factor which appeared in astronomical considerations in the nineteenth century was that due to the motion of the solar system as a whole. After Edmund Halley discovered the proper motion of the fixed stars in 1718, James Bradley, in 1748, was the first to point out the possibility that the apparent change in the position of fixed stars may be due not only to the motion of the stars but also to the motion of the solar system. In 1783 William Herschel, after analysis of the available data for the proper motion of the fixed stars, concluded that the solar system was in motion toward λ Hercules. He tried twice to correct the solar apex in 1805 and 1806. But his conclusion, based on many arbitrary hypotheses, especially concerning the distance to the fixed stars, could not immediately be accepted by his contemporaries. Jean Baptist Biot in 1812 and Bessel in 1818 denied the translatory motion of the solar system. In 1837 Argelander was the first to accept Herschel's conclusion. Not until the middle of the century did Otto Struve and others establish the motion of the solar system and the position of its apex. Since astronomical observations now had to take into account the proper motion of the solar system as well as the orbital motion of the earth, there arose the new problem of determining the path of light rays when not only the terrestrial observer but also the medium and the source of light, the fixed stars, are all in motion. But in the 1850's and 1860's there was still no general theory of light propagation. The wave theory of light had only recently been established, and the general theory of wave propagation, prompted by astronomical need, began to be developed only in the sixties and seventies.

We can see the immaturity of the theory of wave propagation in those days in, for example, the curious conclusion that the director of the Göttingen observatory, Ernst Friedrich Wilhelm Klinkerfues, drew in 1865–1866 from his discussion of the influence of the source of light upon refraction.[47] He concluded that the light emitted from a moving source changes its color without changing its wavelength. He reasoned as follows. Consider the vibrations of ether particles caused by the impact of propagating undulatory disturbances. If the source of light is at rest, the velocity with which the phase of oscillation is transmitted to successive particles—which Klinkerfues called the phase velocity—is the same as the velocity of transmission of the undulatory disturbances—the propagation velocity, as he called it. If the source moves, the rate of change of the phase of a particle will increase or decrease. Consequently, since the color of light is determined by the number of agitations per unit time of the visual nerve and therefore by the rate of change of the phase of the ether particles, the color of light emitted by a moving source should change. On the other hand, since the propagation velocity is determined by the properties of the medium, it does not depend on the motion of the source. Hence the spatial distance between ether particles having the same phase, that is, the wavelength, remains unchanged. Klinkerfues asserted that if one takes this effect of the motion of the source into consideration, it will be possible to explain, without assuming the Fresnel coefficient, the independence of the laws of reflection and refraction of the motion of the earth. However, he inferred that the angle of refraction would be influenced by the motion of the earth, and he attempted to detect the inferred effect. Though the result was negative, he did not give up the attempt to find the change of the angle of refraction caused by the motion of a light source. He also inferred a change in the aberration constant when he used a telescope filled with water, and he carried out an experiment.[48] Again he obtained a negative result, causing him to question the assumption of a stationary ether.

The problem Klinkerfues discussed could have been treated by a purely kinematical method based on Huygens principle. Incapable of

[47]W. Klinkerfues, "Ueber den Einfluss der Bewegung des Mittels und den Einfluss der Bewegung der Lichtquelle auf die Brechbarkeit der Strahls," *Gött. Nachr.* (1865), pp. 157–160, 210; "Weitere Mitteilungen über den Einfluss der Bewegung der Lichtquelle auf die Brechung des Strahls," *Gött. Nachr.* (1865), pp. 376–384; (1866), pp. 33–60; "Untersuchungen aus der analytischen Optik, insbesondere über den Einfluss der Bewegung der Lichtquelle auf die Brechung," *Astron. Nachr., 66* (1866), 337–366.

[48]W. Klinkerfues, *Die Aberration der Fixsterne nach der Wellentheorie* (Leipzig, 1867); "Versuche über die Bewegung der Erde und der Sonne im Aether," *Astron. Nachr., 76* (1870), 33–38.

perceiving this, he tried to derive the law of propagation dynamically, that is, by considering the physical mechanism of the propagation of light. His theory was received with sympathy and even stimulated a number of other investigations. G. B. Airy's famous experiment with a water telescope was performed under the stimulus of Klinkerfues' investigation. In a series of papers published in 1871–1872 Eduard Ketteler carried out a theoretical investigation of the influence of astronomical motions on optical phenomena, which was intended to correct the errors in Klinkerfues' theory.[49] He stated that "elucidation of the theoretical view seems to be desirable since Klinkerfues' conclusions have often been welcomed."[50]

Ketteler discussed reflection, refraction, diffraction, and interference of light on the basis of Huygens principle and showed that these phenomena would not be affected by astronomical motion. In his refutation of Klinkerfues' theory of the Doppler effect, however, Ketteler, too, had recourse to a model of the mechanism of propagation for vibrations proposed by Ernst Mach.[51] The model consisted of an infinitely long chain of metal cylinders connected by steel rings to each other. Ketteler explained that he used this model in order to discuss undulatory motion in an intuitive manner.[52] For us today his argument is rather complicated and turbid precisely because of his recourse to the mechanical model. To Ketteler, the development in terms of a mechanical model seemed to be intuitive and easy to understand because, first, it fit mechanistic modes of thinking, and, second, the lack of a general theory of wave propagation left him no alternative than to discuss the construction of wave fronts and rays in concrete terms.

In 1870 Wilhelm Veltmann gave the first general demonstration of aberration, based not on examinations of separate cases but on a

[49] E. Ketteler, "Ueber den Einfluss der astronomischen Bewegungen auf die optischen Erscheinungen," *Ann. d. Phys., 144* (1871), 109-127, 287-300; *144* (1872), 363-375, 550-563; *146* (1872), 406-430; *147* (1872), 404-429; "Nachträglicher Zusatz zu der Abhandlung über die Aberration," *Ann. d. Phys., 147* (1872), 478-479; "Ueber den Einfluss der astronomischen Bewegungen auf die optischen Erscheinungen. Nachtrag zu den letzten Abhandlungen," *Ann. d. Phys., 148* (1873), 435-448; *Astronomische Undulationstheorie oder die Lehre von der Aberration des Lichtes* (Bonn, 1873).

[50] E. Ketteler, *op. cit.* (note 49), *Ann. d. Phys., 144* (1871), 127.

[51] E. Mach, "Ueber eine Longitudinalwellenmachine," *Ann. d. Phys., 132* (1867), 174-176.

[52] E. Ketteler, *op. cit.* (note 49), *Ann. d. Phys., 144* (1871), 114.

general kinematical consideration.[53] Emphasizing that his theory aims at generality, he stated that his theory "covers not only the given cases but all possible cases, and is therefore exhaustive."[54] He added that "this hypothesis [the Fresnel coefficient] seems to have been utilized only for explaining separate special cases. . . . One obtains neither a clear understanding of the essence of Fresnel's hypothesis nor a general justification for the methods astronomers use in correcting the effect of aberration."[55] It may also be noted that Veltmann distinguished the propagation of light with respect to the ether from that relative to material bodies, calling the former "absolute motion" and the latter "relative motion."[56] At the same time he admitted the possibility that the ether itself moves with respect to space and called the propagation of light with respect to space the "real motion," though he did not discuss this "real motion." Veltmann's nomenclature shows that he did not identify the ether with absolute space in the Newtonian sense.

Veltmann's proof that, if the effects higher than the first order are disregarded, one cannot detect any influence of astronomical motions on optical phenomena is in essence the same as the general proof that H. A. Lorentz gave later. He considered the propagation of light in a system of transparent bodies sharing a common translation of velocity v. Let a ray starting from one point in the system, undergoing reflections and refractions, describe a polygonal path and return to the original point. If we denote the length of each side of the polygon by s_i, and the relative velocity of light with respect to the material body on this side by w_i, the time required for the ray to traverse the whole path will be $\Sigma s_i/w_i$. If the velocity of the ether

[53] W. Veltmann, [a] "Fresnel's Hypothese zur Erklärung der Aberrationserscheinungen," *Astron. Nachr., 75* (1870), 145–160; [b] "Ueber die Fortpflanzung des Lichtes in bewegten Medien," *Astron. Nachr., 76* (1870), 129–144; [c] "Ueber die Fortpflanzung des Lichtes in bewegten Medien," *Ann. d. Phys., 150* (1873), 497–535.

[54] *Ibid.* [c], p. 498: Veltmann states that his theory "nicht blos die vorstehenden, sondern sämmtliche möglichen Fälle umfasst, also wirklich erschöpfend ist."

[55] *Ibid.,* pp. 499–500: "Man scheint jedoch diese Hypothese bisher nur zur Erklärung einzelner specieller Fälle benutzt zu haben. . . . [Man] erhält . . . keine klare Einsicht in das eigentliche Wesen der Fresnel'schen Hypothese und keine allgemeine Begründung des Verfahrens der Astronomen bei der Correction wegen Aberration."

[56] *Ibid.,* p. 501.

relative to the body is u_i, then

$$w_i = c_i + u_i \cos\psi_i,$$

where ψ_i is the angle between the side i and the direction of the translation of the system. By Fresnel's hypothesis $u_i = v/n_i^2$, n_i being the refractive index of the body. Neglecting terms higher than the first order, we obtain

$$\frac{s_i}{w_i} \sim \frac{s_i}{c_i} - \frac{s_i}{c_i^2}\frac{v}{n_i^2}\cos\psi_i = \frac{s_i}{c_i} - \frac{v}{c^2} s_i \cos\psi_i,$$

$$\sum \frac{s_i}{w_i} = \sum \frac{s_i}{c_i} - \frac{v}{c^2} \sum s_i \cos\psi_i,$$

where c denotes the velocity of light in a vacuum. The first term on the right side of the last equation represents the time interval in which the ray traverses the whole path, and the sum in the second term, being the projection of the whole path on the direction of the translatory motion of the system, will vanish for a closed path. Hence the difference in optical path lengths for two rays connecting the same initial and end points remains the same regardless of the motion of the system. Now according to the Fresnel theory, the propagation of light can generally be described as a consequence of interference, and the interference is determined solely by the differences in the optical path lengths. In the approximation to the first order, therefore, no influence of the motion on optical phenomena appears.

Concurrently with Veltmann's theoretical exploration of the general theory of light propagation, experimental investigations led to a provisional accommodation concerning the relation between astronomical motion and optical phenomena. One of these was George Biddell Airy's experiment with a water telescope confirming Fresnel's prediction of 1818.[57] This experiment was motivated by the negative result of Klinkerfues' experiment and by Martinus Hoek's experiment,[58] which was alleged to have confirmed the

[57]G. B. Airy, "On a Supposed Alteration in the Amount of Astronomical Aberration of Light, Produced by the Passage of Light Through a Considerable Thickness of Refractive Medium," *Phil. Mag., 43* (1872), 310–313; "Additional Note to the Paper 'On a Supposed Alteration. . .'," *Phil. Mag., 45* (1873), 306.

[58]M. Hoek, "Détermination de la vitesse avec laquelle est entraînée une onde lumineuse traversant un milieu en mouvement," *Arch. néerl, 3* (1868),

Fresnel coefficient. Airy observed the star γ Draconis with a meridian circle whose tube was filled with water. He made two observations, one in spring and another in autumn after an interval of six months. Then he calculated from the observational results the latitude of the place of observation. If the water in the tube alters the aberration constant, the latitudes obtained by calculation should reveal discrepancies. Airy, however, could not find any. He carried out the experiment in 1871 and 1872, but the results were always negative. Another, far more sweeping, investigation was performed by Éleuthère Élie Nicolas Mascart,[59] who was awarded the 1873 Grand Prix of the Paris Academy of Sciences for this work. The Academy had offered the prize for an experimental investigation of "the modifications produced in the mode of propagation and the properties of light in consequence of motions of the luminous source and the observer."[60]

Mascart confirmed experimentally that no influence of the motion of the earth is observed in any of the examined phenomena such as diffraction by gratings, reflection by mirrors, chromatic polarization by doubly refracting substances, rotation of the plane of polarization by rock crystals, refraction by prisms, Newton rings, and interference by Young's mixed layer. He also repeated Hoek's experiment and confirmed the latter's conclusion. In performing these experiments he analyzed the phenomena by means of the Fresnel theory and the Doppler principle. He analyzed separately the two cases of a terrestrial and a celestial source of light, respectively, that is, the cases with and without relative motion between the observational instrument and the source of light, and predicted that in either case the results will always be negative. His experiments completely confirmed his prediction. Of the various phenomena investigated by Mascart, the propagation of light in a moving doubly refracting substance was the only one that Fresnel had not dealt with. In the analysis of this phenomenon Mascart assumed that the Fresnel coefficient, with

180–185; "Détermination de la vitesse avec laquelle est entraîné un rayon lumineux traversant un milieu en mouvement," *ibid., 4* (1869), 443–450.

[59]E. Mascart, "Sur les modifications qu'éprouve la lumière par suite du mouvement de la source lumineuse et du mouvement de l'observateur," *Ann. de l'École norm., 1* (1872), 157–214; *3* (1874), 363–420.

[60]"Prix décerné. Année 1872.—Prix extraordinaires. Grand prix des sciences mathematique. Rapport lu et adopté dans la séance du 14 juillet 1873," *Comptes rendus, 79* (1874), 1531–1534. Quotation is from pp. 1531–1532.

corresponding refractive indices, is valid for each of the ordinary and extraordinary rays. Having confirmed experimentally the result of his theoretical analysis, he concluded that if one supposes that ordinary rays behave in doubly refracting substances in much the same way as in isotropic substances, then the Fresnel coefficient can be used equally for both ordinary and extraordinary rays.[61] The Prize committee especially appreciated this extension of the Fresnel coefficient to doubly refracting substances. Its report states that "these last results [in the case of doubly refracting substances] have importance and novelty which, in concluding this report, may especially be emphasized."[62]

Mascart's application of the Fresnel coefficient to double refraction drew attention also because of its possible significance for the question of the physical explanation of the coefficient. The question had been discussed by several authors and there were two rival theories. The one supposes, as Fresnel's does, that only the excess of the ether contained in a material body, as compared with the surrounding ether, moves with the same velocity v as the material body, the rest remaining stationary. The other assumes the whole ether within the body to move with the velocity $(1 - 1/n^2)v$.[63] There was also the view that in a moving body the density of the ether is modified.[64] Mascart remarked that if different values of the Fresnel constant are applied for ordinary and extraordinary rays, Fresnel's interpretation of the drag coefficient that the excess ether moves with the body is not tenable.[65] These discussions are concerned with a new physical problem which is independent of the astronomical ones discussed so far in this section. The Paris Academy, explaining

[61] E. Mascart, *op. cit.* (note 59), *Ann. de l'École norm., 3* (1874), 418.

[62] *Op. cit.* (note 60), p. 1534: "Ces derniers résultats ont un caractère d' importance et de nouveauté qu'il convient de signaler d'une manière spéciale en terminant ce Rapport."

[63] For example, A. Beer, "Ueber die Vorstellungen vom Verhalten des Aethers in bewegten Mitteln," *Ann. d. Phys., 94* (1855), 428–434.

[64] J. Boussinesq, "Sur le calcul des phénomènes lumineux produits à l'intérieur des milieux transparents animés d'une translation rapide, dans le cas où l'observateur participe lui-même à cette translation," *Comptes rendus, 76* (1873), 1293–1296.

[65] E. Mascart, *op. cit.* (note 59), *Ann. de l'Ecole norm., 3* (1874), 420: "Il semble résulter de là que, pour calculer l'influence des milieux pondérables, il est nécessaire d'avoir recours à d'autres considérations que celle du transport partiel de l'éther, comme le faisait Fresnel."

the significance of its prize problem, stated that "now that the vibratory motion of light and the existence of the luminiferous ether are universally considered well established it appears of great interest to direct our research toward the properties of this elastic medium and its relation with the ponderable body."[66] In the next stage of the development of the ether problem it was just this relation of the ether with ponderable matter that became the central issue of the ether problem.

4. RELATION OF THE ETHER TO PONDERABLE MATTER–THE PHYSICIST'S PROBLEM

In the latter half of the 1880's physicists became interested in a new aspect of the ether problem. The immediate reason for their interest was the Michelson-Morley experiment, but we cannot forget another contributing factor, namely J. C. Maxwell's electromagnetic theory, which greatly enhanced the status of the ether in physics. In 1886, performing a thorough investigation of "the influence of the motion of the earth on the luminiferous phenomena," H. A. Lorentz wrote: "Examination of this question not only is interesting for the theory of light but has acquired a much more universal importance since it became probable that the ether plays a role in the phenomena of electricity and magnetism."[67] Maxwell, too, had been interested in the ether problem. He even performed an experiment in 1864 to see if the index of refraction was different for light rays moving in opposite directions and confirmed Arago's negative result.[68] In his

[66] *Op. cit.* (note 60), p. 1532: "Aujourd'hui que les mouvements vibratoires de la lumière et l'existence de l'éther lumineux lui-même sont considérés par tous comme des vérités bien établies, il paraît d'un grand intérêt de diriger nos recherches vers les propriétés de ce milieu élastique et ses relations avec la matière ponderable."

[67] H. A. Lorentz, "Over den invloed, dien de beweging der Aarde op de lichtverschijnselen uitoefent," *Versl. Kon. Akad. Wet., 2* (1886), 297–372 (French translation: "De l'influence du mouvement de la terre sur les phénomènes lumineux," *Arch. néerl., 21* (1887), 103–176; *Collected Papers, 4,* 153–214). Quotation from p. 153: "L'examen de cette question n'intéresse pas seulement la théorie de la lumière, il a acquis une importance bien plus générale depuis qu'il est devenu probable que l'éther joue un rôle dans les phénomènes de l'électricité et du magnétisme."

[68] J. C. Maxwell to W. Higgins, 10 June 1867. This letter was published in the latter's paper "Further Observations on the Spectra of some of the Stars and

article "Ether," written for the ninth edition of the *Encyclopaedia Britannica* (1879), after surveying some experimental attempts to find the velocity of the ether relative to terrestrial bodies, he concluded that "the whole question of the state of the luminiferous medium near the earth, and of *its connexion with gross matter,* is very far as yet from being settled by experiment" (italics are mine).[69] Maxwell here already perceived the question of the relation of the ether to ponderable matter.

It was George Francis FitzGerald, the ardent supporter of the Maxwellian theory, who first discussed the influence of the motion of the earth on electromagnetic phenomena in 1882.[70] He argued as follows. Henry Augustus Rowland's experiment had shown that the motion of electric charges produces physical effects. Since an "absolute" motion is meaningless, the motion here should be understood as the motion with respect to the ether. Fizeau's experiment concerning Fresnel's drag coefficient suggests that there exists a relative motion between terrestrial bodies and the ether. Hence one can expect terrestrial electrified bodies to produce electromagnetic effects as a result of the motion of the earth with respect to the ether. On this expectation FitzGerald theoretically examined interactions between terrestrial electric charges and magnets, and between two conducting currents, but he found that effects due to the common translatory motion cancel each other completely. He concluded his discussion by expressing hope for further investigations on possible effects in other, more general cases.

In the previous year, 1881, J. J. Thomson had made an attempt to deduce the drag coefficient from the electromagnetic theory.[71] He extended the Maxwell equations to the case of a moving body and obtained for the case of a dielectric moving with velocity v in the direction of the propagation of light, which is taken as the x-direction, the equation for the x-component of the dielectric

Nebulae. . . ," *Phil. Trans., 158* (1868), 529–564. Maxwell's letter is on pp. 532–535.

[69] J. C. Maxwell, "Ether," *The Scientific Papers of James Clerk Maxwell, 2,* 763–775. The quotation is from p. 770.

[70] G. F. FitzGerald, "On Electromagnetic Effects Due to the Motion of the Earth," *Trans. Roy. Dublin Soc., 1* (1882), 319; *The Scientific Writings of the Late George Francis FitzGerald,* pp. 111–118.

[71] J. J. Thomson, "On Maxwell's Theory of Light," *Phil. Mag., 9* (1880), 284–291.

displacement

$$c^2 \frac{\partial^2 f}{\partial x^2} = \frac{\partial^2 f}{\partial t^2} + v \frac{\partial^2 f}{\partial x \partial t},$$

where c is the velocity of light. If the velocity v of the body lies in the direction of the x-axis, putting $f = a \cos(qt - px)$, the relation for the velocity of light in the medium, q/p, is obtained:

$$\frac{q}{p} = \frac{v}{2} \pm \sqrt{c^2 - \frac{v^2}{4}} \sim \frac{v}{2} \pm c.$$

From this Thomson concluded that when the medium that the light traverses moves, the medium drags the light with half the velocity of the medium, a result that, Thomson asserted, agrees with Fizeau's experiment.

In 1889 Theodor DesCoudres made the first experiment to detect the effect of the motion of the earth on electromagnetic phenomena.[72] He designed the experiment to test if the electromagnetic induction between two coils is affected by the motion. The result was negative. What interests us more than the result of the experiment is the motivation and goal of DesCoudres' investigation. He had reason to hope that an attempt to find the effects of the motion on induction would now be successful: "Although up to now the fundamental assumption of our experiments that the electromagnetic induction propagates itself with a velocity that scarcely differs from that of light had been an unproved hypothesis, Hertz's experiments raised it to the rank of fact."[73] Noting the significance of his attempt, he expected that it would help to determine to "what extent the so-called luminiferous ether partakes in the motion of the ponderable mass of the terrestrial body."[74] These words show that the central interest in the ether problem then was no longer the propagation of light under various conditions nor the determination

[72]T. DesCoudres, "Ueber das Verhalten des Lichtäthers bei der Bewegung der Erde," *Ann. d. Phys., 38* (1889), 71–79.

[73]*Ibid.,* p. 72: "War die bei unserem Experimente gemachte Grundvoraussetzung, dass sich electrodynamische Induction mit einer von der Lichtgeschwindigkeit nicht sehr verschiedenen Geschwindigkeit fortpflanze, bislang eine unbewiesene Hypothese gewesen, so erhoben die Hertz'schen Experimente dieselbe zur Thatsache."

[74]*Ibid.,* p. 72: DesCoudres wanted to determine "inwieweit der sogenannte Lichtäther an den Bewegungen der ponderabelen Massen des Erdkörpers Antheil nimmt."

of the astronomical motion of the earth or the solar system, but the connection between matter and the ether, that is, the behavior of the ether within and around a mass of ponderable matter when the latter is in motion.

The shift of interest indicated above is also reflected in successive papers by Albert Abraham Michelson. In the report of the experiment that he performed in 1881 in Berlin, he stated that the experiment aimed at finding "the velocity of the earth's motion through the ether"[75] and concluded that the hypothesis of a stationary ether is not correct. Then in 1886 he and Edward Williams Morley repeated Fizeau's experiment on the drag coefficient.[76] In this investigation, as is natural in view of its subject, their attention was directed to the connection between ether and matter. Having confirmed Fizeau's result, they concluded that "*the luminiferous ether is entirely unaffected by the motion of the matter which it permeates*" (italics original).[77] By "entirely unaffected" they meant that, since "Fresnel's statement amounts . . . to saying that the ether within a moving body remains stationary with the exception of the portions which are condensed around the particles," that is, with the exception of the excess ether as compared to the surrounding ether, we may say that the remaining ether is entirely unaffected by the motion if we regard each particle of the body and the ether condensed around it as a single body.[78]

Michelson and Morley's most celebrated experiment of 1887 was, according to their words, intended to contribute to solving the following problem.[79] Their experiment in the preceding year had confirmed that the ether contained within a transparent body remains stationary when the body moves. However, in their view this result cannot be extended to the case of an opaque body such

[75] A. A. Michelson, "The Relative Motion of the Earth and the Luminiferous Ether," *Amer. Journ. Sci., 22* (1881), 120–129. For detailed accounts of this experiment and Michelson and Morley's successive experiments the reader is referred to Lloyd S. Swenson, *The Ethereal Aether. A History of the Michelson-Morley-Miller Aether-Drift Experiments, 1880–1930* (Austin and London, 1972). Swenson seems to take it for granted that Michelson viewed his experiment as a quest for absolute motion. I suspect this was not the case.

[76] A. A. Michelson and E. W. Morley, "Influence of Motion of the Medium on the Velocity of Light," *Amer. Journ. Sci., 31* (1886), 377–386.

[77] *Ibid.*, p. 386.

[78] *Ibid.*, p. 379.

[79] A. A. Michelson and E. W. Morley, "On the Relative Motion of the Earth and Luminiferous Ether," *Amer. Journ. Sci., 34* (1887), 333–345, esp. 334.

as the earth. That the ether can penetrate metal is shown by the change of volume of the Torricellian vacuum when a barometer tube is tilted; the ether freely passes through the wall of the tube. But free penetration does not necessarily indicate the absence of resistance. Much less can it be taken for granted that the ether can pass through so extended a body as the earth without resistance. Such an important problem as this should, as Lorentz had stated in his 1886 article, be decided not by supposition, but by experiment. The immediate purpose of Michelson and Morley's experiment was, therefore, to determine the relative motion of the earth and the luminiferous ether, and they concluded that "the ether is at rest with regard to the earth's surface."[80] However, their discussion of the background of the experiment shows that behind this purpose stood the more fundamental concern with the physical problem of the connection between the ether and matter.

Lorentz' 1886 article "On the Influence of the Motion of the Earth on Luminiferous Phenomena,"[81] to which Michelson and Morley referred in their paper, discusses from a unified theoretical point of view the results obtained thus far in the pursuit of the ether problem; it is the fundamental work to which later investigations would always have to refer. In this sense, it, together with the 1887 Michelson-Morley experiment, opened a new epoch in the history of the ether problem. The article clearly identifies "the connection between matter and the ether" as the central issue of the ether problem. Lorentz' article begins with the proof that the two fundamental assumptions of Stokes' theory of the aberration of light, that is, the existence of a velocity potential of the motion of the ether and the null relative velocity between the ether and the surface of the earth, are incompatible. However, of these two assumptions, Lorentz observed, only that of the existence of the velocity potential is indispensable for the explanation of aberration. He therefore examined the possibility of developing a theory in which the assumption of the relative velocity at the surface of the earth is dropped and replaced by Fresnel's hypothesis of the partial dragging of the ether by transparent bodies. He made the following assumptions:[82] first, that the ether surrounding the earth is in motion and that this ether has a velocity potential; second, that the motions of the ether and

[80] *Ibid.*, p. 339.
[81] H. A. Lorentz, *op. cit.* (note 67).
[82] *Ibid.*, § 8.

the earth can be different from each other at the earth's surface; third, that when the ether moves through a transparent body, the elementary waves of light in this body are dragged along the direction of the relative motion of the body with respect to the ether with the velocity kv. Here v denotes the relative velocity of the body to the ether, and $k = 1 - 1/n^2$, n being the refractive index of the body. Finally, Lorentz made no assumptions about opaque bodies. With these assumptions, and neglecting terms higher than the first order of v/c, Lorentz examined the path of light rays with regard to the earth—the relative rays, as he called them—to show that all phenomena occur as if the earth were at rest and the relative rays followed the path of light rays with regard to the ether, that is, the path of the absolute rays. In other words, except for the Doppler effect of light from the stars, there is no detectable effect of the motion of the earth upon optical phenomena. This result, agreeing with the conclusion from Fresnel's theory, accounts for all results of the experiments thus far performed. Since his conclusion depends to a large extent on the drag coefficient, Lorentz declared its experimental confirmation to be of special importance. Fizeau's 1851 experiment had confirmed the coefficient only qualitatively, leaving its numerical value undetermined. But Michelson and Morley's experiment had confirmed the numerical value assumed by Fresnel.

Lorentz' general theory, however, cannot determine whether or not the ether remains stationary, notwithstanding the motion of the earth, because of the second assumption above. Lorentz therefore, in the final part of the 1886 article, tried to approach the problem from another angle by emphasizing the "connection of matter and the ether."[83] He first considered the case of opaque bodies that do not permit the ether to penetrate them. In this case the ether within the telescope tube, together with the telescope, will take part in the motion of the earth. But some experiments suggest that opaque bodies, at least when they are not thick, permit the ether to pass through them freely. For example, when a barometer tube is inclined, the ether contained in the Torricellian vacuum goes out freely through glass and mercury. If we regard the atoms of matter as a local modification of the ether, we may expect that the ether freely penetrates material bodies however thick they might be. Lorentz considered this problem so important that he urged physicists not to be content with considerations of probability or simplicity, but to

[83] *Ibid.*, § 24.

decide on the basis of experiment whether the ether at the surface of the earth is at rest or in motion.[84] He referred to two experiments that had already been performed: Fizeau's 1859 experiment to see if the change of the azimuth of the plane of polarization caused by refraction of polarized light entering into a layer of glass is modified by the motion of the earth and Michelson's 1881 experiment to detect ether drift. The former experiment, Lorentz remarked, seems to have shown that the ether is at rest with respect to the surface of the earth but is not so definitive that it can determine the relative velocity. The latter he showed to be not sufficiently precise when Michelson's overestimation of the effect is corrected.[85]

Thus in his 1886 paper Lorentz reserved his conclusion about the motion of the ether. But there is little doubt that he was inclined to the hypothesis of a stationary ether. He based the theory of optical properties of matter which he had developed since 1875 on the fundamental assumption that the ether exists also in the interior of material bodies, permeating the intermolecular spaces, and can be treated separately from material particles as far as its electromagnetic effect is concerned.[86] Lorentz remarked that his assumption is also supported by the study of the effect of the motion of material bodies on optical phenomena.[87] In 1892 he laid the foundation of the electron theory by adopting the hypothesis of a stationary ether. Characterizing his theory as "the theory of electromagnetic phenomena based on the idea that ponderable matter is completely transparent to the ether and can move without communicating any motion to the latter," he stated that "one can adduce some facts in optics as the ground for this hypothesis."[88] These examples show that

[84] *Ibid., Collected Papers, 4,* 203: "Quoi qu'il en soit, on fera bien, à mon avis, de ne pas se laisser guider, dans une question aussi importante, par des considérations sur le degré de probabilité ou de simplicité de l'une ou de l'autre hypothèse, mais de s'addresser à l'expérience pour apprendre à connaître l'état, de repos ou de mouvement, dans lequel se trouve l'éther à la surface terrestre."

[85] *Ibid.,* §26.

[86] Cf. T. Hirosige, "Origins of Lorentz' Theory of Electrons and the Concept of the Electromagnetic Field," *Historical Studies in the Physical Sciences, 1* (1969), 151–209.

[87] H. A. Lorentz, *Over de Theorie der Terugkaatsing en breking van het licht* (Academisch Proefschrift, Leiden, 1875); *Collected Papers, 1,* 1–192. See p. 87.

[88] H. A. Lorentz, "La théorie électromagnétique de Maxwell et son application aux corps mouvants," *Arch. néerl., 25* (1892), 363–552; *Collected Papers,*

Lorentz' concern for the ether problem was related to his interest in the theory of optical and electromagnetic properties of matter, that is, it was closely connected with the inception and emergence of his electron theory. In turn he could not formulate the electron theory without picturing clearly the connection of material molecules with the ether. It is, therefore, not fortuitous that Lorentz' 1886 paper closes with the suggestion that the ether problem will eventually be reduced to the problem of the connection between matter and the ether, that is, to the question whether or not moving ponderable bodies communicate their motion to the ether within and around them.

Rayleigh also considered the ether problem in the light of the question of the connection between the ether and ponderable matter. In a paper written in 1887 but published in *Nature* in 1892, he stated that the ether problem "must evidently turn upon the question whether the aether at the earth's surface is at rest, absolutely or relatively to the earth."[89] Insofar as effects of the first order in v/c are concerned, Fresnel's theory agrees with all the facts. Michelson's experiment of 1881, as Lorentz noticed, leaves some doubt about its precision. Although the hypothesis of a stationary ether seems, at the present, to be advantageous, Rayleigh agreed, the problem cannot be considered settled. Thus, to decide the problem, he proposed to test experimentally whether or not the path of light passing near a heavy mass moving at high speed is affected by the motion of the mass.[90] If an effect is detected, then moving bodies can more or less communicate their motion to the surrounding ether.

The proposed experiment was carried out by Oliver Lodge.[91] Sending two bundles of light between two horizontal steel disks

---

2, 164–343, esp. 228: "Il m'a semblé utile de développer une théorie des phénomènes électromagnétique basée sur l'idée d'une matière pondérable parfaitement perméable à l'éther et pouvant se déplacer sans communiquer à ce dernier le moindre mouvement. Certains faits de l'optique peuvent être invoqués à l'appui de cette hypothèse. . . ."

[89] Rayleigh, "Aberration," *Nature, 45* (1892), 449–502; *Scientific Papers, 3,* 542–553, esp. 544.

[90] *Ibid., Scientific Papers, 3,* 551.

[91] O. J. Lodge, [a] "Aberration Problems.—A Discussion concerning the Motion of the Ether near the Earth, and concerning the Connexion between Ether and Gross Matter; with some new Experiments," *Phil. Trans., A184* (1893), 727–804; [b] "Experiments on the Absence of Mechanical Connexion between Ether and Matter," *Phil. Trans., A189* (1897), 149–166.

along the same path but in opposite directions, he investigated whether or not putting the steel disks into rapid rotation produces a difference in the speeds of the two bundles of light. He also tested the effect of an electric or magnetic field applied perpendicularly to the disk. The experiment, which he repeated several times from 1891 through 1894, gradually improving the instrument, always gave negative results. What is important for us here is how Lodge viewed the problem situation of the experiment. He said that "the nature of the connexion between ether and gross matter is one of the most striking physical problems which now appear ripe for solution."[92] Explaining his intention he asked if, when a material body moves, the ether within the body moves as a whole with the latter or if only the modifications of the ether produced by the presence of matter travel, the ether itself remaining stationary.[93] Fizeau's 1851 experiment indicates the intermediate case which Fresnel supposed: in Fresnel's theory the ether surrounding material bodies is assumed to be always stationary. All the negative results of the experiments to detect an ether drift can be accounted for by assuming either that the ether is completely connected with matter, or, if FitzGerald's contraction hypothesis is granted, that it is entirely independent of matter. It is therefore desirable, argued Lodge, to test—with his experiment of rotating steel disks—whether the ether outside material bodies remains stationary or not. The conclusion he drew was that "the experiment proves, I think, that by the motion of ordinary masses of matter the ether is appreciably undisturbed, and raises a presumption in favour of the earth's motion being equally impotent."[94]

In 1895 Ludwig Albert Zehnder made an experiment to see whether the ether within opaque solid bodies moves with the bodies or not.[95] He attempted to test by optical methods whether the ether inside an iron cylinder is condensed or not by moving an iron piston back and forth in the cylinder. Having obtained a negative result, Zehnder concluded that solid bodies, like fluid matter, are transparent to the ether. Two years earlier, in 1893, Richard August Reiff had obtained an equation expressing the propagation of light within

[92] *Ibid.*, [a], p. 729.

[93] *Ibid.*, p. 731.

[94] *Ibid.*, p. 753.

[95] L. Zehnder, "Ueber die Durchlässigkeit fester Körper für den Lichtäther," *Ann. d. Phys.*, *55* (1895), 65–81.

moving dielectrics on the basis of Helmholtz' electrodynamics,[96] from which he derived the Fresnel coefficient.[97] In the same year Helmholtz had shown that integration of the Maxwellian stress over a closed surface in the ether gives a finite value and asserted that in an excited electromagnetic field a non-vanishing resultant force acts upon a portion of ether having finite volume, giving rise to a flow of the ether.[98] By bringing up the question of the movability of portions of the ether, Helmholtz' paper seems to have promoted interest in the connection between the ether and matter. A few years later Joseph Larmor, in his historical survey of the ether problem, wrote that Lodge's experiment denied any such motion of the ether as Helmholtz had predicted.[99] In 1897 William Craig Henderson and John Henry performed an experiment to detect directly by means of an interferometer the flow of the ether predicted by Helmholtz' theory to occur when the electric displacement and magnetic force are not zero.[100] They naturally obtained a negative result.

With the exception of Michelson's futile 1897 experiment to detect differences in the relative velocities of the ether and the earth at different altitudes,[101] most of the experiments on the ether problem made in the 1890's thus focused on the connection between the ether and matter. The same tendency is noticeable in theoretical treatments of this period.

It is evidently H. A. Lorentz who, in the period considered, studied most thoroughly the ether problem on the basis of electromagnetic theory. He augmented his investigations after 1892[102] to form the

[96] For Helmholtz' electrodynamics see T. Hirosige, *op. cit.* (note 86), pp. 161-167.

[97] R. Reiff, "Die Fortpflanzung des Lichtes in bewegten Medien nach der electrischen Lichttheorie," *Ann. d. Phys., 50* (1893), 361-367.

[98] H. von Helmholtz, "Folgerungen aus Maxwell's Theorie über die Bewegung des reinen Aethers," *Ann. d. Phys., 53* (1893), 135-143.

[99] J. Larmor, *Aether and Matter. A Development of the Dynamical Relations of the Aether to Material Systems, on the Basis of the Atomic Constitution of Matter, Including a Discussion of the Influence of the Earth's Motion on Optical Phenomena* (Cambridge, 1900), p. 19.

[100] W. C. Henderson and J. Henry, "Experiments on the Motion of the Ether in an Electromagnetic Field," *Phil. Mag., 44* (1897), 20-26.

[101] A. A. Michelson, "The Relative Motion of the Earth and the Ether," *Amer. Journ. Sci., 3* (1897), 475-478.

[102] H. A. Lorentz, "Over de terugkaating van licht door lichamen die zich bewegen," *Versl. Kon. Akad. Wet., 1* (1892), 28-31; "De relative beweging van de aarde en den aether," *ibid., 1* (1892), 74-79; "De aberratietheorie van

monograph *An Essay on the Theory of Electrical and Optical Phenomena in Moving Bodies* published in 1895.[103] In this essay he accounted for the absence of the effect of the motion of the earth by proving the theorem of corresponding states within the first order approximation.[104] I shall discuss Lorentz' theory in section six; in the present section I am concerned with his view of the nature of the problem with which he was wrestling. He began the *Essay* with the words: "The question whether the ether takes part in the motion of ponderable bodies or not has not yet found an answer which satisfies all physicists."[105] He then explained why he had long thought Fresnel's stationary ether to be preferable. His reasons were, first, that the ether cannot be confined within solid or liquid walls and, second, that the Fresnel coefficient had been experimentally confirmed.[106] Lorentz now aimed to develop a theory on the fundamental hypothesis of a stationary ether which would account for all known facts. For this purpose he found the electron theory most suitable, because it enabled him to introduce the penetration of ether into matter into the equations in a satisfactory way.[107] Lorentz' words again show that in his conception the ether problem was almost synonymous with the question of the connection between ether and matter.

Before closing this section, a few words may be devoted to the theory of the propagation of light in moving bodies which Woldemar Voigt developed in 1887 on the basis of the elastic wave theory of light.[108] Voigt's theory is remarkable because of its success in deriv-

---

Stokes," *ibid., 1* (1892), 97-103; "Over den infloed van de beweging der aarde op de voortplanting van het licht in dubbelbrekende lichamen," *ibid., 1* (1893), 149–154. English translations of these papers appear in *Collected Papers, 4,* 215–218; 219–223; 224–231; 232–236, respectively.

[103] H. A. Lorentz, *Versuch einer Theorie der electrischen und optischen Erscheinungen in bewegten Körpern* (Leiden, 1895); *Collected Papers, 5,* 1-138.

[104] For Lorentz' theorem of corresponding states, see T. Hirosige, "Electrodynamics before the Theory of Relativity, 1890–1905," *Japanese Studies in the History of Science,* No. 5 (1966), pp. 1–49, esp. pp. 14–18.

[105] H. A. Lorentz, *op. cit.* (note 103), *Collected Papers, 5,* 1. "Die Frage, ob der Aether an der Bewegung ponderabler Körper theilnehme oder nicht, hat noch immer keine alle Physiker befriedigende Beantwortung gefunden."

[106] *Ibid.,* pp. 1-3.

[107] *Ibid.,* p. 7.

[108] W. Voigt, "Theorie des Lichtes für bewegte Medien," *Gött. Nachr.* (1887), pp. 177–238; *Ann. d. Phys., 35* (1888), 370–396; 524–551.

ing the Fresnel coefficient and other various experimental results by taking into consideration mechanical forces acting between the ether and material particles and by assuming that the ether always remains stationary without sharing the motions of material bodies. We may call it the translation of Lorentz' electron theory into the language of elastic wave theory.

5. APOGEE OF THE ETHER PROBLEM

Physicists studied the connection between the ether and matter with still greater eagerness from the end of the nineteenth century until well into the first decade of the twentieth. This last rise of interest sprang from, among other things, the 1898 Düsseldorf meeting of the Society of German Scientists and Physicians. On this occasion, members of the society organized a special conference to discuss the ether problem. For a few years previous to this, the physics sections of these meetings had held special colloquia on selected topics for which younger scientists were asked to prepare review papers. The preparatory committee of the physics section of the 1898 meeting, consisting of Ludwig Boltzmann, Georg Hermann Quincke, and Emil Gabriel Warburg, adopted "the problem concerning translatory motion of the ether" as the subject of the special discussion.[109] In accordance with this decision, Boltzmann asked Lorentz, who had "deeply studied this problem," to participate in the discussion of the "behavior of the ether in moving media."[110] At the 1897 meeting it had also been decided to invite Dutch scientists to the 1898 meeting, and Lorentz was among those invited.[111] He accepted the invitation with great pleasure.[112] For him this was the first occasion to participate in an international scientific meeting, and he experienced much delight in getting acquainted with German colleagues.[113] The German physicists, too, to judge from the exalted vein of some of their letters, seem to have been greatly stimulated by

[109] F. Klein to Lorentz, 20 October 1897, Algemeen Rijksarchief, den Haag, Lorentz Papers 1.

[110] L. Boltzmann to Lorentz, 13 October 1897, Algemeen Rijksarchief, den Haag, Lorentz Papers 1.

[111] Klein to Lorentz, *op. cit.* (note 109).

[112] Lorentz to Boltzmann, 20 October 1897, Algemeen Rijksarchief, den Haag, Lorentz Papers 1.

[113] G. L. de Haas-Lorentz, ed., *H. A. Lorentz—Impressions of His Life and Work* (Amsterdam, 1957), p. 89.

the discussion of the ether problem with Lorentz. They concluded that it was desirable to repeat experiments on the movability of the ether.[114] DesCoudres was entrusted to repeat Fizeau's experiment on the change of the azimuth of the plane of polarization produced when polarized light is refracted.[115] Stimulated by conversations with Lorentz at the meeting, Max Planck developed the idea of rescuing the Stokes theory of aberration by attributing compressibility to the ether.[116]

Willy Wien prepared the review paper for the discussion. He began his paper with the words: "The questions whether the luminiferous ether takes part in the motion of bodies or not, and whether or not movability can be ascribed to the ether at all have long occupied physicists . . . ."[117] He next discussed successively the questions whether or not, one, the motion of the ether, if it exists at all, expends energy, two, the ether possesses inertial mass, and, three, the motion of solid bodies is communicated to the ether. It is clear that what is of central interest here is not the question of an absolute frame of reference for motion, but the problem of the physical properties of the ether and its connection with ponderable matter. Having examined existing theories and experiments concerning these questions, Wien set forth a program for further research.[118] He noted that there is little chance of achieving a successful theory that is based on the concept of a movable ether without inertia. On the other hand, the use of the notion of a stationary ether leads to a

[114] *Verhandlungen der Gesellschaft deutscher Naturforscher und Ärzte, 70* (1898), 2. Teil, 1. Hälfte, p. 83.

[115] DesCoudres to Lorentz, 18 November 1898, Algemeen Rijksarchief, den Haag, Lorentz Papers 1.

[116] M. Planck to Lorentz, 21 October 1898, Algemeen Rijksarchief, den Haag, Lorentz Papers 1. Planck's idea was not published as a paper but is discussed in some detail in H. A. Lorentz, "De aberratietheorie van Stokes in de onderstelling van een aether niet overal dezelfde dichtheid heeft," *Versl. Kon. Akad. Wet.*, 7 (1899), 523–529. The French translation appears in *Collected Papers, 4,* 245–251.

[117] W. Wien, "Ueber die Fragen, welche die translatorische Bewegung des Lichtäthers betreffen," *op. cit.* (note 114), pp. 49–56. The full paper is published in *Ann. d. Phys., 65* (1898), Beilage, i–xviii. Quotation on p. i: "Die Frage, ob der Lichtäther an den Bewegungen der Körper theilnehme oder nicht, und ob ihm überhaupt Beweglichkeit zuzuschreiben ist, hat die Physiker seit langem beschäftigt und zahllos sind die Annahmen und Vermuthungen, die man für die Eigenschaften des Trägers der electromagnetischen Erscheinungen aufzustellen für nöthig hielt."

[118] *Ibid., Ann. d. Phys., 65* (1898), Beilage, xvii–xviii.

violation of the principle of action and reaction and to theoretical conclusions that disagree with some experimental results. The experimental results contradicting a stationary ether are those of the Michelson-Morley 1887 ether drift experiment, of E. Mascart's experiment on rotation of the plane of polarization by rock crystal, which gave a negative result in contradiction to Lorentz' theoretical prediction that charged condensers do not induce a magnetic field notwithstanding the motion of the earth, and of Fizeau's experiment on the change of the azimuth of the plane of polarization when light is refracted. It is therefore very desirable to repeat these experiments. If they give results refuting the notion of a stationary ether, the only alternative would be to take into account the effect of gravity upon the ether. This is equivalent to bestowing an inertial mass on the ether and therefore will, at the same time, dissolve the difficulty concerning the principle of action and reaction. In this case, the result of Lodge's experiment would have to be disposed of by assuming that a small terrestrial body does not appreciably drag the ether because of its small gravity. The explanation of aberration will be invalidated if the earth puts the surrounding ether in motion by the action of its gravitational force, but the difficulty might be solved by reconsidering the hydrodynamics of a gravitational fluid. The task imposed on theoreticians is to predict cases where motion of the ether is expected to be detected.

Wien's paper was followed by Lorentz' supplementary paper.[119] Lorentz agreed with Wien that the issue was the physical properties of the ether and its connection with ponderable matter. He said: "Ether, ponderable matter, and, we may add, electricity are the building stones from which we compose the material world, and if we could know whether matter, when it moves, carries the ether with it or not, then the way would be opened before us by which we could further penetrate into the nature of these building stones and their mutual relations."[120] As to the movability of the ether, Lorentz

[119]H. A. Lorentz, "Die Fragen, welche die translatorische Bewegung des Lichtäthers betreffen," *op. cit.* (note 114), pp. 56–65; *Collected Papers*, 7, 101–115.

[120]*Ibid.*, p. 56: "Aether, ponderable Materie, und wir wollen hinzufügen Elektricität, sind die Bausteine, aus denen wir die materielle Welt zusammensetzen, und wenn wir einmal wüssten, ob die Materie bei ihrer Bewegung den Aether mit sich fortführe oder nicht, so wäre uns ein Weg gegeben, auf dem wir etwas weiter in das Wesen dieser Bausteine und ihrer gegenseitigen Beziehungen eindringen können."

opposed Wien and defended the stationary ether: the difficulty of the Stokes theory is a kinematical one and can by no means be solved by assuming gravitational action. On the other hand, there are many facts that support the view that material bodies are transparent to the ether. Therefore, "the main question is and will be that of the relation of the theory of a stationary ether, after it has disposed of aberration, to the other facts of the electrical as well as of the optical domain."[121] For Lorentz, who admitted a stationary ether, there doubtlessly existed relative motion between terrestrial bodies and the ether. It is therefore quite natural that he considered it the fundamental problem to find reasons why physical effects of the motion do not reveal themselves. In the paper under consideration, he put forward his plan of explaining the experiments adduced by Wien by means of the theory of electrons. As to the principle of action and reaction, he asserted that since it is a principle obtained within the limits of daily experience it need not be valid for the elementary interaction between the ether and ponderable matter.[122]

Without doubt, the Düsseldorf meeting greatly promoted interest in the ether problem among German speaking physicists. They proposed, discussed theoretically, and carried out various experiments designed to probe into the connections between moving bodies and the ether. The experiment on the change of the azimuth of the plane of polarization produced when light enters a glass layer, which was entrusted to DesCoudres, seems to have been eventually abandoned. There is no report of its result as far as I know. Gustav Mie remarked on the flow of the ether predicted by Helmholtz' theory at the Düsseldorf meeting[123] and later published two papers discussing hydrodynamical motion of the ether on the basis of Helmholtz' theory.[124] The Dutch physicist Hermann Haga, who had attended the meeting, repeated Klinkerfues' 1870 experiment immediately after the Düsseldorf meeting.[125] Klinkerfues'

[121] *Ibid.*, p. 59: "Die Hauptfrage ist und bleibt, wie sich die Theorie des ruhenden Aethers, nachdem sie mit der Aberration abgerechnet hat, zu den sonstigen Thatsachen sowohl auf electrischem wie auch auf optischem Gebiete verhält."

[122] *Ibid.*, p. 64.

[123] *Ibid.*, p. 65.

[124] G. Mie, "Ueber mögliche Aetherbewegungen," *Ann. d. Phys.*, *68* (1899), 129–134; "Ueber die Bewegung eines als flüssig angenommenen Aethers," *Phys. Zeits.*, *2* (1801), 319–325.

[125] H. Haga, "Ueber den Versuch von Klinkerfues," *Arch. néerl.*, *5* (1900), 583–586; "L'expérience de Klinkerfues," *ibid.*, *6* (1901), 765–772.

experiment supposedly had shown a change of the positon of the absorption line of bromine with a change in the relative direction of rays with respect to the motion of the earth; Haga confirmed that such an effect was absent.[126] As to the experiment on the rotation of the plane of polarization by rock crystal, Richard Wachsmuth and Otto Schönrock asserted in 1902 that the instrument used by Mascart had not been perfect and hence the experiment should be repeated.[127] In the same year Egon von Oppolzer proposed to observe the deflection of light from the stars which might be produced by the rotation of the ether that is caused by the diurnal rotation of the earth.[128] Physicists also discussed detection of differences in the intensities of light travelling in different directions by means of a bolometer, a method Fizeau had once proposed.[129] In 1902 H. A. Lorentz and Alfred Heinrich Bucherer debated the feasibility of this experiment and eventually agreed that it would give a negative result.[130] Bucherer then directed his student Paul Nordmeyer to carry out the experiment, and Nordmeyer obtained the predicted negative result.[131] In 1904 W. Wien and Alfred Fritz Schweitzer independently of each other proposed another method of detecting differences in the speeds of light travelling westward and eastward.[132] A. A. Michelson criticized their proposal, indicating that a factor

[126]W. Klinkerfues, *op. cit.* (note 48), *Astron. Nachr., 76* (1870), 33-38.

[127]R. Wachsmuth und O. Schönrock, "Beiträge zu einer Wiederholung des Mascart'schen Versuches," *Verh. Deutsch. Phys. Ges., 4* (1902), 183-188.

[128]Egon v. Oppolzer, "Erdbewegung und Aether," *Ann. d. Phys., 8* (1902), 898-907.

[129]A. H. L. Fizeau, "Constatation du mouvement de la terre par les radiations calorifiques," *Cosmos, 1* (1853), 689-692; *Ann. d. Phys., 92* (1854), 652-655.

[130]A. H. Bucherer to Lorentz, 15 February 1902, 6 April 1902, and 8 December 1902, Algemeen Rijksarchief, den Haag, Lorentz Papers 2. H. A. Lorentz, "The Intensity of Radiation and the Motion of the Earth," *Proc. Roy. Acad. Amsterdam, 4* (1902), 678-681; *Collected Papers, 5,* 167-171. The original Dutch version, *Versl. Kon. Akad. Wet., 10* (1902), 804-808; A. H. Bucherer, "Über den Einfluss der Erdbewegung auf die Intensität des Lichtes," *Ann. d. Phys., 11* (1903), 270-283.

[131]P. Nordmeyer, "Über den Einfluss der Erdbewegung auf die Verteilung der Intensität der Licht- und Wärmestrahlung," *Ann. d. Phys., 11* (1903), 284-302.

[132]W. Wien, "Über einen Versuch zur Entscheidung der Frage, ob sich der Lichtäther mit der Erde bewegt oder nicht," *Phys. Zeits., 5* (1904), 585-586. A. Schweitzer, "Über die experimentelle Entscheidung der Frage, ob sich der Lichtäther mit der Erde bewegt oder nicht," *Phys. Zeits., 5* (1904), 809-811.

they had overlooked in their plan would invalidate the proposed experiment.[133]

From about 1900 English speaking scientists, too, became remarkably active pursuing the inquiry into the ether problem. To Kelvin, speaking at the Royal Institution in 1900, the question of "how could the earth move through an elastic solid, such as essentially is the luminiferous ether," was one of the two clouds hanging over nineteenth century physics.[134] In 1901 Frederick Thomas Trouton performed an experiment which had been proposed by G. F. FitzGerald, who died on 22 February 1901.[135] In the experiment a condenser whose plates are laid in the direction of the motion of the earth is charged with electricity. Since a moving electric charge is equivalent to an electric current, the condenser must acquire, in addition to electrostatic energy, a certain amount of magnetic energy. Since this magnetic energy may come from the kinetic energy of the earth, the condenser must receive an impulse when it is charged with or discharges electricity. Trouton's experiment showed no effect. He then investigated the consequence of another assumption, namely, that the energy in question is supplied by the source of electricity for charging the condenser. If this is the case, a couple which tends to turn the plates of the condenser into the direction perpendicular to the earth's motion should act upon the condenser when it is charged. Trouton and Henry R. Noble tried to find such a couple in 1903 but failed.[136]

In 1902, preceding Wachsmuth's proposal to repeat Mascart's experiment, Rayleigh carried out an experiment on the rotation of the plane of polarization and obtained a negative result.[137] Rayleigh

[133] A. A. Michelson, "Relative Motion of Earth and Aether," *Phil. Mag., 8* (1904), 716–719.

[134] Kelvin, "Nineteenth Century Clouds over the Dynamical Theory of Heat and Light," *Phil. Mag., 2* (1901), 1–40; *Journ. Roy. Inst., 16* (1902), 363–397; reproduced in *The Royal Institution Library of Science, Physical Series* (London, 1970), *5*, 324–358, esp. 324.

[135] F. T. Trouton, "The Results of an Electrical Experiment, Involving the Relative Motion of the Earth and Ether, Suggested by the late Prof. FitzGerald," *Trans. Roy. Soc. Dublin,* 7 (1902), 379–384; *The Scientific Writings of G. F. FitzGerald,* pp. 557–565.

[136] F. T. Trouton and H. R. Noble, "The Mechanical Forces Acting on a Charged Condenser Moving through Space," *Phil. Trans., A202* (1904), 165–181.

[137] Rayleigh, "Is Rotatory Polarization Influenced by the Earth's Motion?" *Phil. Mag., 4* (1902), 215–220; *Scientific Papers, 5,* 58–62.

then attempted to detect the double refraction that would result from the motion of transparent bodies if the Lorentz-FitzGerald contraction was a real effect. This experiment, too, was fruitless.[138] In 1904 D. B. Brace repeated it and concluded from his negative result that the contraction hypothesis was not tenable.[139] Larmor contradicted him, asserting that Brace's result could be accounted for by the theorem of corresponding states.[140] In the following year Brace repeated Fizeau's experiment on the change of the azimuth of the plane of polarization[141] and Mascart's experiment on the rotation of the plane of polarization.[142] Obtaining negative results in both experiments he concluded that the absence of the first order effect was established.

As for the second order effect, E. W. Morley and Dayton Clarence Miller in 1905 repeated Michelson-Morley's 1887 experiment and confirmed the negative result.[143] It should here be noted that their experiment was intended not only to detect an ether drift but, in addition, to test the contraction hypothesis. They reasoned that since the contraction should probably depend on physical properties of the solid body, it would be possible to test the contraction hypothesis by detecting different contractions of mounting beds made of different materials.[144]

All the investigations at the turn of the century thus suggested two conclusions that contradict each other: that the ether is mechanically independent of ponderable matter, and that one cannot in any way

[138] Rayleigh, "Does Motion through the Aether Cause Double Refraction?" *Phil. Mag., 4* (1902), 678-683; *Scientific Papers, 5,* 63-67.

[139] D. B. Brace, "On Double Refraction in Matter Moving through the Aether," *Phil. Mag.,* 7 (1904), 317-329.

[140] J. Larmor, "On the Ascertained Absence of Effects of Motion through the Aether, in Relation to the Constitution of Matter, and on the FitzGerald-Lorentz Hypothesis," *Proc. Phys. Soc. London, 18* (1904), 253-258; *Mathematical and Physical Papers, 2,* 274-280.

[141] D. B. Brace, "The Aether 'Drift' and Rotary Polarization," *Phil. Mag., 10* (1905), 383-396.

[142] D. B. Brace, "A Repetition of Fizeau's Experiment on the Change Produced by the Earth's Motion in the Rotation of a Refracted Ray," *Phil. Mag., 10* (1905), 591-599.

[143] E. W. Morley and D. C. Miller, [a] "On the Theory of Experiments to Detect Aberrations of the Second Degree," *Phil. Mag., 9* (1905), 669-680; [b] "Report of an Experiment to Detect the FitzGerald-Lorentz-Effect," *ibid.,* pp. 680-685.

[144] *Ibid.,* [a], p. 669.

detect the effects of the ether drift that must exist at the surface of the earth if the ether is independent of the motions of material bodies. Lorentz and other theoretical physicists made strenuous efforts to reconcile these conclusions. Lorentz had already in 1895 been able to account for the absence of observable effects of the first order in his *Essay*. As to the Michelson-Morley experiment which was of second order, however, he had had to be satisfied with explaining it by introducing the *ad hoc* hypothesis of contraction. He considerably advanced his theory in 1899.[145] Introducing a transformation of coordinate variables which in its form was identical with the relativistic transformation, he proved that the contraction of moving bodies is necessarily required to secure the theorem of corresponding states in the first order effects. In the same paper he also showed that the mass of any material body will always be altered by its motion.

The "dynamical theory of luminiferous ether" that J. Larmor had developed since 1893 was very similar to Lorentz' theory in many respects. In one of a series of papers on this subject published in 1897, Larmor proved the theorem of corresponding states to the second order by using the same transformation of coordinate variables as Lorentz.[146] Then, in his book *Aether and Matter*, completed by the end of 1898 and published in 1900, he introduced transformations of coordinate variables and field variables of the same form as those used in the theory of relativity.[147] If he had made full use of these transformations, he could have proved the strict correspondence of states between two physical systems of which one is at rest and the other in motion and contracted. Larmor indeed remarked that the electromagnetic equations in the moving system, written in terms of the new variables, have the same form as the Maxwell equations in the stationary system. Nevertheless he called his theory an "approximation carried to the second order"

[145] H. A. Lorentz, "Vereenvoudige theorie der electrische en optische verschijnselen in lichamen die zich bewegen," *Versl. Kon. Akad. Wet.*, 7 (1899), 507–522. The French translation appears in *Collected Papers*, 5, 139–155. For a brief account of the theory developed in this paper see T. Hirosige, *op. cit.* (note 104), pp. 24–27.

[146] J. Larmor, "A Dynamical Theory of the Electric and Luminiferous Medium. Part III: Relations with Material Media," *Phil. Trans.*, *A190* (1897), 205–300; *Mathematical and Physical Papers*, 2, 11–132. For Larmor's theory see T. Hirosige, *op. cit.* (note 104), pp. 10–14.

[147] J. Larmor, *op. cit.* (note 99), pp. 173–179.

and was satisfied with instituting "a correspondence which will be correct to the second order."[148] Larmor's failure to fully exploit his results must be due to his failure to recognize the problem situation that Henri Poincaré had emphasized since 1895. Poincaré occasionally expressed the view that no influence of the earth's motion on optical phenomena will be detected in any order of approximation and that a theory will one day be formulated from which physicists can derive this prediction. At the International Congress of Physics held in Paris in 1900, he praised Lorentz' theory as the most satisfactory one among existing theories but expressed discontent that it had required new hypotheses for each new experimental result.[149] Poincaré's criticism and the negative results of the experiments by Trouton and Noble, Rayleigh, and Brace motivated Lorentz to make fresh efforts to reach, in 1904, a theory that was able "to show, . . . without neglecting terms of one order of magnitude or another, that many electromagnetic actions are entirely independent of the motion of the system."[150] This theory of 1904, refined and augmented by Poincaré in the following year, furnished a satisfactory answer to the question of the "relation. . .[of] the theory of a stationary ether, after it has disposed of the aberration, with other facts in electrical as well as optical domains" which Lorentz had posed at the 1898 Düsseldorf meeting. It may therefore be viewed as the end of the ether problem.

6. CHARACTER OF LORENTZ' THEORY

The Lorentz-Poincaré theory has in recent years been the subject of active discussion by historians of physics. It might therefore be superfluous to describe its content in detail. It is, I believe, nonetheless desirable to reconsider its character in relation to the ether

[148] *Ibid.*, p. 173.

[149] H. Poincaré, "Relations entre la physique expérimentale et la physique mathématique," *Rapports présentés au Congrès international de Physique de 1900* (Paris, 1900), *1*, 1–29; *La science et l'hypothèse* (Paris 1902), Chaps. 9 and 10.

[150] H. A. Lorentz, "Electromagnetische verschijnselen in een stelsel, dat zich met willekeurige snelheid kleiner dan die van het licht bewegt," *Versl. Kon. Akad. Wet. Amst.*, *12* (1904), 986–1009. The English version: "Electromagnetic Phenomena in a System Moving with Any Velocity Smaller than That of Light," *Proc. Roy. Acad. Amsterdam*, *6* (1904), 809-831; *Collected Papers*, *5*, 172–197.

problem, because it will be necessary to evaluate Lorentz' and Poincaré's theoretical efforts against the background of the contemporary problem situation to demonstrate the novelty of Einstein's approach.

Looking back on the history of the ether problem as we have considered it in the foregoing sections, we may safely conclude that the ether problem had not been viewed as a search for an absolute frame of reference for motion. Throughout the whole period considered we can find almost no argument that the ether provides the absolute frame of reference supposed by Newton. The sole exception, it seems, is the view Lodge expressed in 1898. He wrote in a letter to the *Philosophical Magazine,* criticizing W. Sutherland's objection to the Michelson-Morley experiment, that "the whole of this subject [the ether problem] indicates that the aether is a physical standard of rest; and that motion relative to it, which is becoming cognisable by us, is in that sense an ascertained absolute motion."[151] He arrived at his view by discussing the question of the correct expression for the kinetic energy of a moving body.[152] He argued that it seems physically absurd that the amount of kinetic energy of a body, $1/2\ mv^2$, depends on the coordinate system in which the velocity v is measured. Instead, physicists should attach real meaning to an absolute velocity. On the one hand, the ether freely penetrates material bodies and, in turn, exerts no resistance to bodies moving through it, since it lacks viscosity at its boundaries. On the other hand, all interactions between portions of matter are thought to be mediated by the ether. It may therefore be asserted, Lodge declared, that the kinetic and potential energies differ categorically from each other: the former is a property only of matter and the latter only of the ether. Lodge concluded from this that it is reasonable to measure the kinetic energy by taking the ether as the reference of rest. In this sense the ether provides the absolute reference system for velocity. Lodge's argument shows that he based his assertion on the results of contemporary discussion of the connection between the ether and ponderable matter. In other words, his view of the ether as an abso-

[151] O. J. Lodge, "Note on Mr. Sutherland's Objection to the Conclusiveness of the Michelson-Morley Aether Experiment," *Phil. Mag., 46* (1898), 343–344, esp. 344.

[152] O. J. Lodge, "On the Question of Absolute Velocity and on the Mechanical Function of an Aether, with Some Remarks on the Pressure of Radiation," *Phil. Mag., 46* (1898), 414–426.

lute system of reference for motion was a consequence of the pursuit of the ether problem, but not the motive for it. Further, Lodge's view seems to have provoked no serious response.

The foundation of Lorentz' theory is the assumption of a stationary ether and Maxwell's equations describing the electromagnetic state of the stationary ether. Since the interaction between charged particles constituting material bodies and an electromagnetic field depends on the motion of the charged particles with respect to the ether, electromagnetic phenomena occurring in experimental devices that share the motion of the earth are naturally affected by this motion. Lorentz showed that the effects produced compensate each other, thus giving no detectable trace of the influence of the motion. I shall cite an example from Lorentz' article on the electron theory in the *Encyclopedia of Mathematical Sciences* edited by Felix Klein.[153] When a conductor carrying an electric current $I = \rho u$ moves with the earth, it exerts the electric force

$$\mathrm{d}' = \frac{1}{c}\,\mathrm{grad}(\mathrm{w}\cdot\bar{\mathrm{a}}') \tag{1}$$

on an external electric charge. Here w denotes the velocity of the earth, $\bar{\mathrm{a}}'$ is determined by the equation

$$\Delta\bar{a}' - \frac{1}{c^2}\,\ddot{\bar{a}}' = -\frac{1}{c}\,\rho u. \tag{2}$$

(A bar over a letter indicates an averaged value.) Such a force, however, is not observed. Lorentz explained this by noting that force (1) also acts upon electric charges within the conductor. The force induces within the conductor a charge density $\bar{\rho} = 1/c^2\ (wI)$, and this charge in turn gives rise to a scalar potential whose value is $1/c\ (w\bar{a}')$. The contribution of this potential to the electric force cancels out force (1).

It is, however, impossible to prove universally the absence of effects of motion by confirming cancellation in each separate case. Lorentz sought to surmount this limitation by having recourse to the theorem of corresponding states. The theorem allowed him to discuss phenomena occurring in a physical system that is moving

[153] H. A. Lorentz, "Weiterbildung der Maxwellschen Theorie. Elektronentheorie," *Encyklopädie der mathematischen Wissenschaften* (Leipzig, 1904), *5*, Nr. 14. The example is cited from pp. 260–261.

with respect to the ether in terms of phenomena in a system fixed with respect to the ether by carrying out a suitable transformation of variables. But in 1903 the theory was still essentially an approximate theory to the first order. In his *Encyclopedia* article which was finished in December 1903, Lorentz summarized the outlook of theoretical inquiry.[154] If we assume that the molecular forces are modified by motion in the same way as the electric force, then we can derive the contraction of material bodies and account for the result of Trouton-Noble's experiment. But this manner of explanation has the defect that the thermal motion of molecules is entirely neglected. If we admit that material mass, too, undergoes the same change as electromagnetic mass when the body moves, the difficulty might be solved. At the same time Lorentz accepted Poincaré's criticism that the present theory is forced to introduce *ad hoc* hypotheses for each new phenomenon. He urgently hopes, he said, that a theory can be found which can show from fundamental hypotheses and in a general manner that electromagnetic phenomena on the earth are independent of the motion of the earth. He realized this hope himself with his 1904 theory.[155]

Lorentz' 1904 theory is based on the Maxwell equations

$$\left.\begin{aligned} &\operatorname{div} \mathrm{d} = \rho, \ \operatorname{div} \mathrm{h} = 0, \\ &\operatorname{curl} \mathrm{h} = \frac{1}{c}\left(\frac{\partial \mathrm{d}}{\partial t} + \rho \mathrm{v}\right), \\ &\operatorname{curl} \mathrm{d} = -\frac{1}{c}\frac{\partial \mathrm{h}}{\partial t}, \end{aligned}\right\} \tag{3}$$

which are valid in a coordinate system (x, y, z) fixed with respect to the ether, and on the expression of the force exerted on electric charge

$$\mathrm{f} = \mathrm{d} + \frac{1}{c}\,[\mathrm{v}\,\mathrm{h}]\,. \tag{4}$$

In these equations v is the velocity of an electric charge in the stationary coordinate system which is fixed with respect to the ether. Lorentz' aim is to investigate the electromagnetic phenomena that occur in a physical system travelling with a uniform velocity w in the

[154] H. A. Lorentz, *ibid.*, pp. 277–279.
[155] H. A. Lorentz, *op. cit.* (note 150).

direction of the x-axis and to demonstrate that the phenomena do not exhibit any influence of the motion. Lorentz found that to treat the problem by making use of equations (3) and (4) in the stationary system involves enormous complications. To avoid them he introduced the following variables as a mathematical device:

$$x' = kl(x - wt), \; y' = lu, \; z' = lz, \; t' = kl\left(t - \frac{w}{c^2}x\right), \tag{5}$$

where

$$k = \frac{1}{\sqrt{1 - w^2/c^2}}$$

and the coefficient l is to be considered a function of w, whose value is 1 for $w = 0$ and which, for small values of w, differs from unity no more than by an amount of the second order. Further,

$$\begin{aligned} d'_x &= \frac{1}{l^2} d_x, \; d'_y = \frac{k}{l^2}\left(d_y - \frac{w}{c} h_z\right), \; d'_z = \frac{k}{l^2}\left(d_z + \frac{w}{c} h_y\right), \\ h'_x &= \frac{1}{l^2} h_x, \; h'_y = \frac{k}{l^2}\left(h_y + \frac{w}{c} d_z\right), \; h'_z = \frac{k}{l^2}\left(h_z - \frac{w}{c} d_y\right), \end{aligned} \tag{6}$$

and, for electric charge and the relative velocity of the electron with respect to the physical system considered, $u \, (v = w + u)$,

$$\left.\begin{aligned} \rho' &= \frac{1}{kl^3}\rho \\ u'_x &= ku_x, \; u'_y = ku_y, \; u'_z = ku_z. \end{aligned}\right\} \tag{7}$$

By substituting these expressions for variables in (3) and (4), he obtained equations in primed variables that are of nearly the same from as (3) and (4), that is, the Maxwell equations in the rest system of the ether. This is a result which we can easily anticipate because equations (5) and (6) are of the same form as the relativistic Lorentz transformations. That the result is not "exactly the same form" but "nearly the same form" is due to the difference between equation (7) and the relativistic transformations of electric charge and velocity.

Lorentz next expressed the positions of particles constituting the physical system Σ in terms of relative coordinates $x_r = x - wt$, $y_r = y$, $z_r = z$. He assumed system Σ' to consist of particles that are the same as in Σ but at rest relative to the ether and that have the coordinates $x = klx_r$, $y = ly_r$, and $z = lz_r$. The system Σ' may be obtained by enlarging the system Σ by the ratios 1 to lk in the x-direction, and 1 to l in the y- and z-directions. Then electromagnetic phenomena in the system Σ' will be described by the same equations as those obtained from equations (3) and (4) by the transformation above. This result enabled Lorentz to treat electromagnetic phenomena in a physical system moving relative to the ether by considering phenomena in the system Σ' which is at rest with respect to the ether and associated with the system Σ by definite relations. He arrived at the theorem of corresponding states without neglecting any terms of any order of magnitude. However, to settle the problem completely, Lorentz still needed to introduce a certain number of hypotheses. First, he assumed that an electron moving with velocity w relative to the ether is contracted by the fraction $1/kl$ in the direction of the motion and $1/l$ in the directions perpendicular to it. Second, he assumed that intermolecular forces are modified by motion in the same way as the electrostatic force. This hypothesis leads to the contraction of macroscopic bodies. Third, he assumed that the mass of the electron is entirely electromagnetic. From the requirement that the states of the imaginary system Σ' obtained by his transformation are the ones which occur in reality, he then concluded that $l = 1$. The conclusion implies that the contraction of moving electrons and the deformation of physical systems when transformed from Σ to Σ' occur only in the direction of motion. Fourth, Lorentz assumed that the entire mass of all kinds of particles is modified by motion in the same way as the electromagnetic mass of the electron. This hypothesis assures the correspondence of states even in the case when molecules are in thermal motion.

In his 1904 theory Lorentz demonstrated the absence of effects of motion in quite a general manner by means of the theorem of corresponding states rather than showing directly a mutual compensation of effects. The underlying idea, however, was still that the motion does produce certain effects but that they cannot be detected because they cancel each other. Lorentz' own words show this. For example, in *The Theory of Electrons*, which was published in 1909 on the basis of his 1906 lectures delivered at Columbia University, after a detailed description of his 1904 theory Lorentz stated that the "chief difference" between his theory and Einstein's consists in the latter "making us see in the negative result of experiments like those of Michelson, Rayleigh and Brace, not a fortuitous compensation of opposing effects, but the manifestation of a general and fundamental principle."[156]

7. POINCARÉ'S "PRINCIPLE OF RELATIVITY"

Poincaré received Lorentz' 1904 theory with great enthusiasm. In the next year, 1905, Poincaré gave a mathematically refined form to Lorentz' theory, discussed the stability of the deformable electron, and attempted to extend it so as to include gravity.[157] He not only demonstrated that the so-called Lorentz transformation formed a group, but even, though implicitly, used four-dimensional representation.[158] More remarkably, he had for about ten years been proposing "the principle of relativity." He was fascinated with Lorentz' 1904 theory because it seemed to him to embody his principle of relativity. It is because of these facts that Whittaker and his followers regard Poincaré as the founder of the theory of relativity. But what Poincaré called the "principle of relativity," to judge by its purport, cannot be regarded as identical with the principle of relativity as we understand it in terms of the theory of relativity. It is not given the status of a postulate as is the latter in the theory of relativity. From the analysis of the situation of the ether problem as it was understood at the time, Poincaré anticipated the principle of relativity as an empirical law, looking forward to a theory which could explain or prove the principle.

Poincaré's first statement concerning the principle of relativity appears in his 1895 paper dealing with Larmor's electromagnetic theory. There he stated that the conclusions drawn from various empirical facts can be summarized by the assertion that "it is impossible to make manifest the absolute motion of matter, or rather the motion of ponderable matter relative to the ether."[159] In 1899, in

[156]H. A. Lorentz, *The Theory of Electrons and Its Applications to the Phenomena of Light and Radiant Heat* (Leipzig, 1909), p. 230.

[157]H. Poincaré, [a] "Sur la dynamique de l'électron," *Comptes Rendus, 140* (1905), 1504–1508; *Oeuvres de Henri Poincaré, 9,* 489–493. [b] "Sur la dynamique de l'électron," *Rendiconti del Circolo matematico di Palermo, 21* (1906), 129–176; *Oeuvres, 9,* 494–550. The content of paper [b] is carefully analyzed by Arthur I. Miller, *op. cit.* (note 16).

[158]Arthur I. Miller, *op. cit.* (note 16), p. 252.

[159]H. Poincaré, "A propos de la théorie de M. Larmor. (3)," *L'éclairage électrique, 5* (1895), 5–14; *Oeuvres, 9,* 395–413. Quotation is from p. 412: "Il est impossible de rendre manifeste le mouvement absolu de la matière, ou mieux le mouvement relatif de la matière pondérable par rapport à l'éther."

the lecture at the Sorbonne, he said: "I regard it very probable that optical phenomena would depend only on the relative motion of material bodies . . ., and this would be valid not disregarding quantities of the second or third order in the aberration constant, but rigorously. As experiments become more and more exact, this principle will be verified with increasing precision."[160] Poincaré expected that "a well constructed theory should be able to demonstrate the principle at once and with perfect rigor."[161] He expressed the same view again at the Paris International Congress of Physics in the following year: "I do not believe . . . that more exact observations will ever make evident anything else but the relative displacement of material bodies."[162] For all orders "the same *explanation* must be found. . . . [Everything] tends to show that this explanation would serve equally well for the terms of the higher order, and that the *mutual destruction* of these terms will be rigorous and absolute" (italics mine).[163] That Poincaré's desire for rigor motivated and guided Lorentz' efforts toward his 1904 theory is seen from the previously cited conclusion of his *Encyclopedia* article as well as from the introduction of his 1904 paper.[164] At the same time Poincaré's words show that in the physical interpretation of the theory he completely agreed with Lorentz. For him, too, effects of motion exist but do not manifest themselves because of mutual compensation.

[160]H. Poincaré, *Électricité et optique. La lumière et les theories électrodynamiques. Leçon professees à la Sorbonne en 1888, 1890 et 1899* (Paris, 1901), p. 536: "Je regarde comme très probable que les phénomènes optiques ne dépendent que des mouvements relatifs des corps matériels en presence. . . et cela non pas aux quantités près de l'ordere de carré ou du cube de l'aberration, mais rigoureusement. A mesure que les expériences deviendront plus exactes, ce principe sera vérifié avec plus de precision."

[161]*Ibid.:* "Une théorie bien faite devrait permettre de démontrer le principe d'un seul coup dans toute sa rigueur."

[162]H. Poincaré, *op. cit.* (note 149), *Rapports, 1,* 22; *La science et l'hypothèse,* p. 201: "Je ne crois pas, . . . que des observations plus précises puissent jamais mettre en évidence autre chose que les déplacements relatifs des corps matériels." The English quotation is from *Science and Hypothesis* (New York, 1952), p. 172.

[163]H. Poincaré, *ibid., Rapports, 1,* 23; *La science et l'hypothèse,* p. 202: "Il faut trouver une même explication pour les autres, et alors tout nous porte à penser que cette explication vaudra également pour les termes d'ordre supérieur, et que la destruction mutuelle de ces termes sera rigoureuse et absolue."

[164]H. A. Lorentz, *op. cit.* (note 150), *Collected Papers, 5,* 173–174.

The term "principle of relativity" was used by Poincaré for the first time in September 1904 in his address at the International Congress of Arts and Science at St. Louis. He formulated the "principle of relativity" as follows: "The laws of physical phenomena should be the same, whether for an observer fixed, or for an observer carried along in a uniform movement of translation."[165] At first glance, this expression might remind us of Einstein's principle of relativity. But, in fact, it is an "empirical truth"[166] which might some day be denied by an experiment. This address, in which Poincaré acknowledged indications of a crisis in physics,[167] has also attracted attention because it contains a discussion of the synchronization of clocks by means of light signals[168] which resembles that in Einstein's paper on the theory of relativity. But this discussion, too, quite differs in spirit from Einstein's and goes along with Lorentz' attitude. Poincaré wanted to show that, since the velocity of light differs according to its direction of propagation, when the observer is in motion the observer's clock synchronized by means of light signals is advanced or retarded, indicating only the local time of his position, and that he cannot know the disorder of his clock because he has no means other than his clock.

In his article "Dynamics of the Electron" of 1908,[169] Poincaré discussed in detail the connection of the principle of relativity with the Lorentz theory. He first stated the principle in the following form: "Whatever be the method employed, we shall never succeed in disclosing any but relative velocities; I mean the velocities of certain material bodies in relation to other material bodies."[170] After

[165] H. Poincaré, "L'etat actuel et l'avenir de la physique mathématique," *Bulletin des sci. math., 28* (1904), 302-324, esp. 306; *La valeur de la science* (Paris, 1905), pp. 170-211, esp. pp. 176-177: "Le principe de la relativité, d'après lequel les lois des phénomènes physiques doivent être les mêmes, soit pour un observateur fixe, soit pour un observateur entraîné dans un mouvement de translation uniforme."

[166] H. Poincaré, *ibid., Bulletin,* p. 307; *La valeur de la science,* p. 179.

[167] H. Poincaré, *ibid., Bulletin,* p. 302; *La valeur de la science,* p. 171: "Je répugne à donner un prognostic, je ne puis pourtant me dispenser d'une diagnostic; eh bien, oui, il y a des indices d'une crise sérieuse." The English quotation is from *The Value of Science* (New York, 1958), p. 91.

[168] H. Poincaré, *ibid., Bulletin,* p. 311; *La valeur de la science,* pp. 187-188.

[169] H. Poincaré, "La dynamique de l'électron," *Revue gén. des sci., 19* (1908), 386-402; *Oeuvres, 9,* 551-586; *Science et méthode* (Paris, 1908), pp. 215-272.

[170] H. Poincaré, *ibid., Oeuvres, 9,* 563; *Science et méthode,* p. 235: "Quel que soit le moyen qu'on emploie, on ne pourra jamais déceler que des vitesses

demonstrating that in the Lorentz theory the difference between the local and true times exactly cancels out the effect of the contraction of length and that therefore there occurs no manifest change in the velocity of light through motion, he concluded that "it is impossible to escape the impression that the Principle of Relativity is a general law of Nature."[171] In Poincaré's conception, the validity of the principle of relativity, which has been inferred from experience, must be given an explanation,[172] a demand he repeated in his public lecture at the 1909 meeting of the French Association for the Advancement of Sciences. His lecture delivered at the École Supérieure des Postes, Télégraphes et Téléphones shortly before his death in 1912[173] is especially interesting since it shows that to the end of his life he was faithful to the spirit of Lorentz' compensation theory. In this lecture Poincaré illustrated in detail that the Lorentz theory based on the local time and the contraction hypothesis explain "the perfect compensation which is observed in all the experiments of optics," and that "there is compensation equally in electric phenomena. . . ." He "came to believe that the principle of relativity was perfectly exact."[174]

Once physicists had received the Lorentz theory as the long sought satisfactory solution of the ether problem, they turned their attention quite naturally to the question whether the fundamental hypotheses of the Lorentz theory were acceptable or not. We have already seen that Morley and Miller partly intended their 1905 experiment to verify the contraction hypothesis. In the same year D. B. Brace noted that no valid reason had yet been found for the contraction hypothesis and called attention to Fritz Hasenöhrl's 1904

relatives, j'entends les vitesses de certains corps matériels, par rapport à d'autres corps matériels."

[171] H. Poincaré, *ibid., Oeuvres, 9,* 567; *Science et méthode,* p. 240: "Il est impossible d'échapper à cette impression que le principe de relativité est une loi générale de la Nature." The English quotation is from *Science and Method* (New York, n.d.), p. 221.

[172] H. Poincaré, "La mécanique nouvelle," *Revue électrique, 13* (1910), 23–28, esp. 24: "Il faut expliquer pourquoi."

[173] H. Poincaré, "La dynamique de l'électron," *Supplément aux Annales des Postes, Télégraphes et Téléphones* (1913).

[174] *Ibid.,* p. 47: "Ceci explique la compensation parfaite que l'on observe dans toutes les experiences d'optique. On a également la compensation dans les phenomenes electriques. . . . On est arrivé à croire que le principe de relativité était parfaitement exact."

theoretical discussion of the thermodynamics of radiation contained in a moving cavity.[175] Hasenöhrl had shown that one obtained a contradiction of the second law of thermodynamics unless one introduced the contraction hypothesis.[176] That Walter Kaufmann's experiments to determine the velocity dependence of the mass of the electron[177] attracted much attention in the years following 1904–1905 was due, at least partly, to their close connection with the question of the validity of the fundamental physical assumptions of the Lorentz theory. It should also be mentioned in this context that Poincaré paid serious attention to Kaufmann's experiments and on several occasions emphasized the significance of a new "dynamics of the electron."[178] In fact, Poincaré thought of the dynamics of the electron as the theory that could solve the ether problem, as can be seen from the titles of his articles discussing the Lorentz theory, "The New Mechanics" and "The Dynamics of the Electron."

8. THE FUNDAMENTAL PROBLEM FOR EINSTEIN

Einstein's theory of relativity, unlike the Lorentz-Poincaré theory, did not aim at explaining why effects of motion could not be made manifest. Accordingly, in his theory the principle of relativity is not a law to be deduced from the fundamental principles of the theory, but a postulate. In the introduction to his 1905 paper, Einstein stated that we are led to "the conjecture . . . that . . . for all coordinate systems for which the mechanical equations are valid, the same laws of electrodynamics and optics will also be valid. . . . We will raise this conjecture (the substance of which will hereafter be called the 'principle of relativity') to the status of a postulate. . . ."[179] From

[175] D. B. Brace, "The Negative Results of Second and Third Order Tests of the 'Aether Drift' and Possible First Order Methods," *Phil. Mag., 10* (1905), 71-80, esp. 72.

[176] F. Hasenöhrl, "Zur Theorie der Strahlung in bewegten Körpern," *Ann. d. Phys., 15* (1904), 344-370; *16* (1905), 589-592.

[177] For Kaufmann's experiments, see section 12, *op. cit.* (notes 264, 265, and 268).

[178] For example, H. Poincaré, *op. cit.* (notes 169, 172, and 173).

[179] A. Einstein, "Zur Elektrodynamik bewegter Körper," *Annalen der Physik, 17* (1905), 891-921, esp. 891; Einstein speaks of the "Vermutung, . . . dass . . . für alle Koordinatensysteme, für welche die mechanischen Gleichungen gelten, auch die gleichen elektrodynamischen und optischen Gesetze gelten. . . . Wir wollen diese Vermutung (deren Inhalt im folgenden 'Prinzip der Relativität' genannt werden wird) zur Voraussetzung erheben. . . ."

this principle and from the second postulate of the constancy of the velocity of light Einstein logically derived the whole of his theory. The problem that the Lorentz-Poincaré theory set for itself is, of course, also solved by Einstein's theory, or rather it ceases to exist at all, because the unsuccessful experiments associated with the ether problem are only expressions of the fundamental postulate of Einstein's theory. Einstein, indeed, alluded to them only in vague terms as "examples of a similar kind."[180]

If the ether problem as understood by Lorentz, Poincaré, and other contemporary physicists was not the goal of Einstein's theory, then what was the fundamental problem for Einstein when he created the theory of relativity? To answer this question we should first consider Einstein's own statements. Einstein several times expounded the development of his thought which led him to the theory of relativity. His accounts are not consistent to the finest points, but when conflated they are very revealing. I first enumerate them in roughly chronological order of their publication or recording:

[1] Conversation with psychologist Max Wertheimer, which Wertheimer has reported in his *Productive Thinking.*[181] The conversation was held in 1916 or soon after.

[2] Obituary of Ernst Mach which Einstein wrote in March 1916.[182]

[3] "How did I create the theory of relativity,"[183] an improvised account Einstein gave to students of the University of Kyoto on 14 December 1922.

[4] The biography by Anton Reiser published in 1930.[184] Einstein prefaced it with the words: "The author of this book is one who knows me rather intimately in my endeavour, thoughts, beliefs. . . . I found the facts of the book duly accurate. . . ." Anton Reiser is, according to Gerald

180 *Ibid.*

181 Max Wertheimer, *Productive Thinking,* enlarged edition edited by Michael Wertheimer (New York and London, 1959), pp. 213–226.

182 A. Einstein, "Ernst Mach," *Phys. Zeits., 17* (1916), 101–104.

183 Jun Ishiwara, *Einstein Kyôzyu Kôen-roku* (*The Record of Professor Einstein's Lectures*) (Tokyo, 1923), pp. 131–151. Reprint (Tokyo, 1971), pp. 78–88.

184 Anton Reiser, *Albert Einstein. A Biographical Portrait* (London, 1931).

Holton's investigation,[185] the pseudonym of the husband of Einstein's stepdaughter Ilse.

[5] Letters to his friend Michele Besso dated 6 January 1948 and 6 March 1952.[186]

[6] "Autobiographical Notes" published in 1949.[187]

[7] Interviews by R. S. Shankland.[188]

[8] Letters quoted by Carl Seelig in his biography of Einstein.[189]

[9] Letter to Carl Seelig dated 19 February 1955.[190]

[10] Speeches and letters quoted and examined by Holton.[191]

According to these records Einstein took his first step toward the theory of relativity with a conceptual experiment he carried out at the age of sixteen ([1], [6], [7]). While still a student of the cantonal school in Aarau, Switzerland, he asked himself what phenomena would be seen by an observer who ran after propagating light with a velocity equal to that of light. Would he see a standing electromagnetic field which varied only from point to point? From the very beginning it seemed intuitively clear to Einstein that, judged from the standpoint of such an observer, everything would have to happen according to the same laws as for an observer who was at rest ([1], p. 218; [6], p. 52). In other words, he already possessed conception that was destined to develop into the principle of relativity. However, he was at that time "a pure empiricist" who "expected to approach the major questions of physics by observation and experiment" ([4], p. 54). In his second year at the Federal Institute of

[185]Gerald Holton, "Influences on Einstein's Early Work in Relativity Theory," *The American Scholar, 37* (1967-1968), 59–79. A slightly condensed version appears in *Thematic Origins of Scientific Thought,* pp. 197–217, esp. p. 211.

[186]Albert Einstein/Michele Besso, *Correspondence 1903–1955,* translation, notes, and introduction by Pierre Speziali (Paris, 1972), pp. 390–392; 464–465.

[187]Albert Einstein, "Autobiographisches," Paul Arthur Schilpp, ed., *Albert Einstein: Philosopher-Scientist* (New York, 1949 and 1951), pp. 1–95.

[188]R. S. Shankland, "Conversations with Albert Einstein," *Amer. Journ. Phys., 31* (1963), 47–57.

[189]Carl Seelig, *Albert Einstein. Eine dokumentarische Biographie* (Zurich, 1954).

[190]A. Einstein to C. Seelig, 19 February 1955. Published by Seelig in *Technische Rundschau* (1955), and partially quoted by Max Born, *op. cit.* (note 7), p. 193.

[191]Gerald Holton, *op. cit.* (note 19).

Technology in Zurich he planned to perform an experiment to detect changes of the velocity of light caused by the earth's motion ([1], [3], [4]). He designed an experiment to find the difference in the energies carried by two bundles of light travelling in opposite directions by means of thermopiles ([3], p. 79), but "there was no chance to build this apparatus" because "the scepticism of his teachers was too great, the spirit of enterprise too small" ([4], p. 53). Here again it should be noted that, although attempting the experiment, Einstein did not expect it to be successful. "His wish to design such experiments was always accompanied by some doubt that the thing was really so" ([1], p. 214). At this time he did not know the Michelson-Morley experiment, but even when he later got acquainted with it, its "results were no surprise to him, . . . [but] seemed to confirm . . . his ideas" ([1], p. 217; [10]). It may be said that Einstein had prefigured the principle of relativity from the earliest time.

Pondering over the question of the relation of the laws of optical and electromagnetic phenomena to the motion of the observer, young Einstein spent some time trying to modify Maxwell's equations. "If the Maxwell equations are valid with regard to one system, they are not valid in another. They would have to be changed. . . . For years Einstein tried to clarify the problem by studying and trying to change the Maxwell equations. He did not succeed . . ." ([1], p. 216). He tried to treat Fizeau's experiment concerning the drag coefficient with the Maxwell-Lorentz equations and "believed that they were correct and express rightly the facts. That they are valid also in moving coordinate systems indicates the relation of the so-called constancy of the velocity of light. . . . [But] this is not compatible with the law of composition of velocity known in mechanics" ([3], pp. 81–82). "In whatever way he tried to unify the question of mechanical movement with the electromagnetic phenomena, he got into difficulties" ([1], p. 216). Thus he "had to spend nearly one year with fruitless thinking" ([3], p. 82). During that time he even considered the possibility of an emission theory of light ([7], p. 29).

To restate in our terms, the problem with which Einstein was wrestling in these years was to modify Maxwell's theory in such a way that he would obtain a theory of electromagnetic and optical phenomena in which only relative motion had physical meaning. In other words, Einstein was seeking to bring about a unification of mechanics and electromagnetism with regard to the relativity of motion. He had set himself a problem concerned with the form, rather than the content, of theory, but Einstein, empiricist as he was then, did not become clearly aware of this until he came to reflect upon the consequences of Planck's radiation formula. His "Autobiographical Notes" tell us that, looking at Planck's formula, he found that "radiation must . . . possess a kind of molecular structure in energy, which of course contradicts Maxwell's theory. . . . Reflections of this type made it clear to me as long ago as shortly after 1900, i.e., shortly after Planck's trailblazing work, that neither mechanics nor thermodynamics could (except in limiting cases) claim exact validity. By and by I despaired of the possibility of discovering the true laws by means of constructive efforts based on known facts. The longer and the more despairingly I tried, the more I came to the conviction that only the discovery of a universal formal principle could lead us to assured results. The example I saw before me was thermodynamics. . . . After ten years of reflection such a principle resulted from a paradox upon which I had already hit at the age of sixteen" ([6], pp. 51, 53). Again pointing out the relationship between the theory of relativity and his commitment to the quantum theory of radiation, Einstein wrote to Carl Seelig two months before his death: "The insight that the 'Lorentz invariance' is a general condition for any physical theory . . . was for me of particular importance because I had already previously found that Maxwell's theory did not account for the micro-structure of radiation and could therefore have no general validity" ([9], p. 193). As a result of his reflection on the necessity of a reconstruction of physics, Einstein had come to recognize the profound significance of the principle of relativity.

It is of course impossible to reconcile the Maxwell theory with the principle of relativity without modifying traditional notions of time. "The type of critical reasoning which was required for the discovery of this central point was decisively furthered, in my [Einstein's] case, especially by reading of David Hume's and Ernst Mach's philosophical writing" ([6], p. 53; [2], p. 102; [5], pp. 391, 464; [8], pp. 59–60). The first paper of the theory of relativity was completed five to six weeks after he hit upon the modification of the concept of time ([1], pp. 214, 219; [3], pp. 82–83; [8], p. 82).

The most remarkable point in Einstein's first relativity paper of 1905 is that it begins with the discussion of asymmetries with respect to motion involved in the current form of electromagnetic theory. It then proceeds to introduce the principle of relativity as a postulate to remove asymmetries with regard to motion from the electromagnetic theory. By doing so it aims at securing the relativity of motion in electrodynamics as well as in mechanics. The mode of presentation of the 1905 paper completely corresponds to the foregoing presentation extracted from Einstein's writings.

To summarize, Einstein had speculated on the relation between motion and electromagnetic phenomena since as early as the mid-1890's. When he began to ponder the consequences of Planck's radiation theory, he came to consider the problem in broader perspective. He felt the necessity of rebuilding physics on some formal principle. He was especially concerned about formal incongruities between physical theories. The theory of relativity was a fruit of his efforts to eliminate such incongruities. However, the incongruity between electrodynamics and mechanics with respect to the relativity of motion with which he was concerned here was not the only incongruity between these two theories that worried him. Holton has pointed out,[192] Einstein also found a formal incongruity between them in their respective fundamental entities. His other great achievement of 1905, the theory of light quanta, was also intended to remove "a fundamental formal difference"[193] between mechanics and electromagnetism, namely, the difference of having a discrete fundamental entity, the mass point, in mechanics and a continuous one, the field, in electromagnetism. In contrast to Einstein, both Lorentz and Poincaré, and indeed all other contemporary physicists, did not give any consideration to the formal incongruity between physical theories. The allegation that Einstein's theory might have been suggested by Lorentz' 1904 theory has been refuted by Holton.[194] I may add that even if Einstein had been acquainted with Lorentz' 1904 paper the course of events would not have been changed essentially, for he saw the state of physics in those days

[192]Gerald Holton, *op. cit.* (note 12), [a], pp. 629-630.

[193]A. Einstein, "Über einen die Erzeugung und Verwandlung des Lichtes betreffenden heuristischen Gesichtspunkt," *Ann. d. Phys., 17* (1965), 132-148, esp. 132. As early as 1949 Takehiko Takabayasi noted that "the theory of light quanta had been derived from the formal antithesis of the point mechanics and the field theory." T. Takabayasi, "Koten Buturigaku no Hôkai Katei ni tuite" ("On the Process of the Decline of Classical Physics"), *Kagakusi Kenkyu,* No. 11 (1949), pp. 1-9, esp. p. 5.

[194]Gerald Holton, *op. cit.* (note 185), *Thematic Origins of Scientific Thought,* pp. 204-205.

quite differently than Lorentz, Poincaré, and others, and consequently perceived an entirely new problem which none of them had recognized.

Einstein could not have formed his fruitful conception of the unification of mechanics and electromagnetism by a universal formal principle, however, if the two theoretical systems had not been confronting each other on the same footing. The unification of two theories, unlike the reduction of one to the other, presupposes that both theories are equally privileged or, rather, equally unprivileged. To consider mechanics and electromagnetism as two theories having equal status seems quite natural and easy to us today. But such a viewpoint and the problems it set for physicists were just what Lorentz, Poincaré, and others failed to acquire. The difference in this respect between Einstein and Lorentz, Poincaré, and others is so great that it can hardly be considered accidental. It is not unreasonable to seek its roots in the differences between their epistemologies of physics or worldviews. We must consider the influence of Hume and Mach on Einstein which he himself acknowledged.

9. HUME AND MACH

The period beginning in 1902, when Einstein settled in Berne as an officer of the Federal Patent Office, was especially fruitful for the development of his thought because of the evenings he spent with his friends Maurice Solovine and Conrad Habicht reading scientific and philosophical books such as Mach's writings and Hume's *Treatise of Human Nature.* He professed that Hume exerted more direct influence on his work than Mach,[195] but he does not say precisely what he learned from Hume. According to Solovine's recollection, Einstein and his friends "discussed for several weeks David Hume's particularly sagacious criticism of notions of substance and causality."[196] Hume disclaimed the notion of *substantia,* both material and spiritual, and replaced it with bundles of ideas. His criticism of the notion of causality is one of the best known topics in the history of philosophy. Hume asserted that the causal relation merely meant that an object had occurred always in conjunction with another object; it did not express a necessary relation between the two objects. As far as these assertions are concerned, it is difficult to establish a

[195] A. Einstein to M. Besso, 6 January 1948, *op. cit.* (note 186), p. 391.
[196] Albert Einstein, *Lettres à Maurice Solovine* (Paris, 1956), p. viii.

direct, specific connection with Einstein's theory of relativity. Their influence on Einstein would have been a general one.

Hume's ideas of space and time, on the contrary, would probably have had considerable direct influence on the development of Einstein's.[197] Hume stated that "*the idea of space or extension is nothing but the idea of visible or tangible points distributed in a certain order*" (italics original),[198] and that "we have no idea of any real extension without filling it with sensible objects."[199] As to time, it "is always discover'd by some *perceivable* succession of changeable objects" (italics original).[200] We have no "idea of time without any changeable existence. . . ."[201] Asserting that physical theories are "based on the kinematics of rigid bodies"[202] Einstein developed the theory of relativity in his first relativity paper by beginning with definitions of space and time by means of a scale and a clock. This approach to the concepts of space and time immediately reminds us of Hume's assertions that the idea of space is based on an arrangement of tangible objects and that the idea of time is based on a perceptible succession of changeable objects.

However, for Einstein to find fruitful suggestions in Hume's discussion of the notions of space and time, it was necessary that the adaptation of electromagnetic theory to the principle of relativity should first have become the desideratum. The idea of modifying the notions of space and time may have emerged in the effort to solve the problem of adaptation. We are thus led to the question of what conceptual factor enabled Einstein to set himself the problem of adapting the electromagnetic theory to the principle of relativity. To answer this question we must turn to Mach.

As to Mach's influence on Einstein, it has been taken for granted that Einstein's commitment to Machian positivism contributed to the innovations in the concepts of space and time. For example, Philipp Frank has stated: "The definition of simultaneity in the special theory of relativity is based on Mach's requirement that every statement in physics has to state relations between observable quan-

[197]David Hume, *A Treatise of Human Nature* (Oxford, 1888); reprint (Oxford, 1968), pp. 26–68.

[198]*Ibid.*, p. 53.

[199]*Ibid.*, p. 64.

[200]*Ibid.*, p. 35.

[201]*Ibid.*, p. 65.

[202]A. Einstein, *op. cit.* (note 179), p. 892.

tities. . . . Mach's requirement, the positivistic requirement, was of great heuristic value to Einstein."[203] Hans Reichenbach, too, has said: "The physicist who wanted to understand the Michelson experiment had to commit himself to a philosophy for which the meaning of a statement is reducible to its verifiability. . . . It is this positivist, or let me rather say, empiricist commitment which determines the philosophical position of Einstein. . . . He merely had to join a trend of development characterized, within the generation of physicists before him, by such names as Kirchhoff, Hertz, Mach. . . ."[204] Gerald Holton, who has made an interesting, detailed investigation of encounter and deviation in the thought of Einstein and Mach, has found a "Machist component" in the birth of the theory of relativity in Einstein's insistence that the fundamental problems of physics could not be understood before an epistemological analysis, and in his identification of reality with the product of sensations.[205] The Japanese philosopher Wataru Hiromatu has discussed the connection between Mach's philosophical thought and the theory of relativity, especially in their premises of understanding the external world, and has asserted that Mach's ideas, such as the monistic world view, the conception of science as description, the principle of thought economy, and his theoretical investigation in physics as the embodiment of these ideas, pioneered in many respects the theory of relativity, the special as well as the general.[206] Hiromatu's discussion, however, is concerned only with conceptual links, not with any genetic link between Mach's philosophy and the theory of relativity. To find the actual contribution of Mach's thought to the development of Einstein's during the years around 1900 that were crucial for the genesis of the theory of relativity, we have to look at Einstein's own words.

[203]Philipp G. Frank, "Einstein, Mach, and Logical Positivism," P. A. Schilpp, ed., *Albert Einstein: Philosopher-Scientist,* pp. 269–286, esp. pp. 272–273.

[204]Hans Reichenbach, "The Philosophical Significance of the Theory of Relativity," *ibid.,* pp. 287–311, esp. pp. 290–291.

[205]Gerald Holton, "Mach, Einstein, and the Search for Reality," *Daedalus* (1968), pp. 636–673; *Thematic Origins of Scientific Thought,* pp. 219–259, esp. p. 224.

[206]Wataru Hiromatu, "Mach's Philosophy and the Theory of Relativity—In Referring to His Criticism of Newtonian Physics" (in Japanese), in W. Hiromatu and H. Kato, eds. and trans., *Mach: Ninsiki no Bunseki* (*Mach: Analysis of Knowledge*) (Tokyo, 1966); reprint (Tokyo, 1971), pp. 136–173.

Urged by his friend Michele Besso,[207] Einstein in 1897 read Mach's *Mechanics in Its Development, Historically and Critically Described*,[208] which made a strong impression on him. In 1947 Besso asked Einstein if it is permissible to say that Mach's thought played the decisive role in drawing Einstein's attention to "observable quantities—perhaps indirectly to 'clock and scale'."[209] Einstein's answer[210] to Besso's specific question was rather negative; he said that Mach's influence on the development of his thought was surely great, but that it is not clear how much it affected his research directly. However, as to Mach's general influence, his answer was very definite: "I see his great service in that he loosened the dogmatism about the foundation of physics that had been dominant during the eighteenth and nineteenth century. He has tried to show especially in mechanics and heat theory how concepts arose out of experience. He has convincingly advocated the point of view that these concepts, even the most fundamental, receive their justification only from experience, that they are in no way *logically* necessary" (italics original).[211] This statement exactly corresponds to the fol-

[207]Carl Seelig, *op. cit.* (note 189), p. 39.

[208]E. Mach, *Die Mechanik in ihrer Entwicklung historisch-kritisch dargestellt*, 3rd ed. (Leipzig, 1897). It may be convenient to devote a few words to the successive editions of the *Mechanics.* The first edition was published in 1883 and the second in 1888, the text remaining unaltered except for corrections of printer's errors. The third through ninth editions appeared in 1897, 1901, 1904, 1908, 1912, 1921, and 1933, respectively. From the third through the seventh edition Mach made revisions and additions to each new edition. When in 1897 Einstein first read Mach's *Mechanics,* nearly ten years had passed since the publication of the second edition. It would have been difficult for Einstein to purchase a copy of the second edition. The third edition appeared most probably in the first half of 1897, since the author's preface is dated January 1897. It is not unreasonable to assume that Einstein read the newly issued third edition. Upon this assumption I will refer in the following to the third edition. Of course we cannot exclude the possibility that Einstein by some means read the second, or even the first, edition, but the difference between the various editions is not significant for the following discussion, because the difference consists mainly in separate examples and the addition or deletion of references to other authors who discussed related topics in the interim. The fundamental purport is unchanged throughout all editions.

[209]M. Besso to A. Einstein, 12 October, 4 and 23 November, and 8 December 1947, *op. cit.* (note 186), p. 386; Besso is inquiring if "die Mach'schen Gedankengänge entscheidend auf das Beobachtbare hinwiesen—vielleicht eben indirekt, auf 'Uhren und Massstäbe'."

[210]A. Einstein to M. Besso, 6 January 1948, *op. cit.* (note 186), p. 391.

[211]*Ibid.,* pp. 390–391: "Ich [sehe] sein grosses Verdienst darin, dass er den

lowing passage in his "Autobiographical Notes" which was perhaps written shortly before the above letter. There Einstein writes: "Even Maxwell and H. Hertz . . . in their conscious thinking adhered throughout to mechanics as the secured basis of physics. It was Ernst Mach who, in his history of mechanics, shook this dogmatic faith; this book exercised a profound influence upon me in this regard while I was a student."[212] We may assume that Mach's influence upon Einstein consisted essentially in undermining the mechanistic worldview by showing that even the fundamental concepts of mechanics are, in the last analysis, rooted in experience. In physics in general, this was the very goal that Mach himself sought to attain in his *Mechanics* and other writings.[213]

10. CRITICISM OF THE MECHANISTIC WORLDVIEW

To understand the origin of Mach's thought, it is expedient first to examine the essay *History and Root of the Axiom of the Conservation of Work* (hereafter abbreviated as *History and Root*),[214] which is based on a lecture delivered in 1871. Mach stated in his *Mechanics* that the view developed there was first propounded in *History and Root*. The purpose of the essay *History and Root* was to repudiate the mechanistic worldview through examination of the foundation of the principle of conservation of energy. Mach's criticism was directed especially at Hermann von Helmholtz and Wilhelm Wundt.

In the introduction of *On the Conservation of Force* (1847)[215]

---

im 18. und 19. Jahrhundert herrschenden Dogmatismus über die Grundlagen der Physik aufgelockert hat. Er hat besonders in der Mechanik und Wärmelehre aufzuzeigen gesucht, wie die Begriffe aus den Erfahrungen heraus entstanden sind. Er hat überzeugend den Standpunkt vertreten, dass diese Begriffe, auch die fundamentalsten, ihre Berechtigung nur von der Empirie aus erhalten, dass sie in keiner Weise *logisch* notwendig sind."

212 A. Einstein, *op. cit.* (note 187), p. 21.

213 Alfonsina D'Elia analyzes Mach's writings, paying special attention to the latter's criticism of the mechanistic world view, in A. D'Elia, *Ernst Mach* (Firenze, 1971). John T. Blackmore's biographical study *Ernst Mach. His Work, Life, and Influence* (Berkeley, 1972) fails to evaluate Mach's insistence on refuting the mechanistic worldview.

214 E. Mach, *Die Geschichte und die Wurzel des Satzes von der Erhaltung der Arbeit. Vortrag gehalten in der K. Böhm. Gesellschaft der Wissenschaften am 15. Nov. 1871* (Leipzig, 1909). This is a reprint of the first edition which appeared in Prague in 1872.

215 Hermann Helmholtz, *Über die Erhaltung der Kraft, eine physikalische Abhandlung, vorgetragen in der Sitzung der physikalischen Gesellschaft zu Berlin am 23sten Juli 1847* (Berlin, 1847); reprint (Bruxelles, 1966).

Helmholtz formulated the mechanistic worldview with its philosophical foundation by stating that, in view of the principle of sufficient reason, the "ultimate goal of theoretical natural science is to find the ultimate unchangeable causes of the processes in nature."[216] The external objects of science are matter and force which are inseparable from each other, and the ultimate causes that science seeks to discover ought to be motive forces. The actions of the motive forces are determined only by spatial relations between bodies, since motion is the change in the mutual spatial relation between at least two bodies, and consequently the motive force as the cause of motion is deduced only for mutual relations between bodies. Material bodies may be resolved into mass points, and there is no other spatial relation between points than their mutual separations. "The task of physical science, therefore, is defined as consisting in reducing natural phenomena to unchangeable, attractive and repulsive forces."[217] As these words clearly show, Helmholtz not only asserted that all physical phenomena should be explained in terms of material points and central forces, but furthermore reasoned that the reduction of all physical phenomena to mechanics was a necessity on *a priori,* metaphysical grounds.

Helmholtz shared this view with many nineteenth century scientists. One conspicuous proponent of a similar view was Wundt, who, six years before Mach's *History and Root,* tried to establish an epistemological foundation for the necessity of the mechanistic worldview in *The Axioms of Physics and Their Relation to the Causal Principle.*[218] In Mach's words, Wundt was "the proponent of the tendency of modern natural science," and "no objection was raised to Wundt's view."[219] Wundt asserted that all causes in nature are causes of motion for the following reason. The qualitative changes in external objects, judged on the information furnished by

[216]*Ibid.,* p. 2: "Das endliche Ziel der theoretischen Naturwissenschaften ist also, die letzten unveränderlichen Ursachen der Vorgänge in der Natur aufzufinden."

[217]*Ibid.,* p. 6: "Es bestimmt sich also endlich die Aufgabe der physikalischen Naturwissenschaften dahin, die Naturerscheinungen zurückzuführen auf unveränderliche, anziehende und abstossende Kräfte. . . ."

[218]W. Wundt, *Die physikalischen Axiome und ihre Beziehung zum Causalprinzip. Ein Kapitel aus einer Philosophie der Naturwissenschaft* (Erlangen, 1866). See Ernst Cassirer, *The Problem of Knowledge. Philosophy, Science, and History since Hegel* (New Haven, 1950), pp. 87–88.

[219]E. Mach, *op. cit.* (note 214), p. 19.

our senses, is described by the statement that one object has disappeared and another object with partly different qualities has taken its place. Such disappearance and appearance contradict the identity of being and the indestructibility of matter. However, there "is one single case where an object does change before our eyes and yet still remains the same, and this is the case of motion. Here the change consists merely in the alteration of an object's spatial relationships to other objects. . . . [During a change of position, objects] remain identical. . . . We must trace every change back to the only conceivable one in which an object remains identical: motion."[220]

This sort of argument does not appeal to present-day readers, but it appealed to nineteenth century scientists. They thought that it was not accidental and matter-of-fact that every natural phenomenon had to be accounted for by mechanics, but logical and necessary. In their view, the axioms of mechanics were not merely empirical, factual laws, but, like axioms and theorems of geometry, *a priori* or necessary truths. We can find an example in Bernhard Riemann's manuscript "Gravity and Light," inferred to have been written after 1858, which shows from the critic's point of view the current conception. Riemann criticized the attempt to elevate the laws of mechanics to *a priori* truths, remarking especially that the law of inertia cannot be accounted for by the principle of sufficient reason.[221]

In his *History and Root,* Mach, tracing the origins of the principle of conservation of energy to the knowledge that "it is impossible to produce work from nothing," tried to show that this knowledge was rooted far more deeply in human experience over an immensely long period than in modern mechanics. He wanted to assert that the efforts to derive the laws of mechanics *a priori* from the general principle of causality were meaningless. His *Mechanics,* published nearly ten years later, was the fruit of his efforts to widen and deepen the criticism of the *a priori* view of mechanics which he had outlined in *History and Root.* In the preface to the first edition of the *Mechanics,* which he retained throughout all succeeding editions, Mach stated that "the tendency [of this book] is rather an enlightening

220 Quoted by Cassirer, *op. cit.* (note 218), p. 88.

221 B. Riemann, "Gravitation und Licht," *Gesammelte mathematische Werke und wissenschaftlicher Nachlass,* 2nd ed. (1892); reprint (New York, 1953), pp. 532–538. Cf. Yôitu Kondô, *Sin Kikagaku Sisôsi* (*A Conceptual History of Geometry.* Revised) (Tokyo, 1966), p. 189.

one or, to put it more clearly yet, an anti-metaphysical one."[222] In *History and Root* Mach defined metaphysical concepts as concepts of which "we have forgotten how we reached them."[223] Hence, in his *Mechanics* he attempted to elucidate "the questions of the *scientific* content of mechanics, of *how* we obtained it, from what *sources* we have derived it, and to what extent it can be considered our assured possession" (italics original).[224] In other words, Mach's purpose in the *Mechanics* was to put an end to a priority in mechanics and thus to strike a blow at the mechanistic worldview.

Mach's *Mechanics* is divided into five chapters. The first chapter deals with statics. Examining closely the "proofs" of the lever principle by Archimedes and Galileo, the deduction of the theorem of equilibrium of force on an inclined plane by Simon Stevin, the "geometrical proof" of the parallelogram of forces by Daniel Bernoulli, and the derivation of the principle of virtual displacement by Joseph Louis Lagrange, Mach unveiled that behind all these "proofs" are presuppositions of certain intuitive knowledge, which are no more than generalizations of repeated experience obtained in the long history of the human race. The second chapter of the *Mechanics* is devoted to the consideration of dynamics. Analyzing the reasoning by which Galileo inferred the law of inertia (on a horizontal plane), Mach concluded that behind Galileo's reasoning lies the intuitive knowledge that any body having weight never ascends by itself, and also that "it is at all events entirely erroneous to express the inertia as self-evident, or to try to derive it from the general theorem that 'the action of a cause persists'."[225] Similarly Mach pointed to the important part played by intuitive knowledge in Christian Huygens' determination of the center of oscillation of extended bodies. He then proceeded to a detailed discussion of Isaac Newton's conceptions of mass and action and reaction and made it clear that these

[222] E. Mach, *op. cit.* (note 208), p. v: "Ihre Tendenz ist vielmehr eine aufklärende oder, um es noch deutlicher zu sagen, eine antimetaphysische."

[223] E. Mach, *op. cit.* (note 214), p. 2: "Metaphysisch pflegen wir diejenigen Begriffe zu nennen, von welchen wir vergessen haben, wie wir dazu gelangt sind."

[224] E. Mach, *op. cit.* (note 208), p. v, posed "die Fragen . . . worin der *naturwissenschaftliche* Inhalt der Mechanik besteht, *wie* wir zu demselben gelangt sind, aus welchen *Quellen* wir ihn geschöpft haben, wie weit derselbe als ein gesicherter Besitz betrachtet werden kann. . . ."

[225] *Ibid.*, p. 135: "Die Trägheit als selbstverständlich darzustellen, oder sie aus dem allgemeinen Satz 'die Wirkung einer Ursache verharrt' abzuleiten, ist jedenfalls durchaus verfehlt."

two kinds of concepts, mass on the one hand and action and reaction on the other, depend on each other, and that a certain amount of intuitive knowledge and experience underlies the process by which the concepts were formed. This part of the book is written with the greatest ardor and persuasiveness. It is followed by the most famous section of the book, the criticism of Newton's concepts of time and space, which is so well known that we may pass on without giving a detailed account. But one point is worth emphasizing. After illustrating that there is no need to associate the law of inertia with a special absolute space, Mach stresses that "the most important result of our considerations is, however, *that even the apparently simplest laws of mechanics are very complicated in nature, that they rest on unfinished, even never completely terminable experience,* . . . [but] *that they should by no means be regarded as mathematically determined truths, but rather as theorems that not only can be controlled by experience but even need to be*" (italics original).[226] Mach's celebrated opposition to the concepts of absolute space and time must, therefore, be understood in the broader context of his criticism of a priority in mechanics. In the general observation at the end of the chapter Mach stressed that descriptions of Newtonian mechanics should distinguish the parts based on experience from those that are arbitrary convention and that the present form of mechanics is determined by historical contingency. In the third chapter of the *Mechanics,* Mach discussed formal principles of mechanics such as the principle of least action. Here again he rejected a priority. Mach noted that "the largest fault of Descartes, which spoils his study of nature, is that he thinks those propositions to be self-evident and clear that only experience can determine."[227] In the fourth chapter Mach proposed the concept of "economy of scientific thinking," and in the fifth he used the concept as a basis for asserting

[226]*Ibid.,* pp. 231–232: "Das wichtigste Ergebniss unserer Betrachtungen ist aber, *dass gerade die scheinbar einfachsten mechanischen Sätze sehr complicirter Natur sind, dass sie auf unabgeschlossenen, ja sogar auf nie vollständig abschliessbaren, Erfahrungen beruhen,* . . . *dass sie aber keineswegs selbst als mathematisch ausgemachte Wahrheiten angesehen werden dürfen, sondern vielmehr als Sätze, welche einer fortgesetzten Erfahrungscontrolle nicht nur fähig, sondern sogar bedürftig* sind."

[227]*Ibid.,* pp. 274–275: "Der grösste Fehler des Descartes aber, der seine Naturforschung verdirbt, ist der, dass ihm Sätze von vornherein als selbstverständlich und einleuchtend erscheinen, über welche nur die Erfahrung entscheiden kann."

that the mechanistic worldview is unfounded. Mach argued that, from the viewpoint of the economy of thinking, the mechanical hypothesis has no priority over other kinds of hypothesis, and that therefore "we consider as prejudice the view that mechanics should be considered the foundation of all other branches of physics and that all physical processes ought to be explained mechanically."[228]

Since we are preoccupied with the theory of relativity, we are liable to see in Mach's *Mechanics* above all the critical discussion of the concepts of space and time as an early expression of the spirit of the relativity theory. Mach's discussion of space and time could certainly have been suggestive to Einstein in the gestation of the relativistic conception of space and time. Einstein in fact praised Mach's critical mind in his obituary of Mach, quoting at considerable length the passages concerning the concepts of space and time from the *Mechanics.*[229] Nonetheless, Mach's first purpose of the *Mechanics* as a whole was, as he added in later editions, "to convince readers that *properties of nature* cannot be fabricated with the aid of self-evident hypotheses but should be drawn from *experience*" (original italics).[230] Einstein was quite right when he regarded the destruction of the dogmatism of the mechanistic worldview as the greatest merit of Mach's *Mechanics.*

Mach's extended criticism of the mechanistic worldview was not without justification in the late nineteenth century, since the mechanistic worldview was then still the prevalent view. For example, J. C. Maxwell stated in his *Matter and Motion* that "the first part of physical science relates to the relative position and motion of bodies,"[231] because physical science should deal with the simplest and most abstract phenomena in nature and the simplest of all natural phenomena is the change in the arrangement of material bodies. He also asserted that the law of inertia is understandable *a priori.* He said that if a body that is not subject to any influence were to

[228]*Ibid.,* p. 486: "Die Anschauung, dass die Mechanik als Grundlage aller übrigen Zweige der Physik betrachtet werden müsse, und dass alle physikalischen Vorgänge *mechanisch* zu erklären seien, halten wir für ein Vorurteil.

[229]A. Einstein, *op. cit.* (note 182), pp. 102–103.

[230]E. Mach, *Die Mechanik,* 8th ed. (Leipzig, 1921), p. 20: "Mein ganzes Buch verfolgt aber das Ziel, den Leser zu überzeugen, dass man *Eigenschaften* der Natur nicht mit Hilfe sebstverständlicher Annahmen aus den Fingern saugen kann, sondern dass diese der *Erfahrung* entnommen werden müssen."

[231]J. C. Maxwell, *Matter and Motion* (1877); reprinted, with notes and appendices by Joseph Larmor (London, 1920), p. 2.

change its velocity spontaneously, then by the maxim that "the same causes will always produce the same effects"[232] we would be led to a conclusion which "is in contradiction to the only system of consistent doctrine about space and time which the human mind has been able to form."[233] Helmholtz, who in the middle of the century had formulated the mechanistic worldview, later modified his original view of the causal principle,[234] but to the end of his life maintained the view that mechanics occupied the primary place in the whole of physics. In his lecture on mechanics given in 1893–1894, in which he characterized forces as causes that always persist and act according to immutable laws, he declared that "the whole of theoretical physics may be constructed with the aid of the concept of force."[235] A letter by the Japanese physicist Hantaro Nagaoka, written in Berlin in 1893, portrays the state of physics there at that time. Nagaoka wrote that "physicists here seem to believe that it is the modern way to reduce every thing to the mechanical ground."[236] To young Einstein, who "had read with enthusiasm Ludwig Büchner's *Force and Matter*"[237] which emphasizes the inseparability of force and matter and asserts that all forces and actions in nature consist in the conditions or movements of the particles of matter, Mach's radical criticism of the mechanistic worldview must have been a revelation.

11. EPISTEMOLOGICAL STATUS OF MECHANICS IN LORENTZ' AND POINCARÉ'S VIEWS OF PHYSICS

Having considered Mach's criticism of the mechanistic worldview and its general significance for late nineteenth century physics, we now turn to the question of what its specific bearing was on the birth of the theory of relativity. To answer this question we have to return to the scientific thought of Lorentz and Poincaré.

232*Ibid.*, p. 13.

233*Ibid.*, pp. 28–29.

234See the note added in 1881 when "On the Conservation of Force" was included in *Wissenschaftliche Abhandlungen von Hermann Helmholtz* (Leipzig, 1882–1895), *1*, 12–75, on 68.

235H. von Helmholtz, *Vorlesungen über die Dynamik discreter Massenpunkte*, 2nd ed. (Leipzig, 1911; 1st ed. 1898), p. 24.

236Kiyonobu Itakura, Tosaku Kimura, and Eri Yagi, *Nagaoka Hantarô Den* (*A Biography of Hantaro Nagaoka*) (Tokyo, 1973), p. 170.

237Carl Seelig, *op. cit.* (note 190), p. 14.

If "mechanistic worldview" is understood to be a world picture "in which the laws of physics are reduced to those of mechanics,"[238] then Lorentz and Poincaré did not share the "mechanistic worldview." Much less did they attempt to make a mechanical model for electromagnetic phenomena. At the beginning of his *Theory of Electrons* Lorentz rejected such attempts, saying that "we can develop the theory to a large extent and elucidate a great number of phenomena, without entering upon speculations of this kind. Indeed, on account of the difficulties into which they lead us, there has of late years been a tendency to avoid them altogether and to establish the theory on a few assumptions of a more general nature."[239] The ether as Lorentz conceived it was, to use Einstein's words,[240] deprived of all mechanical properties but "rest." Poincaré, too, said: "The end we seek is not the mechanism; the true and only aim is unity."[241] Since the electromagnetic theory can be formulated in such a way that it satisfies the principles of conservation of energy and of least action, mechanical explanation is always and in infinitely many ways possible. We have to be satisfied with the abstract possibility, he asserted. He even declared: "Whether the ether really exists matters little. . . . That, too, is only a convenient hypothesis."[242] Rejecting mechanical explanations, Lorentz and Poincaré sought instead to unify physics under the electromagnetic view of nature.[243] In spite of their denial of the "mechanistic worldview," however, careful examination of their thought reveals that they, too, had not emancipated themselves from the prevalent view that mechanics should occupy the primary position in the logical structure of the edifice of all physics.

Lorentz' greatest contribution to the development of electromagnetic theory is that he took the electromagnetic field, which Maxwell, Hertz, and others had considered a state of the dielectric,

[238] Arthur I. Miller, *op. cit.* (note 16), p. 212, footnote 11.

[239] H. A. Lorentz, *op. cit.* (note 156), p. 2.

[240] A. Einstein, *Aether und Relativitätstheorie* (Berlin, 1920), p. 7.

[241] H. Poincaré, *op. cit.* (note 162), *Rapports, 1,* 26; *La science et l'hypothese,* p. 207: "Le but poursuivi; ce n'est pas le mécanisme, le vrai, le seul but, c'est l'unité."

[242] H. Poincaré, *La science et l'hypothèse,* pp. 245–246: "Peu nous importe que l'éther existe réellement. . . . Ce n'est là aussi qu'une hypothèse commode."

[243] Russell McCormmach, "Einstein, Lorentz, and the Electron Theory," *Hist. Stud. Phys. Sci., 2* (1970), 41–87; Stanley Goldberg, *op. cit.* (note 13); Arthur I. Miller, *op. cit.* (note 16).

as an independent physical reality; he considered it to be the state of the ether.[244] Lorentz' ether is a nonmechanical entity in the sense that its physical state is entirely determined by electromagnetic excitation. It is almost synonymous with the electromagnetic field as we understand it today. It is the substance of electromagnetic phenomena. For Lorentz, however, a substance must be endowed with some mechanical characteristics, however abstract and limited they may be. At the end of his *Theory of Electrons,* defending his theory against Einstein's theory of relativity, he wrote: "I cannot but regard the ether, which can be the seat of an electromagnetic field with its energy and its vibrations, as endowed with a certain degree of substantiality, however different it may be from all ordinary matter. In this line of thought, it seems natural not to assume at starting that it can never make any difference whether a body moves through the ether or not, and to measure distances and lengths of time by means of rods and clocks having a fixed position relatively to the ether."[245] Lorentz believed, in other words, that if something is a substance, then *motion* relative to it can be conceived and this motion must have some physical consequence. He expressed the same thought also in the series of lectures which he delivered in Göttingen in 1910: although the ether in the theory of electrons "is still left with substantiality to such an extent that a coordinate system can be defined thereby," the theory of relativity has attacked even this last substantiality.[246] In other words, in Lorentz' view, if motion or rest cannot be determined relative to the ether, the ether cannot be the substance of electromagnetic phenomena. Lorentz stressed the same point of view in his address at the Royal Academy in Amsterdam in 1915: "To the ether all substantiality is denied to the extent that we cannot speak of rest or motion with respect to it."[247] That substantiality meant for Lorentz the possession of mechanical characteristics

[244]T. Hirosige, "Origins of Lorentz' Theory of Electrons and the Concept of the Electromagnetic Field," *Hist. Stud. Phys. Sci., 1* (1969), 151–209.

[245]H. A. Lorentz, *Theory of Electrons,* p. 230.

[246]H. A. Lorentz, "Alte und neue Fragen der Physik," *Phys. Zeits., 11* (1910), 1234–1257; *Collected Papers, 7,* 205–257. Quotation is from p. 210: "Schliesslich ist ihm nur noch soviel Substantialität geblieben, dass man durch ihn ein Koordinatensystem festlegen kann. Selbst dieser letzte Rest der Substantialität wird durch das *Relativitätsprinzip* angegriffen. . . ."

[247]H. A. Lorentz, "De lichtaether en het relativiteitsbeginsel," *Jaarboek Kon. Acad. Wet.* (1915); *Collected Papers, 9,* 233–243. Quotation is from p. 238: "Aan den aether wordt in die mate alle substantialiteit ontzegd, dat men van rust of beweging to opzichte van hem selfs niet kan spreken."

is apparent also from the following discussion concerning the electromagnetic mass of the electron: "By our negation of the existence of material mass, the negative electron has lost much of its substantiality. We must make it preserve just so much of it, that we can speak of forces acting on its parts, and that we can consider it as maintaining its form and magnitude."[248]

Since the ether, as the substance of electromagnetic phenomena, must possess mechanical characteristics and hence serve as reference system for rest or motion, it is evident that the foundation of electromagnetic theory should be the electromagnetic equations in the coordinate system that is fixed with respect to the ether. Although highly appreciative of Einstein's theory, Lorentz, therefore, held to his theory to the end of his life.[249] He thought that physicists were free to choose either the theory based on the ether or the theory of relativity,[250] and that "each physicist can adopt the attitude which best accords with the way of thinking to which he is accustomed."[251] We must bear in mind that these statements by Lorentz all belong to the period after the advent of the theory of relativity. Earlier, Lorentz had, to my knowledge, never explicitly expressed the idea that the ether determined a physically privileged coordinate system, or that, in other words, the ether furnished an absolute frame of reference, although the ether did play such a role in his theory. The absence of such a statement by Lorentz tends to corroborate my conclusion about the ether problem, namely, that its issue was not finding an absolute frame of reference. Lorentz, I suppose, did not become aware of the kinematical significance of the role played by the ether in his theory until he compared his theory of electrons with Einstein's theory. His statements show that he was unable to reach the theory of relativity and later continued to reject it because he had deeply committed himself to the mechanistic worldview—the view that mechanics should be assigned the primary place in the edifice of physics—without being clearly aware of his commitment.

[248]H. A. Lorentz, *op. cit.* (note 156), p. 43.

[249]Max Born wrote: "When I visited Lorentz a few years before his death, his scepticism [about the theory of relativity] had not changed." M. Born, *Physics in My Generation* (London, 1956), p. 192.

[250]H. A. Lorentz, *op. cit.* (note 247), *Collected Papers, 9,* 241.

[251]H. A. Lorentz, "Considération élémentaire sur le principe de relativité," *Revue gén. des sci., 25* (1914), 179; *Collected Papers,* 7, 147–165. Quotation is from p. 165: "Chaque physicien pourra prendre l'attitude qui s'accorde le mieux avec la façon de penser à laquelle il s'est accoutumé."

The privileged position of mechanics in Poincaré's scientific thought, too, is closely connected with his conception of the principle of relativity as an empirical law rather than a postulate. The principle of relativity here is the principle that it is impossible to experimentally detect motion relative to the ether. Poincaré first formulated the "principle of relative motion" as a general principle of mechanics and then extended it to the "principle of relativity."[252] The principle of relative motion states that "the motion of any system whatever ought to obey the same laws, whether it is referred to fixed axes or to the moving axes drawn by rectilinear and uniform motion."[253] It is, in other words, the principle of relativity as we understand it, within the limit of mechanics. This principle, in Poincaré's view, is not merely a general expression of empirical facts but implies certain elements which transcend experience. It therefore claims a different epistemological status than his "principle of relativity."

As is well known, Poincaré considered axioms of geometry to be conventions.[254] He did not mean, however, that axioms of geometry are an entirely arbitrary invention of the human mind. In the process of selection "our choice among all possible conventions is *guided* by experimental facts" (italics original).[255] The criterion according to which they are chosen is convenience. They are subject to the requirement that they do not contradict each other. From the chosen axioms we then logically construct the whole theory of geometry. By their origin, the axioms are only conventions or "definitions in disguise," but at the same time they are therefore absolutely true. Poincaré thought that nearly the same was true for mechanics: "The principles of this science, although more directly based on experience,

[252]H. Poincaré, [a] "La theorie de Lorentz et le principe de reaction," *Arch. néerl., 5* (1900), 252–278; *Oeuvres, 9,* 464–488. The relevant place is on p. 482. [b] *La science et l'hypothèse,* p. 135. [c] *Op. cit.* (note 169), *Oeuvres, 9,* 552; *Science et méthode,* p. 217. Poincare's distinction between *principe du mouvement relatif* and *principe de relativité* is also noted by Arthur I. Miller, *op. cit.* (note 16), pp. 233–234.

[253]H. Poincaré, *La science et l'hypothèse,* p. 135: "Le mouvement d'un système quelconque doit obéir aux mêmes lois, qu'on le rapporte à des axes fixes, ou à des axes mobiles entraînes dans un mouvement rectiligne et uniforme."

[254]H. Poincaré, *La science et l'hypothèse,* Chaps. 3–5.

[255]*Ibid.,* p. 66: "Notre choix, parmi toutes les conventions possibles, est *guidé* par des faits expérimentaux." The English quotation is from *Science and Hypothesis,* p. 50.

still share the conventional character of the geometrical postulates."[256] In one respect, the principles of mechanics are based on experiments. For an almost isolated system, they can be approximately confirmed by experiment. In the system of mechanics, however, they are generalized so that they become postulates to be applied to the whole universe and are regarded as exactly true, "because they reduce in final analysis to a simple convention that we have the right to make, because we are certain beforehand that no experiment can contradict it."[257]

In the physical sciences other than mechanics, "the scene changes. We meet hypotheses of another kind."[258] A hypothesis in physics "should always be submitted to verification as soon as possible and as many times as possible."[259] It goes without saying that, if it cannot stand this test, it must be abandoned without any hesitation. The most general of the hypotheses that have stood the test are, according to Poincaré, the principle of the conservation of energy, Carnot's principle (the second law of thermodynamics), and the principles of action and reaction, of relativity, of the conservation of mass, and of least action. In mechanics, all but Carnot's principle have the character of a convention and are therefore exactly true. As a consequence of the extensive development of physics during the nineteenth century, the principles have been extended to fields other than mechanics, confirmed there too, and now are considered "experimental truths."[260] Poincaré recognized that, as the other physical principles, the "principle of relativity" "is no longer a convention. It is verifiable, and consequently it need not be verified."[261] In fact, the rapid and unexpected development of physics around the turn of the century seemed to Poincaré to undermine the fundamental principles

[256] *Ibid.*, p. 5: "Les principes de cette science, quoique plus directement appuyés sur l'expérience, participent encore du caractère conventionnel des postulats géométriques." The English quotation is from *Science and Hypothesis*, p. xxvi.

[257] *Ibid.*, pp. 162–163: "C'est qu'ils se réduisent en dernière analyse à une simple convention que nous avons le droit de faire, parce que nous sommes certains d'avance qu'aucune expérience ne viendra la contredire." The English quotation is from *Science and Hypothesis*, p. 136.

[258] *Ibid.*, p. 6.

[259] *Ibid.*, p. 178: "Elle doit toujours être, le plus tôt possible et le plus souvent possible, soumise à la vérification."

[260] H. Poincaré, *op. cit.* (note 166).

[261] H. Poincaré, "L'espace et le temps," *Scientia, 12* (1912), 159–171, esp. 168; *Dernières pensées* (Paris, 1913), p. 105: "Il est vérifiable et par conséquent il pourrait n'être pas verifié."

of physics. He therefore tried, in his St. Louis lecture in 1904, to diagnose the situation in physics and to seek a way out of the crisis of the principles. Although in 1904 he felt that the principle of relativity had been saved by the Lorentz theory, the result of Kaufmann's 1905 experiment showing that the electromagnetic mass of the electron varies according to Abraham's formula rather than Lorentz' once again caused him anxiety. Referring to Kaufmann's result in 1908, he admitted that "it would seem that the Principle of Relativity has not the exact value we have been tempted to give it."[262]

Poincaré's anxiety about the validity of the principle of relativity originated in the very distinction he made between the epistemological status of mechanics and that of other branches of physics. According to his philosophy of science, mechanics provides the frames for describing the processes in nature, which, in mechanics, are conventions and consequently claim strict validity; the theories of other branches of physics are developed so as to conform to these frames. In a lecture delivered in London in the spring of 1912, Poincaré asserted about the most fundamental frame, space, that its definition was reduced to the proposition that the form of the equations of *dynamics* should not be altered by transformations of the coordinate axes.[263] As far as he maintained such a point of view, it must have been inconceivable for him to postulate a universal principle of relativity for both mechanics and electromagnetism that treated the two sciences as equals. Several historians of physics have discussed reasons why Poincaré was not able to reach the theory of relativity. I claim that, as with Lorentz, the most fundamental reason is his mechanistic worldview: as Lorentz, Poincaré believed that mechanics must be assigned primacy in the epistemological structure of the whole of physics.

12. EMANCIPATION FROM THE MECHANISTIC WORLDVIEW AND THE THEORY OF RELATIVITY

The discussion in the preceding section suggests that for the emergence of the theory of relativity a complete emancipation from the mechanistic worldview was the essential prerequisite. Einstein's

[262]H. Poincaré, *op. cit.* (note 169), *Oeuvres, 9,* 572; *Science et méthode,* p. 248: "Le Principe de Relativité n'aurait donc pas la valeur rigoureuse qu'on était tenté de lui attribuer." The English quotation is from *Science and Method,* p. 228.

[263]H. Poincaré, *op. cit.* (note 261), p. 169; *Dernières pensées,* p. 107.

theory of relativity did not intend to reduce either mechanics or electromagnetism to the other, or to assign primacy to one over the other as did Lorentz' and Poincaré's. Einstein reached the theory of relativity by searching for a unification of mechanics and electromagnetic theory at a higher level. For the idea of postulating a universal principle of relativity to arise it was of crucial importance that mechanics and electromagnetism were considered to be of equal standing. Unification of the two theories had to take precedence even over the modification of the concepts of space and time.

The process by which Einstein's theory was gradually accepted during the latter half of the first decade of this century confirms the importance of the complete emancipation from the mechanistic worldview. In fact the purport and significance of Einstein's theory had been misunderstood. Accordingly, physicists did not generally accept it, until they recognized that it was concerned not only with electrodynamics but also with mechanics, that is, that the fundamental postulates of the theory of relativity were universal principles to which mechanics as well as electrodynamics was to be subjected. Such a recognition contradicted the mechanistic worldview.

Walter Kaufmann was the first to cite Einstein's 1905 relativity paper in his article on the mass of the electron. For several years Kaufmann had been engaged in experiments to determine the change of the mass of the electron with change in velocity. In 1901, measuring the electric and magnetic deflection of Becquerel rays, he confirmed that the electron mass increased with its velocity and estimated that the electromagnetic mass of the electron was comparable in its magnitude to the mechanical mass.[264] In the following two papers published in 1902 and 1903, respectively,[265] Kaufmann concluded that the mass of electrons in Becquerel and cathode rays is entirely electromagnetic. These results greatly interested physicists in connection with the electromagnetic view of nature, which was advocated by Wilhelm Wien and Max Abraham.[266] Lorentz, A. H.

[264] W. Kaufmann, "Die magnetische und electrische Ablenkbarkeit der Becquerelstrahlen und die scheinbare Masse der Elektronen," *Gött. Nachr., Math.-phys. Kl.* (1901), pp. 143–155.

[265] W. Kaufmann, "Ueber die electromagnetische Masse des Elektrons," *Gött. Nachr., Math.-phys. Kl.* (1902), pp. 291–296; "Ueber die 'Elektromagnetische Masse' der Elektronen," *Gött. Nachr., Math.-phys. Kl.* (1903), pp. 90–103.

[266] W. Wien, "Ueber die Möglichkeit einer elektromagnetischen Begründung der Mechanik," *Verh. d. Deutsch. Phys. Ges., 8* (1906), 136–141; *Physika-*

Bucherer, and Paul Langevin joined the discussion about electromagnetic mass and the constitution of the electron.[267] Late in 1905 Kaufmann arrived at a definite conclusion about the constitution of the electron. Judging his measurement of the deflections of beta-rays to favor Abraham's rigid sphere electron, he declared that the theory of Lorentz and Einstein was definitely rejected.[268] Kaufmann's conclusion was challenged in the following year by Max Planck.[269] Planck tried to derive the velocity dependence of the electron mass by making use of Einstein's theory and asserted that Kaufmann's result could not refute the Lorentz-Einstein theory conclusively. It was in 1908 that Bucherer for the first time obtained an experimental result in favor of Lorentz' and Einstein's formula of the electron mass,[270] but the experiment was so delicate that the result did not convince all physicists. Disputes continued for a few more years. As late as August 1910 Jakob Johann Laub, in his review article "On the Experimental Foundation of the Relativity Principle," had to admit that there did not yet exist an unequivocal conclusion.[271]

Thus the velocity dependence of the electron mass became one of the topics that was most actively discussed by both experimental and theoretical physicists for nearly ten years after 1905. In textbooks the researches on this subject are often cited as the first experimental verification of the theory of relativity, but, in reality,

---

lische Abhandlungen und Vorträge, 2, 115-120. [b] "Die Kaufmannschen of the electromagnetic view of nature see Russell McCormmach, *op. cit.* (note 243).

[267]A. H. Bucherer, *Mathematische Einführung in die Elektronentheorie* (Leipzig, 1904); P. Langevin, "La physique des électrons," *Revue gén. des sci., 16* (1905), 257-276; *La physique depuis vingt ans* (Paris, 1923), pp. 1-69.

[268]W. Kaufmann, "Über die Konstitution des Elektrons," *Sitzb. preuss. Akad. Wiss.* (1905), pp. 945-956.

[269]Max Planck, [a] "Das Prinzip der Relativität und die Grundgleichungen der Mechanik," *Verh. d. Deutsch. Phys. Ges., 8* (1906), 136-141; *Physikalische Abhandlungen und Vorträge, 2,* 115-120. [b] "Die Kaufmannschen Messungen der Ablenkbarkeit der β-Strahlen in ihrer Bedeutung für die Dynamik der Electronen," *Phys. Zeits.,* 7 (1906) 753-761; *Phys. Abhandlungen und Vorträge, 2,* 121-135.

[270]A. H. Bucherer, "Messungen an Becquerelstrahlen. Die experimentelle Bestätigung der Lorentz-Einsteinschen Theorie," *Phys. Zeits., 9* (1908), 755-762; "Die experimentelle Bestätigung des Relativitätsprinzips," *Ann. d. Phys., 28* (1909), 513-536.

[271]J. Laub, "Über die experimentellen Grundlagen des Relativitätsprinzips," *Jahrb. d. Rad. u. Elekt.,* 7 (1910), 405-463, esp. 462.

contemporary physicists, being concerned primarily with the constitution and mass of the electron, did not ascribe this meaning to their work. By scrutinizing their earlier discussions we find that none of them were conscious of the fundamental difference between Einstein's and Lorentz' theories. Even Planck, who was the first to appreciate and encourage Einstein, did not clearly distinguish the two theories. The only difference he recognized was that Einstein's method was more general than Lorentz'; he spoke of "the 'principle of relativity' recently introduced by H. A. Lorentz and in a more general manner by A. Einstein."[272] Discussing Kaufmann's deflection experiment of 1905, he called the theory of the deformable electron "Lorentz-Einstein's theory."[273] Planck's expression for the principle of relativity, "the postulate that the absolute motion can never be detected,"[274] reminds us rather of Poincaré's definition. Poincaré never mentioned Einstein's name in his many articles referring to the electron mass.[275] His concern in those articles was exclusively with the questions if "the dynamics of the electron" compel us to modify one of the principles of mechanics, the invariability of mass, and how this should be done. Abraham, Kaufmann, Arnold Sommerfeld, and others who defended Abraham's rigid sphere model of the electron, supporting the electromagnetic view of nature as the new mode of physics, refuted Lorentz' and Einstein's formula saying that it upheld the old-fashioned mechanistic view.[276] To our eyes, however, their alleged new mode, the electromagnetic view of nature, is as limited as the mechanistic view of nature because it attempted to substitute electromagnetism for mechanics instead of denying a privileged position to any branch of physics. As the mechanists had tried to reduce the whole of physics to mechanics, so

[272]M. Planck, *op. cit.* (note 269), [a], *Phys. Abhandlungen und Vorträge, 2,* 115; Planck used "das vor kurzem von H. A. Lorentz und in noch allgemeinerer Fassung von A. Einstein eingeführte 'Prinzip der Relativität'."

[273]M. Planck, *op. cit.* (note 269), [b], *Phys. Zeits.,* 7 (1906), 761; this discussion following Planck's paper is omitted from the *Phys. Abhandlungen und Vorträge.*

[274]*Ibid.,* p. 756.

[275]H. Poincaré, *op. cit.* (notes 169, 172, and 173), and "La mécanique nouvelle," in Poincaré, *Sechs Vorträge über ausgewählte Gegenstände aus der reinen Mathematik und mathematischen Physik* (Leipzig, 1910), pp. 49–58.

[276]See the discussion of Planck's paper at the meeting of the Society of German Natural Scientists and Physicians. *Op. cit.* (note 269), [b], *Phys. Zeits.,* 7 (1906), 760–761.

holders of the electromagnetic view of nature tried to reduce physics to electromagnetism. They never thought of subordinating both mechanics and electromagnetism to one universal principle. Accordingly, they regarded the principle of relativity only as a mechanical principle and thus failed to realize the universal significance of Einstein's principle of relativity.

Apart from Einstein, Hermann Minkowski was the first to state clearly that the principle of relativity was to be postulated universally for the whole of physics and, consequently, that the theory of relativity required a radical transformation of the fundamental concepts of physics. On this insight he tried in December 1907 to formulate a relativistic mechanics for an extended body.[277] Minkowski argued that, although many physicists believe that classical mechanics contradicts the postulate of relativity which he adopted in his paper as the foundation of electrodynamics, "it would be quite unsatisfactory if the new concept of time . . . were considered valid only in a limited domain of physics."[278] His recognition of the universal implication of the principle of relativity must have been closely connected with his discovery that Einstein's theory could be expressed in a four-dimensional form. In fact, he almost simultaneously put forth the idea of formulating the theory of relativity in a four-dimensional space. A month before he presented his paper on relativistic mechanics to the Göttingen Academy, on 5 November 1907, he outlined the four-dimensional formulation of the theory of relativity in an address at the Mathematical Society of Göttingen. He began his address with the words: "Starting from the electromagnetic theory of light, a complete transformation seems about to happen in our notions of space and time."[279] No physicist before Minkowski had spoken of a transformation of the notions of space

[277]H. Minkowski, "Mechanik und Relativitätspostulat," appendix to "Die Grundgleichungen für die elektromagnetischen Vorgänge in bewegten Körpern," *Gött. Nachr., Math.-phys. Kl.* (1908), pp. 53–111; *Zwei Abhandlungen über die Grundgleichungen der Elektrodynamik* (Leipzig, 1910), pp. 5–57; appendix, pp. 45–57.

[278]*Ibid.*, p. 45: "Es wäre höchst unbefriedigend, dürfte man die neue Auffassung des Zeitbegriffs, die durch die Freiheit der Lorentz-Transformationen gekennzeichnet ist, nur für ein Teilgebiet der Physik gelten lassen."

[279]H. Minkowski, "Das Relativitätsprinzip," *Ann. d. Phys.*, *47* (1915), 927–938, esp. 927: "Von der elektromagnetischen Lichttheorie ausgehend, scheint sich in der jüngsten Zeit eine vollkommene Wandlung unserer Vorstellungen von Raum und Zeit vollziehen zu wollen. . . ."

and time.[280] Minkowski, being a mathematician, must have been able to focus his attention on the mathematical, formal side of the theory of relativity without being troubled by any special physical view of nature, whether mechanistic or electromagnetic. And because he focused his attention on the formal relations he was able to grasp the full implication of Einstein's theory.

Historical analysis of the reception of Einstein's theory of relativity still remains to be further advanced before we can make a conclusive pronouncement. In such an analysis historians would have to give due consideration to the parts played by the work of Max Planck, Ebenezer Cunningham, A. H. Bucherer, Gilbert Newton Lewis and others.[281] Even without such historical analysis, however, we may reasonably conclude that Minkowski's papers cited above as well as his famous lecture "Space and Time" in 1908[282] played a fundamental part in drawing physicists' attention to the conceptual transformation involved in the theory of relativity. In 1910 Lorentz, in his course of lectures delivered in Göttingen, stated that to deny the existence of a "true" time means to follow Einstein's and Minkowski's thought.[283] Minkowski's work had the remarkable effect of giving a major stimulus to the study of mechanics from the relativistic point of view. Max Born, who had been Minkowski's assistant for a few weeks just before the latter's premature death, developed some relativistic mechanical concepts along the lines pioneered by Minkowski.[284] Philipp Frank, too, was induced by Minkowski's

[280] Of course Einstein here, too, is the exception. In the spring of 1905 he wrote to his friend Conrad Habicht about his current studies: "Die vierte Arbeit liegt im Konzept vor und ist eine Elektrodynamik bewegter Körper unter Benützung einer *Modifikation der Lehre von Raum und Zeit*" (italics mine). Carl Seelig, *op. cit.* (note 189), p. 89.

[281] Papers thus far published that partially meet this need are: T. Hirosige, "Syoki no Sôtaironteki Rikigaku" ("Relativistic Mechanics in its Early Stage"), *Buturigakusi Kenkyû, 4* (1968), 39-54; *5* (1969), 55-70; *6* (1970), 27-61. Stanley Goldberg, "In Defense of Ether: The British Response to Einstein's Special Theory of Relativity, 1905-1911," *Hist. Stud. Phys. Sci., 2* (1970), 89-125.

[282] H. Minkowski, "Raum und Zeit," *Jahresber. d. Deutsch. Math. Verein., 18* (1908), 75-88; *Phys. Zeits., 10* (1909), 104-111.

[283] H. A. Lorentz, *op. cit.* (note 246), *Collected Papers, 7,* 211.

[284] M. Born, "Die träge Masse und das Relativitätsprinzip," *Ann. d. Phys., 28* (1909), 571-584; "Die Theorie des starren Elektrons in der Kinematik des Relativitätsprinzips," *Ann. d.* Phys., *30* (1909), 1-56, 840; "Über die Dynamik des Elektrons in der Kinematik des Relativitätsprinzips," *Phys. Zeits., 10* (1909), 814-817.

paper to attempt a systematization of the theory of relativity comprising both electrodynamics and mechanics. In a paper presented to the Viennese Academy in March 1909[285] he succeeded in showing that, if the principle of relativity is accepted as a universal principle, then from this starting point both electrodynamics and mechanics can be systematically developed. He stated in the introduction to his paper that he had been motivated by the "wealth of ideas" in Minkowski's paper of December 1907. Fróm 1908 on, physicists began to publish many papers in which they applied the theory of relativity to problems in not only electrodynamics or optics but also mechanics. The problems in mechanics included subjects such as the equation of motion of a point mass and relativistic definitions of a rigid body. Needless to say, if physicists discuss mechanical problems from the relativistic point of view, they must assume that relativistic conceptions are valid also in mechanics. The rising interest in relativistic mechanics after 1907, therefore, shows that the universal importance of the theory of relativity was rapidly becoming understood correctly and the theory widely accepted.

Side by side with the general acceptance of the theory of relativity went the recognition that it invalidates the mechanistic worldview. In September 1910 Planck, who now had clearly recognized the fundamental difference between Lorentz' and Einstein's theories and the significance of the modification of the concepts of space and time, declared at the Königsberg meeting of the Society of German Natural Scientists and Physicians: "He who considers the mechanistic view of nature the postulate of the physical way of thinking will never be able to make friends with the relativity theory."[286]

13. CONCLUDING REMARKS

Einstein's achievement and the failure of Lorentz and Poincaré to reach his understanding of relativity raise questions about the growth of science. Scientific research problems are not forced upon

[285] P. Frank, "Die Stellung des Relativitätsprinzips im System der Mechanik und der Elektrodynamik," *Sitzungsb. Wiener Akad. Wiss., 118* (1909), 373–446, esp. 376.

[286] M. Planck, "Die Stellung der neueren Physik zur mechanischen Weltanschauung," *Verh. Ges. Deutsch. Naturf. u. Ärzte, 82* (1910), part 1, pp. 58–75; *Phys. Abhandlungen und Vorträge, 3,* 30–46, esp. 39: "Wer daher die mechanische Naturanschauung als ein Postulat der physikalischen Denkweise ansieht, wird sich mit der Relativitätstheorie nie befreunden können."

scientists automatically by nature. They are the questions that scientists ask nature on the basis of their views of nature and science. The problems of scientific research can be formulated only in correlation with the views of nature and science. Historians, whose view of science is affected by the science of their own time, have not always been able to change their perspective when studying scientific developments that differed greatly from the current circumstances of science. The contrast between most of today's scientific research—the puzzle solving in normal science, as Thomas S. Kuhn has called it[287]—and the creative period in physics around the turn of the century and the first decades afterwards is reflected in the manner in which historians have considered the origin of the theory of relativity.

As the integration of science into the systems of society has advanced, various institutions and the legislation to promote and organize scientific research have been further and further augmented, and under the stimulus of these changes the instruments and facilities for scientific research have undergone rapid technological innovation and an enormous growth in size and complexity. Such recent trends affect the practice of research so that the part occupied by routine work in scientific research is continually on the increase.[288] Project research dominates today's science, and in project research the work is divided into separate parts that are assigned to different scientists. Work on these fragments of projects, performed in by far the majority of cases according to prescription, is not a creative intellectual adventure. In these cases, scientific research is reduced to the manipulation of instruments, data, formulas, and so on, and consequently such factors as philosophical inclination, worldview, and the idiosyncracies of scientists become of little significance. What is important here is only the technical skill of the scientists.

If the history of science is likely to reflect the view of science held

[287]Thomas S. Kuhn, *The Structure of Scientific Revolutions,* 2nd ed. (Chicago, 1970), Chap. 4.

[288]For an introductory discussion of the integration of science into the systems of society and its influence on the qualities of contemporary scientific research, see Tetu Hirosige, *Kagaku no Syakaisi: Kindai Nihon no Kagaku Taisei* (*A Social History of Science: The Social System of Science in Modern Japan*) (Tokyo, 1973), the introductory and final chapters. In English literature I refer the reader to Jerome R. Ravetz' discussion of what he calls the industrialized science: Jerome R. Ravetz, *Scientific Knowledge and Its Social Problems* (Oxford, 1971), esp. Chap. 2.

by the historian, as Alexandre Koyré has suggested,[289] then the modern historian, responding to the trends in science characterized above, will inevitably confine his attention to technical details in the science of the past. Limited by the point of view that corresponds to the characteristics of present-day science, he will neglect to pay due attention to the changes in aspect that give rise to scientific innovation.

The most important feature of scientific innovation is that, to achieve it, the scientist has to bring innovation to the problem itself and to abandon preconceptions. Scientific innovation begins when he perceives a problem to be studied that formerly has not existed as a scientific problem. The theory of relativity, needless to say, was an innovation in the sense just emphasized. It is therefore no surprise that Lorentz and Poincaré, who pursued the ether problem in its traditional formulation, could not create the theory of relativity, eminent though they were in vision and competence. The ether problem did not contain the factor that alone could cause the transformation of the problem structure. For the transformation to be effected, the premise that had made the ether problem the central concern of physicists had to be doubted and abandoned. We have found that premise in the worldview that holds that any physical substance ought to be characterized only by mechanical categories. According to this view, motion of and relative to the ether must always have physical consequences. It was this view that had to be changed.

In view of the close correspondence between the scientist's views of nature and science and his formulation of the problems of scientific research, Mach's refutation of the mechanistic worldview was of crucial importance for the formation of the theory of relativity. Certainly, Mach's criticism of the concepts of absolute space and time, holding that determinations in space and time are no more than the determinations of an event by other events, must have been suggestive to Einstein. But it could be suggestive only after he, viewing the problem situation from a new aspect, had discovered the new problem to be attacked, that is, only after Mach's refutation of the mechanistic worldview had provided him with the new perspective. In this sense I see Mach as having made the most fundamental contribution to the emergence of Einstein's theory of relativity.

[289] A. Koyré, *Étude d'histoire de la pensée scientifique* (Paris, 1966), pp. 71–72 and *passim*.

ACKNOWLEDGMENT

My best thanks are due to Professor Russell McCormmach, without whose kind promptings the present paper would not have been completed until much later. I also wish to express my cordial thanks to Mr. J. Nemoto of the Meteorological Agency who always offers me the convenience of using journals deposited at the library of the Agency. Last, but not least, I am grateful to my colleagues Dr. Sigeko Nisio and Mr. Ichiro Tanaka who have greatly helped me by gathering materials when I was inactive because of illness.

Einstein's Early Scientific Collaboration

BY LEWIS PYENSON*

1. INTRODUCTION

With his earliest work toward the general theory of relativity Einstein joined a direction in physics that, as he saw it, was marked by a prevalent interest in field theory. He recalled in 1933 that, "like most physicists, at this period I endeavoured to find a 'field law,' since, of course, the introduction of action at a distance was no longer feasible in any plausible form once the idea of simultaneity had been abolished."[1] The many field-theoretical studies carried out about 1905 were of two kinds. Some physicists were using field theory in the form of Maxwellian electrodynamics to develop descriptions of the electron as well as a complete system of physics founded on the electromagnetic worldview. Others sought to formulate a field theory of gravitation that could be integrated with electromagnetic theory. Physicists who had serious objections to the program of the electromagnetic worldview tried to answer them by examining gravitation with field-theoretical formalisms.

One way to study Einstein's position in the field-oriented physics of this period is by examining Einstein's early scientific collaboration. By 1913 Einstein had written scientific articles with five other scientists and had corresponded extensively with perhaps a score more. Some of his early contacts are well known. While at Berne, Einstein discussed his ideas with the brothers Paul and Konrad Habicht, with Maurice Solovine, and with Michele Angelo Besso, all of whom were trained in physics but not associated professionally with a university. Then, between 1908 and 1910, several young, established physicists, including Max Laue, Rudolf Ladenburg, and Arnold Sommerfeld, visited Einstein to discuss, for the most part, problems in quantum theory.[2] After 1909 Einstein collaborated with Sommerfeld's student Ludwig Hopf on the quantum theory,

*Institut d'Histoire et de Sociopolitique des Sciences, Université de Montréal.

[1] A. Einstein, *Origins of the General Theory of Relativity* (Glasgow, 1933), p. 6.

[2] See C. Seelig, *Albert Einstein. A Documentary Biography,* transl. M. Savill (London, 1956), pp. 78, 86, 116.

and with Jakob Johann Laub, Walther Ritz, Erwin Finlay Freundlich, and his old friend Marcel Grossmann on the theories of relativity. Unlike Grossmann who helped Einstein with mathematical problems, Laub, Ritz, and Freundlich provided him with the opportunity to discuss his physical ideas on the special theory of relativity and gravitation theory. Laub and Ritz were the first to coauthor papers with him, and, beginning in 1911, Freundlich corresponded extensively with Einstein about astronomical problems in general relativity. Einstein's relationship with these three physical scientists differed greatly from his associations with other scientists in the period before 1914. Through them he was able to keep informed of developments in the physical science community.

In the present study I shall use Einstein's collaboration with Laub, Ritz, and Freundlich to illuminate his relationship to the physical science community in late Wilhelmian Germany. I shall examine to what extent the attitudes toward mathematics of Einstein's collaborators stemmed from the intellectual climate at Göttingen University, where, between 1902 and 1910, each of the three had studied for at least two years. Then I shall discuss their collaboration with Einstein and remark on their importance in the development of Einstein's general relativity. Finally, I shall discuss the reception of the special and general theories of relativity in light of the effects that the collaboration with Einstein had on Laub's and Freundlich's careers.

2. PHYSICAL SCIENCE AT GÖTTINGEN, 1895-1914

During the first decade of the twentieth century Göttingen was one of the universities leading in the fields of pure and applied mathematics, physical chemistry, electrotechnology, applied mechanics, and hydrodynamics.[3] In 1894 several Göttingen institutes accom-

[3] Sources used in assembling information on the institutional evolution of Göttingen physical science include: "Die Physikalischen Institute der Universität Göttingen," *Festschrift im Anschlusse an die Einweihung der Neubauten am 9. Dezember 1905. . .*(Leipzig, 1905); F. Klein and E. Riecke, eds., *Ueber angewandte Mathematik und Physik in ihrer Bedeutung für den Unterricht an den höheren Schulen. Nebst Erläuterung der bezüglichen Göttinger Universitätseinrichtungen* (Leipzig, 1900); K.-H. Manegold, *Universität, Technische Hochschule und Industrie. Ein Beitrag zur Emanzipation der Technik im 19. Jahrhundert unter besonderer Berücksichtigung Felix Kleins* (Berlin, 1970); W. Ebel, ed., *Catalogus Professorum Gottingensium 1734–1962*

modated physics research and instruction. The most important of these was the physical institute with a division for experimental physics under Eduard Riecke and a second division for what was variously called mathematical physics or theoretical physics under Woldemar Voigt. The less important Göttingen observatory under Ernst Schering also treated mathematical physics as well as theoretical astronomy and geodesy. The Göttingen mathematical physics seminar under Felix Klein sought to unify physical science; Klein was joined by Riecke, Voigt, Schering, the mathematician Heinrich Weber, and Wilhelm Schur, director of the division of applied astronomy of the observatory.

Between 1895 and 1898 the institutional configuration of Göttingen physics underwent a transformation. An institute of physical chemistry was created for Walther Nernst, then an associate (*ausserordentlicher*) professor of mathematical physics, and a division for technical physics, that is, for applied mechanics and applied electrical studies, was added to the physical institute. The observatory was reorganized upon Schering's death in 1898. Both Schering's chair and the directorship of the observatory's theoretical division remained vacant. Instead, Martin Brendel became associate professor of applied astronomy, and Emil Wiechert was appointed to the new position of associate professor of geophysics. Although Wiechert's duties included the management of the observatory's earth magnetism laboratory, for all practical purposes he directed a fourth geophysical division of the physical institute.

In 1905, Göttingen physical science underwent its last reorganization of the Wilhelmian period. The technical physics division of the physical institute was given the status of an independent institute for applied mathematics and mechanics under the joint direction of Carl Runge, professor of applied mathematics, and Ludwig Prandtl, associate professor of applied mechanics. Furthermore, Hermann Theodor Simon became director of a new division of the physical institute devoted exclusively to applied electricity. At the same time, Wiechert was elevated to the directorship of an independent institute of geophysics. Finally, the theoretical divisions of the observatory were officially united under the direction of Karl Schwarzschild.

The driving force behind these institutional changes was Felix

---

(Göttingen, 1962); K. J. Truebner *et al.*, eds., *Minerva. Jahrbuch der gelehrten Welt, 1* (1891–92), *et sqq.*

Klein. As the titles of some of the new institutes indicate, Klein sought to promote interdisciplinary studies at Göttingen. He believed that such studies furthered industrial applications of science and that they played a critical role in the development of new scientific ideas.[4] From about 1902 he organized numerous interdisciplinary seminars in the physical sciences, and between 1895 and 1910 he succeeded in establishing nine new senior faculty positions in interdisciplinary physical sciences. Many of the seminars and faculty appointments emphasized technical physics. The technology-oriented fields received the strongest institutional backing, although other interdisciplinary areas without relations to applied science, such as astrophysics and actuarial mathematics, also received support.

Largely as a result of Klein's leadership, mathematics occupied the central position in Göttingen physical science. Beginning in the 1890's, Klein set out to create a new image for Göttingen mathematics by forging links between academic mathematics and industry. He promoted an organization composed of academics and industrialists, the Göttingen Association of Applied Mathematics and Physics, through which he wanted to convince industrialists that university mathematics would be an appropriate investment for industrial capital. Contributions to industry by Göttingen scientists in the period before the First World War included Wiechert's seismic geophysics, Runge's and Prandtl's aerodynamic design, and Nernst's electrochemistry. Klein felt that the industrial connection would be of equal importance for the discovery of new mathematical truths. Applied mathematics, that is, mathematics applied to the sciences and to industry, was for him the source of many new concepts and procedures of pure mathematics; mathematics could not proceed solely by reflection.

Klein preferred intuitive reasoning with models and geometrical pictures to axiomatic, logical exposition. His view of mathematics was a reaction against the trend current at the end of the nineteenth century to separate pure mathematics from applied mathematics and often to include theoretical physics in the latter. Mathematicians generally considered pure mathematics the product of pure reason obtained without any reference to experiment, a "set of *formal*

[4] L. Pyenson, *Göttingen Reception of Einstein's General Theory of Relativity* (diss., Johns Hopkins University, 1973), Chapter 4: "Felix Klein and Mathematical Physics at Göttingen."

implications independent of all content."[5] Applied mathematics, on the other hand, was thought to consist of applications of the relationships revealed by pure mathematics to material facts or physical processes and to depend on an intuitive grasp of physical postulates and models. Analysis, algebra, and logic were part of pure mathematics, whereas geometry and mechanics belonged to applied mathematics.[6]

The division between pure and applied mathematics did not alter the curriculum that pure mathematicians generally claimed. During the period before the First World War, pure mathematicians regularly taught mechanics and geometry to physicists. Some physicists felt that they were not receiving an adequate mathematical education. Carl Runge reported to the Fifth International Congress of Mathematicians in 1913: "Some of my correspondents bitterly complain of the mathematical training of students of physics in consequence of the professors of pure mathematics ignoring some mathematical theorems and methods, that are of greatest importance to the physicist." Runge felt that a closer cooperation between pure and applied mathematicians would do much to alleviate the problem: "On the whole there seems to be no need for special mathematical courses for students of physics, nor does it seem necessary to compel them to attend more mathematical lectures. But a need seems to be strongly felt for mathematicians and physicists to draw closer together. The spirit of the mathematical teaching should be altered, so as to make it more practical and easier to apply to physical problems. At present the gap is very wide and is not tending to close up."[7]

Klein's efforts to bridge the gap between pure and applied mathematics were reflected in the work of other Göttingen mathematicians: they most frequently employed rigid body mechanics and electrodynamics when investigating other problems. David Hilbert,

[5] L. Couturat, *Les principes des mathématiques* (Paris, 1905), p. 4. Emphasis in the original.

[6] Among many other statements, see the distinction made by Marcel Grossmann between pure and applied mathematics for the International Commission on Mathematics Instruction. M. Grossmann, "Der mathematische Unterricht an der Eidgenössischen Technischen Hochschule," published as bulletin 7 of the series *Der mathematische Unterricht in der Schweiz,* edited by H. Fehr (Basle and Geneva, 1911), pp. 29-31.

[7] C. Runge, "The Mathematical Training of the Physicist," *Proceedings of the Fifth International Congress of Mathematicians, 2* (1913), 599.

Wiechert, Max Abraham, and Gunnar Nordström frequently used the Maxwell-Lorentz electromagnetic theory in interpreting both special and general relativity, and Max Born and Gustav Herglotz for some time tried to reconcile special relativity with classical notions of rigid body motion. Their emphasis on mechanics seems to have been related to the strong Göttingen program in technical physics and to Runge's and Prandtl's research on engineering problems associated with hydrodynamics. Interest in electrodynamics at Göttingen was focused on the elaboration of the electromagnetic view of nature. After 1900, the electromagnetic view of nature received widest circulation through the various theories treating the dynamics of the electron.[8] Of the scientists concerned with electron theories between 1900 and 1910 more worked at Göttingen than anywhere else. Alexander Wilkins, Walter Kaufmann, Emil Bose, Schwarzschild, Abraham, and Wiechert published on the electron theory while at Göttingen; Paul Drude and Arnold Sommerfeld, who had been privatdocents at Göttingen in the 1890's, submitted papers on the subject to the Göttingen Scientific Society. Göttingen mathematicians were attracted to the electron theory by its promise for unifying all of physical theory, but they were sometimes not overly careful about giving what physicists would consider realistic formulations of the physical processes they were investigating.

It should be emphasized that the Göttingen *physics* faculty was not sympathetic to Einstein's theory of relativity. None of the six physicists who responded at length to Einstein's theories were senior members of the Göttingen physics faculty. The two most important physicists of the six, Abraham and Born, were privatdocents in mathematical physics. Abraham never accepted either special or general relativity, and Born found the theories difficult to understand. Wiechert, the professor of geophysics, was the only other physics instructor to write on both theories; unfortunately, he never understood either. Three other members of the physics faculty, Kaufmann, Johannes Stark, and Ritz, wrote on special relativity without com-

[8]Among several excellent articles concerning the electromagnetic view of nature, see T. Hirosige, "Electrodynamics Before the Theory of Relativity, 1890–1905," *Japan. Stud. Hist. Sci.*, No. 5 (1966), pp. 1–49; R. McCormmach, "H. A. Lorentz and the Electromagnetic View of Nature," *Isis, 61* (1970), 459–497; R. McCormmach, "Hendrik Antoon Lorentz," *Dictionary of Scientific Biography* (New York, 1973).

menting on general relativity. Kaufmann and Stark, both privatdocents in physics, were excellent experimentalists; Kaufmann initially rejected and Stark never accepted special relativity. Ritz, also a privatdocent, rejected special relativity in favor of an emission theory of light.

In view of the nearly unanimous rejection of relativity among physicists at Göttingen it is not surprising that only three Göttingen physics graduates wrote on special or general relativity before 1919. The three were students of Voigt, who himself referred to Einstein's relativity only in passing.[9] Two of them had previously been educated in Britain: Robert Alexander Houstoun had received a master's degree from Glasgow University in 1902, and Alfred Arthur Robb had obtained a bachelor's degree from Cambridge in 1897. Both Houstoun and Robb studied theoretical optics under Voigt from 1902 to 1905, but their Göttingen studies apparently did not influence their work on special relativity.[10] The third student of Voigt to write on special relativity was Walther Ritz. Other Göttingen physics students who later wrote extensively on special and general relativity were Moritz Schlick and Max Laue,[11] Voigt's students, and Paul Ehrenfest[12] and Johannes Droste. Their work, too, does not appear to have been influenced by Göttingen physicists. On the other hand, Gunnar Nordström, who had come to study physical chemistry in

[9]W. Voigt, "Phänomenologische und atomistische Betrachtungsweise," in E. Warburg's volume *Physik* (Leipzig, 1915), p. 731, which is part of P. Hinneberg's series *Die Kultur der Gegenwart.* See S. Goldberg, "Woldemar Voigt," *Dictionary of Scientific Biography* (New York, 1976), in press.

[10]R. A. Houstoun, *A Treatise on Light* (London, 1915), pp. 465-466. Houstoun here neither rejected nor accepted Einstein's principle of relativity. Robb attempted to put special relativity upon an axiomatic basis by defining rigorously the relations "before" and "after." See A. A. Robb, *Optical Geometry of Motion. A New View of the Theory of Relativity* (Cambridge, England, 1911); *A Theory of Time and Space* (Cambridge, England, 1914). Robb came to reconsider the problem of measuring distance and time through a comment by Joseph Larmor at the British Association Meeting of 1902. A. A. Robb, *Absolute Relations of Time and Space* (Cambridge, England, 1921), p. v. On Robb, see G. Windred, "The History of Mathematical Time," *Isis, 26* (1934), 199-202.

[11]Letter to the author from R. A. Houstoun, 16 December 1972. A copy is located in the Niels Bohr Library of the American Institute of Physics, New York.

[12]On Ehrenfest at Göttingen, see M. J. Klein, *Paul Ehrenfest. The Making of a Theoretical Physicist* (Amsterdam, 1971).

1906, was quickly converted to the mathematical physics of Minkowski.[13] Nordström's switch to mathematical physics was not unusual;[14] Laub, Ritz, and Freundlich were all drawn equally strongly to Göttingen mathematics.

Ritz arrived in Göttingen in the spring of 1901 after studying at the Federal Institute of Technology in Zurich.[15] Ritz was several years behind Einstein in his studies there, and there is no evidence that he met Einstein while they studied in Zurich. Unlike Einstein, Ritz followed attentively the lectures given by Minkowski and Adolf Hurwitz on mechanics and pure mathematics. In 1901, Ritz went to Hilbert in Göttingen with an enthusiastic recommendation from Minkowski.[16] Ritz was strongly attracted to physics and chose to work on a dissertation in optics under Voigt.[17] By the time Ritz left Göttingen in 1903, he had acquired an impressive command of many mathematical tools. He also carried away some knowledge of Wiechert's, Abraham's and Kaufmann's work on the electron theory.

After having studied briefly at the universities of Krakow and Vienna, Laub entered Göttingen University in October 1902 as a mathematics student.[18] He remained until 1905. In October 1903, Hilbert noted that Laub was "diligently" attending his lectures on differential equations.[19] At the same time, Minkowski, who had been

[13]When Nordström arrived in Göttingen, Nernst had just been called to Charlottenburg, and Fritz Dolezalek was the new director of the Göttingen physical chemistry institute. In late 1906 Nordström completed a paper under Dolezalek's direction on Hittorf's transference number for potassium hydrate. Thereafter nearly all of his published work dealt with electrodynamics or Einstein's theories of relativity. H. Tallqvist, "Gunnar Nordström," *Finska Vetenskaps-Societeten Minnestrecknigar och Föredrag* (1924).

[14]P. P. Ewald arrived in Göttingen about the same time to study chemistry under Otto Wallach. He quickly transferred to Hilbert's mathematics. Interview between Ewald and G. Uhlenbeck and T. S. Kuhn, 29 March 1962, Archive for History of Quantum Physics, American Philosophical Society, Philadelphia.

[15]The best source for Ritz's life is Pierre Weiss's biography in P. Weiss, ed., *Walther Ritz Oeuvres* (Paris, 1911), pp. vii–xxii.

[16]Hermann Minkowski to David Hilbert, 11 March 1901, in L. Rüdenburg and H. Zassenhaus, eds., *Hermann Minkowski. Briefe an David Hilbert* (Berlin, 1973), p. 139.

[17]W. Ritz, "Zur Theorie der Serienspektren," *op. cit.* (note 15), pp. 1–77.

[18]J. Laub, *Ueber sekundäre Kathodenstrahlen* (diss., Würzburg University, 1907), "Lebenslauf."

[19]Extract from *Gerd Rosen, 37. Auktion. II. Teil. Bücher und Autographen* (Berlin, 1961), item 2002.

called to Göttingen from Zurich the year before, assigned Laub the task of writing up his introductory lectures on mechanics.[20]

Freundlich studied at Göttingen from 1905 to 1910.[21] Unlike Ritz and Laub, Freundlich came to Göttingen with the familiarity of both technical science and pure mathematics that Klein sought to cultivate. After completing the required studies at the gymnasium in Wiesbaden, Freundlich studied shipbuilding for two semesters at the Polytechnical School in Charlottenburg. He worked for a year and a half as a volunteer in a shipyard in Stettin until ill health forced him to retire. In the spring of 1905 he finally enrolled as a mathematics student at Göttingen University. He was unusually talented in mathematics and shared Klein's philosophy of mathematics education.[22]

Although Ritz, Laub, and Freundlich all had come to Göttingen to prepare for careers in the physical sciences, they also received a thorough exposure to mathematics. They were not the only physicists so influenced. In 1895 Sommerfeld had habilitated in Göttingen under Klein. He recalled later that the experience had profoundly marked his approach to physics.[23] Through their Göttingen education Ritz, Laub, and Freundlich also acquired an appreciation of electrodynamics: Ritz treated it in his dissertation, Laub wrote his dissertation on one facet of electron physics, and Freundlich followed Minkowski's and Born's work in electrodynamics. However, their Göttingen education did not provide them with experimental experience or allow them to acquire sensitivity to experimental results. These they received elsewhere: Ritz carried out experimental work as a postdoctoral assistant to Aimé Cotton in Paris and, later, as an assistant to Friedrich Paschen in Tübingen, Laub earned his

[20] Sheet 4 of a manuscript, "Mechanik I, Göttingen, Winter-Semester 1903/04," box IX, folder 4 of the Minkowski papers located at the Niels Bohr Library of the American Institute of Physics, New York.

[21] E. Forbes, "Erwin Finlay Freundlich," *Dictionary of Scientific Biography* (New York, 1971). E. Freundlich, *Analytische Funktionen mit beliebig vorgeschriebenem unendlich-blättrigem Existenzbereiche* (diss., Göttingen University, 1910), "Lebenslauf."

[22] At the time he was finishing his dissertation Freundlich attended Klein's seminar on philosophy and pedagogy of mathematics. "F. Klein, Material zum psychologischen Seminar, Winter-Semester 1909–1910," Klein Nachlass, XXI, A, Niedersächsische Staats- und Universitätsbibliothek, Göttingen.

[23] T. S. Kuhn, *et al.*, *Sources for the History of Quantum Physics, an Inventory and Report* (*Memoirs of the American Philosophical Society*, Philadelphia, *68* [1967], 139).

doctorate working with Wilhelm Wien at Würzburg on an experimental dissertation, and Freundlich had practical experience in engineering earlier and eventually became an observational astronomer.

3. LAUB AND EINSTEIN

Sometime late in 1905 Laub moved from Göttingen to Würzburg to study under Wilhelm Wien, director of the University's physical institute. Unlike Göttingen scientists, Wien did not assign to mathematics a central, unifying role in physics, but reserved that for theoretical physics. Mathematical physics—as opposed to theoretical physics—was in his view only concerned with developing the mathematical tools for theoretical physics.

Wien felt that theoretical physics was ideally constituted for the task of unifying physics. Far from being the exclusively deductive science of the Kantian point of view, it incorporated inductive as well as deductive method—the two being nearly indistinguishable—and it claimed as its "real and exclusive task" the "establishment of functional connections,"[24] that is, the establishment of laws that govern as many physical processes as possible. Specifically, Wien hoped that theoretical physics would bring unity to physics by establishing an electromagnetic foundation of mechanics. "It is doubtlessly one of the most important tasks of theoretical physics," he wrote, "to link the now completely isolated areas of mechanical and electromagnetic phenomena and to derive the differential equations of each from a common foundation."[25]

Laub began to work on his dissertation under Wien in 1905. An extension of Wien's recent work on secondary radiation,[26] his sub-

[24]W. Wien, "Ziele und Methoden der theoretischen Physik," *Jahrbuch Radioakt. Elektr., 12* (1915), 241–259. The passage cited is on p. 246.

[25]W. Wien, "Ueber die Möglichkeit einer elektromagnetischen Begründung der Mechanik," *Ann. Phys., 5* (1901), 501. "Es ist zweifellos eine der wichtigsten Aufgaben der theoretischen Physik, die beiden zunächst vollständig isolirten Gebiete der mechanischen und elektromagnetischen Erscheinungen miteinander zu verknüpfen und die für jedes geltenden Differentialgleichungen aus einer gemeinsamen Grundlage abzuleiten."

[26]Wien gave Laub an inscribed copy of his paper, "Ueber die Energie der Kathodenstrahlen im Verhältnis zur Energie der Röntgen- und Sekundarstrahlen," *Ann. Phys., 18* (1905), 991-1007. In *Gerd Rosen, Auktion 36, 2. Teil. Bücher und Autographen* (Berlin, 1961), item 4004, Laub wrote that Wien's paper was "very important. . .in connection with Einstein's Nobel Prize work."

ject was an experimental investigation of secondary cathode ray emission, a phenomenon in which a cathode ray beam of electrons is used to generate a secondary stream of cathode rays when it strikes and ionizes a target. Since his dissertation was connected with the problem of a theory of atomic structure,[27] Laub's work was an ideal doctoral thesis: it involved research of great current interest that could be summarized and clarified through new and more precise experiments. At his dissertation defense in November 1906, Laub defended the special theory of relativity to the consternation of his examination committee.[28] Wien, however, was satisfied with his work, and Laub passed. Laub continued to work on secondary cathode ray emission under Wien during the period 1907–1908,[29] but he directed the major part of his work toward Einstein's special theory of relativity.

In June 1907 Laub published his first paper on special relativity, concentrating on a discussion of Fresnel's drag coefficients.[30] (Drag coefficients describe how a ponderomotive body apparently imparts a small portion of its velocity to light, as observed, for example, in the 1851 experiment of H. L. Fizeau, which measured the velocity of light in moving columns of water. By 1895, Lorentz had explained the first order drag coefficient from his equations of electromagnetism by using disturbances in a stationary ether.)[31] Laub derived the first order drag coefficients without making use of the ether, although according to Laue he confused group and phase velocities.[32] Wien encouraged Laub. In a letter to Sommerfeld in June 1908, Wien wrote: "I have recommended that Laub apply the Lorentz transformation to dispersion theory. I did this because the theory of drag coefficients is not rigorous since it only uses the concept of dielec-

[27] J. Laub, *op. cit.* (note 18), p. 86. A shortened version of the dissertation was published under the same title in *Ann. Phys., 23* (1907), 285–300. The thesis was presented by A. Witkowski to the Krakow Academy of Sciences and published in *Bulletin international de l'Académie des Sciences, Cracovie. Classe des sciences mathématiques et naturelles* (1907), pp. 61–87.

[28] C. Seelig, *op. cit.* (note 2), p. 72.

[29] J. Laub, "Ueber die durch Röntgenstrahlen erzeugten sekundären Kathodenstrahlen," *Ann. Phys., 26* (1908), 712–726.

[30] J. Laub, "Zur Optik der bewegten Körper," *Ann. Phys., 23* (1907), 738–744; "Die Mitführung des Lichtes durch bewegte Körper nach dem Relativitätsprincip," *Ann. Phys., 23* (1907), 989–990.

[31] T. Hirosige, "Origins of Lorentz' Theory of Electrons and the Concept of the Electromagnetic Field," *Hist. Stud. Phys. Sci., 1* (1969), 202–205.

[32] M. Laue to A. Einstein, 4 August 1907. Deutsches Museum, Munich.

tric constants and not that of dispersion."[33] A year earlier, at the 1907 Dresden meeting of the Society of German Natural Scientists and Physicians, Wien had argued that the assumption of media that display anomalous dispersion contradicts the second postulate of the relativity theory.[34] The contradiction still puzzled him, and he now asked Laub to investigate the problem from a point of view in sympathy with Einstein's ideas. In 1908 Wien believed that the special theory of relativity was part of the electron theory. By encouraging Laub, Wien acted on his view that theoretical physicists, not mathematicians, could best formulate a new foundation for the electron theory.

In early February 1908, Laub wrote to Einstein, who was then at the patent office at Berne, to inquire if it would be possible for him to spend three months in Berne to work on the theory of relativity. Laub assured Einstein of his great interest in the theory: he considered Einstein's work fundamental not only for electrodynamics but for all of physics.[35] Einstein's reply to Laub is lost, but it must have been favorable. Over the next three months, he and Laub collaborated on two articles on the special theory of relativity.[36] These were the first papers Einstein wrote with another person.

In his letter to Sommerfeld in June 1908, Wien indicated that the notion of force in Einstein's special theory of relativity would have to be modified by future work in dispersion theory. Without doubt,

[33] W. Wien to A. Sommerfeld, 15 June 1908. A microfilm copy is located in the Archive for History of Quantum Physics at the American Philosophical Society, Philadelphia. "Ich habe Laub empfohlen, die Lorentzsche Transformation auf die Dispersionstheorie anzuwenden. Die Theorie der Mitführungskoeffizienten ist nämlich nicht streng, da sie nur den Begriff der dialektr. Constanten nicht aber den der Dispersion benutzt."

[34] W. Wien, "Turbulente Bewegung der Gase," listed in *Phys. Zs.*, *8* (1907), 722. See the discussion between Wien and Sommerfeld in *Phys. Zs.*, *8* (1907), 841–842.

[35] J. J. Laub to A. Einstein, 2 February 1908. Located in the Einstein Archives at Princeton. "Ich versichere Sie, dass mich das grosse Interesse an der Sache nachdem führt, denn ich halte Ihre Untersuchung für fundamental nicht nur für die Elektrodynamik (die ist ja nur ein specielles Anwendungsgebiet) sondern für die ganze Physik."

[36] A. Einstein and J. Laub, "Ueber die elektromagnetischen Grundgleichungen für bewegte Körper," *Ann. Phys.*, *26* (1908), 532–540; "Bemerkungen zu unserer Arbeit," *Ann. Phys.*, *28* (1909), 445–447; "Ueber die im elektromagnetischen Felde auf ruhende Körper ausgeübten ponderomotorischen Kräfte," *Ann. Phys.*, *26* (1908), 541–550.

he had in mind the subject of Laub's recent collaboration with Einstein. Einstein and Laub had taken up the problem of defining force in the special theory of relativity, a critical issue around which would turn many debates in the next three years.[37] Specifically, they criticized Minkowski's 1908 paper on the basic equations for electromagnetic processes, where he first discussed in print his ideas on space-time.[38] In his electromagnetic equations Minkowski required that the electromagnetic force of a particle in motion always be normal to the particle's path in four-dimensional space-time. Einstein and Laub pointed out that this formulation of the electromagnetic force does not allow for current flow in a wire, or, in other words, that Minkowski's force denies that a direct current flowing through a wire in a magnetic field gives rise to a polarization current in addition to the ponderomotive force, an experimentally observable phenomenon. They argued that Minkowski took as the entire ponderomotive force only that component of the force that is normal to the particle's velocity. This conception of force was equivalent to the assumption that the rest mass of the particle remains constant. Einstein and Laub formulated an expression for the ponderomotive (or electromagnetic) force and then sought to demonstrate that their formulation was consistent with the laws of dynamics, in particular with Newton's third law. To show consistency with the latter was important, because Minkowski had attempted to place all of mechanics on an axiomatic foundation without making use of Newton's third law.

The difference between Einstein and Laub's work and Minkowski's is that between a physical and a mathematical theory: Einstein and Laub examined the physical consequences of several possible formulations of electromagnetic force, not including Minkowski's four-dimensional mathematical formulation, to find a theory that integrated several physical laws and phenomena; Minkowski, on the other hand, was concerned primarily with the mathematical consequences of one particular formulation of electromagnetic force and did not investigate whether or not the formulation was consistent with other electromagnetic laws. Laub and Einstein stressed physical arguments, but their understanding of the development of physical theory was not the same as the view that physical theory grows out

[37]Einstein and Laub, "Ueber die im elektromagnetischen Felde . . . ," *ibid.*

[38]H. Minkowski, "Die Grundgleichungen für die elektromagnetischen Vorgänge in bewegten Körpern (1908)," in D. Hilbert, ed., *Hermann Minkowski. Gesammelte Abhandlungen* (Leipzig, 1911), *2,* 352–404.

of or is based on experimental physics. The second paper Einstein and Laub coauthored also illustrates their distinct approach.

Their first paper on the electromagnetic force was followed by a series of treatments by Abraham, Born, Nordström, Ehrenfest, Jun Ishiwara, and Ludwig Silberstein.[39] These physicists shared their objective of clarifying the nature of force in the special theory of relativity. Einstein turned to other problems, in particular to formulating a theory of charged particles, and did not consider the problem of force within the special theory of relativity again after 1908.[40] He realized that any treatment of an electrodynamic force would require the notion of acceleration; to include that he would have to generalize the principle of relativity. Laub shared Einstein's understanding, for although he investigated the special theory of relativity further, he did not address the question of ponderomotive force again.

In the second paper,[41] written in April 1908, Einstein and Laub were concerned with explaining a puzzling effect discovered by H. A. Wilson: when a dielectric is rotated in the gap between the plates of a connected capacitor in the presence of a magnetic field, an equal and opposite charge collects on the plates. For convenience Einstein and Laub assumed that each particle of the dielectric, although actually in rotation, is only in linear motion. They used Minkowski's equations of motion with appropriate boundary conditions to arrive at the observed charge separation on the two capacitor plates. Einstein and Laub claimed that their calculation differs from one based on Lorentz' electron theory and that the Wilson experiment gives identical results for the two theories because the magnetic permeability is unity in each case.[42] Einstein and Laub carried out a theoretical

[39]These papers have been treated in several secondary sources: G. H. F. Gardner, *The Concept of a Rigid Body in Special Relativity* (diss., Princeton University, 1953), pp. 9–14; S. Goldberg, *Early Responses to Einstein's Theory of Relativity, 1905–1911: A Case Study in National Differences* (diss., Harvard University, 1968), pp. 119–129; M. J. Klein, *op. cit.* (note 12), p. 152; L. Pyenson, *op. cit.* (note 4), pp. 234–267. See also L. Silberstein, *Theory of Relativity* (London, 1914), pp. 283 ff.

[40]R. McCormmach, "Einstein, Lorentz, and the Electron Theory," *Hist. Stud. Phys. Sci., 2* (1970), 69–81.

[41]Einstein and Laub, "Ueber die elektromagnetischen Grundgleichungen . . . ," *op. cit.* (note 36).

[42]Subsequently, M. and H. A. Wilson verified Einstein and Laub's theory in their article, "On the Electric Effect of Rotating a Magnetic Insulator in a Magnetic Field," *Proceedings of the Royal Society, London, A 89* (1913), 99–106.

study that appealed to experimental physics. When Max Laue objected to the way in which Einstein and Laub used boundary conditions in the article, for example, they responded by affirming the physical nature of their calculations.[43] In this paper, as in the previous one, Einstein and Laub sought to distinguish the special theory of relativity from the Lorentz electron theory which seemed to many physicists to be identical with it but which in fact was based on a different understanding of physics. Einstein and Laub maintained that the special theory of relativity provided different explanations for certain electromagnetic processes. They did not use experimental propositions to modify the special theory of relativity, and they did not attempt to use their results to decide between the contending theories.

After collaborating with Einstein, Laub continued to work on both theoretical and experimental problems in the special theory of relativity. The initial focus of his work was dispersion theory. In late 1908 Laub formulated an elementary theory of dispersion based on the special theory of relativity. He concentrated on the influence of molecular motion on dispersion in gases.[44] The orientation of Laub's work was probably proposed by Wien, although Einstein was also sympathetic to the attempt to relate the special theory of relativity to molecular theory. In January 1909 Laub used his knowledge of experimental ray physics in his work on dispersion theory.[45] He noted that Wien's as yet unpublished experiments on Doppler-shifted canal ray spectra contradict Stark's so-called photochemical law, which states that energy is transferred only in those collisions between canal ray particles and gas atoms that are above a specific threshold.[46] Arguing against Einstein, Lorentz, Sommerfeld, and Peter Debye, Laub insisted that "all particles contained in radiation bundles share equally in the light processes."[47] From his equations

[43] Einstein and Laub, "Bemerkungen . . . ," *op. cit.* (note 36).

[44] J. Laub, "Ueber den Einfluss der molekularen Bewegung auf die Dispersionserscheinungen in Gasen," *Ann. Phys., 28* (1908), 131-141.

[45] J. Laub, "Zur Theorie der Dispersion und Extinction des Lichtes in leuchtenden Gasen und Dämpfen," *Ann. Phys., 29* (1909), 94-110.

[46] See A. Hermann, *The Genesis of Quantum Theory, 1899-1913,* transl. C. W. Nash (Cambridge, Mass., 1971), pp. 72-77; J. L. Heilbron, *A History of the Problem of Atomic Structure from the Discovery of the Electron to the Beginnings of Quantum Mechanics* (diss., University of California, Berkeley, 1964), p. 178.

[47] J. Laub, *op. cit.* (note 45), p. 95.

Laub was able to indicate second order relativistic effects in dispersion theory. In November 1909 Laub summarized his work in a paper presented to the Heidelberg Academy of Sciences by Philipp Lenard.[48] He described several primary and secondary relativistic effects that, he felt, could be distinguished in canal ray phenomena, and he demonstrated that second order relativistic effects could be distinguished in Zeeman splitting. Laub's work in dispersion theory was not extended by other physicists. The dispersion problem touched on too many critical unresolved issues in quantum theory, statistical mechanics, the electron theory, and the special theory of relativity. Most physicists continued to use Paul Drude's classical dispersion theory until 1921 when, after working on the problem for many years, Rudolf Ladenburg approached the interaction of matter and radiation in dispersion phenomena from the point of view of quantum mechanics.[49]

In 1910 Laub wrote a long review article on the experimental foundation of relativity.[50] He discussed experiments with positive and null results for both first and second order effects in v/c. Laub divided the experiments into "optical" and "electromagnetic" categories. In so doing, he attempted to distinguish phenomena that depend critically on radiation from phenomena that depend on matter. At this time Laub was still not certain how to approach the field-particle antinomy in Einstein's physics. He realized that no *experimentum crucis* could establish the limits within which special relativity was a valid theory. Before Einstein's special theory of relativity could be accepted, physicists would have to refine and reflect upon the material he had presented. Laub concluded his review by asserting the unity and mathematical elegance of relativity theory: "Although it is true that only an experiment can decide if a physical theory is correct, it must still be emphasized that mathematics and its applications are related by an inner harmony."[51] These words did

[48] J. Laub, "Zur Theorie der longitudinalen magnetooptischen Effekte in leuchtenden Gasen und Dämpfen," *Sitzungsberichte der königlichen Akademie, Heidelberg. Mathematische Klasse* (1909).

[49] R. Ladenburg, "On the Quantum-Theoretical Interpretation of the Number of Dispersion Electrons," transl. G. Field in B. L. van der Waerden, ed., *Sources of Quantum Mechanics* (New York, 1968), pp. 139–157. M. Jammer, *Conceptual Development of Quantum Mechanics* (New York, 1966), p. 181.

[50] J. Laub, "Ueber die experimentellen Grundlagen des Relativitätsprinzips," *Jahrbuch Radioakt. Elektr.*, 7 (1910), 405–463.

[51] *Ibid.*, p. 463. "Denn, wenn es auch wahr ist, dass über die Richtigkeit einer physikalischen Theorie nur das Experiment entscheiden kann, so muss

not express Einstein's view of mathematics at the time, for he considered mathematics as a tool to be employed only after the physical situation had been conceptualized. They do suggest that Laub's view of mathematics was influenced by his Göttingen education.

Through Laub, Einstein had access to developments in mathematics and mathematical physics during the period around 1908. Lacking a university appointment, Laub was free to travel throughout German-speaking Europe. He cultivated connections with mathematicians and relayed information to Einstein in Berne. His services as informant were important for Einstein, for Einstein's position at the patent office allowed him little opportunity to consult scientific journals or speak with people in the forefront of science.[52] A letter from Laub to Einstein written in May 1908, soon after they had completed their joint papers, illustrates his concern to keep Einstein informed. He reported that the theoretical physicist Mathias Cantor was not convinced of either the special theory of relativity or Minkowski's space-time because he felt that they were both based on old foundations. "Why the foundations are not correct," Laub commented, "God knows; I don't ask him any longer, and I don't worry about it." Laub was amazed by what Cantor liked about Minkowski's work: Cantor applauded the way space and time coordinates were treated as homogeneous quantities and was impressed "that one can treat that as a rotation." Laub asked Cantor about the physical meaning of time as a fourth spatial coordinate, but he received no answer. Laub concluded from this: "I believe he has let himself be impressed by non-euclidean geometry." Cantor and the professor of mathematics, Eduard von Weber, were planning to introduce Minkowski's work in the physics colloquium at Würzburg. Laub noted that he would add some remarks to their presentation for he was skeptical about Minkowski's work. "Were not your work available," he wrote to Einstein, "we would at best have reached the same position with the Minkowski transformation equation for time (as far as the physical interpretation is concerned) as with Lorentz' 'local time'."[53]

---

doch hervorgehoben werden, dass zwischen Mathematik und ihren Anwendungen ein innerer harmonischer Zusammenhang besteht."

[52] B. Hoffmann, *Albert Einstein. Creator and Rebel* (New York, 1972), pp. 85–86.

[53] Letter from J. J. Laub to A. Einstein, 18 May 1908. Located in the Einstein Archives at Princeton.

Prof. Cantor scheint noch immer "einen Wurm" zu haben. Die Minkows-

4. RITZ AND EINSTEIN

In May 1909 Einstein was appointed associate professor of physics at the University of Zurich. Just before he took up the new position, Einstein sent off a short note to the *Physikalische Zeitschrift* which he had written with Walther Ritz. The note presented a brief clarification of the differences between Einstein's and Ritz's work on the theory of radiation.[54] At the time, Ritz was privatdocent in Göttingen. Since the paper is dated from Zurich, it may have been worked out while Ritz visited Einstein in Switzerland. Einstein's collaboration with Ritz produced no further result; Ritz died four months after the paper was written.[55]

After receiving his doctorate from Voigt at Göttingen, Ritz continued his studies under Lorentz at Leiden and under Aimé Cotton at Paris. After 1903, however, his precarious health interfered with the course of his scientific work and travels. In 1908 Ritz finally habilitated at Göttingen. There he worked intensely to create a new theory of optical spectra and a new electrodynamics, building both upon his previous work in optics under Voigt.[56] Ritz believed in the principle of relativity, but he did not want to include in his theory the result that the mass of a particle depends upon its velocity. Fur-

---

kische Arbeit schätzt er nicht, aber auch unsere nicht, denn es ist noch alles auf den alten Grundlagen. Warum aber die Grundlagen nicht richtig sind, das wissen die Götter; ich frage ihn auch nicht mehr und ich mache mir darüber keine Sorgen. Es ist ganz merkwürdig was dem Cantor an der Minkowskischen Arbeit gefällt. Er schätzt nur die Behandlung der Zeit und Koordinaten als gleichartige Grössen (x_1, x_2, x_3, x_4), dass man das als Drehung behandeln kann. Er schätzt es aus erkenntnistheoretischen Gründen. Auf meine Frage, was das eigentlich physikalisch heisst die Zeit als eine vierte Raumkoordinate (oder $i\ t$) zu behandeln, ist er die Antwort schuldig geblieben. Ich glaube, er hat sich durch die nichteuklidische Geometrie imponieren lassen. Cantor und der Mathematikus v. Weber werden die Arbeit im physikalischen Kolloquium vortragen, woran ich wahrscheinlich anschliessen werde. Ich bin nämlich jetzt noch skeptischer gegen die Minkowskische Abh. gestimmt; wäre nicht Ihre Arbeit vorhanden, so wären wir mit der Minkowskischen Transformations-Gleichung für die Zeit höchstens auf dem selben Standpunkt (was physikalische Deutung betrifft) wie mit der Lorentzschen "local Zeit".

[54] A. Einstein and W. Ritz, "Zum gegenwärtigen Stand des Strahlungsproblems," *Phys. Zs., 10* (1909), 323–324.

[55] See note 15.

[56] On Ritz's spectral theory, see A. d'Abro, *Rise of the New Physics* (New York, 1951), *2,* 181–183, 190. On Ritz's emission theory, see M. J. Klein, *op. cit.* (note 12), p. 155.

thermore, Ritz thought that Einstein was preserving a form of the ether when he postulated that the velocity of light is independent of its source. In the electrodynamics of Maxwell and Lorentz, the ether had provided a mechanical explanation for the constant velocity of light. Ritz sought an "emission" theory of light instead of a Maxwellian field theory; he assumed that the velocity of light particles depends upon the velocity of the source from which they are emitted.

Even before his move to Göttingen Ritz had become increasingly interested in gravitational theory. This interest emerged while he worked at Tübingen during the winter of 1907–1908 in close contact with several physicists who were attempting to frame a new gravitational theory within electrodynamics. The leading figure of the Tübingen group was Richard Gans, who had been a supporter of the electromagnetic view of nature in the years after 1900.[57] Gans had first discussed the problem of gravitation in 1905, when he addressed the Society of German Natural Scientists and Physicians at Meran on "Electricity and Gravitation."[58] His talk was largely a favorable review of Lorentz' and Wien's gravitational theories and contained few of his own ideas. Gans supported the Lorentz-Wien theory until 1912, when he realized that it implied the instability of neutral particles with respect to the energy flux of the gravitational field.[59]

Another physicist at Tübingen who was sympathetic to electrodynamical theories of gravitation was Friedrich Paschen. In 1909 he approved a doctoral dissertation by Fritz Wacker that extended the Lorentz-Wien theory.[60] Since Paschen did not write on gravitation theory himself, we must judge by Wacker's work what kind of argument Paschen found acceptable. Wacker's dissertation was an elaboration of an article he wrote in 1906 on the relationship between electromagnetism and gravitation.[61] Wacker assumed that Abraham's

[57] G. Beck, "Ricardo Gans," *Conferencia pronunciada en la sesión del 10 de Julio de 1954 de la Academia Brasileira de Ciencias en Río de Janeiro.* E. Gaviola, "Ricardo Gans," *Ciencia e Investigación* (Buenos Aires), *10* (1954), 381–384.

[58] R. Gans, "Gravitation und Elektromagnetismus," *Phys. Zs., 6* (1905), 803–805.

[59] R. Gans, "Ist die Gravitation elektromagnetischen Ursprungs?" *Festschrift Heinrich Weber* (Leipzig, 1912), pp. 75–94.

[60] F. Wacker, *Ueber Gravitation und Elektromagnetismus* (Borna-Leipzig, 1909).

[61] F. Wacker, "Ueber Gravitation und Elektromagnetismus," *Phys. Zs., 7* (1906), 300–302.

expressions for the longitudinal and transverse mass of the electron also held for the longitudinal and transverse mass of the planets with respect to their motion in the ether. Wacker then used Lorentz' explanation of the gravitational law to arrive at an expression for the gravitational force on a planet due to the sun. Assuming with Lorentz that the sun was stationary in the ether, Wacker carried out two calculations for the anomalous precession of the perihelion of Mercury. Taking Abraham's model of the rigid electron, Wacker found 5.8 seconds of arc per century; assuming Lorentz' model of the deformable electron, Wacker calculated 7.2 seconds per century. Neither calculation came close to the observed value of around 40 seconds per century.

Immediately after leaving Tübingen, Ritz treated gravitation theory in his long 1908 paper on an electrodynamics based on a particle emission theory.[62] First, Ritz reiterated Maxwell's conclusion that the intrinsic energy of a gravitational field which is constructed in analogy with an electromagnetic field will appear to be a negative quantity. Rejecting Lorentz' 1900 gravitational theory which had been based on hypotheses concerning the relative strengths of electrostatic and repulsive forces, Ritz applied his own electromagnetic theory to gravitation. Assuming that gravitation propagates with the velocity of light and that gravitational force is of the same general form as electromagnetic force, Ritz examined the problem of planetary orbital residues. He was only able to account for a forward precession of 41 seconds of arc per century for Mercury by using a theory that also yielded values of 8 seconds per century for Venus and 3.4 seconds per century for the earth. Ritz admitted that the last two values were unacceptably large, although he did not interpret this discrepancy as a serious threat to the emission theory. Ritz also mentioned his ideas on gravitation in a popular 1908 review on the role of the ether in physics.[63] Finally, in April 1909, Ritz finished a long article devoted exclusively to gravitation.[64] He summarized the limitations of all existing gravitational theories and concluded that the perihelion shift of Mercury and the value of the

[62] W. Ritz, "Recherches critiques sur l'electrodynamique générale," in *op. cit.* (note 15), pp. 419–422. First published in *Annales de chimie et de physique, 13* (1908).

[63] W. Ritz, "Du rôle de l'ether en physique," *ibid.*, p. 455. First published in *Scientia, 3* (1909).

[64] W. Ritz, "La gravitation," *ibid.*, pp. 478–492. First published in *Scientia, 5* (1909).

gravitational constant "will possibly be deduced from the laws of electrodynamics when they are known with more certainty."[65]

In April 1909, that is, at about the same time that he published on gravitation, Ritz also wrote his short note on electrodynamics with Einstein. It may be assumed that Ritz told Einstein of his ideas on gravitation and of those of the Tübingen group. The information may have contributed to Einstein's renewed interest in the electron theory and to a temporary halt in his work on general relativity. A brief review of Einstein's early ideas on gravitation will put the implications of his contact with Ritz at this time into perspective.

Einstein based his work on gravitation on an hypothesis and on propositions that were all experimentally verifiable. In 1907 he postulated as a basic principle the equivalence for all bodies of vertical acceleration during free fall in a gravitational field.[66] As a result of this principle Einstein arrived at the proposition that time in special relativity theory is influenced by gravitation according to the relation $\sigma = \tau[1 + (\phi/c^2)]$. Here τ is the value of time in the absence of a gravitational field, ϕ the potential of a uniform gravitational field, and c the speed of light. The quantity σ, "local time," is time measured in an accelerated frame of reference. Einstein concluded his paper by giving two results of the principle he had formulated: first, the path of light in a uniform gravitational field is curved; second, spectral lines originating from a stationary body are affected by the body's gravitational field. He added that the effects are too small to be observed. When Einstein sent a copy of the article to his friend Konrad Habicht, he mentioned a third physical effect that he was trying to work into his gravitational ideas. Einstein said that he was "busy on a relativistic theory of the gravitational law with which I hope to account for the still unexplained secular changes of the perihelion movement of Mercury," but thus far he had not succeeded.[67]

The details of Ritz's influence on Einstein cannot be established, but I want to suggest that through the exchange of ideas with Ritz Einstein realized that the electron theory, gravitation, and the prin-

[65] *Ibid.*, p. 491.

[66] A. Einstein, "Ueber das Relativitätsprinzip und die aus demselben gezogenen Folgerungen," *Jahrbuch Radioakt. Elektr.*, *4* (1907), 411-462, Section five of the paper was entitled "Relativitätsprinzip und Gravitation."

[67] Letter from A. Einstein to K. Habicht, 24 December 1907, quoted in C. Seelig, *op. cit.* (note 2), p. 76.

ciple of relativity were all integrally connected and that gravitation was best approached by first reformulating the electron theory according to the principle of relativity. Ritz and Einstein both believed in the existence of a connection between gravitation and electrodynamics. Einstein saw the key to the connection in Ritz's demonstration that no classical theory of radiation is possible.[68] In late 1909 Einstein decided that the limitations of classical radiation theory had to be investigated before a connection between electrodynamics and gravitation could be established, and to this end he plunged into the electron theory.[69]

5. FREUNDLICH AND EINSTEIN

In late December 1919 Einstein wrote to the German Minister for Science, Art, and Education to thank the Weimar government for a grant of 150,000 marks for research in general relativity. He felt that the offer was too generous during a period of economic hardship and might "rightfully cause bitter feelings in the public." In any event, Einstein felt that the offer was unnecessary, if only the German astronomical establishment would devote some of its effort to general relativity. He underscored the uncooperativeness of German astronomers by asking the government to provide a special position for the only living German astronomer who had contributed to the general theory of relativity, Erwin Finlay Freundlich.[70]

[68] A. Einstein, cited in L. Infeld and J. Plebański, *Motion and Relativity* (Warsaw, 1960), p. 201.

[69] R. McCormmach, *op. cit.* (note 40), p. 84.

[70] Letter from A. Einstein to the Minister for Science, Art, and Education, 6 December 1919:

> Dieser Tage ist mir ein Bericht des Staatshaushalts-Ausschusses der Verfassunggebenden Preussischen Landesversammlung zugegangen, nach welchem in Aussicht genommen ist, mir zur Unterstützung der Forschungen auf dem Gebiete der allgemeinen Relativitätstheorie aus der Staatskasse 150 000 M zur Verfügung zu stellen. So gross meine Freude und das Gefühl meiner Dankbarkeit über dies wahrhaft grosszügige Entgegenkommen ist, kann ich doch ein schmerzliches Bedenken nicht unterdrücken. Wird nicht in dieser Zeit grösster Not ein derartiger Beschluss mit Recht bittere Gefühle in der Oeffentlichkeit auslösen? Ich glaube, dass wir auch ohne Aufwendung besonderer Staatsmittel die Forschung auf dem Gebiete der allgemeinen Relativitätstheorie wirksam fördern können, wenn nur die Sternwarten und Astronomen des Landes einen Teil ihrer Apparate und ihrer Arbeitskraft in den Dienst der Sache stellen wollen.

Upon receiving a doctorate in 1910, Freundlich was placed by Klein at the Berlin observatory. Freundlich first came into contact with Einstein's work on gravitation at the observatory when Leo Wenzel Pollak, who was then demonstrator in the institute for cosmic physics at the German university of Prague and who was friendly with Einstein, visited the observatory in August 1911. It was Freundlich's task as the junior observer to welcome visitors. Conversation turned to Einstein's work, and Pollak told Freundlich that Einstein wanted astronomers to examine his theory. Freundlich's wife recalls that that very evening her husband wrote to Einstein and offered to collaborate with him.[71] There is no record of Einstein's reply. Rather, for some time Pollak acted as an intermediary between Freundlich and Einstein. Soon after his visit to Berlin Pollak wrote to Freundlich from Prague, sending him the proof sheets of Einstein's latest paper, "On the Influence of Gravity on the Deflection of Light." The paper contained Einstein's first published statement that the deflection of light rays near the sun might be observed during a solar eclipse and that there should be a gravitational red shift in the spectrum of the sun.[72] At the end of the paper Einstein expressed the hope that astronomers would become interested in the questions he posed. Pollak commented: "Prof. Einstein has given me strict orders to inform you that he himself very much doubts that the experiments could be done successfully with anything except the sun. . . . [I beg you] to send further reports to me, or perhaps to Prof. Einstein, on your views on an astronomical verification."[73] Apparently Pollak had asked Freundlich whether or not the deflec-

---

> Bisher war Herr Dr. E. Freundlich am Astro-Physikalischen Institut in Potsdam der einzige deutsche Astronom (neben Schwarzschild), der sich um das Gebiet verdient gemacht hat. Es würde der Sache ein grosser Dienst geleistet werden, wenn dieser Astronom gemäss dem Vorschlage von Herrn Direktor Müller recht bald eine Observatorstelle am Potsdamer Institut erhielte mit dem Auftrage, an der Prüfung der allgemeinen Relativitätstheorie zu arbeiten.

For the full text of the letter see S. Grundmann, "Der deutsche Imperialismus, Einstein und die Relativitätstheorie (1914-1933)," in *Relativitätstheorie und Weltanschauung. Zur philosophischen und wissenschaftspolitischen Wirkung Albert Einsteins* (Berlin-DDR, 1967), pp. 260-261.

[71] K. Freundlich to the author, 29 April 1973. A copy is located in the Einstein Archives.

[72] A. Einstein, "On the Influence of Gravitation [sic] on the Propagation of Light," transl. W. Perrett and G. B. Jeffrey, in *The Principle of Relativity* (London, 1923), pp. 99-108.

tion of light could be measured around other heavenly bodies such as planets or satellites.

During the autumn of 1911 Einstein and Freundlich began collaboration in earnest. A regular pattern emerged in their correspondence: Einstein would ask Freundlich whether or not a particular astronomical observation would be possible, and Freundlich would give an answer which included unexpected facts or judgments. After Pollak's visit Freundlich inquired of Einstein if processes in the solar atmosphere might mask the gravitational deflection. In his first letter to Freundlich on 1 September 1911 Einstein acknowledged the problems of using the sun to observe the effect: "I know very well that to obtain an answer through experience is no easy matter, since refraction in the solar atmosphere might interfere. Nevertheless, one can say with certainty: if no such deflection exists, then the assumptions of the theory are not correct. One must keep in view that these assumptions, even if they seem obvious, are nonetheless rather daring. If only we had a much bigger planet than Jupiter. But nature has not made it her business to make the discovery of her laws convenient for me."[74] The "assumptions" to which Einstein referred concerned the principle of equivalence. In his paper Einstein had noted that the deflection for the planet Jupiter was one hundredth of the deflection for a ray of light grazing the sun. In his letter Einstein also disagreed with a suggestion Freundlich had made, namely, that it might be possible to measure the light deflection using Jupiter.

[73] Letter from L. W. Pollak to E. Freundlich, August 1911, located in the Einstein Archives at Princeton. "Doch hat mir Prof. *Einstein* streng aufgetragen, Ihnen mitzuteilen, dass er selbst sehr bezweifelt, dass die Versuche ausser bei der Sonne, von irgend welchem Erfolge begleitet sein könnten. . . . [Ich bitte Sie,] mir oder vielleicht Herrn Prof. *Einstein* direkt weitere Nachrichten, wie Sie sich zur astronomischen Nachprüfung stellen, zukommen zu lassen. . . ."

[74] A. Einstein to E. Freundlich, 1 September 1911. Photocopy located in the Einstein Archives at Princeton.

> Dass deren Beantwortung durch die Erfahrung keine leichte Sache ist, das weiss ich wohl, weil eben die Refraktion der Sonnenatmosphäre mitspielen mag. Aber eines kann immerhin mit Sicherheit gesagt werden: Existiert keine solche Ablenkung, so sind die Voraussetzungen der Theorie nicht zutreffend. Man muss nämlich im Auge behalten, dass diese Voraussetzungen, wenn sie schon naheliegen, doch recht kühn sind. Wenn wir nur einen ordentlich grösseren Planeten als Jupiter hätten. Aber die Natur hat es sich nicht angelegen sein lassen, mir die Auffindung ihrer Gesetze bequem zu machen.

Three weeks later Einstein wrote to Freundlich about the limits of observational astronomy. He asked whether or not gravitational deflection could be observed during the day using the sun, and he wanted to know why the observed gravitational red shift of the fixed stars should be different from the gravitational red shift of the sun. "I do not grasp how spectral analysis [of the effect] may be useful, since the sun emits all the types of light that the stars we might consider also emit. Or are there enough stars that have no bright lines or even no lines at all where the sun has absorption lines? I am curious how you envision the method."[75] These questions formed the basis of one line of Freundlich's astronomical research during the next several years. After attempting unsuccessfully to measure deflection using photographs of previous eclipses,[76] Freundlich tried to use photographic plates taken by W. W. Campbell at the Lick Observatory to measure the daytime light deflection.[77] Campbell thought that the experiment had no chance of success, and Freundlich eventually found the plates unsuitable.[78] Freundlich then began to look into the stellar gravitational red shift, which had been known for some time through anomalies in the Doppler shifted spectra of various stars.[79]

By 1913 Freundlich had already corresponded with Einstein for two years and yet had never spoken with him. When Einstein was elected to the Berlin Academy of Sciences in that year, Freundlich congratulated him and at the same time told Einstein of his plans to

[75] A. Einstein to E. Freundlich, 21 September 1911. Located in the Einstein Archives at Princeton. "Ich begreife nicht was die spektrale Zerlegung da nützen kann weil ja die Sonne alle Lichtarten emittiert, die die in Betracht kommenden Sterne emittieren. Oder gibt es hinreichend Sterne, die da helle Linien oder gar keine Linien haben, wo die Sonne Absorptionslinien hat? Ich bin neugierig, wie Sie sich die Methode denken."

[76] A. Einstein to Michele Besso, 4 February 1912. P. Speziali, ed., *Albert Einstein-Michele Besso Correspondence, 1903–1955* (Paris, 1972), p. 46.

[77] W. W. Campbell to G. E. Hale, 4 November 1913. Einstein had written to Hale on 14 October 1913 to request such a search, and he had previously contacted Campbell about eclipse measurements. George Ellery Hale Correspondence, located on microfilm in the Niels Bohr Library of the American Institute of Physics, New York. See B. Hoffmann, *op. cit.* (note 52), p. 112.

[78] E. Freundlich, "Ueber einen Versuch, die von A. Einstein vermutete Ablenkung des Lichtes in Gravitationsfeldern zu prüfen," *Astronomische Nachrichten, 193* (1913), 369–372.

[79] E. Freundlich, "Ueber die Gravitationsverschiebung der Spektrallinien bei Fixsternen," *Astr. Nachr., 202* (1915–1916), 17–24.

marry.[80] Einstein wrote back in August 1913 and invited Freundlich and his bride to hear him address the Swiss Society of Natural Sciences at Frauenfeld. In his letter Einstein also commented on two of Freundlich's projects. "I, myself, am fairly firmly convinced that the light rays actually experience a curvature. I am extraordinarily interested in your plan to observe stars near the sun by day. This should be possible if there are no suspended particles in the atmosphere with sizes of the order of the length of light waves, which very slightly deflect the light. I fear that this could wreck your plan. But you will be better informed about these conditions than I. . . ."[81] For several months afterwards Einstein remained hopeful that a daylight photograph of light deflections was possible despite information to the contrary from the Zurich astronomer Julius Maurer; but he worried about effects in the atmospheres of the sun and the earth which might give a negative result.[82] In his letter of August 1913 Einstein also remarked on Freundlich's work with double stars. Willem de Sitter had recently written a short note in which he claimed that double star data could be used to disprove Ritz's emission hypothesis.[83] Freundlich had noted that current astronomical evidence could not support de Sitter's claim.[84] Einstein emphasized the importance of the problem for his theories: "I am also very curious about the outcome of your examination of double stars. If the velocity of light depends even to a very tiny part upon the velocity of the light source, then my whole theory of relativity, including my gravitational theory, is false."[85]

[80] K. Freundlich, *op. cit.* (note 71).

[81] A. Einstein to E. Freundlich, August 1913. Photocopy in the Einstein Archives. "Ich bin im Stillen ziemlich fest überzeugt, dass die Lichtstrahlen thatsächlich eine Krümmung erfahren. Ausserordentlich interessiert mich Ihr Plan, sonnennahe Sterne bei Tage zu beobachten. Dies müsste möglich sein, wenn es nicht überall in der Atmosphäre suspendierte Körperchen von der Grössenordnung der Lichtwellenlängen gibt, die das Licht nur wenig ablenken. Ich fürchte, dass daran Ihr Plan scheitern könnte. Aber Sie werden über diese Verhältnisse besser orientiert sein als ich. . . ."

[82] A. Einstein to E. Freundlich, 27 October 1913. Photocopy in the Einstein Archives.

[83] W. de Sitter, "Ein astronomischer Beweis für die Konstanz der Lichtgeschwindigkeit," *Phys. Zs.*, *14* (1913), 429.

[84] E. Freundlich, "Zur Frage der Konstanz der Lichtgeschwindigkeit," *Phys. Zs.*, *14* (1913), 835-838.

[85] A. Einstein to E. Freundlich, *op. cit.* (note 81). "Sehr neugierig bin ich auch auf die Ergebnisse Ihrer Untersuchungen über die Doppelsterne. Wenn

By the time of the Frauenfeld conference Freundlich began to encounter opposition to his work with Einstein. He had been attempting to organize an expedition to the Crimea to test Einstein's deflection prediction during the August 1914 solar eclipse.[86] However, Freundlich's superiors at the Berlin observatory were not sympathetic to his plans. Freundlich wrote to Einstein soon after their Frauenfeld meeting and explained that the director of the Berlin observatory, Hermann Struve, would not provide the funds for Freundlich's expedition. Einstein replied in early December 1913: "After receiving your last letter I immediately wrote to Planck, who has made serious efforts for the matter and has undertaken to talk the matter over with Schwarzschild. *I will not write to Struve.* If the [Berlin] Academy does not want [to give], then we will get that little bit of money from private sources."[87] Einstein concluded the letter by saying that he would try to get money for the expedition with the help of Fritz Haber, and, if that failed, he would contribute to the expedition from his own pocket. As Einstein feared, the Berlin Academy refused to support the expedition. Freundlich finally contacted the chemist Emil Fischer, who "gave him 3000 marks for the journey and supplied him with another 3000 marks from Krupp."[88]

As the expedition approached, Einstein's support for it weakened. To Besso, Einstein wrote in March 1914: "Now I am fully satisfied, and I no longer doubt the correctness of the whole system, whether the observation of the eclipse will succeed or not. The reasonableness of the matter is too evident."[89] All of Freundlich's plans, however, were aborted by the outbreak of the First World War. His equipment

---

die Lichtgeschwindigkeit auch nur im Geringsten von der Geschwindigkeit der Lichtquelle abhängt, dann ist meine ganze Relativitätstheorie inklusive Gravitationstheorie falsch."

[86] Freundlich first announced plans to observe the eclipse in *op. cit.* (note 78), p. 372. The article was dated January 1913.

[87] A. Einstein to E. Freundlich, 7 December 1913. Photocopy in the Einstein Archives. "Nach Empfang Ihres letzten Briefes habe ich sofort an Planck geschrieben, der sich wirklich ernsthaft für die Sache verwandt hat und es übernahm, mit Schwarzschild über die Sache zu reden. *An Struve werde ich nicht schreiben.* Wenn die Akademie nicht. . .will, dann kriegen wir das bischen Mammon von privater Seite."

[88] K. Freundlich, *op. cit.* (note 71).

[89] *Einstein-Besso Correspondence, op. cit.* (note 76), p. 53. "Nun bin ich vollkommen befriedigt und zweifle nicht mehr an der Richtigkeit des ganzen Systems, mag die Beobachtung der Sonnenfinsternis gelingen oder nicht. Die Vernunft der Sache ist zu evident."

was impounded by the Russians and he was not able to make any photographic plates. After being interned briefly in Russia, he was exchanged for several Russian officers held in Germany and returned to Berlin in early September 1914.[90]

6. THE REACTION OF THE SCIENTIFIC ESTABLISHMENT

Neither Laub nor Ritz nor Freundlich were members of the German scientific establishment when they collaborated with Einstein, nor did they become members later. Ritz died soon after working with Einstein. The unsuccessful careers of Laub and Freundlich invite inquiry. For both, the collaboration with Einstein led to a confrontation with a university chairholder over issues in Einstein's development of the relativity theory.

Laub

In 1909 Einstein wrote to Laub, then in Würzburg, urging him to accept a post as assistant to Philipp Lenard at Heidelberg. "Do tolerate Lenard's whims, as many as he may have. He is a great master, an original thinker!"[91] Lenard was indeed a brilliant experimentalist. He was on less secure grounds as a theoretician, but he held his theoretical work in high regard. Lenard had devoted much effort to investigating the photoelectric effect; he had demonstrated by early 1905 that the number of electrons emitted from a photoelectric metal was proportional to the energy of the incident light and that the electron velocity—and hence the kinetic energy of the emitted electron—varied inversely with the wavelengths of the incident light. Several months later Einstein revolutionized physics by interpreting the photoelectric effect in terms of a quantum hypothesis. Lenard had won a Nobel Prize in 1905 for his work in experimental physics, but he never forgave Einstein for overshadowing him with respect to the work on the photoelectric effect.[92] Furthermore, as Laub recalled many years later, Lenard was "no great partisan of 'mathematical' physics."[93] Early in 1912, when the Heidelberg faculty had

[90] K. Freundlich, *op. cit.* (note 71).

[91] A. Einstein to J. Laub, n.d. Photocopy in the Einstein Archives; the letter itself is in the E. T. H. Library, Zurich. "Ertragen Sie Lenards Schrullen, soviel er nun haben mag. Er ist ein grosser Meister, ein origineller Kopf!"

[92] *Nobel Lectures in Physics, 1901-1921* (Amsterdam, 1967), p. 137.

[93] J. Laub to C. Seelig, 10 September 1959. E. T. H. Library, Zurich. "[Er] war kein grosser Anhänger der 'mathematischen' Physik."

the opportunity of creating a new associate professorship of theoretical physics, Lenard objected strenuously. Friedrich Pockels described Lenard's reaction in a letter to Laub: "He [Lenard] wanted to deny theoretical physics any right to an independent existence [saying] that he was enough of a theoretician himself, etc. For example, he wanted to refuse on principle [to accept] purely theoretical workers as doctoral candidates."[94]

Soon after his arrival in Heidelberg Laub found himself involved in a serious dispute with Lenard. The conclusion of Laub's paper, presented by Lenard in 1909 to the Heidelberg Academy of Sciences, that relativistic effects could be observed in gaseous dispersion[95] was difficult to verify and had been challenged by one of his colleagues.[96] Lenard was not happy with this situation. As Laub related in letters to Johannes Stark, his position required an inordinate expenditure of time.[97] Nevertheless, throughout 1910 Laub worked on a long review for Stark's *Jahrbuch,* summarizing all the experimental evidence in favor of the special theory of relativity.[98] During the same time his relations with Lenard deteriorated. In June 1910 Laub complained to Stark that he was having difficulty in getting along with Lenard.[99] It was becoming increasingly apparent that Lenard would not allow Laub to habilitate. On 25 November 1910 Laub wrote to Stark that he was looking for a position in America, most probably referring to one at the Massachusetts Institute of Technology.[100] Laub told Stark that his references were Runge, Einstein, Wien, Pockels, and Lenard. At this point he was evidently still speak-

[94] F. Pockels to J. Laub, 12 March 1912. Deutsches Museum, Munich. "Er wollte der theoret. Physik jede selbständige Existenzberechtigung bestreiten, da er selber genügend Theoretiker sei, u.s.w. Z.B. wollte er rein theoretische Arbeiter als Doktorarbeiter grundsätzlich ablehnen."

[95] J. Laub, *op. cit.* (note 48).

[96] Most probably August Becker. F. Pockels to J. Laub, 5 November 1911. Deutsches Museum, Munich.

[97] J. Laub to J. Stark, 15 March 1910. Staatsbibliothek Preussischer Kulturbesitz, Berlin.

[98] J. Laub, *op. cit.* (note 50).

[99] J. Laub to J. Stark, 13 June 1910. Staatsbibliothek Preussischer Kulturbesitz, Berlin.

[100] J. Laub to J. Stark, 25 November 1910. Staatsbibliothek Preussischer Kulturbesitz, Berlin. See also G. N. Lewis to J. Stark, 24 January 1911, Deutsches Museum, Munich, where Lewis asks Stark for his opinion of Laub, who has just applied for a job. I thank Stanley Goldberg for pointing out the Lewis letter to me.

ing with Lenard. However, a few weeks later, in a letter to Einstein which is now lost, Laub reported that he had broken with Lenard.

When Einstein learned of the final altercation, he was furious. "Lenard is really a perverted fellow! Put together entirely of bile and intrigue! However, you are considerably better off than he. You can get away from him, but he must keep house with the monster until he bites the dust. I will do what I can to obtain an assistant's post for you."[101] After writing to Anton Lampa and Walther Nernst, Einstein hoped to place Laub in a position in Chile.[102] Finally, in early 1911, Laub found a job as associate professor of geophysics in the new National University of Argentina at La Plata. He arrived in April 1911[103] and joined a growing circle of German physical scientists and mathematicians.

Laub's personal beliefs estranged him further from the German-speaking physics establishment. Like Einstein during this period, Laub was a pacifist sympathetic to socialism. In 1916 Laub publicly expressed his moral and political views in the conclusion of a report on the physics institute of the National Institute for Secondary School Teachers in Buenos Aires, of which he had been director for the preceding three years. To Laub the First World War was a tragedy which exemplified all that was morally reprehensible in the world. The war threatened to destroy culture and civilization by fomenting antagonisms between different groups. His report continued: "The spiritual end of science consists in this: to strive toward the highest ideal of our life, the one which rises above the individual and the na-

[101] A. Einstein to J. Laub, 11 November 1910. Photocopy at the Einstein Archives from original in the library of the E. T. H., Zurich. "Das ist wirklich ein verdrehter Kerl, der Lenard! So ganz aus Galle und Intrigue zusammengesetzt! Aber Sie sind doch erheblich besser daran als er. Sie können von ihm weggehen, aber er muss mit dem Scheusal wirtschaften, bis er ins Gras beisst. Ich will thun, was ich kann, um Ihnen eine Assistentenstelle zu verschaffen."

[102] A. Einstein to J. Laub, 15 November 1910, quoted in *op. cit.* (note 19), item 1978. Ellipses and brackets given in text. "Ich habe an Lampa und Nernst geschrieben und ich habe einem mir sehr gut bekannten Herrn, der sehr einflussreiche Verbindungen in Chile hat und dorthin gestern abgereist ist, den Auftrag gegeben, Ihnen dort eine Stelle zu verschaffen. Lassen Sie also L[enard] ruhig schüfteln. Sie sind mit einem Fuss schon Machtsphäre entronnen. . .[sic]."

[103] J. Laub to P. Ehrenfest, 24 April 1911. Located on microfilm in the Ehrenfest Correspondence, Archive for History of Quantum Physics, American Philosophical Society, Philadelphia.

tion, the one that has to be sacred to everyone and can be expressed in one simple word: humanity. We confess to be convinced that science will heal our misery, eliminate the artificial hatred between different peoples, and subordinate the petty and oppressing significance of capital to the reign of the free spirit."[104] Laub believed that science could secure the material needs of mankind both by freeing it from the bonds of capitalist economic structure and at the same time by providing a spiritual unity. Statements concerning the importance of spiritual goals in science were not unusual among German-speaking scientific academics at this time,[105] but only a few, among them Einstein, associated themselves with pacifism or alluded to socialist ideas.

Laub continued to produce studies on many experimental and theoretical subjects between 1911 and 1917. In early 1917 the institute where Laub was employed collapsed as a result of political and economic turmoil.[106] Laub moved to Spain and then to Berlin after the war. During this period, Laub's politics veered even farther left. In 1919, together with Rudolf Grossmann of the Ibero-American Institute in Hamburg, he translated a 1917 address given by Horacio Oyhanarte on the necessity of Argentine neutrality.[107] Grossmann was a highly respected scholar who, like several others in this period,

[104] J. Laub, "El Departamento de Física y su Enseñanza," in *El Instituto Nacional del Profesorado Secundario en la primera década de su existencia* (Buenos Aires, 1916), pp. 335-336. "El fin espiritual de la ciencia consiste en esto: llegar al más alto ideal de nuestra vida, al ideal que se levanta sobre el individuo y sobre la nación y que debe ser sagrado para todos y que se expresa en la sencilla palabra: Humanidad. Confesamos y estamos convencidos que la ciencia aliviará nuestra miseria, eliminado las enemistades artificiales entre las diferentes naciones y subordinando el valor del capital mezquino y oprimente bajo el dominio del expíritu libre."

[105] Two physicists who often phrased their work in terms of spiritual goals at this time were Gustav Mie and Max Planck. G. Mie, "Naturgesetz und Geist," *Deutsche Revue, 41* (1916), 150-163. S. Goldberg, "Max Planck's Philosophy of Nature and His Elaboration of the Theory of Relativity," *Hist. Stud. Phys. Sci.,* 7 (1976), 125-160.

[106] See, for example, the recent study by D. G. Woodbury, *The Argentine Socialist Party in Congress: The Politics of Class and Ideology, 1912-1930* (diss., Columbia University, 1971).

[107] Horacio B. Oyhanarte, *Argentiniens Neutralität. Rede, 24.-25. Sept. 1917,* transl. J. Laub and R. Grossmann (*Auslandspolitische Schriften des Ibero-Amerikanischen Instituts,* No. 1, Hamburg, 1920). Laub's preface is dated Charlottenburg, 20 January 1920.

published communist political ideas under a pseudonym.[108] In 1920 Laub found permanent employment with the Argentine foreign service in Europe. He served in various posts for nineteen years, and he was Argentine ambassador to Poland when Hitler crossed the Polish corridor in September 1939.[109] Subsequently, Laub spent the Second World War in Argentina. Throughout these years, he retained personal ties with the scientific world, vacationing with Nernst[110] and corresponding extensively with C. A. Krukow on the development of high-frequency radio transmission.[111] In 1945 Laub tried to rejoin the European scientific community, although both Einstein and Laue told him that the scientific world had changed and that he would have difficulty adjusting to it.[112] Finally, in 1953, Laub obtained the position of associate in the physical institute at Fribourg, Switzerland. In 1962 he died, unknown and without means.[113]

Freundlich

Several months after his attempt to observe the 1914 eclipse Freundlich began to consider the anomalous perihelion precession of Mercury. He first examined the attempts of Hugo von Seeliger to provide a classical explanation for the shift. In February 1915 Freundlich questioned Seeliger's hypothesis that the presence of interplanetary dust accounts for anomalous planetary residues in general.[114] Freundlich's article may have been prompted by Einstein's as

[108] Under the name Pierre Ramus, Grossmann wrote *Generalstreik und direkte Aktion im proletarischen Klassenkampfe* (Berlin, 1910); *Jahrbuch der freien Generation; Volkskalender und Dokumente der Weltanschauung des Sozialismus-Anarchismus* (Paris, 1910); *Die Neuschöpfung der Gesellschaft durch den kommunistischen Anarchismus* (Wien-Klosterneuburg, 1923).

[109] The Argentine Diplomatic Service to the author, 12 March 1973. A copy is located in the Einstein Archives at Princeton.

[110] W. Nernst to J. Laub, 25 July 1934, *op. cit.* (note 19), item 2035.

[111] *Op. cit.* (note 26), item 3968. A transcript of a manuscript by Laub, "Die Herkunft des hochfrequenten Drahtfunks und seine Vorteile gegenüber dem üblichen drahtlosen Rundfunk" (Fribourg, Switzerland, n.d.), is located at the Leo Baeck Institute in New York.

[112] A. Einstein to J. Laub, 2 June 1946, 16 April 1947, 18 July 1953. M. Laue to J. Laub, 21 July 1947. *Op. cit.* (note 26), item 2366; *op. cit.* (note 19), items 1979, 1980, 2021.

[113] A. Jaeggli to the author, 11 December 1972. A copy is located in the Einstein Archives at Princeton.

[114] E. Freundlich, "Ueber die Erklärung der Anomalien im Planetensystem durch die Gravitationswirkung interplanetarer Massen," *Astr. Nachr., 201* (1915), 49-55.

yet unpublished conjecture or calculation showing that general relativity is able to account for these anomalies. Perhaps because he did not have an actual calculation from general relativity in hand, Freundlich decided to attack Seeliger on Seeliger's own ground. In his critique Freundlich made no use of non-Newtonian gravitational theories. Rather, he attempted to demonstrate, by using the results of classical astronomy, that Seeliger's theory could not yield correct values for the residues.

In criticizing Seeliger, Freundlich was challenging the doyen of German astronomy who, for the past twenty years, had been working on alternative gravitational theories and explanations of anomalies in planetary motion.[115] Seeliger's power in the German astronomical community was unquestioned. His conservatism on political and scientific matters was well known, and a direct challenge to his scientific standing could have been expected to receive little encouragement from German astronomers.[116] Furthermore, Seeliger was a veteran of many scientific debates and knew from experience how best to defend his own integrity while at the same time discrediting that of an opponent. Seeliger's reply to Freundlich came

[115]H. v. Seeliger, "Ueber Zusammenstösse und Theilungen planetarischer Massen," *Abhandlungen der mathematisch-physikalischen Classe der königlich-bayerischen Akademie der Wissenschaften, München, 17* (1891), 457–490; "Ueber das Newtonsche Gravitationsgesetz," *Astr. Nachr., 137* (1895), 129–136; "Ueber das Newtonsche Gravitationsgesetz," *Sitzungsberichte der mathematisch-physikalischen Classe . . . München, 26* (1896), 373–400; "Kosmische Staubmassen und das Zodiakallicht," *ibid., 31* (1901), 265–292; "Das Zodiakallicht und die empirischen Glieder in der Bewegung der innern Planeten," *ibid., 36* (1906), 595–622; "Die empirischen Glieder in der Theorie der Bewegungen der Planeten Merkur, Venus, Erde und Mars," *Vierteljahrsschrift der Astronomischen Gesellschaft, 41* (1906), 234 ff. See A. S. Eddington, "Hugo von Seeliger," *Monthly Notices of the Royal Astronomical Society, London, 85* (1925), 316. M. Jammer, *op. cit.* (note 49), pp. 127–128; A. Einstein, "Considerations on the Universe as a Whole," in M. K. Munitz, ed., *Theories of the Universe* (New York, 1957), p. 276; J. D. North, *The Measure of the Universe* (Oxford, 1956), p. 48.

[116]O. Struve and V. Zebergs, *Astronomy of the Twentieth Century* (New York, 1962), pp. 39–40. In 1910 Max Planck considered Seeliger, along with Rayleigh, van der Waals, and Arthur Schuster, as one of the "old men" who would not be interested in devoting the first Solvay Congress to problems of quantum physics. M. Planck to W. Nernst, 11 June 1910. Quoted by Jean Pelseneer in "Historique des instituts internationaux de physique et de chimie Solvay depuis leur fondation jusqu'à la deuxième guerre mondiale." A copy of Pelseneer's manuscript is located in the Archive for History of Quantum Physics at the American Philosophical Society, Philadelphia.

five months later. It was a credible defense of the dust hypothesis, and it made Freundlich's attack appear ridiculous.[117]

Freundlich now continued his investigation of the astronomical consequences of general relativity from another point of view. While Seeliger was drafting his reply, Freundlich reconsidered the red shift in the Fraunhofer lines of certain classes of fixed stars.[118] He examined the predictions of a red shift for massive bodies of Einstein's general relativity theory and of Nordström's euclidean gravitational theory,[119] and he referred to Schwarzschild's 1914 paper on an experimental measurement of the solar red shift.[120] Freundlich's primary observational data were Campbell's measurements of the stellar red shift in Class B stars.[121] Making use of his previous work on double stars, Freundlich calculated that, if Hans Ludendorff's result for the mass of spectroscopic Class B double stars is correct, then Einstein's and Nordström's theories both predict enormous mass densities for those stars. When Freundlich's paper appeared, Seeliger decided he could use it to further discredit his opponent. In January 1916 Seeliger showed that Freundlich's relativistic formula for the determination of stellar densities was wrong.[122] Furthermore, Seeliger insisted that Campbell's and Ludendorff's measurements disproved the stellar gravitational red shift. Freundlich

[117]H. v. Seeliger, "Ueber die Anomalien in der Bewegung der innern Planeten," *Astr. Nachr., 201* (1915), 273-283.

[118]E. Freundlich, "Ueber die Gravitationsverschiebung der Spektrallinien bei Fixsternen," *Astr. Nachr., 202* (1915), 1.

[119]Nordström produced several different gravitational theories during the period 1912-1915 which were expressed in terms of four-dimensional euclidean space-time. For reviews of Nordström's work see M. Abraham, "Neuere Gravitationstheorie," *Jahrbuch Radioakt. Elektr., 11* (1914), 497-508; M. Laue, "Die Nordströmsche Gravitationstheorie," *ibid., 14* (1917), 263-313; A. L. Harvey, "A Brief Review of Lorentz-Covariant Theories of Gravitation," *American Journal of Physics, 33* (1965), 449-460; L. Pyenson, *op. cit.* (note 4), pp. 279-289.

[120]K. Schwarzschild, "Ueber die Verschiebungen der Bänder bei 3883 Å im Sonnenspektrum," *Sitzungsberichte der königlichen preussischen Akademie der Wissenschaften zu Berlin* (1914), pp. 1183-1213. See E. Forbes, "A History of the Solar Red Shift Problem," *Annals of Science, 17* (1961), 140.

[121]W. W. Campbell, "On the Motions of the Brighter Class B Stars," *Lick Observatory Bulletin, 6* (1911), 101-124; "Some Peculiarities in the Motions of the Stars," *ibid.*, pp. 125-135.

[122]H. v. Seeliger, "Ueber die Gravitationswirkung auf die Spektrallinien," *Astr. Nachr., 202* (1916), 83-86.

immediately wrote a two line defense of his own calculations, and he resolved to await further quantitative evidence.[123]

Freundlich's exchange with Seeliger displeased his superior at the observatory, Hermann Struve. Struve had descended from a long line of Russian astronomers and since 1904 had been director of the Berlin observatory.[124] For a generation, Struve had calculated masses and densities of planetary moons by working with orbital residues.[125] He was incensed by Freundlich's attack on Seeliger, for he thought that Freundlich lacked astronomical evidence to support his statements. Late in 1915 he had Freundlich removed from his position at the observatory.[126]

Einstein was shaken by Struve's action. He felt that Seeliger had over-reacted in his initial reply to Freundlich. In early December 1915 Einstein wrote bitterly to Sommerfeld in Munich: "Tell your colleague Seeliger that he has a horrible temper. I enjoyed it recently in a reply in which he corrected the astronomer Freundlich."[127] Two months later Einstein was even more concerned, for Seeliger's second rebuttal had appeared. In a letter to Sommerfeld Einstein observed:

> It should always be credited to Freundlich that he has invented the statistical method that allows fixed stars to be used to answer the question of the line shift. Even if he has made that dreadful arithmetical error . . . still the value of the whole calculation should not be forgotten for that reason. Mistakes can be corrected and are, indeed, always corrected with time. [The achievement] lies in discovering a path and making it passable.
>
> From my point of view the matter looks like this. Freundlich was the only colleague who effectively supported me up to now in my

[123] E. Freundlich, "Bemerkung zu meinem Aufsatz in A. N. 4826," *Astr. Nachr., 202* (1916), 147-148.

[124] F. W. Dyson, "Karl Hermann Struve," *Month. Not. R. Astr. Soc. London, 81* (1921), 270-272.

[125] L. Courvoisier, "Hermann Struve," *Astr. Nachr., 212* (1920), 33-38.

[126] E. Forbes, *op. cit.* (note 21), writes that Freundlich was fired in 1913 over his publication of relativistic calculations concerning the perihelion shift of Mercury. Forbes has agreed by personal communication that my chronology is more likely.

[127] A. Einstein to A. Sommerfeld, 9 December 1915, in A. Hermann, ed., *Einstein-Sommerfeld Briefwechsel* (Basle, 1968). "Sagen Sie Ihrem Kollegen Seeliger, dass er ein schauerliches Temperament hat. Ich genoss es neulich in einer Erwiderung, die er an den Astronomen Freundlich richtete."

> endeavors in the area of general relativity. He has devoted to the problem years of thought and work as far as that was possible with his fatiguing and dull duties at the observatory.
>
> Freundlich has yet a second achievement. I will not speak of the refutation of Seeliger's theory of the perihelion movement of Mercury, since that work might be considered carrying coals to Newcastle. But Freundlich has shown that modern astronomical devices suffice to prove light deflection around Jupiter, something I had not thought possible. . . .[128]

Einstein alluded to Freundlich's incorrect formula for the relativistic mass density, an error Seeliger had criticized. Freundlich's reputation suffered from his error for many years. When Freundlich produced calculations of the solar light deflection in 1931, the exchange with Seeliger made his work immediately suspect. Campbell remarked in a letter to R. J. Trumpler:

> Confidentally I am also telling Dr. Pritchett that Dr. Freundlich was compelled to resign from the Berlin-Babelsberg Observatory about the year 1915 by its Director Hermann Struve because of Freundlich's trickery and dishonesty in scientific matters, as shown by Seeliger. . . . To give you the right angle on Freundlich, I advise

[128] A. Einstein to A. Sommerfeld, 2 February 1916, *ibid.*
Es darf Freundlich nicht vergessen werden, dass er die statistische Methode erdacht hat, die gestattet, Fixsterne heranzuziehen bei der Beantwortung der Frage der Linienverschiebung. Wenn ihm auch der üble Rechenfehler unterlaufen ist . . . (Dichtebestimmung), so darf deshalb der Wert der ganzen Sache nicht vergessen werden. Fehler können berichtigt werden und werden stets mit der Zeit berichtigt. . . . [Das Verdienst] liegt darin, dass man einen Weg entdeckt und soweit ebnet, dass er passierbar wird.

Von meinem Standpunkt aus betrachtet sieht die Angelegenheit so aus. Freundlich war der einzige Fachgenosse, der mich bis jetzt in meinen Bestrebungen auf dem Gebiete der allgemeinen Relativität wirksam unterstützte. Er hat dem Problem Jahre des Nachdenkens und auch der Arbeit gewidmet, soweit dies neben dem anstrengenden und stumpfsinnigen Dienst an der Sternwarte möglich war. . . .

Freundlich hat noch ein zweites Verdienst–Ich will nicht von der Widerlegung von Seeligers Theorie der Perihelbewegung des Mercur reden, da diese That vielleicht als Einrennen einer offenen Thür zu bezeichnen ist. Aber Freundlich hat gezeigt, dass die modernen astronomischen Hilfsmittel ausreichen, um die Lichtablenkung am Jupiter nachzuweisen, was *ich* nicht für möglich gehalten hatte. . . .

> you to read [his] article, if you have not already done so. The whole incident was described to Dr. Moore and others here a fortnight ago by Professor Georg Struve, an astronomer in Berlin University and Babelsberg Observatory, who has been making observations with the Lick telescope during the past three months. He also remarked that Freundlich has no standing whatever with German astronomers.[129]

Freundlich's subsequent attempt to find a position in German astronomy is worth considering briefly. In 1917 or 1918 Einstein arranged a meeting between Planck and Struve to discuss Freundlich's habilitation at Berlin. Struve told Planck in no uncertain terms what he thought of Freundlich.[130] Nevertheless, Einstein continued to intervene on Freundlich's behalf. In 1918, after meeting with the Prussian Minister for Culture and Education, Einstein wrote to Freundlich: "The man is massive and not unsympathetic. He agreed with all I said, but promised nothing, so that I don't know whether my visit was of any use. But I made an impression. I warned him that S. will react with excuses to which no importance should be attached and asked him to consult with Planck."[131] "S." was evidently either Struve or Seeliger. Finally, early in 1918, Einstein was able to obtain support for Freundlich from the newly formed Kaiser Wilhelm Institute for Physical Research. By March 1919 Einstein had succeeded in placing Freundlich as an observer at the Potsdam Astrophysical Observatory. Largely because the Weimar Ministry of Culture wanted to create a showcase for pure science at any cost Einstein was successful in obtaining for Freundlich a key position in the Einstein Institute at Potsdam.[132]

[129] W. W. Campbell to R. J. Trumpler, 21 October 1931. Photocopy in the Einstein Archives.

[130] A. Einstein to E. Freundlich, no date. Photocopy in the Einstein Archives.

[131] A. Einstein to E. Freundlich, no date. Photocopy in the Einstein Archives. "Der Mann ist massiv und nicht unsympathisch. Er ging auf alles ein, was ich sagte, versprach aber nichts, sodass ich nicht weiss, ob mein Gang direkt genützt hat. Ich hatte aber Eindruck. Ich bereitete ihn darauf vor, dass S. mit Ausflüchten reagieren werde, denen kein Gewicht beizulegen sei und bat ihn, auch mit Planck Rücksprache zu nehmen."

[132] See S. Grundmann, *op. cit.* (note 70), pp. 210–215, for the founding of the Kaiser Wilhelm Institutes and their support of Freundlich's work in general relativity, as well as for the construction of the Einstein Institute at Potsdam.

7. CONCLUSION

Laub, Ritz, and Freundlich all expressed a similar understanding of theoretical physics. They held that applied mathematics and experiments were important for developing the implications of theory, but in their view theory did not grow out of either mathematics or experiment. Laub supported special relativity, Ritz the emission theory, and Freundlich general relativity against alternative possibilities because in their opinion these theories expressed clearer pictures of the physical world. Each was thoroughly educated in applied mathematics at Göttingen, but none made extensive use of complicated or unusual mathematics in his work on relativity and gravitation. All were proficient in experimental and observational physics, yet none sought to base theory upon observed results.

Their approach to theoretical physics was similar to that of Einstein, who himself had an opportunity to become familiar with mathematics and experimental physics as a student at the Zurich Polytechnic. Between 1896 and 1900 Einstein registered for thirty-eight required courses and eleven electives;[133] of the required courses eleven were considered pure mathematics.[134] However, Einstein did not regularly attend the mathematics courses.[135] Heinrich Weber's physics laboratory attracted Einstein most, but he never seriously considered experimental physics as a career.[136] From the beginning Einstein's work was that of a theoretical physicist with a driving concern for developing a unified picture of the physical world. The most characteristic intellectual quality of Einstein's work is a creative independence of mind. As has been suggested in several recent studies, his receptive attitude toward new ideas is also expressed in his admiration for the iconoclasm inherent in Ernst Mach's philosophy.[137] Einstein was open to stimulating ideas wherever he found

[133] As recorded in Einstein's matriculation record at the E. T. H. in Zurich.

[134] According to the classification of M. Grossmann, *op. cit.* (note 6).

[135] G. Holton, "Mach, Einstein, and the Search for Reality," *Thematic Origins of Scientific Thought* (Cambridge, Mass., 1973), p. 221.

[136] R. McCormmach, "Editor's Introduction," *Hist. Stud. Phys. Sci.*, 7 (1976), ix–xxv.

[137] H. Steinberg, "Grundzüge der philosophischen Auffassungen Albert Einsteins," in *Relativitätstheorie und Weltanschauung. Zur philosophischen und wissenschaftspolitischen Wirkung Albert Einsteins* (Berlin-DDR, 1967), esp. pp. 46–58; G. Holton, *op. cit.* (note 135).

them, whether the author was director of a laboratory or an old schoolmate. Einstein's independence also led him to favor an uncomplicated way of life. It has been maintained that Einstein's independence of mind and bohemian predilection placed him in a philosophical anti-establishment milieu centered in German-speaking Switzerland.[138] In view of his early scientific collaboration, we may ask whether or not his independent course can be considered part of a larger anti-establishment current. Some material seems to speak against this proposition. Einstein associated with other young theoretical physicists outside the physics establishment who sought radical approaches for unifying physics, but the group cannot be called an intellectual movement, for the radical theoretical physicists subscribed to no single program or worldview. One finds few ideological congruences in the lives and works of those, such as Laub, Freundlich, and perhaps Abraham, who would have to be included in such a current. Some establishment physicists actively sought new nonmechanical unifications of physical science that were similar in many respects to the ideas of younger physicists.

To consider whether or not Einstein and his collaborators might have been part of an anti-establishment current in Wilhelmian physics, it is important to know if their nonacademic and, in the cases of Laub and Ritz, non-German background and their social and political attitudes were shared by a distinct group outside the mainstream of physics who had corresponding scientific views. Preliminary studies seem to indicate that so many German physics students of the period had a similar background, that information of this type, taken by itself, allows no conclusions. Over fifteen percent of all doctorates awarded in the physical sciences (excluding organic chemistry but including mathematics) in Germany circa 1900 were earned by non-German Jews; this percentage is much higher than that for organic chemistry, the social sciences, or the humanities.[139] Also, the large number of non-Europeans obtaining doctorates in

[138]L. S. Feuer, *Einstein and the Generations of Science* (New York, 1974), esp. pp. 26–46. See the review of Feuer's book by J. L. Heilbron, *Science, 185* (1974), 776–778.

[139]These statistics are drawn from a study of all German Ph.D. dissertations published in 1899. L. Pyenson and D. Skopp, "La dynamique du doctorat à l'époque de Guillaume II," *Bulletin du Centre Interuniversitaire d'Etudes Européennes, 2,* no. 16 (1 May 1975), 10–19.

mathematics and physics implies that, like Einstein's collaborators, a substantial fraction of young mathematicians and physicists were not members of the German academic elite. For the time being the precise outlines of intellectual currents in physics have to remain conjecture. We are only beginning to understand how to define and delineate the physics community in German-speaking Europe during the period before 1914. We lack detailed information concerning the way various minor currents emerged to occupy a central role within the theoretical physics discipline. We have not yet charted the evolution of interdisciplinary formations such as astrophysics, cosmic physics, physical chemistry, geophysics, and we have not yet begun to examine networks of scientific communication. Several excellent analyses have recently appeared which bring into view the broad outline of theoretical physics in Wilhelmian Germany.[140] The road is open for a detailed examination of scientific productivity and career patterns in theoretical physics within all of German-speaking Europe, the Low Countries, Scandinavia, and Russia which would bring us closer to a balanced understanding of the rise of contemporary physics in its perhaps most important setting.

Einstein's early scientific collaboration places in clearer perspective his revolutionary approach to physics. Although many physicists in late Wilhelmian Germany were sympathetic with attempts to unify physics, only a few exceptional physicists in established positions, such as Planck and Laue, paid serious attention to Einstein's physical interpretation of the theories of relativity. If Einstein did not share the neglect that his collaborators suffered, it was only because, by 1909, the latitude and genius of his physics made it extraordinarily difficult for other physicists to damage his reputation by personal vindictiveness. His young collaborators became victims of the rejection of his revolutionary physical thought by the Wilhelmian physical science establishment.

[140] R. McCormmach, "On the Growth of the Physics Discipline in the Nineteenth Century," address published as "Editor's Foreword" in *Hist. Stud. Phys. Sci., 3* (1971), ix–xxiv; R. McCormmach, "On Academic Scientists in Wilhelmian Germany," *Daedalus* (Summer, 1974), pp. 147–171; R. McCormmach, "Wilhelmian Theoretical Physics and the Physical World Picture," paper presented at the first meeting of the Joint Atlantic Seminar on the History of Physics, University of Montreal, 22 March 1974; P. Forman, J. L. Heilbron, S. Weart, "Physics *circa* 1900. Personnel, Funding and Productivity," *Hist. Stud. Phys. Sci., 5* (1975).

ACKNOWLEDGMENTS

For their comments on earlier drafts of this paper, I want to thank Eugene Frankel, Stanley Goldberg, Gerald Holton, Karl Hufbauer, Russell McCormmach, Otto Nathan, and Brigitte Schroeder-Gudehus. Miss Helen Dukas of the Institute for Advanced Study kindly helped me navigate through the Einstein Archives. I am grateful to the Estate of Albert Einstein for permission to cite Einstein's correspondence. A portion of this research was made possible through a grant from the Deutschen Akademischen Austauschdienst.

Max Planck's Philosophy of Nature and His Elaboration of the Special Theory of Relativity

STANLEY GOLDBERG*

1. INTRODUCTION

For more than three years after the publication of Einstein's first paper on the theory of special relativity Max Planck was nearly the only physicist of importance who received the new theory with the attention it deserved[1] and who immediately and without regard for the generally hesitant response to it by his colleagues applied himself to the exacting task of its elaboration. Almost all of his original contributions to the theory of special relativity appeared during this brief period. Planck defended the theory although it did not yet agree with experimental results, he encouraged his students to examine applications of the theory to a wide range of physical processes, and he wrote several papers in which he brilliantly demonstrated the heuristic power of the theory.

Planck, then in his late forties, received the theory of special relativity with a youthful—if critical—open-mindedness that he did not have for all of Einstein's new ideas. It is well-known, for example, that he long remained skeptical of Einstein's theory of the photoelectric effect.[2] His receptiveness for special relativity does not mark him as a revolutionary, however. In the present study I shall show that Planck's reaction and early contributions to the theory of special relativity were entirely in keeping with his conservatism, with his views of the nature of physical reality, and with his chief goal in physical research, namely, to uncover and confirm the absolutes of physics.

Planck frequently stated in public his metaphysical, epistemological, and ethical views. However, almost all of the addresses and

*School of Natural Science, Hampshire College, Amherst, Mass. 01002.

[1]M. Planck, *Wissenschaftliche Selbstbiographie* (Leipzig, 1948), p. 31. The essay has been translated as M. Planck, *Scientific Autobiography and Other Papers,* trans. Frank Gegnor (New York, 1949).

[2]Cf. M. J. Klein, "Einstein and the Wave Particle Duality," *The Natural Philosopher, 3* (1964), 1–50.

essays containing them were written after 1908, that is, they were not published concurrently with his principal achievements in physics. The historian interested in the relationship between Planck's work in physics—in the present instance, in his work on special relativity—and his philosophical outlook is therefore forced to assume that Planck's philosophical views were essentially the same throughout his career. In the case of Planck's elaboration of the theory of special relativity, this assumption of continuity in Planck's philosophical thought allows us to assert a close correlation between the philosophical views Planck held at the time of publication of Einstein's theory and the nature of Planck's response to special relativity. In the present study I bring together and discuss in detail the elements needed to demonstrate such a correlation: a full account of the physical researches resulting from Planck's interest in the new theory, followed by an examination of the philosophical beliefs that motivated his involvement in its development.

2. PLANCK'S ELABORATION OF THE THEORY OF SPECIAL RELATIVITY

The Early Response to Einstein's Theory

Einstein submitted his first paper on relativity to the *Annalen der Physik* at the end of June 1905; it was published in September.[3] Max Planck and Wilhelm Wien were immediately impressed by Einstein's paper. Wien, the physicist Jakob J. Laub has pointed out, must be considered the "first of the great physicists" to appreciate the profundity and originality of Einstein's theory of special relativity. When Einstein's article appeared in the *Annalen,* Wien asked Laub, who was then studying under Wien at Würzburg, to prepare a colloquium talk on the subject,[4] but Wien did not support the

[3]A. Einstein, "Zur Elektrodynamik bewegter Körper," *Ann. d. Phys., 17* (1905), 891–921. The article has been translated a number of times; see A. Einstein, "On the Electrodynamics of Moving Bodies," in H. A. Lorentz et al., *The Principle of Relativity,* trans. W. Perrett and G. B. Jeffrey (New York, n.d.), pp. 35–65.

[4]J. J. Laub to C. Seelig, 11 September 1959. The letter is contained in the Seelig papers in the library of the Eidgenössischen Technischen Hochschule, Zurich, Switzerland, to whom I am indebted for permission to read and make use of the material; in particular I acknowledge the valuable help given to me by Dr. A. Jaeggli.

Very little attention has thus far been paid to Laub. He seems to have been

theory and remained skeptical until 1909.[5] He was unable to accept the second postulate of the theory, the invariance of the velocity of light.[6] Planck has also been called "Einstein's earliest patron in scientific circles."[7] The first lecture Planck gave in the colloquium series at Berlin University in the fall of 1905 was a review of Einstein's relativity paper. Max von Laue, who was Planck's assistant in 1905, was so impressed by the lecture that he used his next vacation to visit Einstein in Berne.[8]

On the whole, however, the response to the new theory during the first three or four years after the publication of Einstein's paper was not very great.[9] Most of it came from physicists in Germany, Austria, and other German speaking countries and appeared in German language publications. Only a handful of physicists came to grips directly with the issues that Einstein had presented. Of these, Laub was the most penetrating and the most perceptive.[10] Max Abraham did not accept the theory, but he appreciated the logic of Einstein's arguments and the basic premises of the theory.[11] Hermann Minkowski, on the other hand, failed to appreciate the significance of Einstein's approach, even though he cast the theory of special relativity into a formalism of great heuristic power. His emphasis was on creating a theory of matter rather than a theory of

---

particularly instrumental in convincing Wilhelm Wien of the meaning and power of Einstein's theory. The early relationship between Einstein and Laub is discussed in Carl Seelig, *Albert Einstein: Eine dokumentarische Biographie* (Zurich, 1952), pp. 85-87.

[5] W. Wien, *Über Elektronen* (Leipzig, 1906; 2nd ed. 1909).

[6] *Ibid.*, 2nd ed., p. 28. The second edition contains a number of new footnotes, one of which (note 2, p. 28) extends over nearly five pages and gives a brilliant account of the fundamental aspects of Einstein's theory of relativity, thereby rendering meaningless a good part of the text to which it refers.

[7] G. Holton, "Mach, Einstein, and the Search for Reality," *Daedalus, 97* (1968), 636-673; on p. 642.

[8] M. von Laue to C. Seelig, 13 March 1952. The letter is contained in the Seelig papers (note 4).

[9] S. Goldberg, *Early Response to Einstein's Theory of Relativity, 1905-1911: A Case Study in National Differences* (diss. Harvard University, 1969); "Poincaré's Silence and Einstein's Relativity," *Brit. J. Hist. Sci., 5* (1970), 73-84; "In Defense of Ether; The British Response to Einstein's Theory of Relativity, 1905-1911," *Hist. Studies Phys. Sci., 2* (1970), 89-125.

[10] See especially J. J. Laub, "Zur Optik der bewegten Körper," *Ann. d. Phys., 23* (1907), 738-744; *25* (1908), 175-184.

[11] S. Goldberg, "The Abraham Theory of the Electron: the Symbiosis between Theory and Experiment," *Arch. Hist. Exact Sci.,* 7 (1970), 9-25.

measurement.[12] H. A. Lorentz, by his own account, did not understand the difference between his own electron theory and Einstein's kinematical theory based on the behavior of rigid rods, perfect clocks, and light signals until after 1909.[13]

If physicists were reluctant to accept Einstein's theory, it was often because they did not grasp the significance of his approach, not because they were wary of the elements of the theory, which were known to them and had individually been discussed since 1902.[14] Laub was struck by his colleagues' inability to understand the logic of Einstein's theory;[15] Arnold Sommerfeld found it impossible to convince Wien of the second postulate.[16]

Some physicists were so convinced of the truth of existing theory that they rejected Einstein's theory outright. For example, in 1909 Fritz Hasenöhrl began a review article on the literature on the concept of the inertia of energy by stating that he would only discuss the development of the concept "from the standpoint of the true Lorentz theory."[17] Hasenöhrl accredited recent attention to Einstein's principle of relativity to "its blinding elegance."

The First Experimental Tests of Einstein's Theory

The first physicist to subject the predictions of the theory of special relativity to a comparison with experimental results was Walter

[12] Since Minkowski's untimely death cut off his important contributions, it is difficult to give a definitive judgment of it. His work was, in a sense, carried on by his last assistant, Max Born. In the years 1908–1910 Born published a series of papers which reveal a gradual transformation of his point of view; abandoning Minkowski's interpretation, he came to appreciate fully the kinematical character of Einstein's theory. See S. Goldberg, *Early Response . . . , op. cit.* (note 9), chap. 2.

[13] S. Goldberg, "The Lorentz Theory of the Electron and Einstein's Theory of Relativity," *Amer. J. Phys., 37* (1969), 982–994; G. Holton, "On the Origins of the Special Theory of Relativity," *Amer. J. Phys., 28* (1960), 627–636; H. A. Lorentz, *The Theory of Electrons* (New York, 1952), p. 321, fn. 72*. This book was originally published in 1909. The starred footnotes were added to the second edition, published in 1915.

[14] S. Goldberg, "Henri Poincaré and Einstein's Theory of Relativity," *Amer. J. Phys., 35* (1967), 934–944; G. Holton, *ibid.*

[15] J. J. Laub to C. Seelig, 11 September 1959 (note 4).

[16] A. Sommerfeld, "Ein Einwand gegen die Relativtheorie und seine Beseitigung," *Phys. Zs., 8* (1907), 841–842; see especially the discussion between Sommerfeld and Wien.

[17] F. Hasenöhrl, "Die Trägheit der Energie," *Jahrb. Rad. u. Elek., 6* (1909), 485–502; esp. pp. 485–486.

Kaufmann.[18] Late in 1905 Kaufmann completed a series of experiments on the changes in the "transverse" mass of an accelerating electron, that is, on the changes in the mass of an electron when the electron is accelerated in a direction perpendicular to the direction of motion. Kaufmann had carried out his investigation by photographically determining the magnetic and electric deflections of β-rays emitted by a kernel of radium bromide. He had obtained photographed deflection curves each point of which corresponded to a different velocity and mass of the electron. Kaufmann, who had been making measurements of this kind since 1901, now used his experimental technique to test the validity of the various electron theories, that is, Abraham's theory of the rigid electron, Bucherer's theory of the deformable electron of constant volume, Lorentz' theory of the deformable electron of variable volume, and Einstein's theory relating changes in the specific charge of the electron to the problem of simultaneity. If he could determine the velocity that corresponded to any point of the photographed curves from the constants of his experimental setup, then, Kaufmann reasoned, a comparison between this value and any of the velocities derived from the curve by applying one or the other of the three theories would show which, if any, of the theories agreed best with experience. More importantly, if the measurements confirmed the validity of Lorentz' and Einstein's theories—their predictions were the same[19]—then one would have to assume an "internal potential energy" of the electron to account for the work required in the deformation of Lorentz' electron, and it would be proven that a purely electromagnetic foundation of the mechanics of the electron as well as of mechanics in general is impossible.[20]

Planck took up the defense of the theory of special relativity in 1906. In March of 1906 Planck read a paper before the German Physical Society in which he investigated the theoretical implications

[18] W. Kaufmann, "Über die Konstitution des Elektrons," *Ann. d. Phys., 19* (1906), 487-553.

[19] S. Goldberg, *op. cit.* (note 11). In fact, Einstein's first prediction for the transverse mass of the electron differed formally from Lorentz' prediction due to Einstein's idiosyncratic definition of force in his first paper on relativity. Lorentz' expression, as almost everyone immediately recognized, was implied by Einstein's. Cf. A. I. Miller, "On Lorentz' Methodology," *Brit. J. Phil. Sci., 25* (1974), 29-45, esp. p. 42.

[20] W. Kaufmann, *op. cit.* (note 18), p. 494.

of Einstein's work.[21] In September, at the Stuttgart meeting of the Society of German Natural Scientists and Physicians, he presented the results of his examination of Kaufmann's experimental techniques and analysis, for Kaufmann had concluded in favor of Abraham's and Bucherer's theories and against Lorentz' and Einstein's, particularly against the admissibility of the relativity principle.[22]

Planck took up work on the theory of special relativity not because he was convinced that it was correct, but because he believed that a physical idea of such simplicity and generality deserved to be investigated in more than one way. His first theoretical task was to find the equations of motion that would have to replace Newton's if the principle of relativity was to be generally valid. Planck applied the Lorentz transformations, Newton's second law, and the transformation equations for electric and magnetic field strength as given by Einstein in his first 1905 paper on relativity to a mass point having a charge *e* and derived the equations of motion for the particle that are valid for any inertial frame of reference under the influence of an electromagnetic field. By comparing these equations with the Lagrangian equation of motion, he then derived the kinetic potential and, by another simple step, the kinetic energy of the mass point. Finally Planck derived the least action formulation and Hamilton's canonical equations. He concluded that all of these relations also apply to any other coordinate system related by the Lorentz transformations.[23] Planck's paper was important because it

[21] M. Planck, "Das Prinzip der Relativität und die Grundgleichungen der Mechanik," *Verh. d. Deutsch. Phys. Ges.*, *8* (1906), 136–141.

[22] M. Planck, "Die Kaufmannschen Messungen der Ablenkbarkeit der β-Strahlen in ihrer Bedeutung für die Dynamik der Elektronen," *Phys. Zs.*, 7 (1906), 753–761; *Verh. d. Deutsch. Phys. Ges.*, *8* (1906), 418–432.

[23] In 1905, H. Poincaré had already considered many of the problems taken up by Planck in this paper. He had, among other things, correctly formulated the transformation equations for force and had considered the least action formulation for the electromagnetic Lagrangian. Planck's approach and point of view had little in common with Poincaré's. Planck began from the principle of relativity and the Lorentz transformations and deduced the consequences. Poincaré, on the other hand, used the Lorentz transformations to explain why it was not possible to detect absolute motion. H. Poincaré, "Sur la dynamique de l'électron," *Rend. Cir. mat. di Palermo, 21* (1906), 129–176. Cf. H. Poincaré, *Oeuvres d'Henri Poincaré* (Paris, 1934–1954), *9*, 494–550. See H. M. Schwartz, "Poincaré's Rendiconti Paper on Relativity," *Amer. J. Phys., 39* (1972), 1287–1294; *40* (1973), 862–872, 1282–1287. Schwartz has modern-

showed that one could use the principle of relativity to express the laws of motion in closed form and that one did not need quasi-stationary approximations such as the one Abraham had been forced to use because of his strict adherence to an electromagnetic worldview.[24]

Planck's first purpose in examining Kaufmann's work was to find out how great a distance there is between the individual measured deflection and the corresponding deflections calculated according to the different theories with the constants of the experimental setup.[25] He chose not to assume—as Kaufmann had—that the observed deflections are at once reducible to infinitely small deflections and, hence, that the simultaneous electric and magnetic deflections are independent of one another.[26] In spite of his different method of calculation, Planck found that his results agreed with Kaufmann's. Nonetheless, he insisted that the results of the experiment do not justify Kaufmann's conclusion. He pointed out that the deflections predicted by the theories are closer to each other than the prediction of any one theory to the observed deflection. Dismissing as inconclusive whatever objections might be made to Kaufmann's work, Planck saw no alternative but to assume that there was still a "significant gap" in the theoretical interpretation of the measured quantities which would have to be filled before Kaufmann's measurements could be used to decide between Abraham's theory and the relativity theory.[27]

In the ensuing discussion Kaufmann, Abraham, and Bucherer pressed Planck to decide for one of the theories on the basis of Kaufmann's experiments, but Planck refused. Abraham then fell back on theoretical arguments. He insisted that his theory is correct because it is the only one of the theories that has been built on a purely electromagnetic basis. That, he said, is more convincing than the experiments. Lorentz' theory requires nonelectromagnetic forces to ensure the stability of the electron and Einstein's theory says

---

ized all notation and draws conclusions that do not flow from a historical study of the context of Poincaré's paper. See also A. I. Miller, "A Study of Henri Poincaré's "Sur la Dynamique de l'Electron'," *Arch. Hist. Exact Sci., 10* (1974), 207-328.

[24] S. Goldberg, *op. cit.* (note 11).

[25] M. Planck, *op. cit.* (note 22), p. 419.

[26] W. Kaufmann, *op. cit.* (note 18), p. 524.

[27] M. Planck, *op. cit.* (note 22), p. 428.

nothing about the electromagnetic nature of the electron. In reply, Planck pointed out that the electromagnetic basis of Abraham's theory is as much a postulate as the postulate of the impossibility of detecting absolute motion of the Lorentz and the Einstein theories. Planck added that he preferred the relativity postulate.[28]

Within six months of the Stuttgart meeting A. Bestelmeyer established a new value for the rest mass of the electron.[29] Bestelmeyer had used electrons produced by X-rays of various intensities to determine to what extent the ratio of charge to mass might be a function of the intensity of the primary radiation. Rather than using parallel electric and magnetic fields to deflect the charged particles as Kaufmann had done, Bestelmeyer introduced the now familiar crossed field velocity filter arrangement. Bestelmeyer's new value for the specific charge of the resting electron was approximately nine percent lower than Kaufmann's; used in conjunction with Kaufmann's data and Kaufmann's method of analysis, the new value led to results that were so close to the predictions of both Abraham's theory and Lorentz' and Einstein's theories that they could not be used to differentiate between the theories.

In the meantime others, chiefly H. Geiger in England and J. Stark in Germany, were investigating the ionization in gases caused by β-radiation. Planck had come to question the accuracy of the constants of Kaufmann's experimental setup since they yielded a velocity greater than that of light for the least deflection radiation. He now attributed part of the failure of Kaufmann's experiment to a miscalculation of the electric field.[30] He argued that, since the recent experiments have shown that β-radiation ionizes the air remaining between the condenser plates in Kaufmann's apparatus, the electric field between the plates can no longer be assumed to be homogeneous as Kaufmann had assumed. Planck repeated his and Kaufmann's calculations again, taking into account the changed view of the electric field and Bestelmeyer's new value. This time, although his result still did not clearly confirm either Abraham's or the relativity theory, Planck concluded that it "increased the chances" of the latter. Kaufmann denied that his experimental results could be so

[28]M. Planck, *op. cit.* (note 22), pp. 759-761.

[29]A. Bestelmeyer, "Spezifische Ladung und Geschwindigkeit der durch Röntgenstrahlen erzeugten Kathodenstraheln," *Ann. d. Phys., 22* (1907), 429-447.

[30]M. Planck, "Nachtrag zur Besprechung der Kaufmannschen Ablenkungsmessungen," *Verh. d. Deutsch. Phys. Ges., 9* (1907), 301-305.

grossly in error and continued to dispute the question with Planck and Stark through the year 1907.[31] Eventually, however, his contributions in this area stopped.

During all of this time Einstein refrained from entering the dispute, but he remarked that even though Kaufmann's data seemed to favor the predictions of the Abraham theory, his theory was the only one that comprehended a wide variety of phenomena.[32]

Thermodynamic Theories of Stationary Radiation in Moving Blackbodies

In 1904 Fritz Hasenöhrl had taken up the problem of bringing the laws governing stationary radiation in moving bodies into agreement with thermodynamics.[33] Hasenöhrl had arrived at results that contradicted the second law of thermodynamics and he had chosen to correct the situation by adopting the Lorentz contraction hypothesis. The following year Kurd von Mosengeil had investigated the same problem at Planck's suggestion and concluded that Hasenöhrl had based his work on an incorrect argument and that a correctly constructed theory of stationary radiation in moving blackbodies did not need the contraction hypothesis. When Mosengeil died early in 1906 before the publication of his paper in the *Annalen der Physik*, Planck edited his student's work and saw it through publication.[34]

[31] *Ibid.*; W. Kaufmann, "Bemerkungen zu Herrn Planck: Nachtrag zur Besprechung der Kaufmannschen Ablenkungsmessungen," *Verh. d. Deutsch. Phys. Ges., 9* (1907), 667-673; J. Stark, "Bemerkung zu Herrn Kaufmann's Antwort auf einen Einwand von Herrn Planck," *Verh. Deutsch. Phys. Ges., 10* (1908), 14-16. W. Kaufmann, "Erwiderung an Herrn Stark," *Verh. d. Deutsch. Phys. Ges., 10* (1908), 91-95.

[32] A. Einstein to J. Stark, 1 November 1907. Einstein thanked Stark for pointing out to him Planck's work on Kaufmann's experiments. He wrote that he had known nothing of Planck's research: "Es ist gut, dass Sie mich auf die Planck'sche Arbeit über die Kaufmann'schen Versuche aufmerksam gemacht haben. Ich wusste nichts von einer Untersuchung des Herrn Planck über diesen Gegenstand." I am indebted to the Handschriftenabteilung, Staatsbibliothek, Preussischer Kulturbesitz, Berlin (Dahlem), for permission to quote the letter from Einstein to Stark. The letter is contained in the collection Nachlass Stark and is designated Autogr. 1/223. Cf. G. Holton, *op. cit.* (note 7), pp. 651-652.

[33] F. Hasenöhrl, "Zur Theorie der Strahlung in bewegten Körpern," *Ann. d. Phys., 15* (1904), 344-370; "Ueber die Veränderung der Dimensionen der Materie infolge ihrer Bewegung durch den Aether," *Sitzungsber. Akad. Wiss. Wien, 113* (1904), 469-492.

[34] Kurd von Mosengeil, "Theorie der stationären Strahlung in einem gleichförmig bewegten Hohlraum," *Ann. d. Phys., 22* (1907), 867-904. An

It is instructive to consider Hasenöhrl's and Mosengeil's work in some detail in the present study. Like Planck, Hasenöhrl and Mosengeil tried to construct theories that would answer the questions raised by recent work in the area of heat radiation. Unlike Planck, however, they sought their answers only in the careful examination and combination of accepted theories and concepts. Other physicists, Hasenöhrl noted, had attempted to develop a theory of heat radiation in moving bodies from the point of view of the electromagnetic theory. Hasenöhrl and Mosengeil used thermodynamic concepts instead, supplementing them by electromagnetic ones where necessary. As a consequence of their cautious adherence to proven modes of physical thought they obtained only specialized results. When Planck came to treat the same problem, he achieved through his emphasis on general principles—the principle of least action and the principle of relativity—a general new dynamics of moving systems, which included heat radiation theory only as a special case.

Until physicists began to study the problem of heat radiation in moving bodies, it had been possible for them to treat the energy of a body in such a way that the kinetic energy of the body was completely divorced from any consideration of the body's internal state. That separation ceased to be possible for moving, radiation-filled cavities; it was therefore necessary to account for the energy of such a body in a new way. Hasenöhrl chose a thermodynamic approach, even though he realized that the laws of thermodynamics were inadequate for his purpose.[35] Thermodynamics failed to provide a value for the radiation pressure on a moving surface and Hasenöhrl used the value that Abraham had derived from the Lorentz theory: the radiation pressure was equal to the ratio of the total relative radiation (to be defined below) to the velocity of light.

After giving a summary of the geometric relations between absolute and relative radiation velocity and direction, Hasenöhrl first defined the kinds of radiation, that is, "absolute radiation," "total relative radiation," and "true relative radiation." He defined absolute

---

earlier version of this work appeared with the subtitle *Berliner Inaugural-Dissertation* (Berlin, 1906). Since Planck made significant alterations in the article, the editors of Planck's collected papers included it in the collection. See M. Planck, *Physikalische Abhandlungen und Vorträge* (Braunschweig, 1958), henceforth referred to as *PAV*; in *2,* 138–175.

[35] F. Hasenöhrl, "Zur Theorie der Strahlung . . . ," *op. cit.* (note 33).

radiation as the quantity of energy crossing a unit surface at absolute rest in unit time, total relative radiation as the quantity of energy crossing a unit surface moving with absolute velocity v in unit time, and true relative radiation as the total relative radiation decreased or increased by the work done by the radiation pressure or against the radiation pressure, respectively. Hasenöhrl next calculated the radiation density in the moving cavity, one part of which is equivalent to the work required to put the system into motion. Poynting and others had stated that mechanical work is directly transformed into radiation energy and vice versa purely as a consequence of the energy law and independently of any special assumptions about the nature of radiant heat. Hasenöhrl applied this point of view to his investigation and showed that the changes in the radiation process due to the motion of the body are reversible. He identified the mechanical work as an apparent mass change due to the inertia of the radiation.

Hasenöhrl concluded his analysis by examining the effects on the temperature of a radiation-filled cavity of a closed cycle of velocity changes under adiabatic conditions. His calculations showed that at the end of the process the net work was zero and the energy density the same as at the beginning of the process, but that the temperature was higher, a result that contradicted the second law of thermodynamics. He considered two explanations. One would require the assumption that the emission capacity of a blackbody changes as a function of its velocity, that is, that the internal energy of a radiating body is a function of its motion. If this were so, he said, then his work, including his derivation of radiation pressure, would be wrong. The hypothesis did not appear probable to him, but he admitted that it was possible. The other possible explanation required the Lorentz-Fitzgerald hypothesis that the dimensions of matter depend on their absolute velocity.[36] The alteration in the energy density produced by the change in volume in this case was, according to Hasenöhrl's calculations, just sufficient to compensate for the change in temperature.

Mosengeil claimed that his analysis differed greatly from Hasen-

[36] Lorentz' 1904 paper "Electromagnetic Phenomena in a System Moving with any Velocity less than that of Light" [reprinted in H. A. Lorentz et al., *The Principle of Relativity* (New York, n.d.), pp. 9–34] in which he placed the Lorentz-Fitzgerald contraction into the framework of his transformation equations had not yet been published.

öhrl's.[37] He believed that Hasenöhrl attributed greatest significance to the apparent support for Lorentz' theory that had been the result of his investigation. Mosengeil could not accept Hasenöhrl's result, since he believed that Hasenöhrl had made an error in his analysis by using the Lambert cosine law to calculate the energy density in a moving blackbody. He wanted to show that an accurate theory of stationary radiation in blackbodies would have different results.

In the first part of his paper Mosengeil assumed a moving cavity and a resting observer. He first showed that the intensity of the stationary radiation in moving blackbodies is independent of the nature of the emitting substance. He then derived expressions for the pressure, entropy, energy, and momentum of radiation in a moving cavity in terms of the temperature, volume, and speed of the cavity. He did not need the Lorentz contraction hypothesis but was able to obtain the desired expressions by using the well-known relationship between energy and momentum of electromagnetic radiation, and by defining the necessary relationship between the temperatures of blackbodies at rest and in motion.[38] In the next part of the paper, Mosengeil again derived these results by now assuming an observer who is moving relative to a blackbody and by applying the Lorentz transformation equations. He concluded that one did not need the Lorentz contraction hypothesis to show that the laws of stationary radiation in moving blackbodies are compatible with both electrodynamics and thermodynamics. As far as the validity of the principle of relativity was concerned, that was still an "open question."[39]

Hasenöhrl found Mosengeil's analysis without error. However, he rejected Mosengeil's criticism by correctly pointing out that, if the Lambert relationship applied to systems at rest, then it should also apply to systems in motion, since, according to Lorentz, uniform motion should have no effect on the phenomena.[40] Later Hasenöhrl attributed the difference in their analyses to Mosengeil's assumption of a change in temperature in the cavity with changing velocity rather than to the use of the Lambert relationship. Hasenöhrl proved

[37]K. von Mosengeil, *op. cit.* (note 34), p. 867.

[38]*Ibid.*, pp. 871-895.

[39]*Ibid.*, p. 904.

[40]F. Hasenöhrl, "Zur Theorie der stationären Strahlung in einem gleichförmig bewegten Hohlraume," *Ann. d. Phys.*, *22* (1907), 791-792.

his point by deriving results identical with Mosengeil's with the Lambert relationship.[41]

Given the structure of Mosengeil's paper—the introduction criticizing Hasenöhrl's work, the examination of the laws governing radiation in blackbodies from the point of view of a moving blackbody and resting observer, and then a rederivation of the same results from the point of view of relative motion using the Lorentz transformations—Mosengeil's unwillingness to commit himself to the principle of relativity appears to be a nonsequitur. If we remember, however, that Mosengeil's paper was edited and corrected by Planck, then Mosengeil's remarks may be seen as the—reflected or direct—expression of Planck's wariness with regard to the validity of Einstein's theory which continued through 1908.

Dynamics of Moving Systems

Several months after the appearance of Mosengeil's work, Planck presented a paper before the Prussian Academy of Sciences in which he developed a general theory of dynamics. He included in it not only Mosengeil's study of the laws governing stationary radiation in moving cavities, but also his own earlier analysis of the dynamics of ponderable matter.[42] Planck believed that it had become necessary to reevaluate certain concepts and laws of theoretical physics that had long been taken for granted.

> Recent experimental as well as theoretical investigations in the area of heat radiation have led to the common result that any system that has been stripped of all ponderable matter and consists entirely of electromagnetic radiation obeys both the fundamental equations of mechanics and the two laws of thermodynamics so completely that, so far, none of the conclusions drawn from these laws have proven inadequate. As a consequence it has become necessary to revise the foundations of a number of conceptions and laws which up to now have commonly been used as basic fixed and almost self-evident prerequisites of all theoretical speculations

[41] F. Hasenöhrl, "Zur Thermodynamik bewegter Systeme," *Sitzungsber. Akad. Wiss., Wien, 116* (1907), 1391-1405; *117* (1908), 207-215.

[42] M. Planck, "Zur Dynamik bewegter Systeme," *Sitzungsber. Preuss. Akad. Wiss.* (1907), pp. 867-904; also in *Ann. d. Phys., 26* (1908), 1-34; and in *PAV, 2,* 176-209. All references are to the version that appeared in the *Annalen der Physik.*

> in these fields. A closer look at them shows that some of the simplest and most important among them in the future can no longer claim exact validity but are no more than widely applicable and, as far as applications are concerned, very important approximations.[43]

Planck discussed three of the fundamental concepts that needed to be examined. He first considered the accepted definition of energy. It was no longer permissible, he asserted, to define the energy of a moving body as the sum of kinetic and potential energy. Every ponderable body contains a finite quantity of energy in the form of radiant heat; when the body is set in motion, the radiant heat is also set in motion. Although the energy of this radiant heat is a function of the velocity, it is impossible in the case of radiant heat to separate the kinetic from the potential energy, and, therefore, it is impossible to divide the total energy of the body into kinetic and potential energy. Planck next showed that the concept of inertial mass of a ponderable body similarly needed to be redefined. From the time of Newton, he said, the concept of mass as an absolutely unchanging property of matter, independent of physical or chemical action, had been the first building block of almost every physical world system. It is easy to show, however, that the mass of a body is not constant but a function of the temperature of the body. It is most directly defined in terms of the kinetic energy, and since it is impossible to completely separate the energy of motion from the internal energy one may say that a constant with the properties of inertial mass in the classical sense does not exist. Attempts to distinguish between "true" and "apparent" mass and to assign to the first the properties of an absolute constant merely masked the problem, Planck believed, and did not solve it. If the "true" mass were taken to be an invariant, the concepts of momentum and kinetic energy would lose their usual meaning. Finally Planck considered the question of the identity of inertial and ponderable mass. The radiant energy in a totally evacuated space, bounded by mirrored walls, contains inertial mass. Planck found it necessary to ask whether or not radiant energy then also contains ponderable mass, for, if it does not, then the generally accepted hypothesis of the identity of inertial and ponderable mass no longer holds. In the face of this state of affairs Planck found it very important to seek out and emphasize

43*Ibid.*, p. 1.

those of the laws of general dynamics that had proven to be absolutely exact for the results of recent researches, too, and to distinguish them from those that were now found to be only useful approximations.

Of all the well-known laws Planck found only the principle of least action still generally valid: it embraced mechanics, electrodynamics, and the two laws of thermodynamics, as well as the laws of blackbody radiation, as Planck was about to show. But the principle of least action alone was not a sufficient basis for a general dynamics, since it did not contain a substitute for the division of energy into kinetic and potential energy. Hence Planck proposed combining the principle of least action with the principle of relativity and developing the consequences to which such a combination might lead, not only for ponderable bodies, but for cavity radiation as well.

As in his other discussions of the theory of relativity, Planck noted that there was as yet no direct confirmation of the theory of relativity except for the results of the Michelson-Morley experiment. But, he added, there was also nothing known to prevent one from ascribing general and absolute validity to the principle. Planck argued that the principle was so decisive and fruitful that it should be given as thorough an investigation as possible; the investigation he proposed was to take the form of an examination of the consequences of the principle.[44]

Planck began his analysis of the consequences of combining the principle of least action with the principle of relativity by discussing blackbody radiation in a vacuum, because blackbody radiation is the only physical system whose dynamic, electrodynamic, and mechanical properties can be stated with absolute precision and independently of conflicting special theories. He expressed the entropy, energy, pressure, and momentum of blackbody radiation as functions of the independent variables velocity, volume, and temperature, making use of Mosengeil's equations.

Planck next wrote general expressions for energy, momentum, and kinetic potential. He noted that the general equations also apply to blackbody radiation. Then, in accordance with his introductory remarks, he substituted the principle of relativity for the usual analysis of the kinetic potential. He applied the Lorentz transformations to his general equations and derived the relationships between the variables of a body in different reference systems. Planck concluded

[44] *Ibid.*, pp. 1–5.

the first part of his analysis by listing the many properties and relations that are invariant in a transformation from one reference system to another.[45]

In the last section of the paper Planck applied to specific cases the general dynamical relations he had derived. He pointed out that the most important consequence of these general relations involved the dependence of the physical state of a body on its velocity. The special relations resulting from this dependence could be summarized in a single differential equation that represented the general expression for the application of the relativity principle to the kinetic potential. Then he showed that for the case of blackbody radiation his general results reduced to the specific equations Mosengeil had derived. Next Planck redefined inertial mass. He showed that the most general expression for inertial mass was equivalent to the differential equation he had derived. It stated that "every increase or decrease of heat changes the inertial mass of a body in such a way that the increase of the mass is always equal to the quantity of heat that is absorbed during an isobaric change of the body divided by the square of the velocity of light in a vacuum."[46] H. E. Ives, an American physicist, later referred to Planck's result as the "first valid and authentic derivation" of the relationship between mass and energy.[47] Planck concluded from his theory that every body contains a colossal quantity of internally stored energy—Planck called it "latent" energy—which ordinary physical and chemical processes hardly affect. He added that to pursue the implications of this result one could no longer use the concepts of kinetic gas theory and treat the chemical atom as a rigid body or as a material point or consider inertial mass as something given, nor could one still assume equipartition of energy in statistical equilibrium. Planck suggested that the relationship between mass and energy might prove to be verifiable by experimental tests.

Shortly after Planck had published his paper, Hasenöhrl responded with a modification of his earlier work of 1904 and 1905 in which he derived essentially all of Planck's results by using the hypothesis that no experiment could reveal the difference between absolute rest and inertial motion. The use of this hypothesis required the

[45] *Ibid.*, p. 23.

[46] *Ibid.*, p. 29.

[47] *Ibid.*, p. 27. Cf. H. Ives, "Derivation of the Mass-Energy Relation," *J. Opt. Soc. Amer.*, *42* (1952), 540–543.

introduction of the Lorentz-Fitzgerald contraction. At the end of his analysis, Hasenöhrl, obviously puzzled and hurt, remarked that, although he had introduced the notion of temperature dependent mass among other things in his earlier work, Planck had not acknowledged his contributions.[48]

Planck had demonstrated the heuristicity of the theory of relativity, but he was well aware that the theory was not yet fully established. In a paper delivered at the 1908 meeting of the Society of German Scientists and Physicians at Cologne[49] he noted that Einstein's theory differed from other similarly applicable theories only with respect to very small terms, and that except for those terms it could be considered correct. In the same paper he continued to develop the theoretical consequences of the theory of relativity in connection with his constant and primary concern, the unification of the different fields of physics. His remarks dealt with the role of the principle of action and reaction in electrodynamics. Lorentz had denied the principle general validity in his electrodynamics. Max Abraham had contradicted Lorentz and maintained the validity of the principle provided that mechanical momentum be supplemented by electromagnetic momentum. Abraham had justified the new concept by establishing an analogy between the conservation of momentum and that of energy. Reluctant to accept a generalization of momentum that was based on an analogy between the universal physical concept of energy and the specifically mechanical concept of momentum, Planck proposed instead to derive a suitably generalized definition of momentum from general dynamics. He noted that a definition that united within itself the mechanical and electromagnetic form was possible if one assumed the validity of Einstein's theory of relativity.

In the theory of relativity momentum can be defined in terms of a vector representing energy flow. Specifically, Planck stated that the momentum per unit volume at a point in space is equal to the ratio of the energy flow per unit surface area per unit time (a vector) to the square of the speed of light. He found that this definition of momentum provided the desired new insight into the "real signifi-

[48]F. Hasenöhrl, "Zur Thermodynamik bewegter Systeme," *op. cit.* (note 41), esp. *117,* 215.

[49]M. Planck, "Bemerkungen zum Prinzip der Aktion und Reaktion der allgemeinen Dynamik," *Verh. Deutsch. Phys. Ges., 10* (1908), 728-732; also in *PAV, 2,* 215-219.

cance of the reaction principle": the principle of the equality of action and reaction would now have general application as the "inertia law of energy."[50] Planck noted further that it is possible to speak of a momentum flow as one speaks of energy flow. Energy is a scalar and energy flow a vector. By analogy, since momentum is a vector, momentum flow must be a triple tensor. When Planck examined this tensor, he recognized that it was identical with the well-known Maxwell tension, a concept that had hitherto resisted physical interpretation.

After his 1908 paper Planck made few new contributions to the theory of relativity.[51] Instead, he frequently referred to it in his philosophical and popular addresses and essays. Planck continued to have reservations about the theory of relativity because of the lack of direct experimental confirmation. By 1908 Kaufmann's results had been discredited, and although A. H. Bucherer claimed that his new experimental results supported the theory of relativity he did not gain the confidence of many physicists.[52]

3. PLANCK'S PHILOSOPHICAL RESPONSE TO THE THEORY OF RELATIVITY

Planck's early theoretical work in relativity reflects the large philosophical purpose that directed all of his work in physics: "It has always seemed to me that the most important thing, the goal that guided all of my scientific endeavor, is the greatest possible simplification and unification of the physical worldview and that the first means of reaching this goal is the reconciliation of opposites through mutual fertilization and amalgamation."[53] His efforts to achieve a unified physical world picture took the form of a search for absolutes. The theory of relativity was based on an absolute, the

[50] *Ibid.*, p. 218.

[51] Planck did make a significant contribution to the dispute over the problems related to the relativistic definition of a rigid body. See S. Goldberg, *Early Response . . .* , *op. cit.* (note 9), pp. 111–148.

[52] C. E. Guye and C. Lavanchy, "Verification expérimentale de la formule de Lorentz-Einstein par les rayons cathodiques de grade vitesse," *Comptes Rendus, 161* (1915), 52–55.

[53] M. Planck, "Erwiderung auf die Ansprachen vom 26. April 1918 zu Max Planck's 60. Geburtstag in der Deutschen Physikalischen Gesellschaft," *PAV*, *op. cit.* (note 34), *3,* 327–330. Planck was replying to speeches by E. Warburg, A. Sommerfeld, M. von Laue, and A. Einstein. Unless otherwise specified all translations in this essay are by the author.

measure of the space-time continuum, and it made the velocity of light into an invariant. Planck was attracted to the theory of relativity by these aspects of it, for he quickly comprehended that the implications of such a theory might prove to be important for his search for a physical world picture.[54] The concern with invariants requires an understanding of the nature of the relative, of experience, to Planck the basis of all knowledge. In physics, experience consists of measurements. In my discussion of Planck's philosophy of nature as it found expression in his response to relativity, I shall consider his views of the role of measurement and the nature of experience and knowledge, his great concern with absolutes, and finally his physical world picture.

The Role of Experience in the Development of Physical Theories

In 1908 Planck gave the first exposition of his views on the relationship between the world of experience, the theoretical structures of physics, and the "real world" in an address on "The Unity of the Physical World Picture."[55] In this address—and in his later writings—Planck emphasized the fundamental importance of experience and measurement for science. Only through experience can man know anything about the world:

> The source of all knowledge and therefore the origin of each science is personal experience. It is the immediately given, the most real thing that we can think of, and the first point to which we connect the thought processes that constitute science. For the material with which we work in every science we receive either directly through our sense perceptions or indirectly through accounts from others, from our teachers, from writings, from books. There are no other sources of knowledge.
>
> In physics we are dealing with the experiences that are given to us by our senses of inanimate nature and that find expression in more or less exact observations and measurements. The content of

[54] M. Planck, *Wissenschaftliche Selbstbiographie, op. cit.* (note 1); see also *Scientific Autobiography . . . , op. cit.* (note 1).

[55] M. Planck, "Die Einheit des physikalischen Weltbildes," *Phys. Zs., 10* (1909), 62–75; also in *PAV, op. cit.* (note 34), *3,* 6–29; M. Planck, *Wege zur physikalischen Erkenntnis,* 4th ed. (Leipzig, 1944), pp. 1–24; cf. M. Planck, "The Unity of the Physical Universe," trans. R. Jones and D. H. Williams, in *A Survey of Physical Theory* (New York, 1960), pp. 1–40; S. Toulmin, ed., *Physical Reality* (New York, 1970).

that which we see, hear, feel is the immediately given and consequently incontestable reality.[56]

With the development of physical science, Planck continued, the laws of physics become divorced from their origin in experience and undergo a gradual deanthropomorphization. As the analytical techniques of physics improve and the understanding of physical processes deepens, anthropomorphisms disappear in two ways. First, physical knowledge is gradually unified: concepts and laws that were developed for specific phenomena are generalized until they apply to seemingly disparate fields of physics and thus unify the fields. Secondly, successive abstractions gradually remove explanatory patterns from the realm of direct experience: the physical concept of force no longer corresponds to the exertions of men and animals, the definition of color no longer describes the sensation of color.[57]

However, Planck did not agree with Ernst Mach that all physical theories are merely an attempt to organize physical experience in an economical way.[58] Planck rejected Mach's assertion that the process of deanthropomorphization in physics was due to an inclination toward "economy of thought," that is, that knowledge is gathered and organized so that a single mind can accommodate vast quantities

[56]M. Planck, "Positivismus und reale Aussenwelt," in *Wege . . . , op. cit.* (note 55), pp. 201-218, on p. 202. In translating the essay James Murphy divided it into two essays: "Is the External World Real?" and "The Scientist's Picture of the Physical World." See M. Planck, *Where is Science Going?* (New York, 1932), pp. 64-106. For other similar statements by Planck on the fundamental role of experience see *Der Kausalbegriff in der Physik* (Leipzig, 1932); "The Concept of Causality in Physics," trans. Frank Gegnor, in *Scientific Autobiography . . . , op. cit.* (note 1), pp. 121-150; "Physikalische Gesetzlichkeit im Lichte neuerer Forschungen," *Naturwissenschaften, 14* (1926), 249-261; also in *PAV, op. cit.* (note 34), *3,* 159-171; *Wege . . . , op. cit.* (note 55), pp. 156-178; "Vom Relativen zum Absoluten," *Naturwissenschaften, 13* (1925), 52-59; also in *PAV, op. cit.* (note 34), *3,* 145-158; *Wege . . . , op. cit.* (note 55), pp. 142-155; "From the Relative to the Absolute," in *Where is Science Going?*, pp. 170-200; *Neue Bahnen der physikalischen Erkenntnis* (Leipzig, 1914); also in *PAV, op. cit.* (note 34), *3,* 65-76; *Wege . . . , op. cit.* (note 55), pp. 42-53. Cf. "New Paths to Physical Knowledge," *A Survey . . . , op. cit.* (note 55), pp. 45-55; "Die Einheit . . . ," *op. cit.* (note 55); "Zwanzig Jahre Arbeit am physikalischen Weltbild," *Physica, 9* (1929), 193-223; also in *PAV, op. cit.* (note 34), *3,* 179-208.

[57]M. Planck, "Die Einheit . . . ," *op. cit.* (note 55), pp. 3-4; see also articles referred to in note 56. Planck frequently discussed deanthropomorphization in connection with the importance of sense data.

[58]For an analysis of Mach's theory of economy see J. T. Blackmore, *Ernst Mach: His Work, Life and Influence* (Berkeley, 1972), pp. 173-179.

of experience in the course of a lifetime. Mach had suggested that such a process was valuable for human survival. Planck emphatically denied that men like Copernicus, Kepler, Newton, Huygens, or Faraday had been motivated by a desire for economy of thought when they created the laws that bear their names. Instead, they were driven by their belief in the reality of the picture they created, Planck maintained.[59] He argued that the principle of economy was of little use to the practicing physicist who could not possibly know without hindsight which point of view would be the most economical, and he recommended that the principle be given a somewhat less conspicuous position. "If the physicist wants to promote his science, he must be a realist, not an economist; that is, in the changes of phenomena he must above all search for and separate out that which is permanent, unchanging, independent of the human senses. The economy of thought serves him in this as a means but not as an end. It has always been like this and will always remain so in spite of E. Mach and his supposed antimetaphysics."[60]

Planck's view of the importance of experimental data explains his caution in admitting the validity of the theory of relativity even while he was working on its elaboration. Planck found the theory too attractive and important to reject it on the basis of Kaufmann's experimental results, but his commitment to measurement and experiment was great enough to motivate the immense effort he put into reassessing Kaufmann's experiment and independently recalculating the results from Kaufmann's data. His commitment to experimental certainty similarly supported him in his refusal, in 1906, to admit to Abraham, Bucherer, and Kaufmann that Kaufmann's results had decided the issue by experiment.[61] Finally, Planck also revealed his commitment to the importance of measurement in his 1907 paper on general dynamics in repeated references to the possibility of an experimental decision on the validity of the theory of relativity through investigations of the relationship between mass and energy.[62]

Planck's work on relativity also reflects his view that measurement and experimental data are only the starting points of physical knowledge. While Planck insisted on experimental confirmation of a

[59] M. Planck, "Die Einheit . . . ," *op. cit.* (note 55), p. 23.

[60] M. Planck, "Zur Machschen Theorie der physikalischen Erkenntnis. Erwiderung," *Phys. Zs., 11* (1910), 1186–1190, on 1190.

[61] See above, p. 131.

[62] M. Planck, "Zur Dynamik bewegter Systeme," *op. cit.* (note 42), pp. 24–33.

new theory, he recognized other means of testing it. The logical development and internal consistency of a theory and the degree to which the theory is capable of revealing that which is "permanent, unchanging, independent of the human senses" were equally important to him. When Planck took up the task of producing the generalized laws of motion that were to replace Newton's laws,[63] he looked forward to further investigations that might reconcile the principle of relativity with observation, but he also outlined a logical method of investigation: "A physical idea that exhibits the simplicity and generality of the principle of relativity deserves to be tested in more than one way and if it is incorrect it should be driven *ad absurdum;* that can be done in no better way than by an investigation of the consequences to which it leads."[64] This manner of testing theories was not a new idea to Planck. In 1897, Planck wrote a letter to his friend Leo Graetz in which he discussed whether or not it was possible to reconcile the second law of thermodynamics with mechanics. Planck's assistant Ernst Zermelo had argued that no mechanical proof was possible, while Ludwig Boltzmann had taken the opposite position. Planck saw "only one way to reach a definitive conclusion about the question. One must embrace one of the two positions in advance and see how far one proceeds towards the light or towards the absurd."[65] Seeing "how far one proceeds towards the light or towards the absurd" was an important element in Planck's notion of the physical world picture.

Invariants and the Real World

Planck discussed the difference between the world of experience and the real world in almost every essay he wrote after 1908 on the nature of scientific knowledge. He held that reality consists of the constant elements of physical world pictures which are independent of all individual intellectual characteristics.[66] Planck was able to believe in an absolute, invariant real world because the development of physics had led to the discovery of universal physical constants such as the velocity of light, the charge and rest mass of an electron, or

[63] M. Planck, "Das Prinzip der Relativität . . . ," *op. cit.* (note 21).

[64] *Ibid.,* p. 137.

[65] Max Planck to Leo Graetz, 23 May 1897. The original letter is in the Deutschen Museum, Munich. The translation is by T. S. Kuhn. See T. S. Kuhn, *Blackbody Theory and Quantum Discontinuity, 1894–1912* (forthcoming). Cf. H. Kangro, *Vorgeschichte des Planckschen Strahlungsgesetzes* (Wiesbaden, 1970), esp. pp. 128–131.

[66] M. Planck, "Die Einheit . . . ," *op. cit.* (note 55), pp. 20–22.

the elementary quantum of action[67] and of universal physical laws. The universal constants were the result of a great number of measurements, and one could be certain that future measurements would give the same numerical values for them within the limits of experimental error. They were independent of individual experience. Planck found it impossible that any "real physicist," knowing of the existence of the universal physical constants, could espouse positivism. "Physical science demands the assumption of a real world that is independent of us and that we can never, to be sure, know directly, but that we can always perceive only through the eyeglasses of our sense perceptions and through the measurements that they allow us to make."[68] The universal physical constants may never be precisely known or may some day have to yield to higher absolute concepts, Planck noted, but the quest for invariants is still the ideal, indeed the highest, purest motivation for doing physical science. It is the search for the absolute.[69]

Planck's concern with absolutes did not conflict with his efforts on behalf of the theory of relativity. Planck held that every relative is necessarily connected to an absolute. The denial of the absolute nature of space and time in the theory of relativity does not, he argued, eliminate the absolute; it locates the absolute beyond space and time in the metric of the four-dimensional manifold in which space and time have been fused into a uniform continuum by means of the velocity of light.[70] The metric, Planck claimed, possesses a transcendental character entirely independent of all arbitrary choices such as measuring processes or reference systems that determine space and time.[71]

[67]M. Planck, "Die Stellung der neueren Physik zur mechanischen Naturanschauung," *Phys. Zs., 11* (1910), 922–932; also in *PAV, op. cit.* (note 34), *3,* 30–46; *Wege . . . , op. cit.* (note 55), 25–41; cf. "The Place of Modern Physics in the Mechanical View of Nature," in *A Survey . . . , op. cit.* (note 55), pp. 27–44.

[68]M. Planck, *Religion und Naturwissenschaft* (Leipzig, 1938); also in *Wege . . . , op. cit.* (note 55), pp. 291–307, on p. 300; cf. "Religion and Science," *Scientific Autobiography . . . , op. cit.* (note 1), pp. 151–187, on p. 173.

[69]M. Planck, "Vom Relativen . . . ," *op. cit.* (note 56), 158.

[70]M. Planck, "Vom Relativen zum Absoluten," *op. cit.* (note 56), p. 153.

[71]*Ibid.* See also "Kausalgesetz und Willensfreiheit," (1923), in *Wege . . . , op. cit.* (note 55), p. 128. This essay was translated as two separate essays by James Murphy in M. Planck, *Where is Science Going?* (New York, 1932): "Causation and Free Will: The Problem Stated," pp. 107–140, and "Causation and Free Will: The Answers of Science," pp. 141–169.

Planck considered the invariants in the theory of relativity of overriding importance. In his 1907 paper on general dynamics Planck concluded the central part of the paper by listing the invariants that his theory had produced.[72] Again, in his 1909 lectures on theoretical physics at Columbia University, Planck devoted a considerable part of the lecture on relativity to a discussion of the physical quantities that the theory showed to be invariant: the action, the Lagrangian density, the entropy, and the pressure of a system. Other physicists, for example Poincaré, had noted the existence of such invariants,[73] but only Planck accorded them so much prominence.

The second class of constituents of the real world that Planck postulated, the true laws of the universe, were as absolute and unchanging to him as the universal physical constants. Empirical laws were to him merely approximations to these absolute laws which were exact laws and universally valid with respect to place and time. The first such law, Planck once said, which he knew with certainty to possess absolute independent validity was the law of the conservation of energy. Even as an old man, Planck still recalled his gymnasium teacher's concrete account of it.[74]

Of all the quantities that remained invariant under the theory of relativity, perhaps the most important to Planck was the principle of least action:

> The most brilliant achievement of the principle of least action is shown by the fact that Einstein's theory of relativity, which has robbed so many theorems of their universality, has not disproved it, but has shown that it occupies the highest position among physical laws. The reason for this is that Hamilton's "action" . . . is an invariant with respect to all Lorentz transformations, that is, it is independent of the system of reference of the observers. . . .
>
> As in the case of the principle of least action, the principle of the conservation of energy has also a special position in the theory of relativity. Energy is, however, not an invariant with reference to the Lorentz transformations any more than it was earlier with respect to the Galilean transformations. . . . The principle of least action stands superior to both [conservation of energy and con-

[72] M. Planck, "Zur Dynamik bewegter Systeme," *op. cit.* (note 42), p. 23.

[73] H. Poincaré, *op. cit.* (note 23).

[74] M. Planck, *Wissenschaftliche Selbstbiographie, op. cit.* (note 1), pp. 7–8; *Scientific Autobiography . . . , op. cit.* (note 1), pp. 13–14.

servation of momentum], even when considered together, and it appears to govern all reversible processes in Nature. . . .[75]

Planck showed that one can derive the conservation of energy principle from the principle of least action, but that one cannot do the converse. The reason for this is that the conservation of energy principle provides only one equation, while the principle of least action provides as many equations as there are variables.[76]

Planck's position on the validity of the theory of relativity gradually changed from early noncommittal statements[77] to an expression of hope for future experimental confirmation in 1910.[78] However, Planck was drawn to the theory by its invariants,[79] and it is safe to assume that it was these, and especially the invariance of the principle of least action, that most inspired Planck with the belief in the eventual validation of the theory of relativity.

Planck's Physical World Picture

Planck first discussed the physical world picture in his 1908 Leiden lecture on "The Unity of the Physical World Picture."[80] These early remarks lack the clarity with which he defined the con-

75 M. Planck, "Das Prinzip der kleinsten Wirkung," *Kultur der Gegenwart,* Part 3, Division 3, Vol. 1, E. Warburg ed., *Physik* (Leipzig, 1915), pp. 692–702; also in *Wege . . . , op. cit.* (note 55), pp. 68–78; *PA V, op. cit.* (note 34), *3,* 91-101. All references will be to *Wege. . . .* The quote is on pp. 77–78. Cf. "The Principle of Least Action," *A Survey . . . , op. cit.* (note 55), pp. 69–81, esp. pp. 80–81. The translation is by R. Jones and D. H. Williams. See also M. Planck, *Acht Vorlesungen über Theoretische Physik* (Leipzig, 1910), pp. 97–98.

76 *Ibid.*

77 M. Planck, "Bemerkungen . . . ," *op. cit.* (note 49).

78 M. Planck, "Die Stellung . . . ," *op. cit.* (note 67), p. 929.

79 M. Planck, *Wissenschaftliche Selbstbiographie, op. cit.* (note 1), p. 32; *Scientific Autobiography . . . , op. cit.* (note 1).

80 Planck discussed the concept of the physical world picture in the following essays: "Die Einheit . . . ," *op. cit.* (note 55); "Die Stellung . . . ," *op. cit.* (note 67); "Physikalische Gesetzlichkeit . . . ," *op. cit.* (note 56); "Die Kausalität in der Natur," in *Wege . . . , op. cit.* (note 55); "Erwiderung . . . ," *op. cit.* (note 53); *Der Kausalbegriff . . . , op. cit.* (note 56); "Vom Relativen zum Absoluten," *op. cit.* (note 56); *Neue Bahnen . . . , op. cit.* (note 56); "Zwanzig Jahre . . . ," *op. cit.* (note 56); "Theoretische Physik," *PA V, op. cit.* (note 34), *3,* 209–218; "Sinn und Grenzen der exakten Wissenschaft" (1941), in *Wege . . . , op. cit.* (note 56); pp. 323–339; see also "The Meaning and Boundaries of Exact Science," in *Scientific Autobiography . . . , op. cit.* (note 1), pp. 80–112.

cept later, but they show that Planck was thinking of an ultimate intellectual picture of the physical universe. The physicist, he explained, derives his knowledge from experience and from "physical thinking": he uses his observations to draw conclusions that can never be tested by other direct observations. Through physical thinking, through the generalization of experience, the physicist arrives at knowledge that is not verifiable in experience, but that nonetheless corresponds to what Planck called reality. The physical world picture, Planck asserted, was of the nature of such knowledge. It was not an arbitrary invention of the human intellect, but the reflection of real processes in nature that are independent of man.[81] Planck's assertion of the validity of the physical world picture rested on his belief that the same laws that govern external nature also govern the workings of the mind. Planck frequently referred to this argument and he employed the strategy it supported in his theoretical work on the theory of relativity.

The principal characteristic of the physical world picture that Planck sought was its unity. All observed physical phenomena had to be included.[82] Planck did not claim that any of the competing worldviews of the period represented the ultimate solution. The "constant, unified world picture" was still only the goal "that true natural science approaches through all its changes."[83] World pictures had succeeded each other through the centuries according to the same process of change:

> It must be noted that the continual displacement of one world picture by another is dictated by no human whim or fad, but by an irresistible force. Such change becomes inevitable whenever scientific research hits upon a new fact in nature for which the currently accepted world picture cannot account. To cite a concrete example, such a fact is the velocity of light in empty space and another is the part played by the elementary quantum of action in the regular occurrence of atomic processes. These two facts, and many more, could not be incorporated in the classical world picture, and consequently its framework had to be destroyed and a new world picture was introduced in its place.
>
> This in itself is enough to make one wonder. But the circumstances which call for even greater wonderment, because it is not

[81] M. Planck, "Die Einheit . . . ," *op. cit.* (note 55), p. 25.
[82] *Ibid.*, p. 24.
[83] *Ibid.*, p. 27.

> self-evidently a matter of course by any means is that the new world picture does not wipe out the old one, but permits it to stand in its entirety and merely adds a special condition to it. . . . In fact, the laws of classical mechanics continue to hold satisfactorily for all processes in which the velocity of light may be considered to be infinitely great, and the quantum of action to be infinitely small.[84]

Planck noted the contradiction that, although new observations had led to improvements and simplifications of the physical world picture, it still continuously moved further away from the world of sense perceptions, that is, it became less and less anthropomorphic. This move from the world of experience was nothing other than a move toward the real world.[85]

In 1908 Planck believed that the ultimate physical world picture was close at hand, perhaps because he still shared in the *fin de siècle* feeling that the major task of physics was completed.[86] The deanthropomorphized world picture that he anticipated might appear "colorless and drab," especially when compared with vivid earlier world pictures, but it would have the desirable "unity of all separate parts of the picture, unity of space and time, unity of all investigators, nations, and cultures."[87]

Following his 1908 paper and over the next twenty years Planck repeatedly introduced the concept of the physical world picture in his public lectures. In 1929 Planck returned to the theme with great clarity and sureness. The evidence suggests that Planck did not change his mind very much during this twenty year period; rather he now gave his formerly intuitive and incomplete ideas a precise formulation.[88]

As in his earlier work Planck asserted the existence of three separate worlds: the world of sense perceptions, the inaccessible real world, and the physical world picture. The first two of these were worlds

[84] M. Planck, "The Meaning and Boundaries . . . ," *op. cit.* (note 80), pp. 97–98.

[85] M. Planck, "Zwanzig Jahre . . . ," *op. cit.* (note 56), pp. 184–185.

[86] On attitudes toward the state of physics at the end of the nineteenth century see Lawrence Badash, "Completeness of Nineteenth Century Science," *Isis, 63* (1972), 48–58, and the literature cited there. With regard to Planck see especially pp. 54–55.

[87] M. Planck, "Die Einheit . . . ," *op. cit.* (note 55), p. 18.

[88] M. Planck, "Zwanzig Jahre . . . ," *op. cit.* (note 56).

outside the control of man, but the physical world picture was for Planck a conscious, purposeful creation of the human spirit and as such changeable and subject to development. The physical world picture had the two-fold task of making known the real world as completely as possible and of describing the world of sense perceptions in as simple terms as possible. Planck added that it would be useless to decide for the one or the other of the two aspects of the task; each, taken by itself, is unsatisfactory. On the one hand physicists can never have direct knowledge of the real world, and, on the other hand, they will never be able to decide which description of natural phenomena is the simplest. The main thing, he insisted, is that the two aspects of the task that he had assigned to the physical world picture never contradict each other, but rather complement each other.[89]

Depending on whether physicists inclined toward the first or toward the second aspect of the task they were metaphysicians or positivists, Planck claimed. Physicists who were neither but who sought instead to uncover the inner completeness and the logical structure of the physical world picture Planck identified as "axiomatizers."[90] Each of these groups of physicists became influential at different times in the history of science. In periods when the physical world picture displays a stable character, such as, in Planck's view, the second half of the nineteenth century, physicists may conclude that they are closer to understanding the real world. During periods of uncertainty and change positivism comes to the fore, and the world of experience is seen as the only secure basis for physical research.[91]

The transformations of physical world pictures did not take place in rhythmical oscillations, Planck held, but in a determined direction marked by a constant increase of the content of man's world of experience, of his understanding of this world, and of his control over it.[92] Theoretical physics might give "the impression of an old and honored building which is falling into decay with parts tottering one after the other and its foundation threatening to give way," Planck noted in 1910, but that conception was erroneous and, although great changes were in fact taking place, they were only extensions and perfections of the structure.[93] Planck's historical reconstructions of

[89] *Ibid.*, p. 182.
[90] *Ibid.*, p. 183.
[91] *Ibid.*
[92] *Ibid.*, p. 184.
[93] M. Planck, "New Paths . . . ," *op. cit.* (note 56), p. 45.

the development of the theory of relativity in which he insisted on an evolutionary rather than a revolutionary development reflect the same view of the nature of change in physical world pictures. Moreover, Planck's analysis of the logic of the theory of relativity and of its logical relationship to other theories stressed the view that the introduction of the theory of relativity was one more step in the evolution of the physical world picture.

> The principle of relativity by no means simply disintegrates and demolishes—it only throws aside a form that through the continuous enlargement of science was already ruptured—but to a much greater degree it orders and builds. In place of the old building that has become too small it erects a new, more comprehensive and durable one which contains all of the treasures of the old . . . in changed, clearer arrangements. . . . It removes from the physical world picture the inessential ingredients arbitrarily introduced by our . . . habits and thereby cleanses physics of the anthropomorphical impurities originating from the individuality of the physicist. . . .[94]

The application of the principle of relativity to mechanics led to a modification of Newton's laws of motion, which were no longer invariant in four-dimensional coordinate systems, and thus to a generalization and simplification of Newtonian mechanics.[95]

When Planck spoke of the unity of the physical world picture, he meant the reduction of all physical phenomena to one basic law. Planck believed that that law was the principle of least action.

> For as long as physical science has existed it has had before it as its highest, worthiest goal the solution of the problem of uniting all observed and still to be observed natural phenomena in a single, simple principle which permits us to calculate both past and especially future processes from present ones. It is in the nature of things that this goal has not yet been attained nor ever will be completely attained. But it is indeed possible to approach it more and more, and the history of theoretical physics shows that in this way a large number of important successes could be obtained which clearly speak in favor of the view that the ideal problem is not a purely utopian, but rather an eminently fruitful one, and

[94] M. Planck, "Die Stellung . . . ," *op. cit.* (note 67), p. 932.
[95] M. Planck, "Theoretische Physik," *op. cit.* (note 80), p. 214.

that it therefore deserves to be always kept in mind, especially with regard to application.

Among the more or less general laws that mark the achievements of physical science in the development of the last centuries, the principle of least action today is probably that which by form and content may claim to come closest to the ideal goal of theoretical research.[96]

In 1906, in an exchange with Bucherer relating to the theory of relativity, Planck first indicated how much importance he attached to the principle of least action in this connection. He advised Bucherer to investigate whether or not the principle could be accommodated by Bucherer's new theory, since the equations of motion of electrons would be reduced to those of general mechanics through the Lagrangian form.[97] A year later, in a letter to Einstein written shortly after the presentation of his new general dynamics based on the principle of least action and the principle of relativity, Planck sought Einstein's advice on "urgent" questions about the principle of least action and its bearing on the admissibility of Einstein's principle of relativity. Planck informed Einstein that Bucherer had privately told him without proof or explanation that the principle of relativity was incompatible with the principle of least action. Planck was pleased to learn that Einstein did not share Bucherer's view.[98] In his 1929 address on the physical world picture Planck concluded that the introduction of the theory of relativity into the physical world picture was "one of the most important

[96] M. Planck, "Das Prinzip der kleinsten Wirkung," *op. cit.* (note 75), p. 68.

[97] M. Planck, "Die Kaufmann'schen Messungen . . . ," *op. cit.* (note 22).

[98] Max Planck to A. Einstein, Grunewald [Berlin], 6 July 1907. This letter is in the possession of the Einstein Archives, Princeton University, to whom I am indebted for permission to use it. I am particularly grateful to Helen Dukas who generously provided me with a typescript copy of the handwritten original.

Bucherer had a penchant for making strong assertions with little or no documentation, not only in his private letters, but in his published papers as well. This led him into some rather dramatic, public, polemical debates, notably with E. Cunningham in England during 1907–1908, a period when Bucherer rejected the theory of relativity on any grounds he could find: it did not conform to experiment, it was a misguided attempt to modify Maxwell's equations, it violated electromagnetic assumptions, and, as in this case, it was at odds with the principle of least action. In 1908 he changed his mind suddenly, apparently solely on the basis of his own experiments. Ironically, challenges to his experimental results by A. Bestelmeyer led him into yet another acrimonious public debate. Cf. S. Goldberg, *Early Response . . . , op. cit.* (note 9), pp. 97ff.

steps toward its unification and completion."[99] The contribution of the theory of relativity toward unity lay in no small part in confirming the principle of least action as the most general of all physical principles and thus establishing one of the unifying elements of Planck's vision of the ultimate physical world picture.

In 1908 Planck believed that the developing physical world picture now consisted of only two remaining major divisions: matter and ether. He was hoping that the two divisions would soon merge as the physical world picture neared completion.[100] Between 1907 and 1910 the problem of matter and ether caused him considerable agitation. In his letters to Lorentz he constantly came back to the question. On 1 April 1908 Planck wrote to Lorentz that given that experience tells us that there exists a true equilibrium for the finite distribution of energy between matter and ether, the electron theory must include a hypothesis that prevents the transformation of all energy from matter into ether over a period of time. He repeated what he had written to Einstein in July 1907: the simple Maxwell equations have to be retained for the free ether.[101] A year later[102] Planck wrote to Lorentz that, if the quantum of action is indeed a constant of the ether, the question arises whether or not Maxwell's equations are still applicable to free ether. The answer is no. Therefore Planck concluded once again that the quantum of action has nothing to do with processes in free ether. Planck continued to be concerned with Einstein's suggestions about the quantization of light energy. Again and again the problem of the possible interaction between the ether and the quantum resonators came up in Planck's letters to Lorentz.[103]

99M. Planck, "Zwanzig Jahre . . . ," *op. cit.* (note 56), p. 187.

100M. Planck, "Die Einheit . . . ," *op. cit.* (note 55).

101M. Planck to H. A. Lorentz, 1 April 1908. I am indebted to Stephen Brush, University of Maryland, who kindly provided me with a microfilm copy of selected correspondence of Lorentz which is available at the Hague. A copy of the same microfilm is on file with the Niels Bohr Library at the American Institute of Physics in New York.

102Planck to Lorentz, 16 June 1909.

103Besides the letters already cited, the reader is directed in particular to the letter from Planck to Lorentz dated 7 October 1908. In his letter Planck considers Lorentz' suggestion that the quantum resonators are not the same as ordinary matter and that the role of the molecules of ordinary matter is to transmit energy between the resonators. Planck was not happy with the idea. In the absence of ordinary matter the ether would play the role of transmitter and this conception did violence to the absolutely fixed ether which obeyed Maxwell's equations "exactly."

Planck's letters to Lorentz reveal his casual use of the concept of the ether and his resistance to achieving a more profound understanding of the problems involved. Planck understood the impossibility of action at a distance. He was not willing to blend the notions of the ether and the quantum of action, as attractive as the idea might be for a unified physical world picture. Lorentz and Planck agreed that it was highly unlikely that light quanta could maintain their individuality while propagating through free ether.[104] Furthermore, Planck argued that adopting the quantum hypothesis for light would invalidate the theories of interference, refraction, and diffraction. He encountered great difficulties in reconciling the notion of a coupling constant between, say, ether and electrons with the traditional concept of "free ether." He even toyed with the assumption that energy exchanged between electrons and the free ether could only occur in units of $h\nu$. Writing to Lorentz in July 1909, Planck remarked on the differences in their interpretation of h. Lorentz, Planck pointed out, interpreted h in such a way that the degrees of freedom of the ether are limited, so that every degree of freedom of the ether can assume energy only in multiples of h. Planck claimed exactly the opposite, namely that the electrons give up their energy in multiples of h. "From your point of view [Planck wrote], the unyielding point is the ether, from my point of view, it is the electrons."[105]

In 1907 Planck was still seriously considering the possibility of measuring the motion of the earth relative to the ether. In a letter to Lorentz dated 19 October 1907, six months after he had delivered his paper on general dynamics, Planck stated that he was occupied with the investigation of a method for measuring the influence of the motion of the earth on the intensity of radiation of a blackbody. He had little hope that the experiment would be practical: "If only $(v/c)^2$ were not so absurdly small! It is a real pity."[106]

By 1910 he had abandoned such hopes not only on practical grounds, but on theoretical ones as well. The principle of relativity, he now said, forced the abandonment of the concept of the rest ether as a substantive carrier of electromagnetic waves. The ether had to be

[104] Planck to Lorentz, 16 June 1908.

[105] Planck to Lorentz, 10 July 1909. "Bei Ihnen liegt . . . die Hartnäckigkeit beim Aether, bei mir bei den Elektronen."

[106] Planck to Lorentz, 19 October 1907. "Wenn nur $(v/c)^2$ nicht so unsinnig klein wäre! Es ist ein wahrer Jammer."

replaced by the concept of an absolute vacuum without physical properties in which electromagnetic energy is continuously propagated. The speed of propagation of the electromagnetic energy could not be considered a property of the absolute vacuum, but rather a property of the electromagnetic energy itself. "Where there is no energy, there exists no propagation velocity either."[107]

Planck was not rejecting the concept of the ether but replacing it by the concept of an absolute vacuum. His substitution was similar to the transformations of the ether concept by Lorentz and Poincaré. Planck often used the terms "ether" and "vacuum" interchangeably.[108] He sometimes thought it desirable to relinquish the ether concept entirely because of the difficulties it presented, but as late as 1946 he was only willing to reject the study of the mechanical properties of the ether, not the ether itself.[109]

4. CONCLUSION

Although Planck was intimately associated with the two major innovations in physics in the twentieth century, quantum physics and relativistic physics, he never intended to participate in a revolution in physics. As M. J. Klein has pointed out with regard to quantum physics, Planck tried throughout his career to reconcile the quantum hypothesis with classical physics.[110] He took the same approach to the theory of relativity.

Planck expressed his sense of continuity in the first sentence of his 1906 paper on relativistic generalization of the laws of dynamics where he described the principle as "recently introduced by Lorentz and in a more general form by Einstein.[111] In his reassessment of the Kaufmann data Planck referred to the "Lorentz-Einstein theory in which the principle has exact validity." He added that for the sake of convenience he would call this theory the theory of relativity in the remainder of the article, but during the discussion following the presentation of the paper Planck again referred to the "Lorentz-Einstein theory" which takes as a basic postulate that no absolute

[107]M. Planck, *Acht Vorlesungen . . . ,*" *op. cit.* (note 75), pp. 116–117. The quotation is on p. 117.

[108]*Ibid.*, pp. 110 ff.; see also, "Die Stellung . . . ," *op. cit.* (note 67).

[109]M. Planck, *Scheinprobleme der Wissenschaft* (Leipzig, 1947), p. 7.

[110]M. J. Klein, *op. cit.* (note 2).

[111]M. Planck, "Das Prinzip der Relativität . . . ," *op. cit.* (note 21).

translation is detectable.[112] His first reference to Einstein's theory of relativity occurred in his analysis of the principle of action-reaction in 1908.[113] In 1910 Planck described Lorentz, Einstein, and Minkowski as among the "pioneers" who worked with the new concept of relativity: Lorentz discovered the concept of relative time and introduced it into electrodynamics without, however, drawing any radical consequences from it, Einstein proclaimed the relativity of time as a universal postulate, and Minkowski then fashioned a consistent mathematical system of relativity theory.[114] Planck described the origin of the theory of relativity as typical of the origin of all scientific theories in the sense that it had been the result of a contradiction between theory and experiment. Such a contradiction represents progress, for it leads to changes or improvements in existing theory that may affect other parts of physics as well and even affect the development of physics far more than could have been foreseen.[115]

Planck was equally conservative in his use of the theory of relativity in his early work on the subject. He almost never referred explicitly to the second postulate of the theory, the invariance of the speed of light; in fact, he never spoke of the invariance of the speed of light as a postulate. Rather, he considered it a consequence of the principle of relativity.[116] In 1926 Planck remarked that the work of elaborating the theory of relativity, of developing its con-

[112]M. Planck, "Die Kaufmannschen Messungen . . . ," *op. cit.* (note 22), pp. 756, 761.

[113]M. Planck, "Bemerkungen . . . ," *op. cit.* (note 49).

[114]M. Planck, "Die Stellung . . . ," *op. cit.* (note 67), p. 927.

[115]M. Planck, "Sinn und Grenzen . . . ," *op. cit.* (note 80), pp. 336ff.

[116]Planck gave his first full account of his view of the import of the invariance of the speed of light in his Columbia University lectures, *Acht Vorlesungen . . . , op. cit.* (note 75).

There has been considerable discussion in the literature over the years of how Einstein first conceived of the theory of relativity and how in the development of the theory he saw the relationships between the two postulates and their consequences, the Lorentz transformation equations, and the various kinematical relationships that the transformations entail. Historians obtained most of the evidence employed in that discussion from Einstein's first paper on relativity in 1905 and his recollections in the autobiographical sketch he wrote much later. Cf. A. I. Miller, *op. cit.* (note 19), esp. pp. 41–43 and the literature cited there. Without wanting to reconstruct how Einstein created the theory, I recommend his *Relativity: the Special, the General Theory* (New York, 1918) as a good source for his view of the logical relationship of the parts of the theory.

sequences for the purpose of testing its validity—work in which he had had a large share—was made easier because the assertions made by the theory fitted classical physics perfectly. He went on to state that, if he were not stopped by historical considerations, he would not hesitate for a second to consider the theory of relativity a part of classical physics.[117]

In several of his addresses Planck spoke of the close correspondence between universal physical laws and ethics. His remarks there prove that his conservative approach to the theory of relativity as well as his loyalty to it stemmed not only from his scientific and philosophical convictions but also from his ethical beliefs. Unfortunately, the problem of the connections between scientific achievements and ethical or psychological motivations is too elusive and too complex to be developed here. We catch a glimpse of these inner connections in Einstein's words at the celebration of Planck's sixtieth birthday:

> Man tries to make for himself, in the fashion that suits him best, a simplified and intelligible picture of the world; he then tries to some extent to substitute this cosmos of his for the world of experience and thus to overcome it. This is what the painter, the poet, the speculative philosopher, and the natural scientist do, each in his own fashion. Each makes this cosmos and its construction the pivot of his emotional life, in order to find in this way the peace and security that he cannot find within the all-too-narrow realm of swirling personal experience. . . .
>
> The supreme task of the physicist is to arrive at those universal elementary laws from which the cosmos can be built up by pure deduction. There is no logical path to these laws; only intuition, resting on sympathetic understanding, can lead to them. . . . The longing to behold [cosmic] harmony is the source of the inexhaustible patience and perseverance with which Planck has devoted himself . . . to the most general problems of our science. . . . The state of mind that enables a man to do work of this kind is akin to that of the religious worshipper or the lover; the daily effort comes from no deliberate intention or program, but straight from the heart.[118]

[117]M. Planck, "Physikalische Gesetzlichkeit . . . ," *op. cit.* (note 56), p. 167.

[118]Quoted in Banesh Hoffmann, *Albert Einstein: Creator and Rebel* (New York, 1972), on p. 222. Hoffmann remarks that these words also reveal a great deal about Einstein. Cf. G. Holton, "Mach, Einstein and the Search for Reality," *op. cit.* (note 7).

ACKNOWLEDGMENTS

I am indebted to Albert Stewart, Antioch College, Herbert Bernstein, Hampshire College, and Lynn Miller, Hampshire College, for comments on an earlier draft of this paper. I am particularly indebted to Gerald Holton, Harvard University, under whose direction I began the work which led to this paper, and to Russell McCormmach, Johns Hopkins University, who made many valuable suggestions.

The research for the present study was carried out, in part, under a grant from the National Science Foundation.

The Concept of Particle Creation before and after Quantum Mechanics

BY JOAN BROMBERG*

"One of the most important results in the recent development of electron theory," Victor Weisskopf wrote in the mid-thirties, "is the possibility of transforming electromagnetic field energy into matter. A light quantum, for example, in the presence of other electromagnetic fields in empty space, can be absorbed and transformed into matter, with the creation of a pair of electrons with opposite charge."[1] By "recent," Weisskopf meant the period since P. A. M. Dirac's theory of the electron, which was published in 1928.[2] Today, physicists by and large still believe that the concept of the creation of fundamental particles, as well as that of the inverse process of their destruction, is a product of quantum mechanics.

In actuality, creation and annihilation concepts antedate quantum mechanics. The concept of the annihilation of pairs of oppositely charged, elementary particles,[3] for example, dates from the turn of the twentieth century. It became important in astrophysics about 1924, and continued to be through the early thirties. The annihilating pairs were first positive and negative electrons, later protons and electrons, and finally, starting in 1931, electrons and anti-electrons. For the physicists who were building quantum mechanics, the concept was familiar and acceptable, if minor, and they included it in their theories where convenient. In Dirac's "hole" theory of 1930, for example, pair annihilation was neither novel nor central. Dirac's object was to deal with a difficulty inhering in relativistic electron

*Center for the Philosophy and History of Science, Boston University, Boston, Mass. 02215.

[1]"Über die Elektrodynamik des Vakuums auf Grund der Quantentheorie des Elektrons," *Kongelige Danske Videnskabernes Selskab, Math. fys. Medd., 14* (1936), No. 6, on p. 3; reprinted in Julian Schwinger, ed., *Selected Papers on Quantum Electrodynamics* (New York, 1958), pp. 92-128, on p. 92.

[2]"The Quantum Theory of the Electron," *Proc. Roy. Soc., A 117* (1928), 610-624, and *118* (1928), 351-361.

[3]Whenever the participating particles are charged, the principle of charge conservation demands that creation and annihilation involve oppositely charged particles. I therefore use the terms "pair" and "particle" creation and annihilation interchangeably in this paper.

theory, which was that the theory allowed electrons to make transitions to states of negative energy. He achieved his object by interpreting electrons in states of negative energy as unobservable, and empty negative-energy states, or "holes," as protons. As a by-product, when an electron jumped into a vacant negative-energy state, an electron and a proton disappeared together into radiation. Since pair annihilation was already an accepted concept, this by-product was admissable.

All of this is easy to show, and I believe I do in section 2 of this paper. The question that arises next is harder: if creation and annihilation concepts were already known, how were they transformed by the advent of quantum mechanics? I suggest, first, that there was a change in the logical relation between the processes of creation and annihilation. Before quantum mechanics, a belief in annihilation did not entail a belief in creation; after quantum mechanics, it did. Second, I suggest that there was a change in the argument for the plausibility of the two processes. In sum, I argue that pair production and annihilation retained a steady position within physics as acceptable, if peripheral, concepts, but that they changed their relation, and their justification.

To begin to substantiate these suggestions I examine, in section 1, the situation before quantum mechanics. I analyze the writings of James Jeans, Walther Nernst, and A. S. Eddington on pair creation and annihilation within astrophysics, showing that these authors all started with models that pictured particles as aether structures and light as waves in the aether. I emphasize the dissimilarity between light and matter that is embodied in their aether models; light is motion, whereas matter is generally an arrangement like an aether strain or a configuration of aether particles. The possibility that the potential energy in the strain or configuration constituting the particle can be released, giving rise to wave motions, is more or less obvious.

Later, the justification for creation and annihilation processes shifted from aether conceptions to Einstein's mass-energy equivalence, $E = mc^2$. Physicists interpreted the latter in a variety of ways: some understood it in the energeticist sense to mean that matter was also energy, while others saw matter and energy as manifestations of the same underlying substance or world-condition. Both the aether models and these later generalizations, however, have in common that they give the processes a reductionist or unitary underpinning. They therefore embody a common justification: matter and light can convert into each other because they represent different conditions, or different manifestations, of the same thing. When the psychological habits of physicists are also taken into account, the opposition between generalized and aetherial justifications of pair annihilation and creation is not so clear-cut. Some physicists who renounced the use of models on philosophical grounds in the early 1920's continued to use them as tools in thinking, treating creation and annihilation in both mass-energy and aether-model terms.

Quantum mechanics removed the reductionist underpinning to a large extent, as I argue, with special reference to Pascual Jordan and Dirac, in sections 3 and 4. The introduction of matter waves promoted the view that light and matter are similar—light is quantum and wave, matter is particle and matter-wave—and this similarity made reductionist theories less interesting. There is no logical reason why a quantum-mechanical electron cannot be reduced to a structure in the electromagnetic field or aether, but the new symmetry between electromagnetic fields and matter fields made the reduction more arbitrary and less appealing. In addition, the emerging physics provided strategies for circumventing the use of reductionist models in the treatment of annihilation and creation.

The weakening of the reductionist interpretations of light and matter undermined the old justifications for creation and annihilation, and new ones were needed. One such, as I illustrate in the work of Jordan, was grounded in the similarity between light and matter itself. Similar objects have similar properties; light quanta can be created and destroyed, and since matter is like light, pair production and annihilation are readily intelligible. Dirac provided a different argument by suggesting that the most fruitful strategy for theorists was to extend the formalism of quantum mechanics and to link the new features that arise in it with experimentally known processes or entities. In this view, creation and annihilation functioned as a justification for the new feature in Dirac's theory, that is, of transitions into and out of negative energy states. Through the association, however, they also became interpreted in terms of these transitions.

In Dirac's theory, the postulation of pair annihilation demands the admission of pair creation, for the transition of an electron into a negative energy state, which is annihilation, requires that that state has previously been emptied, for example, through the transfer of its previous occupant into a positive energy state with the creation of a particle pair. Jordan's approach leads to the same result. The laws that predict the absorption of light also predict its emission, so that a theory that develops the annihilation of matter as an analog to the absorption of light must necessarily allow for the creation of matter. It was therefore precisely the new justification for the creation and annihilation processes that gave rise to the new relation between them.

If the physicists are wrong in believing that creation and annihilation processes were first conceived of by Dirac, it is nevertheless not hard to see how they came by their belief. To begin with, Dirac substituted a mathematically precise handling of these processes for qualitative speculations. Then, in 1932, the discovery of the positron endued his theory with unexpected importance. Hole theory was seen to be central to quantum field theory. To go back before the discovery of the positron, to Dirac's first paper, seemed to physicists a sufficient step into the past to understand the genesis of the concepts of pair creation and annihilation. From their viewpoint the present paper, which goes no further than 1931, may be called a prehistory of pair creation and annihilation.

- 1 -

Stars send energy into space.[4] The contraction hypothesis, which was generally accepted at the turn of this century, had postulated that this radiant energy arises from the gravitational potential energy that is lost as a star diminishes in bulk. This hypothesis gave an age for the sun of some tens of millions of years. When the new phenomena of radioactivity were applied to the dating of terrestrial rocks, however, the age of the earth was estimated to be very much greater. The emission of gamma rays and fast particles from radioactive substances could not, in turn, be substituted as a source of

[4] For histories of theories of stellar energy, see Reginald L. Waterfield, *A Hundred Years of Astronomy* (New York, 1938); A. S. Eddington, *The Internal Constitution of the Stars* (Cambridge, 1926), principally Ch. XI, "The Source of Stellar Energy"; and Bengt Strömgren, "On the Development of Astrophysics During the Last Half Century," and "The Growth of Our Knowledge of the Physics of the Stars," in J. A. Hynek, ed., *Astrophysics* (New York, 1951), pp. 1-11 and 172-258.

stellar radiant energy. The rate at which the known radioactive elements liberate energy was insufficient.

In 1920 measurements of atomic masses by the British physicist F. W. Aston created a flurry of excitement among astrophysicists.[5] At that time, nuclei were thought to be composed of electrons and protons. Aston showed that the atomic weight of helium is 4.00 and that of hydrogen 1.008,[6] with the implication that when four protons and two electrons combined to form helium almost one percent of the mass of the constituents was radiated away. Assuming this process in stars, and given reasonable estimates as to the amount of hydrogen contained in stars, one can arrive at a lifetime for the sun of about one billion years. This span is, in Eddington's phrase, "cramped" but sufficient.[7]

After 1920, fusion came to be regarded as a possible source of stellar energy. Eddington, for example, advocated hydrogen fusion[8] in a long review article on his theory of stars, published in 1921 in the *Zeitschrift für Physik*. The attraction of fusion was that it gave a common explanation for stellar energy and the evolution of the elements. For Eddington compatibility with evolution was one of the touchstones of any theory.

In the same article, Eddington referred to another possible energy source, the "annihilation of electron pairs [paarweise Vernichtung der Elektronen]." He included it there only for the sake of completeness, however, as the second of the two mechanisms through which sufficient energy could be made available.[9] James Jeans also

[5] *Internal Constitution*, pp. 295–296.

[6] *Ibid.*, p. 292.

[7] *Ibid.*, p. 293.

[8] "Das Strahlungsgleichgewicht der Sterne," *Zeit. f. Phys.*, 7 (1921), 351–397. This discussion is on pp. 392–396.

[9] Eddington had first considered the possibility that stellar energy was provided by pair annihilation in 1917, according to A. Vibert Douglas, *The Life of Arthur Stanley Eddington* (New York, 1957), pp. 60–68. I have not tried to trace the development of his views on sources of stellar energy from 1917 to 1924. It is sufficient for my themes to use him as an example of a supporter of fusion in 1921, and to examine his reasons for championing pair annihilation after 1924. Douglas, p. 68, credits Eddington with being the first to apply annihilation to stellar phenomena and takes Jeans to task for claiming priority for himself. That this dispute arose at all is evidence of how naturally pair annihilation follows from any aether theory of matter. In worrying over priorities, I believe Douglas also misses the close connection between radioactivity and particle annihilation, which I mention below.

mentioned pair annihilation in discussions of gaseous stars. In 1919, for example, he introduced the "annihilation of pairs of positive and negative electrons" as "the only source of energy of sufficient power" to substitute for gravitational energy of contraction as the origin of stellar radiant energy. He rejected annihilation, however, in favor of gravitational contraction for astronomical reasons.[10]

Beginning about 1924,[11] new results on the relation between the masses of stars and their luminosities gave added plausibility to pair annihilation. Eddington found that the luminosity, or total energy radiated per unit time, is determined by the mass of a star. Consequently, if the mass remains more or less constant as the star evolves, its luminosity should remain fairly constant. A star is also characterized by its spectral type, or the distribution of the intensity of the radiated energy over the various frequencies. The spectral type is determined by the star's surface temperature according to Planck's law for black body radiation. Now a correlation exists—the Russell-Hertzsprung diagram—between luminosity and spectral type for a large number of stars. In the diagram stars cluster around two lines. One line shows the luminosity constant as the effective surface temperature increases; the second, intersecting line—the main sequence—shows luminosity decreasing as the surface temperature decreases. Before 1924, these lines had been interpreted as revealing the life-history of stars. Diffuse, giant stars contract at constant mass, their luminosity tending to decrease with decreasing surface areas; at the same time, they move through spectral types representing increasing surface temperatures, corresponding to the increase in temperature of an ideal gas with contraction. The two tendencies keep their luminosity roughly constant as they trace out the first line of the Russell-Hertzsprung diagram. When they contract to a point where the ideal gas laws cease to be applicable, they turn to

[10] *Problems of Cosmogony and Stellar Dynamics* (Cambridge, 1919), pp. 286-287. The book was the Adams Prize Essay of 1917. See also "The Internal Constitution and Radiation of Gaseous Stars," *Roy. Astron. Soc., London, Monthly Notices, 79* (1919), 331-332.

[11] Eddington's investigations into the mass-luminosity relation started in 1916. The earliest results seemed to apply only to giant stars. That such a relation could obtain for stars of the main sequence only became apparent about 1924, when it was realized that the high degree of ionization inside stars caused them to behave as gases at unexpectedly high densities. See Strömgren, "The Growth of Our Knowledge," *op. cit.* (note 4), p. 254, and Douglas, *Eddington, op. cit.* (note 9), Ch. 7.

follow the second line, decreasing both in luminosity and surface temperature as they contract further.

Eddington's results made this interpretation impossible. If the main sequence of the diagram were to continue to represent stellar evolution, it was necessary to assume that stars lose considerable mass as they move along it. The amount lost would exceed that which occurs in the fusion of hydrogen, which suggested that this loss of mass accounts for stellar energy; that is, the energy was seen to represent the conversion of mass into radiation.

After 1924, both Eddington and Jeans favored pair annihilation in stars. Walther Nernst emerged among the opponents of pair annihilation. In the rest of this section, I shall discuss these three authors and show how annihilation, or creation, was originally rooted in one or another hypothesis on the reduction of matter to aether. I shall also try to give some indication of the more abstract justification each author gave to the annihilation and creation processes later in the twenties.

Long before Jeans came to advocate pair annihilation in stars, he had already used the idea in another connection. In a letter to *Nature* in 1904, he offered it as an explanation for radioactivity,[12] basing his suggestion on the supposition of a particulate aether.[13] He represented molecules as composed of J. J. Thomson's "corpuscles" or "ions" (positive and negative electrons), which in turn are composed of aether particles. Molecular agitations are responsible for heat, ionic or corpuscular agitations for spectra. He suggested that agitations of the particles of aether give rise to that disruption of the atom that constitutes radioactivity: "It is possible that the atomic instability . . . must be traced to the agitation of the ultimate constitutents of these ions or corpuscles." His choice of a particulate aether permitted him to apply the concepts of kinetic theory here.

At this point, Jeans invoked a specific hypothesis of aether constitution as a "definite mechanical illustration." It was that offered by Osborne Reynolds in his Rede Lecture at Cambridge in 1902.[14]

[12] "A Suggested Explanation of Radio-activity," *Nature, 70* (1904), 101.

[13] Barbara Giusti Doran characterizes this as a minority view: "The conception that dominated . . . was that of a *continuous* aether" ("Origins and Consolidation of Field Theory in Nineteenth Century Britain," *Hist. Studies Phys. Sci., 5* [1975], 133–260, on p. 161).

[14] *On an Inversion of Ideas as to the Structure of the Universe* (Cambridge, 1903). See also *The Sub-Mechanics of the Universe* (Cambridge, 1903). Doran

Reynolds postulated the fundamental aether particle to be a spherical "grain" and matter to be composed of certain strained and abnormal configurations of these grains.[15] Jeans wrote that "a grain moving with exceptionally high velocity may . . . effect a rearrangement of the adjacent ether structure. . . . It seems probable that the rearrangement would consist of the combination and mutual annihilation of two ether strains of opposite kinds, i.e. in the coalescence of a positive and negative ion, and would therefore result in the disappearance of a certain amount of mass. . . . [T]he process of radioactivity would consist in an increase of material energy at the expense of the destruction of a certain amount of matter." "Material energy" means here the kinetic energy of the emitted alpha particles. The aether model obviously facilitated the framing of the annihilation hypothesis by reducing the conversion of mass into energy into the familiar conversion of potential energy of strain into kinetic energy.

When, after the first world war, Jeans connected pair annihilation with stellar radiation, he was transforming an explanation of the process of the radioactive emission of high speed alpha particles into an explanation of a different subatomic process, leading to the production of gamma rays. The two processes were closely related, as Jeans conceived them in 1904 and again in 1924.[16] In 1904 the two were both rearrangements: pair annihilation was a rearrangement of the primal aether particles, and radioactivity was a kind of depolymerization, in which the secondary particles, the corpuscles forming the atom, rearranged themselves in such a way that a group of them became detached. In 1924 the two both generated energy through the release of binding energy: "An electron at rest must be thought of as having some energy, for it consists of electric charges which have been brought very close to one another in opposition to their mutual repulsions. . . . [If spherical its] electrostatic energy is $3/4mC^2$. . . . There must be further energy $1/4mC^2$ of unknown type. . . . Presumably the whole of this would be set free if the electron could be persuaded to explode and scatter to infinity."[17]

---

delineated Reynolds' place among British aether theorists in footnote 68 of the paper cited in the note above.

[15] *Structure of the Universe,* pp. 6 and 17-23.

[16] Gerald Holton first alerted me to a continuing connection between pair annihilation and radioactivity in twentieth century physics.

[17] "The Ages and Masses of the Stars," *Nature, 114* (1924), 828-829.

By this time, radioactivity was a nuclear process; some of the electrons and protons, bound into the nucleus by forces that were still mysterious, could disperse, setting their binding energy free.

In this same article of 1924, Jeans continued: "Some years ago I suggested that the source of stellar radiation was to be found in an actual destruction of matter in a star's interior, the mechanism probably being that positive and negative electric charges fell together and annihilated one another." The whole question then looked speculative: "It takes on an entirely new complexion in light of the definite results recently obtained by Eddington." Taking it for granted that the lines of the Russell-Hertzsprung diagram traced the life story of stars, Jeans illustrated the new situation by examining the case of Sirius. In evolving from its present luminosity to the much smaller luminosity of our sun, it will lose sixty percent of its total mass, according to the Eddington relation. This loss is enormously greater than the total of what Jeans called its "super-electronic" energy. Jeans concluded that the "main part of its loss of mass will of necessity be 'sub-electronic'"; that is, it will arise from the disappearance of the mass of some of the electrons and protons.[18] His discussion of "sub-electronic" energy reveals that at the time of this article, Jeans no longer based particle annihilation on a conception of the electron as an aether structure. Instead, the basis is the more general principle that the mass of a body represents its energy. In the 1924 edition of his *Report on Radiation and the Quantum Theory,* he wrote that "any attempt to refer back the atomicity of *e* to the structure of a supposed ether simply discloses the fact that the fundamental equations of the ether are not fully known."[19] In Jeans's work we see a clear case of the origin of the idea of pair annihilation within an aether theory of matter and its subsequent abstract underpinning.

Walther Nernst thought electron-proton annihilation a "highly fantastic hypothesis."[20] He also opposed fusion as a source of stellar energy, believing that the temperatures at which it would occur were

[18] "The energy mC^2 . . . may be called 'sub-electronic' energy, any further energy which electrons may have in virtue of their motion or positions in space being 'super-electronic' energy" (*ibid.*).

[19] This was the second edition, published by the Physical Society of London, 1924. Quotation on p. 79.

[20] "Physico-chemical Considerations in Astrophysics," *Jour. Franklin Inst., 206* (1928), 135–142, on p. 136.

too high to be realized in stellar interiors.[21] Instead he espoused his own theory, formulated before fusion and annihilation processes had been attributed to stars, of hyper-uranium elements, as yet unknown, yielding an energy surpassing that of known radioactive elements. The stars contained these elements, which provided their energy source.[22]

To this theory Nernst added another ingredient: the requirement that the universe be in a steady state. The reasons he presented for the requirement were not exclusively scientific; he spoke of it as an "intellectual necessity."[23] He had doubted the heat death of the universe, which his requirement contradicted, from the time he had first heard of it as a student. It was "so unlikely, that any theory that leads to this consequence must be characterized as extremely improbable and therefore incomplete."[24] This conclusion led him to introduce into his theory a process that was, in a broad sense, a converse of particle annihilation. He assumed a *creation* of matter out of the light aether; specifically he assumed a creation of the hyper-uranium radioactive atoms.[25]

In a 1912 version of his theory, Nernst described the aether as the "Ursubstanz," the stuff of which all things are made. He illustrated the processes in it by using the analogy of a gas. In contrast to Jeans, however, Nernst used the kinetic theory model to elucidate the fabrication of aether structures, not their disruption. "We suppose that, indeed, in the course of time, all the elements in the universe become dissolved in an *Ursubstanz*, which we must identify with the so-called light aether, that, however, in this medium, as in a gas as conceived in the kinetic theory, all possible constellations, even of the most improbable kind, can arise, and that in this way from time to time an atom of some element (indeed, most probably of a heavy element) reconstructs itself."[26] Just as aether models facilitated

[21] *Ibid.*, p. 141.

[22] *Ibid.*, pp. 137–138, and *Das Weltgebäude im Lichte der neueren Forschung* (Berlin, 1921), pp. 4–5.

[23] "The assumption appears at the outset to be wholly arbitrary and so it is to a certain degree, but nevertheless it appears that hitherto every prominent investigator has accepted this assumption simply as an intellectual necessity" ("Considerations," p. 136).

[24] *Weltgebäude*, pp. 1 and 12–13.

[25] *Weltgebäude*, pp. 2–5 and 32–34; "Considerations," p. 141.

[26] *Weltgebäude*, p. 4. I use the long quotation Nernst gives here of his earlier theory as my source for it.

Jeans's thinking about annihilation, they also helped Nernst imagine a mechanism for creation. I should point out that Nernst's model led to a significant asymmetry between annihilation and creation: the dissolution of the ordered structure of an atom is thermodynamically much more probable than the converse process of synthesis.

In the 1921 version of his theory, Nernst placed the creation of atoms on an energeticist basis. He represented the aether as matter, even to the extent of being capable of absorbing and dispersing radiation.[27] But, he argued, Einstein's mass-energy equivalence shows that matter itself is energy: "The existence of mass is determined solely by a heaping up of energy . . . the atoms of the various elements [appear to us] solely as energy concentrations."[28] The conversion of matter and aether into each other thus came to represent a transformation of energy from one form to another.

Nernst's hypothesis of the creation of radioactive elements was opposed by Eddington on two grounds. First, the idea of the creation of elements as complex as, or more complex than, uranium struck Eddington as open to "an obvious *a priori* objection. . . . Surely we must rather look on the stars as the crucibles where these unstable compounds are brewed out of simpler matter. . . . Surely it is an anti-evolutionary theory to postulate that [intensely complicated and unstable elements are] the form in which matter first started."[29] Second, Nernst's own a priori objection to the heat death of the universe was not shared by Eddington, who believed in "an inexorable running down of the universe."[30] Eddington therefore had no need for a creation hypothesis, even one which, in keeping with the spirit of evolution, might postulate creation of the simplest material particles.

When he wrote *The Internal Constitution of the Stars,*[31] Eddington had not yet come to accept light quanta. Light and matter were, respectively, wave and particle. Experimental and theoretical results, he explained, "strongly suggest that free radiation has no atomicity of constitution. The modern quantum theory appears to incline to

[27] *Ibid.,* pp. 40 and 31–32.

[28] *Ibid.,* p. 23.

[29] "Sub-Atomic Energy," *Memoirs and Proceedings of the Manchester Lit. and Phil. Soc., 72* (1927–1928), 111.

[30] *Ibid.,* p. 117. See *Internal Constitution, op. cit.* (note 4), p. 316.

[31] It was written between May 1924 and November 1925, with "further amendments in proof" to March 1926 (*Internal Constitution,* p. v).

this view that the quantum is only called into being in the process of interaction of radiation and matter."[32] Like Nernst, he possessed a highly asymmetric picture of the process of annihilation; its converse, the gathering together of electromagnetic waves into a pair of localized particles, strained the imagination. Unlike Nernst, he had no reasons for invoking the converse process. It was therefore easy for Eddington to hold that the "creation of matter . . . may be denied without striking too heavily at our sense of the fitness of things."[33]

Despite his rejection of Nernst's hypothesis, Eddington's own speculations had some resemblance to it. Acknowledging that the mechanism of electron-proton annihilation was completely unknown, he nevertheless thought it useful to "have a scheme in mind." He suggested that "the process consists in evolving certain kinds of nuclei which are self-destroying. The destruction occurs spontaneously some time after the formation of the nucleus."[34] This suggestion is also one in which energy comes from as yet unknown radioactive elements. What is new about these elements is not that they emit alpha, beta, or gamma rays of a peculiarly high energy, but that they exhibit a novel form of radioactive decay.

As a philosopher, Eddington rejected visualizable models. "Our scientific information is summed up in measures," he wrote. "Science has at last revolted against attaching the exact knowledge contained in these measurements to a traditional picture-gallery of conceptions which convey no authentic information of the background and obtrude irrelevancies into the scheme of knowledge."[35] Indeed, his position was even more strict. The "background"—that is, the things in themselves—could not be an object for *any* kind of knowledge. Only the relations among things are knowable, as we deduce them from the relations among the pointer-readings, or measurements, which constitute our information.[36]

All the same Eddington did use aether models in describing pair annihilation. In *The Internal Constitution of the Stars* he pictured

[32]*Ibid.*, p. 57. Jeans held the same view (*Report on Radiation, op. cit.* [note 19], p. 80).

[33]*Internal Constitution*, p. 305.

[34]*Ibid.*, p. 306.

[35]*The Nature of the Physical World* (Cambridge, 1928), p. xi.

[36]John W. Yolton, *The Philosophy of A. S. Eddington* (The Hague, 1960), especially Ch. III, "The Doctrine of Structure." The "background," although unknowable, does exist; thus, according to Yolton, Eddington combined an "operational attitude" with realism.

the proton and electron as coalescing, "their positive and negative charges cancel and nothing is left but the energy which, released from all constraint, spreads out through the aether as an electromagnetic ripple. Or, instead of considering the two charges, we may fix attention on the field of force between them which involves something of the nature of a tube of discontinuity in the aether; this tube may slip back, healing the discontinuity, and at the same time starting a wave of radiant energy."[37] He gave a similar picture in his later Joule lecture. The electron and proton are each "a kind of nucleus of twist in the aether"; the "opposite twists undo one another, releasing the energy like a splash in the aether."[38]

Eddington's model was based on Larmor's work:

> The rudiments of the idea that the mass of ordinary matter is an index of the presence of energy which might conceivably be set free, can be traced back to 1881 when J. J. Thomson showed that the electric field of a charged body possesses inertia or mass. The discovery of the electron and the tendency to regard its mass as residing in its electrical field strengthened the belief in large quantities of field energy bound, but perhaps not permanently, in the constitution of matter. How far this conception had advanced by 1900 may be seen for example in J. Larmor's *Aether and Matter*. . . . There the generation of a positive and negative electron by rotating the walls of a tube with respect to an inner core is described; and the possibility that the walls may ultimately slip back annihilating the electrons and releasing the energy is guardedly touched on. The subject of the intrinsic energy of matter was made clearer and more precise by Einstein who showed the identity of mass and energy; that is to say, mass and energy are the modes in which the same underlying condition manifests itself in different types of experiment. . . . The intrinsic energy of structure of any given mass of matter thus became known. . . .[39]

[37] *Internal Constitution*, p. 293.

[38] "Sub-Atomic Energy," *op. cit.* (note 29), p. 103.

[39] *Internal Constitution*, p. 294. Larmor's conjecture goes back to Faraday and Maxwell in the sense that each of them thought of (equal and opposite) charges as termini of lines of force. Maxwell, in particular, thought of them as termini of strained portions of the aether. However, such "creation and annihilation" had, for Maxwell, nothing to do with the properties of elementary particles. On the contrary, the elementary constituents, or molecules, of the elements are permanent, and their permanence and sameness constitute one of the proofs of the existence of God (J. C. Maxwell, "Molecules," in *Scientific Papers, 2* [New York, 1965], p. 376). In this connection, it is not

Eddington saw the idea of annihilation as arising through the mediation of aether theories of matter. If we also take account of his continuing use of Larmor's model as a "mechanical illustration," I believe we can read this quotation as an approximation to a history of his own thinking about annihilation. We can add him to the list of those whose conceptions were rooted in aether models.

Later Eddington justified the annihilation concept by a formulation of the mass-energy equivalence conforming to his philosophical position: mass and energy are "the modes in which the same underlying condition manifests itself." However, aether models seem to have conditioned his thinking even where he makes no attempt to give an illustration or a history of his own thinking. Phrases like the "potential energy concealed in matter," "the intrinsic energy of structure of any given mass of matter," and the "energy of constitution of . . . the electrons" recur often in his writings.[40] The justification of annihilation in terms of a formal conception of $E = mc^2$ seems, for Eddington, to be complemented by a kind of pictorial, psychological justification in terms of aether models.

- 2 -

Throughout the twenties, the "mutual suicide" of a proton and an electron continued to be regarded as a possible cause of stellar energy. Its leading rival remained the fusion hypothesis.[41] At that time it was thought that fusion probably took place in one step, which would give rise to the release of a single quantum of an energy of 27 million electron volts. The quantum produced by particle annihilation would have an energy of about 940 million electron volts.[42]

---

surprising that Eddington unites speculation about the annihilation of elementary particles with the conviction that the elements are products of evolution.

40The first quotation is from *Space, Time and Gravitation* (Cambridge, 1920), p. 61. The second and third are from *Internal Constitution*, pp. 294 and 292, respectively.

41See Waterfield, *Astronomy, op. cit.* (note 4), pp. 228ff. Waterfield writes: "Up until about 1930 the majority of astronomers were on the whole inclined to favor the theory of annihilation" (pp. 235–236). "Mutual suicide" is Eddington's phrase.

42*Internal Constitution*, pp. 315–316. The numbers are from Ernest Rutherford, "Address of the President . . . at the [Royal Society] Anniversary Meeting, Nov. 30, 1928," *Proc. Roy. Soc., A 122* (1929), 12–13. Present

The astrophysical hypotheses were linked to the concerns of physics proper in several ways. Of interest here are those links that served to bring the concept of particle annihilation to the attention of theoretical physicists. Cosmic ray physics was probably the most important of these and is the one I will discuss here. Cosmic radiation, which was just beginning to be explored in the twenties, was thought to consist of electromagnetic quanta from extraterrestrial sources. The energy of these quanta was estimated to be much higher than the 3 to 4 million electron volt gamma rays which were the most energetic quanta given off in the transmutations of radioactive nuclei. Consequently, cosmic rays were assumed to be produced in an intra-atomic process more energetic than any yet seen in the laboratory.[43] The two astrophysical hypotheses—particle annihilation and hydrogen fusion—were now pressed into service as possible explanations of cosmic ray quanta.[44] Conversely, an accurate determination of the energy of these quanta would bear on the astrophysical controversy: energies close to 27 million electron volts would suggest that hydrogen fusion was going on somewhere in the universe, whereas energies approaching 940 million electron volts would make the Jeans-Eddington theory more probable.

Determinations of the energy of the impinging cosmic quanta depended upon a question of great interest to theoretical physicists, that of the proper theory of Compton scattering. Radiant energy is ordinarily calculated from measurements of the extent to which the rays are absorbed in matter. At such high energies, however, there is little true absorption, and the reduction in intensity that cosmic rays exhibit after traversing air or water is caused by Compton scattering from atomic electrons. Different theories of scattering give different relations between the energy of the impinging quanta and the number scattered. Compton's old-quantum-theory formula had been superceded by work of Dirac and Walter Gordon in 1926. In 1928 Oskar

---

theory favors fusion via two, multistep, catalyzed processes (Bernard Cohen, *The Heart of the Atom* [New York, 1967], Ch. X).

[43] R. A. Millikan, "High Frequency Rays of Cosmic Origin," *Proc. Nat'l. Acad. Sci., 12* (1926), 48–55. Rutherford adhered to the gamma ray theory in late 1928 as "natural," but not proven ("Address of the President," p. 14). The gamma ray energies cited are his.

[44] Other processes were also suggested. For example, Millikan suggested the capture of orbital electrons by nuclei, in the article cited in the previous note. Nernst thought the rays might be the gammas given off in transmutations by his hyper-uranium radioactive elements ("Considerations," *op. cit.* [note 20], pp. 137–138).

Klein and Y. Nishina, working in Copenhagen, published a newer formula based upon Dirac's relativistic theory of electrons, which implied still a different estimate of the energy of the cosmic rays.[45]

All of this suggests that contemporary physicists could not have failed to take notice of the hypothesis of particle annihilation. Rutherford's Presidential Address at the Royal Society in 1929 indicates that it was noticed in conspicuous settings. In this address, Rutherford discussed fusion and annihilation as physical processes, albeit processes that do not take place on the earth. He discussed the bearing of measurements of cosmic ray absorption for the existence of such processes, pointing out "that the absorption coefficient of the most penetrating type of radiation, deduced by [R. A.] Millikan and [G. H.] Cameron from their experiments, is in excellent accord with that to be expected on the Klein-Nishina theory for a quantum of energy of 940 million volts—the energy demanded for the transformation of the internal energy of the proton into radiation."[46]

When Bohr and Dirac exchanged letters at the end of 1929 on Dirac's theory of electrons and its problems, they treated the idea of the annihilation of an electron and proton pair. In a letter of 26 November Dirac outlined the suggestion he was writing up at this time for dealing with the difficulties of the relativistic theory of the electron, remarking that his suggestion incorporated pair annihilation. Bohr commented that Dirac's hypothesis would not give quantitative results fitting the astrophysical data. "As regards the problem of annihilation of electrons and protons which you mention in this connection it appears to me that the astrophysical evidence is of a very conflicting nature. Thus Eddington's theory of the equilibrium of stars seems to indicate that the rate of energy production per unit of mass ascribed to such annihilation is larger in the earlier stages of stellar evolution . . . than in the later stages. . . . As far as I can see any views like yours would claim a variation . . . in the opposite direction."[47]

[45] See Rutherford, "Address of the President," *op. cit.* (note 42), and O. Klein and Y. Nishina, "The Scattering of Light by Free Electrons According to Dirac's New Relativistic Dynamics," *Nature, 122* (1928), 398–399, and *Zeit. f. Phys., 52* (1929), 853–868.

[46] "Address of the President," *op. cit.* (note 42), pp. 15 and 12–17. The energy contribution of the annihilated electron is so small in comparison with that of the proton that it is ignored.

[47] Dirac to Bohr, 26 Nov. 1929; Bohr to Dirac, 24 Nov. 1929 and 5 Dec. 1929 (Bohr Scientific Correspondence, Archive for Quantum Physics, Copen-

Dirac's written paper, "A Theory of Electrons and Protons,"[48] makes it clear that his primary purpose was to deal with the negative energy difficulty, and his secondary purpose—one which took shape in the course of the work—was to present a theory of protons. It also makes it clear that the chief novelty he introduced was the identification of the proton with the absence of an electron, whereas the concept of pair annihilation was not a novelty at all. He began by stating the difficulty: relativistic theories of the electron all yield solutions in which the electron has a negative total energy, and quantum mechanical relativistic theories, in addition, permit the electron to make transitions from states of positive energy to these states of negative energy. He then argued for the basic premise that these states, and the transitions to them, cannot be disregarded as nonphysical, but are to be assumed to exist, and to be interpreted in a way that agrees with experiment.[49]

To arrive at such an interpretation, Dirac invoked the exclusion principle, which says that two electrons cannot occupy the same state. He then postulated that "*all the states of negative energy are occupied except perhaps a few of small velocity*."[50] Leaving a few states empty allowed him to create a theory of protons, for empty states, or "holes," exhibit the properties of positively charged, positive energy particles. This interpretation opens the possibility of postulating "only one fundamental kind of particle, instead of the two, electron and proton, that were previously necessary."[51] Dirac devoted two paragraphs to defending the suggestion that vacant electron states can be regarded as particles. At the end of the second of these, he dedicated a single sentence to the proposition of pair annihilation: "When an electron of positive energy drops into a hole

---

hagen). I am indebted to Professor Aage Bohr for permission to quote from these letters here and in note 83. Bohr, for his part, preferred to make use of the unsettled state of explanations of stellar radiation to work out an idea of his own; namely, that energy is not conserved in systems of nuclear dimensions. See my paper, "The Impact of the Neutron: Bohr and Heisenberg," *Hist. Studies in the Phys. Sci., 3* (1971), 309-323.

[48] "A Theory of Electrons and Protons," *Proc. Roy. Soc., A 126* (1930), 360-365; received 6 Dec. 1929.

[49] *Ibid.*, pp. 360-361.

[50] *Ibid.*, p. 362.

[51] *Ibid.*, p. 363. Other physicists were also trying to reduce the two known, fundamental particles to one in these years. As an example, see the unpublished attempt by Heisenberg to reduce electrons and protons to a single particle, described in "The Impact of the Neutron," *op. cit.* (note 47), pp. 324-326.

and fills it up, we have an electron and proton disappearing together with emission of radiation."[52] That is the only mention of annihilation in the entire paper; it is clearly not in need of justification, and not the focus of the paper. Dirac's commentators saw his theory in the same way, as, for example, Oppenheimer's articles make clear.[53]

For the sake of details that will be useful later, as well as for additional evidence, it is worthwhile to look at Dirac's presentation of these ideas in the first, 1930, edition of his *Principles of Quantum Mechanics,* in a somewhat shortened version. From the Dirac wave equation for an electron moving in an electromagnetic field, a wave equation can be derived governing a particle of the same mass, and of equal but opposite charge, moving in the same field. The first equation may be written like this:

$$\left\{\left(\frac{W}{c}+\frac{e}{c}A_o\right)+\alpha_x\left(p_x+\frac{e}{c}A_x\right)+\cdots+\alpha_m\, mc\right\}\psi_e=0, \qquad (1)$$

where e is the charge of the electron and is a negative number, $(W+eA_o)$ is the electron's kinetic energy, A_o, A_x, ... are the electromagnetic potentials, and ψ_e is the wave function which gives the state of the electron. The dots represent terms for y and z analogous to that for x.[54]

The second equation is:

$$\left\{\left(\frac{W}{c}-\frac{e}{c}A_o\right)+\alpha_x\left(p_x-\frac{e}{c}A_x\right)+\cdots+\alpha_m\, mc\right\}\psi_p=0. \qquad (2)$$

Here $-e$ is a positive number. To each eigenfunction, ψ_e, representing a solution of the first equation for a *negative* value of $(W+e\,A_o)$, there corresponds a ψ_p, which is the complex conjugate of ψ_e, and which is an eigenfunction of equation (2) corresponding to a positive eigenvalue of $(W-eA_o)$. Since equation (2) contains the positive charge, $-e$, whereas equation (1) contains the negative electron charge, ψ_p describes a particle that moves in the external electro-

52 "Electrons and Protons," p. 363.

53 "On the Theory of Electrons and Protons," *Phys. Rev., 35* (1930), 562–563; "Two Notes on the Probability of Radiative Transitions," *ibid.,* pp. 939–947.

54 The other symbols are as follows: p_x ... are the components of the momentum, c is the velocity of light, and m is the mass. The α_x... are four operators introduced by Dirac, and can be represented by four by four matrices. The function ψ_e must then be taken as a four-component vector.

magnetic field, A_o, like a positive particle possessing positive energy. This is the formalism.[55]

The formalism suggests that the negative energy solutions have something to do with the motion of positive particles. The hypothesis had been made previously, as he pointed out in "Electrons and Protons," that an electron in a negative energy state *is* a proton in a positive energy state. There were obvious objections to this, among them that it would violate charge conservation: "A transition of an electron from a state of positive to one of negative energy would be interpreted as a transition of an electron into a proton."[56] To avoid this objection Dirac introduced the idea that a *vacancy* in a negative energy state is a positive energy proton.

Equation (2) governs a particle having the same mass as the electron, whereas the proton is almost two thousand times more massive. Since Dirac recognized this difficulty from the beginning,[57] why did he nonetheless identify the particle as a proton?[58] I have already suggested one reason: the desire to reduce the two fundamental particles of matter to one. A second reason was undoubtedly weightier. The proton was known to exist, and Dirac preferred to identify the constructs in his theory with entities that were known.

[55] *The Principles of Quantum Mechanics* (Oxford, 1930), pp. 255-257.

[56] "Electrons and Protons," *op. cit.* (note 48), pp. 361-362.

[57] *Ibid.*, p. 364.

[58] I should like to point out the differences between my interpretation and that which N. R. Hanson gives in *The Concept of the Positron* (Cambridge, 1963). In the section, "The Dirac Particle," Hanson correctly points out that the negative energy difficulty predates Dirac's theory of the electron, and that Dirac recognized it from the start. Hanson then maintains that Dirac sought to correct it by interpreting electrons with negative energy as protons (pp. 146-147); that the difficulties of asymmetry of the proton-electron mass made this interpretation untenable (p. 147); that the idea of vacancies in the continuum of negative energy states now began to be taken seriously (p. 149); that these began to be regarded as positive electrons (p. 150); that "this shift of ideas percolated only slowly into . . . physical thought in 1931" (p. 150); and that by 1931 Dirac had finally arrived, though reluctantly, at the theoretical prediction of positive electrons, as "holes" in the sea of negative energy electrons, in his "Quantized Singularities" (p. 152). Hanson's reconstruction is based on the assumption that Dirac made use of the identification of negative energy electrons as protons. This is where his error lies. He fails to recognize that Dirac's theory of protons *was* a hole theory, which from its inception took seriously the idea of vacancies, and that it was only after the idea of the hole as proton was made implausible that Dirac suggested the hole as the yet undiscovered positive electron.

It was in this spirit that he appealed to pair annihilation, a process which, if not known, was at least probable. In *Quantum Mechanics*, his reference to it, again brief, is in the penultimate paragraph: "It will . . . be possible for [a positive energy] electron to drop into an unoccupied state of negative energy. In this case we should have an electron and proton disappearing simultaneously, their energy being emitted in the form of radiation. Such processes probably actually occur in nature."[59]

In contrast to its minor role in the *construction* of Dirac's interpretation of the negative energy states, electron-proton annihilation played an important role in its *refutation*. The probability of annihilation was too large. In Oppenheimer's calculations, the longest average lifetime for protons turned out to be of the order of 5×10^{-9} seconds, so that they would scarcely be expected to be observed.[60] Oppenheimer's reaction to this and other difficulties was to reject that part of Dirac's suggestion that constituted a theory of protons, and to assume that *all* the negative energy states for electrons were filled, so that *all* transitions into these states could be ruled out.[61] Dirac accepted this disconfirmation of his theory of protons.[62] The effect of these refutations was that the theory was rendered improbable,[63] and from mid-1930 until the discovery of the positron, the sea of negative energy electrons together with its occasional positive holes was largely abandoned.

Hanson finishes the section by implying that the creation of pairs was a new feature. Dirac advanced "one of the boldest conjectures of all time" (p. 152). "A spectacular departure from classical theory is the process of particle-creation, first described in the early 1930's. That particles could 'materialize' out of radiation is an idea for which twentieth-century physics had no preparation" (p. 56). With this, of course, I also disagree.

59 *Quantum Mechanics, op. cit.* (note 55), p. 257.

60 "On the Theory of Electrons and Protons," p. 563, and "Two Notes," pp. 939–943. *Op. cit.* (note 53). The mechanism Oppenheimer assumed was one in which a free electron dropped into a hole with the emission of two quanta.

61 *Ibid.*, p. 563.

62 "Quantized Singularities in the Electromagnetic Field," *Proc. Roy. Soc., A 133* (1931), 61. Dirac did not follow Oppenheimer in rejecting the idea of holes, however, but merely their interpretation as protons.

63 In a letter of 11 March 1975, Professor Dirac pointed out to me that he was very much influenced by Hermann Weyl's statement that the mass of the "holes" must be that of the electrons, as it appeared, for example, in *Gruppentheorie und Quantenmechanik,* 2nd edition (Leipzig, 1931), p. 234.

- 3 -

In the last section, I traced the way in which the well-known concepts of annihilation and creation were taken from astrophysics into Dirac's electron theory. In this section and the next, I examine the new content given to the concepts when they were handled within the frame of the new ideas of wave and matrix mechanics; I draw on the work of Jordan and Dirac for examples. A central feature of quantum mechanics is the wave-particle duality and its consequence that light and matter are similar entities. Jordan described this feature in 1928 in the following way: "Quantum physical reality is *simpler* in a marked way than the system of ideas through which classical theories sought to represent it. In the classical system of representation, wavelike and corpuscular radiations [Strahlungen] are two fundamentally different things; in reality, however, there is instead only one single type of radiation, and *both* classical representations give only a partially correct picture of it."[64]

The *similarity* of light and matter suggests a different relation between the two than that which I examined in section 1. There I showed that wave-like light and particle-like matter were related in that both could be *reduced* to the same substratum. For Jordan, however, they were related in that both presented different *special cases* of a general entity which is both wave and particle, a "single type of radiation." A consequence of this change in the general outlook on the nature of matter and light was an important change in methodology: the mathematical-physical formalism used to describe matter became routinely applied to light, and conversely. This fluidity in formalism in turn served to deepen the similarity perceived between light and matter, since it continually gave rise to new analogies between them.

The analysis I have just given distorts history in one respect, and a correction of this distortion is pertinent to my argument. It was, of course, only after the matter wave—that is, after 1925—that the similarity of light and matter appeared as a consequence of the wave-particle duality. Historically, the former preceded and gave rise to the latter. The concept of the light quantum or light particle gave rise to a whole spectrum of attempts to assimilate it to the notion of particle in general. One such attempt, Louis de Broglie's application

[64] "Der Charakter der Quantenphysik," *Naturwissenschaften, 41* (1928), 765-772, on p. 771.

of Einstein's formula for the relativistic energy of the matter particle to the light particle, was, as is well known, a starting point for wave mechanics. A very different attempt—which helps to illustrate the spectrum—was G. N. Lewis' "photon," a particle "which is not light," but which is a carrier of radiant energy in the same way that a material particle carries kinetic energy, and which has the same permanence as a material particle.[65]

This older perception of the similarity of light and matter is embodied in two papers by Jordan; the first created a formalism providing a natural setting for pair production, and the second placed the latter within that formalism. For Jordan, the idea of the wave-particle duality was of crucial importance, for it was the similarity of light and matter that helped convince him of light quanta.[66] However, this pair of papers treats only particulate aspects of radiation and matter, because they connect with researches by Wolfgang Pauli and others reaching back to a time, 1923, before the wave aspect of matter became part of physics. In 1923, Pauli published an investigation of the behavior of light and electrons interacting by the mechanism of scattering. He assumed that the free electrons obeyed the classical, Maxwell-Boltzmann statistics. In 1924 and 1925, as is well known, Einstein extended to material atoms the statistics that J. C. Bose had just introduced for light quanta. Following this, Jordan undertook to revise Pauli's work by

[65]"The Conservation of Photons," *Nature, 118* (1926), 874–875. This counts as a pre-quantum mechanical paper because Lewis builds on Compton, but takes neither de Broglie nor matrix and wave mechanics into account. Roger Stuewer describes Lewis' paper, and some of his antecedent and subsequent researches, in "G. N. Lewis on Detailed Balancing, the Symmetry of Time, and the Nature of Light," *Hist. Studies Phys. Sci., 6* (1975), 469–511.

[66]In his doctoral thesis, "Zur Theorie der Quantenstrahlung," *Zeit. f. Phys., 30* (1924), 297–319, submitted in November 1924, Jordan held that the existence of the light quantum was not proven. By the time of the article on "Thermodynamic Equilibrium," cited in note 70, and submitted in July 1925, Jordan had accepted both quanta and matter waves. Several events intervened between November and July, including the Bothe-Geiger experiments, which provided strong reasons for the acceptance of quanta. However, the use Jordan made of the symmetry in the nature of light and matter as a science-creating device in the July paper, together with the symmetry inhering in the picture of light and matter waves interacting in ordinary space, which underlay his program for second quantization, support the interpretation I give here. For Jordan's program, see his "Zur Quantenmechanik der Gasentartung," *Zeit. f. Phys., 44* (1927), 473–480, and his and O. Klein's "Zur Mehrkörperproblem der Quantentheorie," *ibid., 45* (1927), 751–765.

assuming Bose-Einstein statistics for the interacting electrons. This was in the summer of 1925, just before matrix mechanics. In that same summer and in the fall, Otto Stern published papers on the behavior of light and material particles interacting through the mechanism of the interconversion of matter and radiation. Jordan saw that he could extend his revision of Pauli's paper to include this interaction, and in early 1927 he published the second paper showing how this could be done.

Pauli's paper[67] made use of light quanta, but in a noncommittal way. The problem he attacked was this: a rigorously classical treatment of the interaction of free electrons and radiation leads to results contradicting Planck's radiation law and the equipartition law of classical kinetic theory. It was therefore necessary to import a quantum mechanical component into the treatment, which Pauli did by taking Compton scattering as the mechanism by which light and matter interact. He assumed light to possess the quantized energy hf and the quantized momentum hf/c (where f and c are the frequency and velocity of light, respectively, and h is Planck's constant) in accord with Compton's treatment, but at the same time he incorporated Bohr's caveat against drawing any consequence as to the existence of light quanta. The transference of the quantized momentum hf/c in a light-matter interaction cannot be doubted, "even though it is probably not justified . . . to draw definitive conclusions on the nature of radiation from this, by applying the concepts of classical electrodynamics."[68]

Einstein's papers in 1924 and 1925 on gas theory[69] had called attention to de Broglie's matter waves, but had made use of the mathematical formalism of particle physics. Jordan's follow-up article[70] was dominated by his acceptance of the spirit of the wave-particle duality, but here too the mathematical analysis was entirely in terms of particles. The relevant parts of the article are its first two

[67] "Über das thermische Gleichgewicht zwischen Strahlung und freien Elektronen," *Zeit. f. Phys., 18* (1923), 272–286.

[68] "Wenn es auch vielleicht nicht berechtigt ist, von hier aus mittels einer Anwendung der Begriffe der klassischen Elektrodynamik auf das dem einzelnen Elementarprozess entsprechende elektromagnetische Feld entgültige Schlüsse über die Natur der Strahlung zu ziehen . . ." (*ibid.*, p. 274).

[69] "Quantentheorie des einatomiges idealen Gases," *Sitzungsberichte der Preuss. Akad. der Wissen.* (1924), pp. 261–267, and *ibid.* (1925), pp. 3–14.

[70] "Über das thermische Gleichgewicht zwischen Quantenatomen u. Hohlraum-Strahlung," *Zeit. f. Phys., 33* (1925), 649–655.

and final four pages.[71] Jordan opened by considering general particles, "light quanta *or* ideal [gas] atoms,"[72] with the property of obeying Bose-Einstein statistics. He then exhibited two special cases. In the first the rest mass of the particles is zero and the total number of particles is allowed to vary; these particles are to be interpreted as light quanta, and they can be shown to obey Planck's radiation law. In the second the rest mass of the particles is finite and their total number is constant; these are the particles of Einstein's ideal gas, and they obey the Bose-Einstein distribution law. At the end of the paper, Jordan pointed out that light quanta interacting with free electrons by means of the Compton effect did not obey Planck's law. Such quanta were a third, special case of his general Bose-Einstein particles, one in which the rest-mass of the particles is zero, but the total number of quanta remains constant. The procedure he used at the start of the paper made it possible to derive the distribution law for this case also, and Jordan wrote it down, exhibiting its deviation from the Planck formula.

The two notes of Otto Stern, upon which Jordan based his second paper,[73] treated the universe as a gigantic *Hohlraum,* or cavity, containing matter and radiation in equilibrium. Stern's purpose was to see what conclusions follow from the assumption that equilibrium is maintained by the conversion of particles into radiation and back again. He devoted the first paragraph of each of the papers to a defense of this assumption. The first paper starts with the statement that mass and energy "are equivalent and connected through the fundamental equation $U = mc^2$. This equation leads immediately to the question of whether there are processes in which material masses (atoms, electrons) pass over into electromagnetic radiation, and conversely."[74] The way in which this question is posed suggests that Stern viewed Einstein's mass-energy equation as an analog of $W = JQ$, the thermodynamic equation connecting heat, Q, and work, W. The thermodynamic equation expresses the interconvertibility of heat

[71] The most interesting and original part is the middle, where Jordan uses the similarity of light and matter to obtain an expression for the probability of their interaction, which depends on the final as well as the initial state of the particle. See note 66.

[72] "Über das thermische Gleichgewicht," *op. cit.* (note 70), p. 649. My italics.

[73] "Über das Gleichgewicht zwischen Materie und Strahlung," *Zeit. f. Elektrochem., 31* (1925), 448–449, and "Über die Umwandlung von Atomen in Strahlung," *Zeit. f. Phys. Chem., 120* (1926), 60–62.

[74] "Über das Gleichgewicht," p. 448.

and work; likewise, Einstein's equation leads to an expectation that material mass and energy are interconvertible. If Stern did think in terms of this comparison, then, just as heat and work are usually pictured as interconvertible by virtue of both being forms of energy, so he may have pictured both mass and energy as forms of energy, or of some other fundamental entity. There is good reason to suppose that mass can be transformed into radiation, Stern continued: "The huge quantity of energy radiated by a star in the course of its development can scarcely be provided through any other hypothesis. On the other hand, to save the world from the heat death, Nernst once made the hypothesis that atoms of high atomic number arise out of the radiation filling space (*Weltraumstrahlung*)." Hence, science seems to answer the question Stern had posed with a "yes."

In his second paper, Stern began with the assumption of Eddington's theory that particles can annihilate each other to give radiant energy. He treated particle creation as an immediate consequence: "If one makes this hypothesis [of annihilation], one must naturally assume the inverse process, the conversion of radiation into material mass, as also present."[75] The creation of matter was less securely based in experience than annihilation, but the principle that microscopic processes are reversible was too fundamental and too natural to leave the former in much doubt. Stern then went on to analyze the equilibrium state, deriving a formula for the number of electrons, or protons, per unit volume of the cavity-universe as a function of temperature.[76]

Jordan had been concerned with exactly such expressions for numbers of particles in his paper of 1925. He saw that Stern's formula could be assimilated to his own earlier discussion and made it his fourth special case; namely, the case where the rest-mass is finite, while the total number of particles is variable. Considering the average number of particles per unit of phase space and assuming the expression for entropy given by Einstein's theory, Jordan considered the following cases:

A. The rest energy of the particle is zero (light quanta);
B. The rest energy is positive (Einstein gas).

[75] "Über die Umwandlung," p. 60.

[76] "Über das Gleichgewicht," pp. 448–449, and "Über die Umwandlung," p. 62. The results were unsatisfactory. Stern's formula gave a sufficient density of matter only at unreasonably high temperatures.

And in addition:

a. The total number of particles is variable;

b. The total number is constant. . . .

Case Bb is the normal Einstein gas theory; Case Ba, by contrast, leads directly to Stern's formula.[77]

In deriving Stern's formula, Jordan nowhere mentioned the mass-energy equivalence. It was not needed; creation and annihilation of mass particles were here independent of their transformation into radiant energy. Indeed, Jordan introduced creation and annihilation several pages before he took up the processes by which particles transform into light.[78] Thus, from Stern's paper to Jordan's, the basis for creation and destruction of matter was transformed. From the mass-energy equivalence, it became the view that light and matter are special cases of a generalized "particle." This view expressed the idea that light and matter are deeply similar entities.

Dirac's 1927 paper, "The Quantum Theory of the Emission and Absorption of Radiation," provides an example of an argument from the similarity of light and matter that is an inverse to Jordan's.[79] The state of a free mass particle can be specified by giving its momentum and rest mass. One particular state is that of zero momentum. Dirac applied this to light quanta; the state of zero momentum is then also a state of zero energy.[80] In this zero state "the light-quantum has the peculiarity that it apparently ceases to

[77]"Über die thermodynamische Gleichgewichtskonzentration der kosmischen Materie," *Zeit. f. Phys., 41* (1927), 711–717. The number Jordan considered was actually the average number of particles per unit cell in phase space (p. 711). He made a similar analysis for a gas of particles satisfying Fermi statistics (p. 712).

[78]The general treatment of variable numbers of mass particles is on pp. 711–712, and a specific derivation of distribution laws for mixtures of negative and positive electrons, under the condition that the total charge remains constant, is on pp. 712–713. Later in the paper, Jordan went on to consider the situation in which the number of mass particles varies when the particles are transformable into an electron and a quantum. This mechanism was suggested by G. E. M. Jauncey and A. L. Hughes in a paper on cosmic rays, "Radiation and the Disintegration and Aggregation of Atoms," *Proc. Nat. Acad. Amer., 12* (1926), 169–173.

[79]"The Quantum Theory of the Emission and Absorption of Radiation," *Proc. Roy. Soc., A 114* (1927), 243–265; reprinted in *Selected Papers on Quantum Electrodynamics,* pp. 1–23.

[80]Since $E = c\sqrt{p^2 + m_o^2 c^2}$ and $p = m_o = 0$. The spin of the particle, or the polarization of the photon, must be taken into account for a full specification.

exist. . . ." "When a light-quantum is absorbed," Dirac continued, "it can be considered to jump into this zero state, and when one is emitted it can be considered to jump from the zero state to one in which it is physically in evidence, so that it appears to have been created . . . there are an infinite number of light-quanta in the zero state."[81]

Jordan started from equations for an Einstein-Bose ideal gas. He exhibited the attributes that are possible for the gas particles and the way in which differing combinations of attributes give differing individual cases. Finally, he combined attributes in a novel way by combining finite mass with variable total number of particles. This combination corresponded to the relatively unfamiliar gas of mass particles whose members could be brought into existence or annihilated. Dirac started from the equation of relativistic mechanics, $E^2 = m^2c^4 + p^2c^2$, for a generalized particle. He also implicitly dealt with the attributes such particles could possess; they might have finite rest mass (electrons or protons) or zero rest mass (quanta), and they might have any momentum including zero (electrons or protons) or the frequency-dependent momentum $p = E/c = hf/c$ (quanta). Like Jordan he introduced, implicitly, a novel combination: zero rest mass together with zero momentum. This combination corresponded to the new conception of quanta with permanent existence. The result of both Jordan's and Dirac's work was an extension of the similarity between light and matter through an increase in the number of analogous processes they undergo. Jordan's view of mass particles as capable of being created or absorbed was analogous to the established view of quanta as capable of being emitted or absorbed. Dirac's view of the emission and absorption of quanta as a change of state was analogous to an established view of processes that mass particles undergo.

Dirac's hole theory may have been, in part, yet another instance of the transfer to matter of concepts that had been created originally for light. The infinite sea of unobservable negative energy electrons of 1929 is analogous to the infinite set of unobservable zero energy photons of 1927.[82] The transition of an electron from an observable

[81] "Quantum Theory . . . of Radiation," pp. 260–261.

[82] For the electron sea, however, he needed the additional assumption that any physical condition that was perfectly uniform throughout space and time would be unobservable ("Theory of Electrons and Protons," p. 362). The ad hoc assumption rendered his theory of electrons more vulnerable. Thus, Bohr

state to a vacant, unobservable state parallels the transition of a quantum from an observable state to the unobservable state of zero momentum (absorption). To the extent that echoes of his earlier paper were in Dirac's mind, it was natural that he should see the electron that had jumped into a negative energy hole as annihilated.[83]

The assimilation of the creation and destruction of particle pairs to the change of state of an electron[84] finally placed creation on an equal footing with annihilation. For every electron that falls into oblivion another must become observable together with its positive hole. The same kind of parity followed from Jordan's approach, since he modelled the creation and annihilation processes for matter on those for quanta, and for quanta emission and absorption have equal status. Dirac's and Jordan's work contrasts with Nernst's kinetic theory conception, which made the creation of matter much less probable, and with Eddington's, which did not admit creation at all.

- 4 -

In 1931, Dirac discussed the methodology that he thought his hole theory embodied:

> There are at present fundamental problems in theoretical physics . . . the solution of which problems will presumably require a more

commented: "Your idea is indeed very fascinating, but I must confess that . . . before all we [Bohr and Klein] do not understand, how you avoid the effect of the infinite electron density in space" (Bohr to Dirac, 5 Dec. 1929 [Bohr Scientific Correspondence]).

[83]These connections are not mentioned by Dirac, but are my interpretation of his work. In justifying them, I would point out it is characteristic of Dirac that he is not alert to his sources. I would hypothesize that he works from an exceptionally large store of ideas of various origins, and that he incessantly transforms and forges new connections among them. His papers show ideas reappearing in subtly altered versions, and in a multiplicity of different relations. If this is one source of Dirac's creativity, a consciousness of the sources of his concepts might have been somehow rejected as hindering the fluent progress from one idea to another.

[84]This assimilation also appears in Jordan's "Gleichgewichtskonzentration der kosmischen Materie," *op. cit.* (note 77) in a different way. The general transformation of a particle of type k into one of type l is there particularized in various ways. One way is the transformation of electrons and protons into quanta, and another is the transformation of a particle in state k into the same particle in state l.

> drastic revision of our fundamental concepts than any that have gone before. Quite likely . . . it will be beyond the power of human intelligence to get the necessary new ideas by direct attempts to formulate the experimental data in mathematical terms. . . . The most powerful method of advance that can be suggested at present is to employ all the resources of pure mathematics in attempts to perfect and generalise the mathematical formalism that forms the existing basis of theoretical physics, and *after* each success in this direction, to try to interpret the new mathematical features in terms of physical entities (by a process like Eddington's Principle of Identification). ["The Theory of Electrons and Protons"] by the author may possibly be regarded as a small step according to this general scheme of advance.[85]

The new feature in the theory of electrons was the existence of electrons in negative kinetic energy states. The addition of the exclusion principle converted this feature into the existence of an infinite "sea" of negative energy electrons with an occasional unfilled state, or "hole" in the sea. To interpret the new feature of the theory in terms of experimentally known physical entities, Dirac invoked an analogy. The problem of vacant negative energy states, he wrote in "Electrons and Protons," "is analogous to that of the X-ray levels in an atom with many electrons. According to the usual theory of the X-ray levels, the hole that is formed when one of the inner electrons of the atom is removed is describable as an orbit and is pictured as the orbit of the missing electron before it was removed. This description can be justified by quantum mechanics, provided the orbit is regarded, not in Bohr's sense, but as something representable, apart from spin, by a three-dimensional wave function."[86] In his Oppenheimer Memorial Prize Lecture, Dirac compared the wave functions representing holes in his relativistic electron theory to those describing electrons missing from the outermost, valence energy levels in many electron atoms.[87] Using this analogy, he now

[85] "Quantised Singularities," *op. cit.* (note 62), pp. 60–61. See also Dirac's Bakerian Lecture. "The Physical Interpretation of Quantum Mechanics," *Proc. Roy. Soc., A 180* (1942), pp. 1–40, especially p. 5. I am indebted to Professor Mario Bunge for the last reference, and for suggestive comments on Dirac and Eddington.

[86] "Electrons and Protons," *op. cit.* (note 48), p. 363.

[87] *The Development of Quantum Theory* (New York, London, and Paris, 1971), p. 50. I am grateful to Barbara Guisti Doran for pointing out this reference.

employed "Eddington's Principle" and identified the hole with the presence of a proton. In the same way, he identified the transition of an electron into (or out of) a hole with pair annihilation (or creation).

In his interpretation of his theory, Dirac invoked no model of an elementary particle.[88] It is only the presence or absence of particles and their representation that are at issue. Certainly, there is nothing like the abnormal piling of the granular aether of Reynolds, the strains of the continuous aether of Larmor, or the configurations and motions of the ideal gas aethers of Jeans and Nernst. Dirac took a strong stand against such visualizations. In the preface to the 1930 edition of *Quantum Mechanics*, he wrote that nature's "fundamental laws do not govern the world as it appears in our mental picture in any very direct way, but instead they control a substratum of which we cannot form a mental picture without introducing irrelevancies."[89] Although these words are reminiscent of Eddington's,[90] there was an important difference between Dirac and his elder compatriot. Dirac not only repudiated visualizable models, as Eddington did, but he found it possible to think about creation and annihilation without them, as Eddington did not.

We have seen that both Jordan and Dirac handled creation and annihilation processes without aether models of matter or light, and therefore without the justification for these processes that inhere in these models. Indeed, they handled the subject without any model for matter or light at all, but they did use conceptual models for the *mechanism* of creation and annihilation processes. Their mechanisms differed entirely from those associated with aether models, as well as from those associated with the conception of light and matter as

[88] I use the characterization of model in Dorothy M. Emmet, *The Nature of Metaphysical Thinking* (London, 1945), p. 87. "The 'models' of physics are . . . hypotheses symbolizing a possible way in which things may be connected [and] are drawn from types of relation which seem intelligible to us in some more familiar setting."

[89] Preface, *Quantum Mechanics, op. cit.* (note 55).

[90] See p. 172 above. Eddington used the word "substratum" with the same meaning Dirac gave it in *Space, Time and Gravitation, op. cit.* (note 40), p. 187. In pointing out these similarities, I do not evaluate the extent to which Eddington was an influence on Dirac. However, it would be worthwhile to have a systematic comparison of Dirac's ideas with the philosophy embedded in Eddington's *Mathematical Theory of Relativity* (Cambridge, 1923), and it would also be good to have a group study of the philosophies of the scientists and mathematicians who were working in Cambridge during the twenties.

manifestations of energy or of some other common substance. Jordan's model is an analogy with the emission and absorption of light. Dirac's is the quantum mechanical transition of an electron between states of different energies. Their new ways of conceiving of the processes of creation and annihilation led on to novel results. Jordan, for example, arrived at new formulas for the density of particles; Dirac arrived at a theory of protons. Thus, the early history of pair creation and annihilation is also part of the history of heuristics in physics over the first thirty years of this century.

ACKNOWLEDGMENTS

I am grateful to Erwin N. Hiebert and Gerald Holton for making the Harvard libraries accessible to me, and to Joseph Agassi, Barbara Giusti Doran, John L. Heilbron, and Russell McCormmach for valuable suggestions and criticisms.

Chemistry as a Branch of Physics: Laplace's Collaboration with Lavoisier

BY HENRY GUERLAC*

Few collaborations in the history of science can equal, for the eminence of the two participants, that between the coryphaeus of the Chemical Revolution, Antoine-Laurent Lavoisier, and his colleague in the Royal Academy of Sciences, the mathematician and physicist, Pierre-Simon de Laplace. We remember this collaboration chiefly, if not solely, for the famous investigation on calorimetry which resulted in one of the acknowledged classics of science,[1] the joint "Mémoire sur la chaleur," published in the *Histoire et Mémoires* of the Academy of Sciences in 1784. Yet no one has tried to determine when the partnership began, what brought it about, how long it lasted, or what effect it may have had on the scientific philosophy of both men. To find answers, where possible, to some of these questions is the purpose of this study. Perhaps, too, this inquiry may help us understand why it was the physicists and mathematicians of the Academy, rather than the chemists, who were among the first to enlist under the banner of the new Antiphlogistic Chemistry.[2]

*The Society for the Humanities, Andrew D. White House, 27 East Avenue, Cornell University, Ithaca, New York 14853.

An abstract of part of this paper was read at the XIIe Congrès International d'Histoire des Sciences, Paris, 1968.

[1] Often reprinted, it may be conveniently consulted in *Oeuvres de Lavoisier, publiées par les soins du Ministre de l'Instruction Publique, 2* (1862), 283–333; in the *Oeuvres complètes de Laplace, 10* (1894), 147–200; in Charles Richet's collection "Les maîtres de la science," Solovine's series "Les maîtres de la pensée scientifique"; and in German translation, notably in *Ostwald's Klassiker der exakten Wissenschaften, No. 25.* It has been translated into Spanish, but there is as yet no English translation. For further bibliographical details see Denis I. Duveen and Herbert S. Klickstein, *A Bibliography of the Works of Antoine Laurent Lavoisier, 1743–1794* (London, 1954), pp. 54–56.

[2] "Je ne m'attends pas que mes idées soient adoptées tout d'un coup. . . . En attendant, je vois avec une grande satisfaction que les jeunes gens qui commencent à étudier la science sans préjugé, que les géomètres et les physiciens qui ont la tête neuve sur les vérités chimiques, ne croient plus au phlogistique dans le sens que Stahl l'a présenté, et regardent toute cette doctrine comme un échafaudage plus embarrassant qu'utile pour continuer l'édifice de la science chimique." (*Oeuvres de Lavoisier, 2,* 655.)

See also the remark of Chaptal: "La chymie moderne doit une partie de ses

LAVOISIER AND THE PHYSICISTS

That Lavoisier, by native disposition and early training, should have found satisfaction in associating with the physicists and mathematicians among his colleagues is not surprising. His early investigations—notably the experiments on street lighting and on hydrometry—have more the character of physical inquiries than of chemistry as it was then understood.[3] Indeed Lavoisier seems at times to have thought of himself as a *physicien*—the term *physique* was broadly and loosely used in the eighteenth century—rather than as a chemist.[4] As early as 1766, when he was only aspiring to membership in the Academy of Sciences, he attempted to persuade its members to create a new division or class of "physique expérimentale," clearly with the idea of making room for himself in this august and rigidly constituted body. When the Academy was founded, this important subject, he wrote, "sorti de l'obscurité des laboratoires des anciens chimistes, et maniée par les savantes mains des Huyghens, des Mariotte, des Perrault" began to take on a new character, but it

---

succès à une classe d'hommes chez qui l'habitude d'une étude profonde des sciences exactes a fait une nécessité de n'admettre que ce qui est démontré, et de ne s'attacher qu'à ce qui est susceptible de l'être. Lagrange, Condorcet, Vandermonde, Monges [*sic*], Laplace, Meusnier, Cousin, les plus célèbres mathématiciens de l'Europe, se sont intéressés tous aux progrès de cette science, et l'ont enrichie de leurs découvertes." (*Elémens de chymie,* 3rd ed. [Paris, AnV–1796], *1,* 1.)

[3] Even his earliest scientific observation, of an aurora borealis observed at Villers-Cotterets when he was only twenty, shows a knowledge of stellar astronomy and a quantitative habit of thought that was doubtless the fruit of his studies with the Abbé de Lacaille.

[4] Lavoisier not infrequently referred to the subject of his interest as "physique," and to himself as a "physicien." Lavoisier's first book bore the title: *Opuscules physiques et chimiques.* In a review of this work in *Hist. Acad. Roy. Sci.,* Année 1774 (1778), pp. 77 f., Lavoisier is described as bringing chemistry and physics closer together. See also *Oeuvres de Lavoisier, 2,* 95–96. In describing to Benjamin Franklin the new scheme he has adopted for his *Traité élémentaire de chimie,* Lavoisier says that "la chimie s'est trouvée beaucoup plus rapprochée qu'elle ne l'était de la physique expérimentale." See René Fric, "Une lettre inédite de Lavoisier à B. Franklin," *Bulletin Historique et Scientifique de L'Auvergne,* 2ème série, No. 9 (Septembre 1924), pp. 145–152, esp. p. 149. Lavoisier's last book, published posthumously by his widow, bore on the half-title, which served as the title page, the caption *Mémoires de Chimie.* But as William Smeaton has shown, the heading of the first page of both volumes is given as *Mémoires de physique et de chimie,* which Lavoisier evidently considered the real title of the work. See his review of Duveen and

had been inexplicably neglected during the Academy's reorganization of 1699.[5]

After Lavoisier's admission to the Academy in 1768, the first colleague, and the first physicist, with whom he became actively associated was Mathurin-Jacques Brisson.[6] In 1768 and 1769 the two men reported on a number of proposals and inventions submitted for the Academy's approval.[7] These involved, for the most part, questions of applied physics and mechanics.[8] In the years 1772–

Klickstein's *Bibliography of the Works of Antoine Laurent Lavoisier* (London, 1954) in *The Library Transactions of the Bibliographical Society*, 5th series, *2* (London, 1956), 133. Duveen and Klickstein (*op. cit.*, p. 201) do remark that the first volume of the *Mémoires* "is in reality a monograph on physics."

[5] *Oeuvres de Lavoisier—Correspondance recueillie et annotée par René Fric, 1* (1955), 7–12. Fric published unsigned drafts, attributed to Lavoisier, of letters addressed to Mignot de Montigny (President of the Academy in 1766) and to Grandjean de Fouchy (the Perpetual Secretary), urging the creation of a "classe de physique expérimentale." Because Lavoisier was not yet a member of the Academy, these drafts have puzzled scholars, but they were evidently supposed to be signed by one or more of Lavoisier's supporters and transmitted to the two officials. Nothing came of this effort (the letters were probably never sent) and Lavoisier was not admitted to the Academy until 1768. But it is significant that in the reorganization of 1785, which was largely the work of Lavoisier, a "classe de physique générale" was in fact established. For the Academy of Sciences and Lavoisier's role in the reorganization, see Roger Hahn, *The Anatomy of a Scientific Institution—The Paris Academy of Sciences, 1666–1803* (Berkeley, Los Angeles, London, 1971), pp. 98–101.

[6] Brisson, born in Fontenay-le-Comte on 30 April 1723, had been an assistant of Réaumur and like the latter was both naturalist and physicist. In 1760 he had published his six-volume *Ornithologie.* He was a disciple of the Abbé Nollet, whom he succeeded as teacher of physics at the Collège de Navarre. In 1771 he brought out, with anti-Franklinist notes, a French translation of Priestley's *History of Electricity.* In 1781 he published his *Dictionnaire raisonné de physique* in two volumes. Disillusioned by early attempts to study chemistry, he was drawn to it after his association with Lavoisier, when he saw it becoming a "new science." He shared with Lavoisier an interest in accurate hydrometry, and his most successful work was his *Pesanteur spécifique des corps* (Paris, 1787).

[7] In 1768 Brisson and Lavoisier reported on a table of specific gravities of solid substances, submitted by a M. Thévenard, a port official of Lorient. Later that summer the two men evaluated a new kind of hydrometer submitted by a certain Cartier. In December they reported on a new design for street lamps, and in July 1769 on a proposal for a nonsmoking fireplace.

[8] To the same period belongs the unfavorable report by Lavoisier and Macquer on the theory of color of a M. de La Folie. It discloses, as did his work on street lighting, Lavoisier's grasp of optical problems and his familiarity with Newton's work on light and color. (*Oeuvres de Lavoisier, 4,* 21–27.)

1774, Lavoisier joined Brisson, Louis-Claude Cadet de Gassicourt, P.-J. Macquer, and other academicians in subjecting a wide variety of mineral substances to the intense heat of great burning glasses,[9] an investigation as much physical as, properly speaking, chemical.

The first months of 1776 were the occasion for a quite different sort of collaboration of Lavoisier with physical scientists of the Academy. January was the coldest month France had experienced since 1709. Three academicians—the mathematicians Etienne Bezout and Charles-Auguste Vandermonde, together with Lavoisier—were charged with collecting temperature readings and comparing these figures with temperatures reported for the year 1709. To make this comparison possible, Brisson put at their disposal a thermometer made by Réaumur about 1730 on which fortuitously the low reading for 1709 had been marked. The problem of checking this thermometer, and comparing it with those used by the several observers (including a mercury one built by Cappy and used by Laplace), must have been Lavoisier's introduction to the problems of accurate thermometry; through the work of Réaumur and from the recent book of the Genevan Jean-André Deluc (1727–1817), a *membre correspondent* of the Academy, he could learn the current state of the art.[10]

Early in 1777 Lavoisier undertook with two physicists, Vandermonde and Gaspard Monge, to investigate the problem of heat conduction. On 4 February they carried on an experiment, reminiscent of an earlier one of Isaac Newton, to determine the relative rates of cooling of hot bodies suspended in air and *in vacuo.*[11] Benjamin Franklin, who was greatly interested in the subject of heat and who had only recently arrived in Paris as representative of the rebelling American colonies, may have been present at these experiments.[12]

The first trace of Laplace's association with Lavoisier—apart from

[9] See the report on these experiments in *Oeuvres de Lavoisier, 3,* 274–283.

[10] "Expériences faites par ordre de l'Académie, sur le froid de l'Année 1776," *Oeuvres de Lavoisier, 3,* 355–377; and "Second mémoire sur le froid de 1776," *ibid.*, pp. 378-386. See also *Procès-verbaux,* 1777, f. 181v.

[11] René Taton, *L'oeuvre scientifique de Monge* (Paris, 1951), p. 321. Taton derived this information from papers in the archives of the Baron de Chaubry.

[12] Joseph Black (*Lectures on the Elements of Chemistry,* ed. John Robison, 2 vols. [Edinburgh, 1803], *1,* 24) wrote that such experiments were performed by "Dr. Franklin and some of his friends in Paris." For Newton's (or Desaguliers') experiment, see H. Guerlac, "Newton's Optical Aether," *Notes and Records of the Royal Society of London, 22* (1967), 45–57.

his minor contribution to the Academy's investigation of the cold of the year 1776—was his presence, on 27 February 1777, when Lavoisier formally repeated, before a crowd of observers in his new laboratory at the Arsenal, certain of Joseph Priestley's experiments on gases.[13] The serious collaboration of Laplace with Lavoisier began not long after, in the early months of 1777, with experiments carried on in concert on the vaporization of fluids.

THE BEGINNING OF THE COLLABORATION

Laplace was six years younger than Lavoisier and his junior in Academic rank. Four years earlier, when Laplace entered the Academy as "adjoint mécanicien," Lavoisier had already reached the higher grade of "associé chimiste." In 1777, despite the brilliance and daring of Laplace's early papers—papers that dealt with the two problems that were to occupy him throughout his career: the mathematical theory of probability and the gravitational physics of the solar system—he was far from enjoying the eminence he was later to attain, and he stood well below Lavoisier in scientific prestige. The partnership, as he later acknowledged, was distinctly to his professional advantage.

Lavoisier had already made his historic discovery of the chemical rôle of atmospheric air in combustion and calcination, and his find-

[13]*Oeuvres de Lavoisier, 2,* 785 (reprinted from Lavoisier's posthumous *Mémoires de Chimie*). Lavoisier's note gives the date as 1776, but there are good reasons for putting it a year later. Lavoisier could hardly have forgotten Franklin's presence, when he recalled the event years later, but Franklin did not arrive in Paris until December 1776. Moreover the experiments were doubtless performed in Lavoisier's new laboratory at the Arsenal, which he did not occupy until the late spring of 1776. There is, moreover, the suggestion, in a note of Lavoisier to Franklin dated 8 June 1777, that the early experiments had not been too successful. Lavoisier invited Franklin to come to the Arsenal laboratory on 12 June to observe a repetition of the chief experiments of Priestley on different kinds of air: "Je souhaite bien sincèrement que vous puissiez accepter cette proposition; nous n'aurons que M. le Veillard, M. Brisson et M. Bezout. Le trop grand nombre de personnes n'étant point en général favorable au succès des expériences." (*Oeuvres de Lavoisier—Correspondance, 3* [1964], 601 [no. 344].)

Laplace, it should also be remembered, had been present, together with Lavoisier, Louis Clouet, and other observers, in November 1776 when tests were made at the Arsenal of the Chevalier d'Arcy's method of evaluating the strength of gunpowder. (Marcelin Berthelot, *La Révolution Chimique—Lavoisier* [Paris, 1890], p. 297.)

ings had been published in his first book, the *Opuscules physiques et chimiques* of 1774. In July of 1775 he entered upon his duties as one of the *régisseurs,* in effect the scientific director, of the gunpowder administration (*Régie des poudres et salpètres*) established by Turgot, the economist and *philosophe* who in 1774 had become Controller General of Finance; in the spring of 1776 he took up his residence at the Arsenal where a portion of his apartment was made into a spacious laboratory. Only a few months before (on 13 February 1776) he had confirmed Priestley's discovery of a new kind of air—Priestley's "dephlogisticated air" or oxygen gas—and recognized it as the effective aerial agent in combustion and calcination. By this time Lavoisier had formulated in his mind his new theory of combustion, but he had not come out openly against the generally accepted phlogistic hypothesis, chiefly, it would appear, because he believed that one important feature of his theory needed experimental verification. It was to this end, it would seem, that he sought the assistance of Laplace.

Precisely when, in the late winter or early spring of 1777, the two men carried out their first joint experiments we do not know, but they must surely have been performed in Lavoisier's new laboratory, which he was busy equipping with the finest apparatus obtainable. Our earliest reference to these experiments is from the *Procès-verbaux* or minutes of meetings of the Academy where we learn that on 9 April 1777, at its Easter public session of that year (the *rentrée publique*), Laplace read a memoir on "la nature du fluide qui reste dans le récipient de la machine pneumatique," a description anything but informative.[14] No mention is made of the fact that this was a report on experiments performed jointly with Lavoisier.

But this was clearly the case; and from a number of allusions to these experiments in Lavoisier's later papers, and from a recently discovered unpublished letter of Lavoisier, we can reconstruct, at least in a general way, what they were about. The experiments were

[14]*Procès-verbaux,* 1777, f. 256. On this same occasion Lavoisier first reported his discoveries on animal respiration. The reading was finished on May 3 (*ibid.,* f. 328 v); it was published in *Mém. Acad. R. Sci.,* Année 1777 (1780), pp. 185-194, and may be consulted in *Oeuvres de Lavoisier, 2,* 174-183. The substance of this paper is also given in his "Mémoire sur la combustion en général," read to the Academy on 12 November 1777 and published in *Mém. Acad. R. Sci.,* Année 1777 (1780), pp. 592-600. See *Oeuvres de Lavoisier, 2,* 225-233.

designed, it appears, to determine the conditions under which various fluids can be vaporized. Water, ether, and alcohol were subjected to varying pressures in the receiver of an air pump to determine the temperature and the air pressure at which each substance vaporized. As Lavoisier wrote not long after: "Nous avons observé, M. Laplace et moi, dans un mémoire lu à la rentrée dernière de l'Académie, que le passage d'un fluide à l'état d'expansibilité tenait à deux causes: 1° au degré de chaleur communiqué à ce fluide; 2° au poids de l'atmosphère dont sa surface est chargé."[15] For each fluid, Lavoisier continues, there is a particular dependence of the "moment of vaporization" upon the temperature and weight of the atmosphere.[16]

There can be little doubt that it was Lavoisier who proposed the experiments and sought the assistance of Laplace. In the "Avertissement" to his *Opuscules,* written in 1773, Lavoisier outlined several subjects he had already investigated or proposed to examine further, and which he expected to include in a projected second volume of the work.[17] Two of these relate to heat: the ebullition of fluids in a vacuum and the related phenomenon of evaporative cooling. Both foreshadow the experiments with Laplace, and the second refers to a phenomenon, known to him through the published literature, that strongly supported his ideas about vaporization.

[15]"Expériences et observations sur les fluides élastiques en général et sur l'air de l'atmosphère en particulier," *Oeuvres de Lavoisier, 5,* 272. According to Edouard Grimaux, editor of this volume, Lavoisier read this memoir, in slightly different form and under a different title, at a meeting on Saturday, 10 May 1777, in the presence of Emperor Joseph II, who was present incognito as the Comte de Falkenstein. See *Procès-verbaux,* 1777, f. 342, where it is merely recorded concerning Lavoisier's paper that he read "un mémoire sur le gas." A fuller account is given in Bachaumont, *Mémoires secrets, 10* (1784), 130, under the date 12 May 1777.

[16]A more detailed account, setting forth the same principle, is given by Lavoisier in his first published version of his theory of vaporization: "De la combinaison de la matière du feu avec les fluides évaporables," *Mém. Acad. R. Sci.,* Année 1777 (1780), pp. 420-432. Still another reference to these experiments occurs in a later memoir: "Sur quelques fluides que l'on peut obtenir, dans l'état aériforme, à un degré de chaleur peu superieur à la temperature moyenne de la terre," *Mém. Acad. R. Sci.,* Année 1780 (1784), pp. 334-343. See also *Oeuvres de Lavoisier, 2,* 261-270.

[17]*Oeuvres de Lavoisier, 5,* 267-270. No second volume was ever written, although Lavoisier revived the project years later.

LAVOISIER'S COVERING THEORY

At this point we may well ask what impelled Lavoisier to explore with Laplace the behavior of fluids during vaporization, and what significance these experiments had for his general notions of chemistry and his theory of combustion. Some manuscript notes reveal that as early as 1766 he had begun to speculate on heat and the nature of vapors.[18] By the summer of 1772 he had worked out in its general lines an interesting theory of the classical Four Elements. Three of these, he believed, could exist in either of two states, fixed or free. For the doctrine of "fixed air" he was chiefly indebted to Stephen Hales; water too, as he had learned from his teacher, Guillaume-François Rouelle, could exist not only free, but combined with salts as "water of crystallization," a term of Rouelle's coinage. Fire, or the "matter of fire" as Lavoisier called it—a fine particulate, elastic fluid able to pervade the pores of all material bodies—could also exist either free or in the state of *feu fixé*. When combined with bodies, air and fire lose their characteristic properties of "elasticity" and "expansibility," and fire no longer affects the thermometer.

What interested Lavoisier especially was the presumed ability of these elements to combine together; most striking was the apparent combination of the "matter of fire" with water and with air. Fire is the chief agent of *physical* change: combined with water, or with the "base of air," it imparts to them new elastic or "aeriform" properties. Thus at least by 1772 he was convinced that aeriform fluids were nothing but combinations of a particular material fluid with the "matter of fire." Indeed the physical state of a body—whether solid, liquid, or "aeriform"—depends only on the amount of the "matter of fire" with which it is combined. In principle, any body can assume successively all three of these states, a notion already suggested by Turgot.

[18] These speculations were set down by Lavoisier in short notes inspired by reading a memoir on the elements published by the German chemist and physiologist Johann Theodor Eller in the *Mémoires* of the Berlin Academy. Lavoisier's notes were first studied, and their source identified, by my student, J. B. Gough, "Lavoisier's Early Career in Science—An Examination of some New Evidence," *The British Journal for the History of Science, 9* (1968), 52-57. See also Gough's Cornell doctoral dissertation *Foundations of Modern Chemistry: The Origin and Development of the Concept of the Gaseous State* (1971), and Robert Siegfried, "Lavoisier's View of the Gaseous State and Its Early Application to Pneumatic Chemistry," *Isis, 63* (1972), 59-78.

The essentials of what I shall call his "covering theory" were set down before August 1772 in a draft document that he later referred to as his "System on the Elements."[19] This truncated draft deals largely with air, with its "fixability" and the cause of its elasticity when in the free state; fire, however, is treated only insofar as it accounts for the production of airs and water vapor. Presumably—or so I believe—he intended to complete the paper with a discussion of the two states of water, and to treat more fully the dual role of fire. Instead, in two draft memoirs written in the spring of 1773, rather than enlarge upon his "System on the Elements," he dealt at some length with his theory of vaporization, the causes for the state of "expansibility," and his notion of the three states or modes of matter.[20] In the previous autumn he had discovered, in the experiments recorded in his famous *pli cacheté*, the role of air in combustion and calcination. This deflected him from his original purpose and turned his energy and enthusiasm into new paths. He came to perceive that his "théorie singulière" (that air is not a simple substance, but a "fluide particulier combiné avec la matière du feu") could clarify his understanding of the processes of combustion and calcination.[21]

[19]See my *Lavoisier—The Crucial Year* (Ithaca, 1961), Chapter 3 and Appendix IV.

[20]René Fric, "Contribution à l'étude de l'évolution des idées de Lavoisier sur la nature de l'air et sur la calcination des métaux," *Archives internationales d'histoire des sciences, 12* (1959), 137-168. J. B. Gough has produced evidence that both the words *vaporisation* and *expansiblité* were first used by the *philosophe* and economist Turgot in Tome 6 of Diderot's *Encyclopédie.* Lavoisier's notion that vaporization is the act of solution in fire or the matter of heat is likewise derived from Turgot. See Gough's "Nouvelle contribution à l'étude de l'évolution des idées de Lavoisier sur la nature de l'air et sur la calcination des métaux," *Archives internationales d'histoire des sciences, 22* (1969), 267-275. For a fuller account see his doctoral dissertation, *op. cit.* (note 18), pp. 79-93. As a mere conjecture, Lavoisier outlined briefly his theory of vaporization in 1774 in his *Opuscules, op. cit.* (note 4). See *Oeuvres de Lavoisier, 1,* 612-613.

[21]The third of the MSS published by Fric (see above note 20) entitled "Sur une nouvelle théorie de la calcination et de la réduction des substances métalliques" seems almost certainly to have been his first public announcement of his revolutionary theory of the role of air in these processes, delivered at the Easter public session of the Academy of Sciences in April 1773. It is surely of some interest that, with unaccustomed generosity, Lavoisier attributes the earliest ideas leading to his new theory to his friend and collaborator Trudaine de Montigny.

Lavoisier's "covering theory" has two notable consequences. In the first place it led him to distinguish, as we do today, between the purely *physical* and the purely *chemical* (or combinatorial) properties of substances; in the older chemistry these had been habitually confused or intermingled.[22] But more immediately significant was what the theory of vaporization contributed to understanding the mystery of combustion. If it could be shown that air is a compound of a "base" with the "matter of fire," then when air is fixed in the process of combustion there should be a release (*dégagement*) of the "matter of fire," for it is from the elastic properties of the fire that air acquires its "expansibility." If so, it must be the fire-matter originally combined with the "base of air" which, now that it is liberated, yields the heat and flame associated with combustion. In other words, it is not the phlogiston, or fixed fire, of combustible substances that accounts for the heat and light (as the later partisans of the phlogistic theory believed), but the "matter of fire" freed from its combination with the air. It was evidently to determine by experiment that substances when they vaporize do in fact combine with the "matter of fire," and under what conditions this occurs, that Lavoisier undertook those first experiments with Laplace.

Such an explanation can help us answer a persistent question: why did it take Lavoisier so long to come out publicly against the phlogiston theory and set forth his own rival theory? This he did not do until the autumn of 1777. In 1772 he had discovered the role that air plays in these processes, and in February 1776 he had identified Priestley's new gas, the "dephlogisticated air," as that portion of the atmosphere responsible for the effects he had discovered. Why did he delay until late in 1777 before making his own theory public and openly challenging the phlogistic hypothesis? To be sure, there was perhaps an understandable element of caution, a hesitation to question a widely-held theory. But he probably also believed he had not fully supported his case. Why had he not? To those of us who were taught that combustion is only the combination of oxygen with a combustible substance, he would seem to have all the evidence he needed. Yet he surely did not. The strictly *chemical* aspect of the problem Lavoisier had brilliantly demon-

[22] Lavoisier was quite clear about this distinction. Later, in discussing aeriform fluids, he wrote: "Si ces airs ont beaucoup de rapport avec l'atmosphère par les qualités qu'on peut regarder comme physiques, ils en diffèrent tous essentiellement par leurs qualités chimiques," *Oeuvres de Lavoisier, 5,* 272.

strated, but the *physical* aspect—the production of heat and light, which after all is the striking feature of combustion—remained unexplained, unless, as Macquer was to do,[23] one sought to combine the phlogistic theory with the newly discovered role of "dephlogisticated air." According to the phlogistic hypothesis the heat and light produced during combustion were nothing but manifestations of the phlogiston suddenly freed from the burning substance. To complete his theory, Lavoisier had to account in plausible fashion for these physical effects. If he could show experimentally that all fluids, on changing to the vapor state, were combined with a notable amount of heat, it would be this heat that is released from "dephlogisticated air" during combustion; his theory would now have a semblance of finality. At this time—unfamiliar with Black's success in measuring the latent heat of vaporization—he could only attempt to determine directly the various conditions under which fluids vaporized. This was what he set out to do with the assistance of Laplace. With the experiments completed to his satisfaction, he was free to make his theories public without delay.

Laplace had much to offer a man like Lavoisier, even though the latter was ill-equipped to follow, much less savor, the stream of highly technical papers that Laplace read before the Academy. Yet the breadth of Laplace's scientific interest was far greater than one might have gathered from a casual glance at these early memoirs. The wide range of his knowledge was noted by Anders Johan Lexell, the Swedish mathematician who visited Paris in 1780-1781. In writing his impressions of the Academy of Sciences in Paris, he described Laplace as the most gifted mathematician in France (a man, also, well aware of his superiority): "Il a des connaissances encore dans les autres sciences, mais il me semble qu'il en fait un abus, car à l'Académie il veut décider de tout."[24]

It is certain that Laplace showed a reach of scientific understanding, a depth of philosophical insight, unequalled among his colleagues. Even in his first papers he made explicit a unified view of natural knowledge that went far beyond the specific problems he was treating. As early as 1773, he set forth his belief in an absolute

[23]*Dictionnaire de Chimie,* 2nd ed., *1* (1778), 389-399.

[24]Lexell to Johann-Albrecht Euler (7 January 1781) in Arthur Birembaut, "L'Académie Royale des Sciences en 1780 vue par l'astronome suédois Lexell (1740-1784)," *Revue d'histoire des sciences et de leurs applications, 10* (1957), 148-166.

determinism in nature.[25] The subject of his special interest, physical astronomy, he took to be the model which all the other sciences should emulate; of all the sciences it is the one "qui fait le plus d'honneur à l'esprit humain."[26] Yet he was well aware that elsewhere in physical nature one cannot expect to find the same elegance and apparent simplicity. Many causes, and these largely unknown, interact to produce an event. Our knowledge of the laws of nature, of the causes that operate, can in most cases be obtained only through the use of the calculus of probability. This developing branch of mathematics is applicable to all the "sciences physico-mathématiques." For example, when we measure the intensity of illumination, the different degrees of heat in a body, the forces and resistances of bodies, we are concerned to determine the physical causes of our sensations, not the sensations themselves. We can compute the mathematical expectation that certain causes are operating, and this is quite different from moral expectation, which merely relates to a state of mind and lies beyond the realm of calculation.[27]

Our lack of certitude, moreover, applies not only to the causes of events, but to the events themselves, that is, to the observations and measurements that we make. Even our most careful measurements disagree to some extent. What values, then, should we accept? The notion that there must exist a "law of error" which dictates the "best mean" had been suggested in 1775 by Thomas Simpson who perceived that observational errors should be thought of "not as unrelated happenings, but as properties of the measurement process itself and the observer involved."[28] In any process of measurement there must be a probable distribution of errors so that the mathematics of probability can be applied. In 1772, while still teaching at the Ecole Militaire, Laplace took up the problem and read a preliminary paper to the Academy. At first he deemed the results of little practical value. But later—after learning that Daniel Bernoulli and Lagrange were interesting themselves in the problem—he appended his preliminary thoughts on error theory to his

[25] Roger Hahn, "Laplace's First Formulation of Scientific Determinism in 1773," *Actes du XIe Congrès International d'Histoire des Sciences,* Warsaw-Cracow, 1968, *2,* 167-171.

[26] *Oeuvres de Laplace, 8,* 144.

[27] *Ibid.,* p. 147.

[28] Churchill Eisenhart, "The Meaning of 'Least' in Least Squares," *Journal of the Washington Academy of Sciences, 54* (1964), 28-29.

"Mémoire sur la probabilité des causes par les événements," published in 1774.[29]

Interest in the problem of accurate measurement and "theory of error" increased in succeeding years. It had become a major concern in the Academy of Sciences in the very year, 1777, that brought Lavoisier and Laplace together. In February, a memoir on the "mean" by the Marquis de Condorcet came to the attention of the Academy.[30] Later that month, at a regular meeting of that body, Laplace read the first part of a long memoir on the method of taking a mean of several observations. On 8 March he finished reading this memoir, which is fully transcribed in the *Procès-verbaux* but was never published.[31] The space given to the transcription of Laplace's memoir testifies to the importance the Academy gave to the subject. Lavoisier, who, as we shall see, had in mind experiments that would require considerable precision, was present when Laplace read his long and sophisticated memoir. He must have been impressed by the interest the Academy's mathematicians displayed in the evaluation of experimental results. Whether or not he hoped to make use of their methods, he could well have seen the advantage of working with a man like Laplace, sensitive to the theoretical difficulties of accurate experimentation.

A succinct account of these early joint experiments of Lavoisier and Laplace, and indeed the earliest to be published, is given in a small book of J.-B. Bucquet (1746–1780), a physician and chemist with whom Lavoisier was actively collaborating in this same year, 1777.[32] The first section of Bucquet's book is entitled "Histoire abrégée des différens Fluides aériformes ou Gas" and contains a passage, not least interesting for its early use of the word "gas," which reads as follows: "Tous les corps susceptibles de se volatiliser à un degré de chaleur modéré, peuvent se mettre dans l'état de gas, & subsister sous cette forme tant qu'ils restent exposés à la même température." Messrs. Lavoisier and Laplace, he goes on, have found another way of turning fluids into the gaseous state, i.e., by diminishing the atmospheric pressure upon them. While it is well known that

[29] *Oeuvres de Laplace, 8,* 27–65.

[30] *Procès-verbaux,* 1777, f. 47v.

[31] *Procès-verbaux,* 1777, fols. 86 and 121v–142.

[32] *Mémoire sur la manière dont les animaux sont affectés par différens Fluides Aériformes, Méphitiques, & sur les moyens de remédier aux effets de ces Fluides. Précédé d'une Histoire abrégée des différens Fluides Aériformes ou Gas* (Paris, 1778).

water boils in a vacuum at a temperature below its ordinary boiling point

> Mrs. Lavoisier & de la Place ont été plus loin, & par une trés-belle suite d'expériences dont les détails seront publiées dans le volume de l'Académie des sciences de 1777, ces Savans ont démontré que tous les fluides volatils, pris même au terme de zéro du thermomètre de M. de Réaumur, entroient en expansion, & se réduisoient sous forme de gas lorsqu'on les plaçoit, soit dans le vide de la machine pneumatique, soit dans celui du baromètre, et que dans cet état ils avoient la propriété de soutenir le mercure à une hauteur déterminée propre à chacun d'eux.[33]

The year 1777 was one of the most active and productive of Lavoisier's scientific career. Besides working with Laplace, he had experimented intensively with Bucquet. On 5 September the two men registered with the Academy (*pour prendre date*) no less than twenty-six memoirs, representing work carried out independently or in concert.[34] One of these, entitled "De la combinaison de la matière du feu avec les fluides évaporables, et de la formation des fluides élastiques aériformes,"[35] was not only Lavoisier's first formal

[33] *Ibid.*, p. 11. Although Lavoisier seems to have coined the term "aériforme," it was picked up by Bucquet who used it before Lavoisier in a paper delivered at the Société Royale de Médecine late in January 1777. I owe this information to Dr. J. B. Gough.

[34] *Procès-verbaux*, 1777, fols. 527–528. For a partial list, see Maurice Daumas, *Lavoisier théoricien et expérimentateur* (Paris, 1955), pp. 39–41. Fourteen of these memoirs were joint efforts; two were by Bucquet alone and ten by Lavoisier. See E. McDonald, "The Collaboration of Bucquet and Lavoisier," *Ambix, 13* (1966), 77. Several of these memoirs dealt with heat; one of these, on the rate at which various fluids absorb heat, was obviously related to the experiments with Laplace.

[35] Initialled by Condorcet on 5 September 1777 (Lavoisier Papers, Archives of the Academy of Sciences, dossier 1315), this memoir was read to the Academy on 18 July 1778 (*Procès-verbaux*, 1778, f. 251) and published in *Mém. Acad. Roy. Sci.*, 1777 (1780), pp. 420–432. See *Oeuvres de Lavoisier, 2*, 212–224. McDonald (*ibid.*, p. 78) suggests that Lavoisier may have used in this memoir conclusions drawn from his work with Bucquet. For a discussion of this paper, referring in passing to the collaboration of Lavoisier and Laplace, see John F. Fulton, Denis I. Duveen, and Herbert S. Klickstein, "Antoine Laurent Lavoisier's 'Réflexions sur les effets de l'éther vitriolique et de l'éther nitreux dans l'économie animale," *Journal of the History of Medicine and Allied Sciences, 8* (1953), 318–323. The full import of these experiments with Laplace (incorrectly spoken of as carried out "prior to 1777") escaped these

exposition of his covering theory but gives some details about the joint experiments with Laplace. Although it was not published until 1780, it is the earliest account Lavoisier gave in print of these experiments. After setting forth his belief that fire (*fluide igné* or *matière du feu*) is a subtle fluid which penetrates all bodies, an opinion held by "le plus grand nombre des anciens physiciens," and asserting that, like water, it can exist in both free and fixed forms (*feu libre* and *feu combiné*), he offers evidence to show how the combination of the fire-matter with fluids turns them into elastic vapors. Compelling evidence is drawn from the reports of Georg Richmann, William Cullen, Dortous de Mairan, and Antoine Baumé on the phenomenon of evaporative cooling.[36] These experiments seem to show that when one observes the drop in temperature of a thermometer whose ball is moistened with a volatile liquid, it is only because when the liquid turns into a vapor it has taken up heat or the matter of fire from the surroundings. The more volatile the fluid, the greater the cooling effect. In general, aeriform substances are made up of certain fluids combined with the matter of fire.

But, Lavoisier goes on, the more rapid the evaporation, the greater the drop in temperature; moreover one can enhance this evaporative cooling by placing the fluids in circumstances capable of promoting and accelerating the formation of vapors, and a way of doing this is to carry out the process in the vacuum of an air pump. This important additional support for his theory, which he proceeds to describe, is drawn from the "travail très-considérable, entrepris en commun par M. de La Place et par moi, et dont l'Académie a déjà connaissance, d'après le mémoire qui a été lu à la séance publique de Pâques dernier."[37] He then goes on to describe an experiment to prove three things: (1) that the weight of the atmosphere is a resistance to be overcome, a force which opposes the vaporization of a fluid; (2) that when this confining force is withdrawn, evaporable fluids expand and change into "fluides élastiques aéri-

---

authors who were chiefly interested in physiological problems; but they correctly write that Lavoisier and Laplace "demonstrated with ether that the gaseous state is a modification of bodies dependent on temperature and pressure" (p. 320). I owe this reference to Diana Long Hall.

[36]For the influence on Lavoisier of the work on evaporative cooling, see J. B. Gough, "Nouvelle contribution à l'étude de l'évolution des idées de Lavoisier sur la nature de l'air . . . ," *op. cit.* (note 20), p. 270; and for more detail, his doctoral dissertation, *op. cit.* (note 18), pp. 53–67.

[37]*Oeuvres de Lavoisier, 2,* 217.

formes, en espèces d'airs"; and (3) that the transformation of ordinary liquids into the state of elastic fluids is accompanied by an absorption of matter of fire at the expense of surrounding bodies.

To demonstrate these conclusions they filled a small flask, phial, or simple glass tube with ether, covering the phial or tube with a moistened bladder, wrapped tightly around with a string. The tube is so completely filled with ether that no air remains between the liquid and the bladder. The flask or tube so disposed is placed in the receiver of a good air pump, the top of which is equipped with a leather cover, pierced by a rod to which is affixed an awl or other pointed instrument capable of cutting open the bladder at the proper moment. When the air is sufficiently pumped out, as shown by the barometer attached to the air pump receiver, the bladder is pierced. Immediately the ether begins to boil and vaporizes with astonishing rapidity. It is transformed into an elastic fluid capable of exerting in winter a pressure of eight to ten inches of mercury and twenty to twenty-five inches in the hottest days of summer. If one inserts a small thermometer into the flask or tube containing the ether, the temperature drops rapidly during evaporation, because of the great amount of heat (*feu libre*), which combines with the ether when it turns into a vapor.

The experimenters then let the air back in, bringing the pressure in the receiver up to the ordinary twenty-eight inches of mercury; they were astonished to note an unexpected effect: "Ce qui est très-remarquable, l'éther ainsi mêlé d'air atmosphérique ne se condense pas pour cela: il reste dans l'état de fluide élastique permanent, et forme une espèce particulière d'air inflammable, que je n'ai pas eu le temps d'examiner."[38]

The same experiment, he reports, succeeds with other evaporable fluids: with alcohol and even water, but with this difference, that the vapor of alcohol produced in the vacuum of the air pump gives only the pressure of an inch of mercury in winter and of four or five inches in summer. There is, he concludes, less fluid vaporized in the case of alcohol than with ether, hence less *fluide igné* is absorbed and the cooling is less pronounced. Lavoisier believed that he had demonstrated in these experiments with different fluids that the cooling effect is nearly proportional to the quantity of fluid vaporized.

[38]*Ibid.*, p. 218.

Lavoisier then reports that less striking effects were noted when the evaporable fluid was placed in an open vessel in the receiver of the air pump. A small quantity of alcohol is put in an open dish at the temperature of fifteen degrees of their thermometer. When the pump is activated and the barometer which indicates the inside pressure has dropped nineteen lines below its level, the alcohol begins to boil; but the ebullition is not continuous as in the earlier experiment. It stops as soon as the pumping ceases because, Lavoisier says, the alcohol when vaporized turns into an elastic fluid which forms a sort of atmosphere pressing on the surface of the fluid and opposing the vaporization, and also because a part of the *feu libre* passes into the state of *feu combiné*[39] and the cooling effect slows down the ebullition.

He sums up his theory: from vaporization in a vacuum, it seems to be proved that the transformation of liquids into elastic, aeriform fluids "est soumise à deux lois dont l'effet est opposé: d'une part, le degré de chaleur auquel ils sont exposés tend à les vaporiser; de l'autre, la pression de l'atmosphère met obstacle à leur vaporisation, de sorte qu'il sont ou dans l'état d'élasticité, ou dans celui de liquidité, suivant que l'une de ces deux forces l'emporte sur l'autre."[40] There are really no details of experimental measurements, no references to attempts to measure what we would call vapor pressure, and no specific data as to the temperatures at which, under ordinary atmospheric pressure, the various fluids vaporize. Yet he concludes: "Au reste, toute cette théorie deviendra beaucoup plus claire d'après les expériences dont nous nous occupons, M. de la Place et moi, et dont nous rendrons compte dans ce même volume ou dans le suivant."[41] The promised paper never appeared. In a much later memoir by Lavoisier he gives us some information as to the sort of thermometer used and specifies the boiling points of the several liquids.[42] At one point he writes: "Nous avons déjà établi, M. de Laplace et moi, dans un mémoire lû à l'Académie en 1777,

[39]The notions of "free fire" and "combined fire" make their first appearance in print in this paper, although they already appeared in Lavoisier's unpublished notes and drafts.

[40]*Oeuvres de Lavoisier, 2,* 220–221.

[41]*Ibid.,* p. 221.

[42]"Mémoire sur quelques fluides que l'on peut obtenir, dans l'état aériforme, à un degré de chaleur peu supérieur à la température moyenne de la terre," *Oeuvres de Lavoisier, 2,* 261–270, first published in *Mém. Acad. Roy. Sci.,* 1780 (1784), pp. 334–343.

que l'éther se vaporisait à une temperature de 32 à 33 degrés d'un thermomètre de mercure divisé en quatre-vingt parties, depuis la glace fondante jusqu'à l'eau bouillante, le baromètre étant à 28 pouces de hauteur."[43] Later on in this paper Lavoisier writes that alcohol vaporizes at seventy-one or seventy-two degrees of their thermometer and that water "ne se vaporise et ne prend l'état aériforme qu'à 85 degrés complet, et même un peu au delà."[44]

Some rather tantalizing suggestions emerge from an undated letter Lavoisier wrote to Trudaine de Montigny, probably in the spring or summer of 1777. It begins as follows: "Je regrette infiniment de n'avoir point encore pu terminer les expériences dont nous nous occupons, M. de la Place et mois, mais j'espère qu'elles seront acheveés sous peu de jours, et aussitôt je ferai emballer la machine pneumatique et je la ferai partir pour Montigny."[45]

This at least tells us of the interest Lavoisier's friend and occasional collaborator took in the experiments we are considering. Possibly he had lent the air pump (*machine pneumatique*) to Lavoisier, but Lavoisier's remark that its use needs no detailed explanation suggests that it was Lavoisier who was lending it to Trudaine. What is most interesting is that Lavoisier describes in detail, with an accompanying sketch, an adaptation of this air pump which seems intended to measure vapor pressures. The contrivance has at least a family resemblance to what physical chemists know as a McLeod gauge. Inside the receiver of the air pump is suspended what he calls an English tube (*éprouvette angloise*), the upper part of which consists of a narrow, finely graduated glass tube, closed at the upper end, the lower portion being expanded into an ovoid bulb open at the bottom. By means of a simple mechanical contrivance this English tube can be suspended just above, or lowered into, a vessel two-thirds filled with mercury placed in the receiver. This is held above the mercury when the pump is in operation. When the vacuum has been produced, and one wishes to know the extent of the evacua-

[43]*Ibid.*, p. 262.

[44]*Ibid.*, p. 267. It is in this paper (p. 269) that Lavoisier clearly states his doctrine in print "que *solidité, liquidité, élasticité*, sont trois états différents de la même matière, trois modifications particulières par lesquelles presque toutes les substances peuvent successivement passer, et qui dépendent uniquement du degré de chaleur auquel elles sont exposées, autrement dit de la quantité de fluide igné dont elles sont pénétrées. . . ."

[45]Archives de Chabrol. This letter is missing from the *Correspondance de Lavoisier, 3* (1964), where it should have appeared.

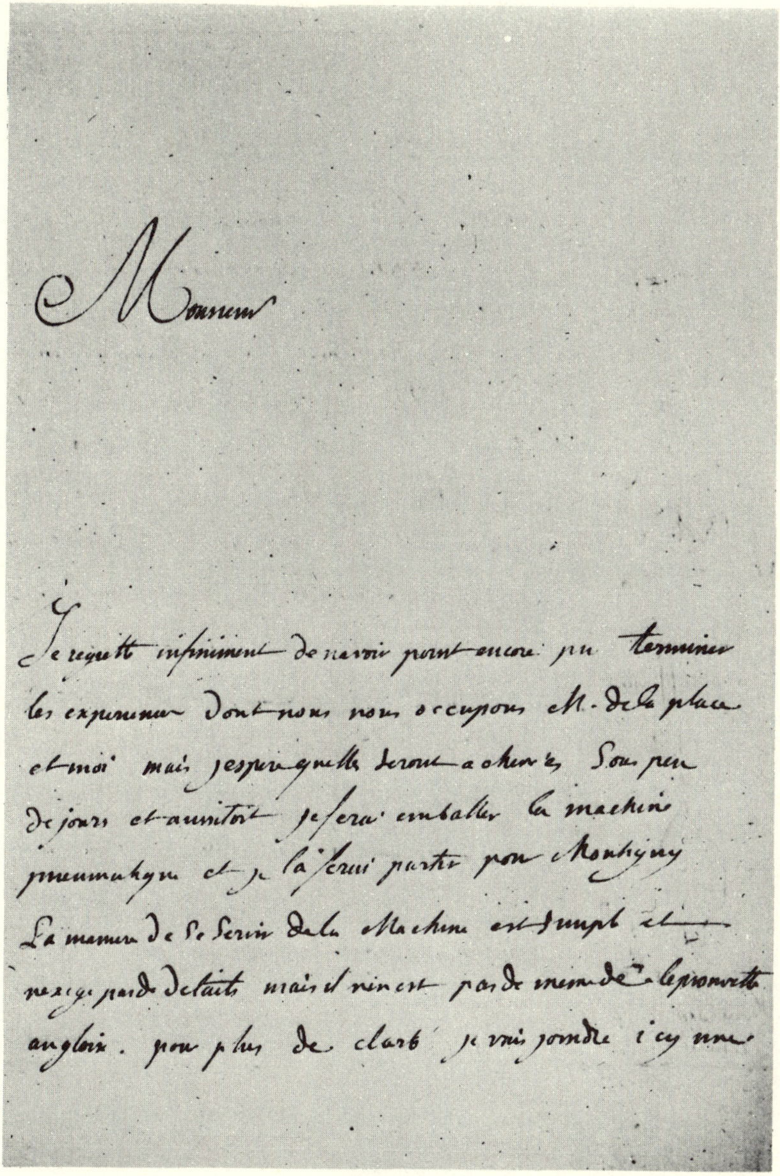

Monsieur

Je regrette infiniment de n'avoir point encore pu terminer
les experiences dont nous nous occupons M. de la place
et moi mais jespere quelles seront achevées sous peu
de jours et aussitot je ferai emballer la machine
pneumatique et je la ferai partir pour Montigny
La maniere de se servir de la Machine est simple et
n'exige pas de details mais il n'en est pas de meme de l'eprouvette
anglaise. pour plus de clarté je vais joindre icy une

Figure 1

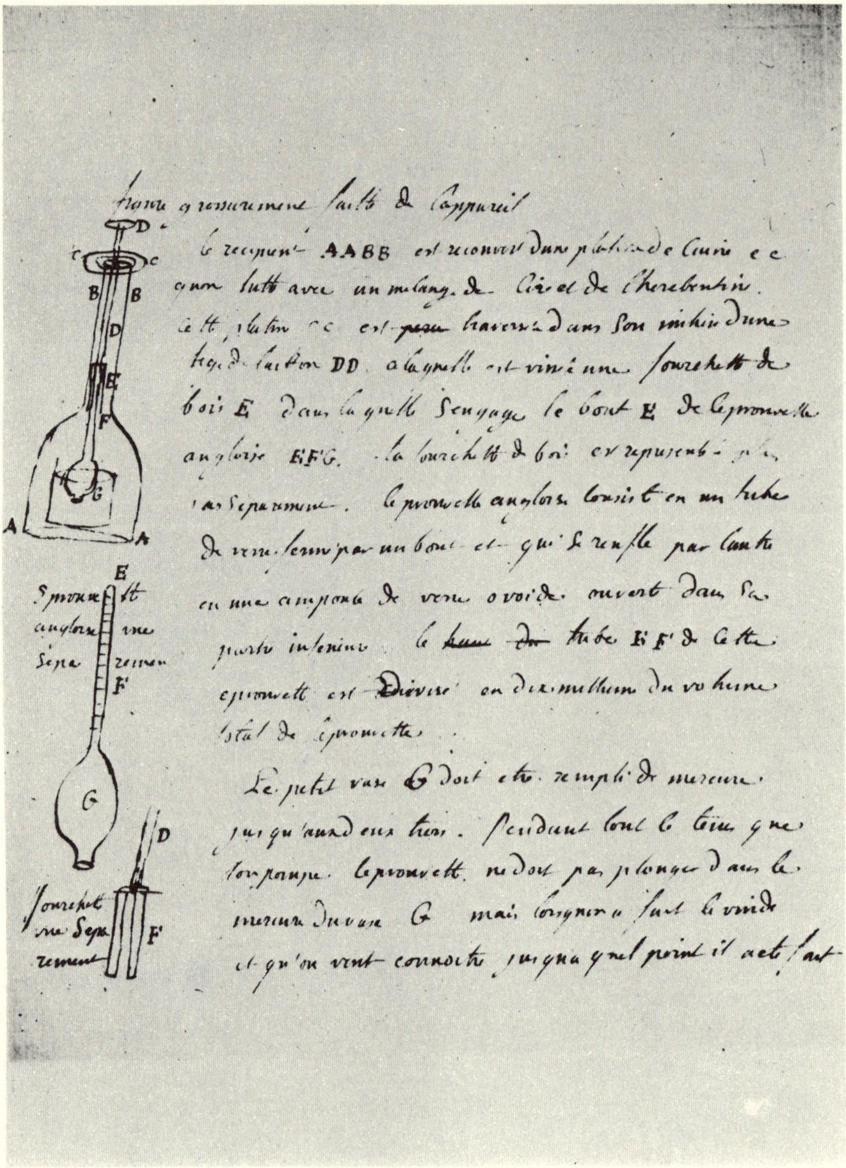

figure grossierement faite de l'appareil

le recipient AABB est recouvert d'une platine de cuivre c c qu'on lutte avec un melange de cire et de therebentine. cette platine c c est ~~percée~~ traversée dans son milieu d'une tige de laiton DD a laquelle est vissée une fourchette de bois E dans laquelle s'engage le bout E de l'eprouvette angloise EFG. la fourchette de bois est representée [illegible] separement. l'eprouvette angloise consiste en un tube de verre fermé par un bout et qui se renfle par l'autre en une ampoule de verre ovoide ouvert dans sa partie inferieure. le ~~haut~~ ~~du~~ tube EF de cette eprouvette est ~~divisé~~ divisé en dix millièmes du volume total de l'eprouvette.

Le petit vase G doit etre rempli de mercure jusqu'aux deux tiers. pendant tout le tems que l'on pompe l'eprouvette ne doit pas plonger dans le mercure du vase G mais lorsqu'on a fait le vuide et qu'on veut connoitre jusqu'a quel point il a été fait

Eprouvette angloise vue separement

fourchette vue separement

Figure 1 (Continued)

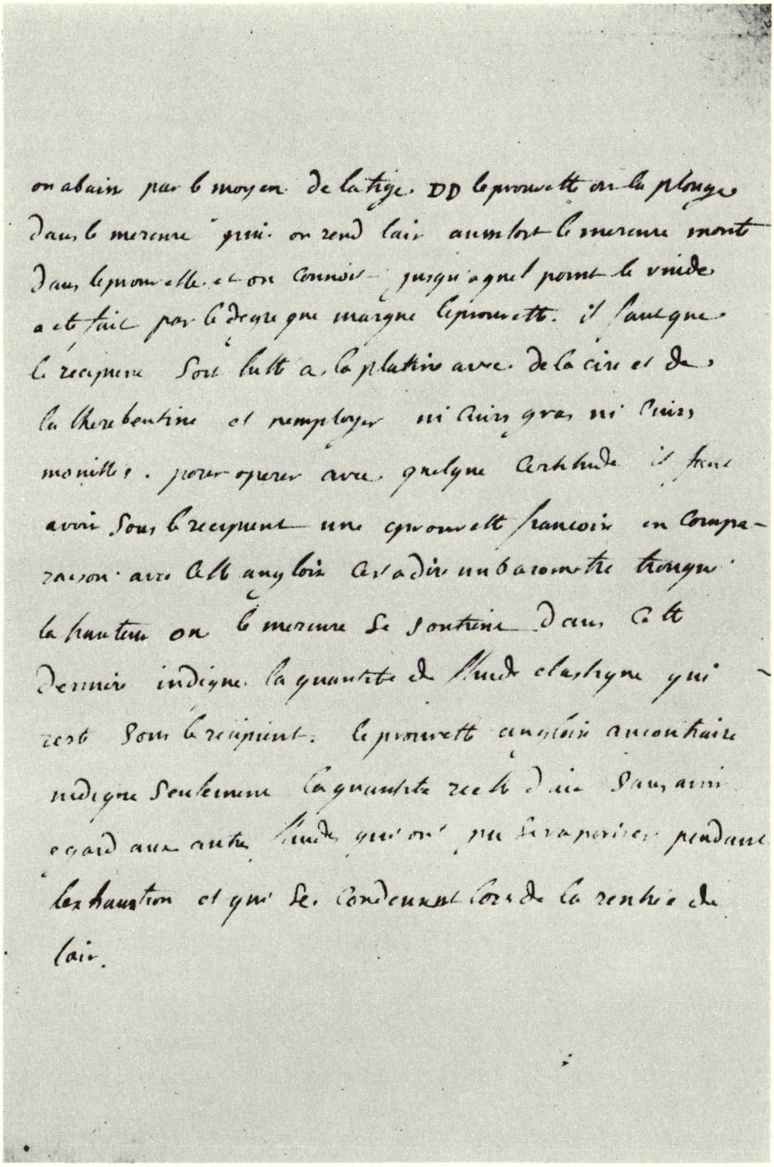

on abaisse par le moyen de la tige DD l'eprouvette on la plonge
dans le mercure puis on rend l'air aussitot le mercure monte
dans l'eprouvette et on connoit jusqu'a quel point le vuide
a été fait par le degré que marque l'eprouvette. il faut que
le recipient soit lutté a la platine avec de la cire et de
la therebentine et n'employer ni cuirs gras ni cuirs
mouillés. pour operer avec quelque certitude il faut
avoir sous le recipient une eprouvette francoise en compa-
raison avec celle angloise c'est a dire un barometre tronqué
la hauteur ou le mercure se soutient dans cette
derniere indique la quantité de fluide elastique qui
reste sous le recipient. l'eprouvette angloise au contraire
n'indique seulement la quantité reelle d'air sans avoir
egard aux autres fluides qui ont pu se vaporiser pendant
l'exhaustion et qui se condensent lors de la rentrée de
l'air.

Figure 1 (Continued)

tion, the open end of the tube is lowered into the mercury, then air is allowed to re-enter the receiver. The extent the mercury rises in the English tube indicates the total pressure, that of the air and of any vapor that has been produced (the diagram does not show, and there is no explicit mention of, any vessel containing an evaporable fluid). As Lavoisier describes matters, to make an accurate comparison there should also be placed in the receiver what he calls an *éprouvette françoise,* which is nothing but a small barometer. The height of the mercury in the latter indicates the quantity of elastic fluid that remains in the receiver. The English tube indicates only the amount of air in the receiver, without regard, Lavoisier writes, to the other fluids which have been vaporized during the evacuation and which are condensed as the air re-enters. There is no record of the successful use of this piece of equipment. If it performed as Lavoisier expected, the results obtained would probably have been inconsistent and unreliable, which perhaps explains why there is no reference to it in any of Lavoisier's published, or for that matter unpublished, writings.

To sum up: Lavoisier and Laplace were in fact correct that temperature and atmospheric pressure are the two variables determining the vaporization of a fluid. They had doubtless discovered—but not accurately measured—the characteristic vapor pressures of different liquids. But they misunderstood the role played by atmospheric pressure. They conceived of the atmosphere as acting like a force of compression and believed that if this force is increased a vapor should be totally condensed. They were unaware, of course, of the independent pressures exerted by the constituents of a mixture of gases, a matter that was left for John Dalton to elucidate. This explains Lavoisier's astonishment that when air was let back into the receiver of the vacuum pump, the ether vapor did not completely condense. Nor did they realize that the partial pressures of evaporable fluids depend only on the temperature. Atmospheric pressure acts, we now know, only to determine the *rate* at which the vapor-liquid equilibrium is reached.

Although no joint paper on these vaporization experiments ever appeared, as Lavoisier had promised, the two men did continue, for a time, to occupy themselves with this question. An undated and apparently insignificant paper published by J.-B. Dumas, the first editor of the *Oeuvres de Lavoisier,* evidently refers to a modified approach to the problem they were exploring.[46] While Laplace's name does not appear, the paper is written in the first person plural and makes clear reference to the earlier experiments. Having carried out (it reads) a series of experiments made with the air pump and shown the influence of heat and air pressure on the vaporization of fluids, "nous avons pensé" that it would be important to push forward similar experiments so as to establish tables giving the barometric pressures and temperatures at which different fluids vaporize. Experiments of this kind performed in the receiver of an air pump would be "longues et pénibles." A better plan is to use the Torricellian vacuum in the form of a series of mercury thermometers. Into three of these they proposed to introduce, on top of the mercury, a small layer of water, ether, or alcohol. These experiments may never have been tried, and it would seem that the two men never did more than to determine the possibility of making barometers sufficiently accurate and comparable to be useful in such experiments. Later, in passing, Lavoisier referred to this work: "Quelques expériences d'ailleurs que nous avons faites, M. de Laplace et moi, dans une autre vue, nous avaient appris qu'il était possible, en prenant des tubes égaux et en employant toutes les précautions convenables, de construire des baromètres rigoureusement d'accord entre aux."[47]

We can hardly doubt the importance that Lavoisier attached to these early experiments with Laplace. Crude though they were, they provided experimental support for the theory in terms of which, henceforth, he set forth his new view of chemistry. Sometime in 1778, encouraged by these results, Lavoisier revived his plan of preparing a second volume of his *Opuscules* and set down on paper an "Introduction et plan" for this projected volume.[48] He intends, he writes, to eliminate historical matters (such as had filled so much space in the volume that had appeared in 1774), for they are "rarement utile au physicien." Instead, he continues, he has tried to follow, as much as the subject permits, "la méthode des

[46] "Appareil pour la mesure de l'élasticité des vapeurs à diverses températures," *Oeuvres de Lavoisier, 3,* 749–752.

[47] "Mémoire sur la construction des baromètres à surface plane," *Oeuvres de Lavoisier, 3,* 761.

[48] "Introduction et plan d'un deuxième volume des Opuscules Physiques et Chimiques," *Oeuvres de Lavoisier, 5,* 267–270.

géomètres, méthode précieuse qui conduirait toujours à l'évidence, si nous pouvions toujours partir en physique de données sûres et démontrées." What he has in mind is something quite different from the miscellany of experiments and observations he had earlier envisaged for a second volume of the *Opuscules.* What he now proposes is to set forth a "système général" he had long been meditating (clearly his "covering theory") and to expound it in a rational and deductive fashion—for this is the "méthode de géomètres." So as to present matters to the reader in the most direct fashion, "j'ai commencé par les expériences de l'évaporation des fluides dans la machine pneumatique, faites par M. de Laplace et par moi, quoique ces expériences, dans l'ordre chronologique, soient les dernières faites."[49]

The outline of fourteen projected chapters begins with reference to an introduction (some "réflexions générales sur la chimie vaporeuse") followed by two chapters elaborating his theory of vaporization and discussing the nature of airs "résultant de la combinaison de la matière du feu avec une base quelconque." Then come headings for chapters on the formation of acids and on combustion.

THE OPEN ASSAULT ON PHLOGISTON

On 12 November 1777—the occasion was the autumnal *rentrée publique* of the Academy of Sciences—Lavoisier read before a distinguished audience, which included Benjamin Franklin who was greeted by a tumultuous ovation, one of his most important papers. Entitled (in its published form) "Mémoire sur la combustion en général," it announced his own theory of combustion and included his first public criticism of the phlogistic hypothesis.[50]

[49]*Ibid.*, p. 268. This method of presentation anticipates the scheme followed later in his *Traité élémentaire de chimie* of 1789, where the first chapter is entitled "Des combinaisons du calorique et de la formation des fluides élastiques aériformes," and the second, "Vues générales sur la formation & la constitution de l'atmosphère de la terre." Not surprisingly he refers back to the early joint experiments which Laplace read to the Academy in 1777. See *Oeuvres de Lavoisier, 1,* 23–25.

[50]*Oeuvres de Lavoisier, 2,* 225–233. The minutes of the Academy's meeting on that occasion (*Procès-verbaux,* 1777, f. 542v) simply record that "M. Lavoisier a lu un mémoire sur la théorie de la combustion." For Franklin's presence, and the ovation he received, see Alfred Owen Aldridge, *Franklin and his French Contemporaries* (New York, 1957), p. 65. Lavoisier's memoir was

For some time it had been rumored that Lavoisier planned just such an open attack. This was of particular concern to two of the most persistent partisans of phlogiston, Macquer and his disciple, the Dijonnais lawyer and chemist, L.-B. Guyton de Morveau.[51] Both followed Lavoisier's progress with interest and no little concern. In January 1774, Lavoisier had sent Guyton a complimentary copy of his *Opuscules,* expressing his admiration for Guyton's experimental work, but making clear that he could not agree as to the cause of the increased weight of metallic calxes, and suggesting that in a second volume of the *Opuscules* he may write against Guyton.[52] Guyton, in his reply, courteously answered that he was pleased to hear that Lavoisier deemed him a worthy adversary.[53] Later the two men met for the first time in the early spring of 1777, when Guyton—on a visit to Paris—carried out some experiments (on iron oxide) in Lavoisier's furnace.[54] Early in 1777 (or late in 1776) when Guyton was seeing through the press the first volume of his *Elémens de chymie théorique et pratique,* and before sending to the printer its

given a second reading at a regular meeting of the Academy on 11 December 1779 and the reading was continued on 18 December (*Procès-verbaux,* 1779, fols. 305v and 309). The results were published in *Mém. Acad. R. Sci.,* Année 1777 (1780), pp. 592–600. This was Lavoisier's first published criticism of the phlogistic hypothesis. It is almost certain that Lavoisier was not the author of the two anonymous attacks on phlogiston published in 1773 and 1774 in Rozier's *Observations sur la physique* as has often been claimed. See Carleton Perrin, "Early Opposition to the Phlogiston Theory: Two Anonymous Attacks," *The British Journal for the History of Science, 5* (1970), 128–144.

[51] See Guyton's letter to Macquer, published by Denis I. Duveen and H. S. Klickstein (*Osiris, 12* [1956], 345–346). Maurice Daumas and I independently pointed out to the authors that they had misread the name of the recipient of this letter, who was not Louis-Charles-Henry Macquart (1745–1808), as they believed, but Guyton's patron in the Academy of Sciences, the eminent chemist P. J. Macquer. The authors acknowledged their error, publishing my detailed argument, with only minor verbal changes, in *Isis, 49* (1958), 73–74.

[52] *Oeuvres de Lavoisier—Correspondance, 2,* 404–406 (No. 212).

[53] This letter, offered for sale at the Hotel Drouut in December 1965, was purchased for the Academy of Sciences; it may be consulted in the Academy's Archives.

[54] Berthelot, *Révolution chimique,* p. 268. On 25 April 1775 Guyton wrote to Lavoisier from Dijon expressing his pleasure at having made Lavoisier's acquaintance "pendant mon séjour à Paris." I consulted this letter at the Château de La Canière near Riom in 1939 and took notes from it. Although now in the Archives de Chabrol, this letter—like all too many others—escaped the notice of M. René Fric and was not included in fasc. *2* of the Lavoisier *Correspondance.*

second chapter ("Des dissolutions par l'air"), where Lavoisier's work is referred to, he wrote Lavoisier "pour le presser de me développer toute son idée," but the reply only convinced him that Lavoisier "se mettroit d'attaquer notre phlogistique."[55]

In the last months of that same year Guyton read accounts of Lavoisier's paper presented at the *rentrée publique*, and they first confused and then disturbed him. A letter from Macquer, his patron and official correspondent, dated 15 January 1778, gave him a clearer picture of what had transpired. Macquer writes with jocose exaggeration:

> M. Lavoisier m'éffrayoit depuis long-temps par une grande découverte qu'il reservoit *in petto*, & qui n'alloit pas moins qu'à renverser de fond en comble toute la théorie du phlogistique ou feu combiné: son air de confiance me faisoit mourir de peur. Où en aurions-nous été avec notre vieille Chymie, s'il avoit fallu rebâtir un édifice tout différent?[56]

Then with respect to Lavoisier's recent memoir, he expresses his relief that the danger seemed not as great as he had feared: "Heu-

[55]This correspondence seems not to have come to light; it is mentioned, however, in Guyton's letter to Macquer (22 January 1778) referred to in Note 51. In a letter of 25 March 1777, Guyton informs Lavoisier that he is sending a complimentary copy of the first volume of his *Elémens de Chymie;* Lavoisier will discover in it "votre propre bien" and Guyton expresses the hope that his correspondent will feel he made good use of it. Despite several references to Lavoisier's discoveries, Guyton expounds the phlogistic theory of calcination; his interpretation of recent discoveries of pneumatic chemistry is along phlogistic lines. For Guyton's views about phlogiston and pneumatic chemistry, and his gradual conversion to Lavoisier's doctrine, see William A. Smeaton, "Guyton de Morveau and the Phlogiston Theory," in *Mélanges Alexandre Koyré*, eds. I. Bernard Cohen and René Taton, 2 vols. (Paris, 1964), *1*, 522–540.

[56]*Encyclopédie méthodique—Chimie, 1* (Paris, 1786), 628. Cited by Duveen and Klickstein, *op. cit.* (note 51), p. 347. Macquer's reference to a "grande découverte" that Lavoisier "réservait in petto" may possibly explain an item in the *Procès-verbaux* (19 April 1777): "M. Lavoisier demande des commissaires pour une découverte qu'il a faite, et sur laquelle il demande le secret. MM. de Montigny, le chevalier d'Arcy et le chevalier de Borda ont été nommés" (*Procès-verbaux,* 1777, f. 289). Daumas (*op. cit.*, p. 38) comments: "Aucune indication n'est donnée par la suite dans les procès-verbaux sur la nature de cette découverte." It may refer to the discoveries on respiration, which Lavoisier described to the Academy on 3 May, but it was probably a version of his theory of vaporization or the substance of his "Mémoire sur la combustion en général."

reusement M. Lavoisier vient de mettre sa découverte au jour, dans un Mémoire lu à la dernière Assemblée publique; & je vous assure que depuis ce temps j'ai un grand poids de moins sur l'estomac."[57] Then describing with precision the main drift of Lavoisier's memoir he writes rhetorically: "Jugez si j'avois sujet d'avoir une si grande peur." To this Guyton answers in the same spirit that Macquer's account, and his evaluation, had greatly pleased him: "Je dois même vous remercier de ne m'avoir pas laissé apercevoir votre frayeur, j'avais assez de la mienne et l'assurance du chef est un point bien important."[58]

Macquer was unduly sanguine, for what he had heard at the Academy's public session was the earliest considered assault by Lavoisier on the fortress of "notre veille Chymie." Nor did he perceive, or relish, the significance of Lavoisier's rather pedantic opening, an implicit criticism of the state of chemistry, and the advocacy of a "systematic," or perhaps as we would say a theoretical, approach to chemistry. Only confusion will result, Lavoisier wrote, if the scientist merely piles up facts, observations, and experiments, forgetting that they are only the materials of the edifice. One should, he argued, attempt to order the facts in terms of a conceptual scheme (as we might call it) which subordinates some facts to others and places the whole in a meaningful relationship.[59]

The ideal is exemplified in the body of the paper by Lavoisier's exposition of his hypothesis, as he cautiously calls it, of aeriform fluids and combustion. He begins by listing four phenomena associated with combustion. The experiments—for the most part his own—have verified: that in all combustions there is a "dégagement de matière du feu ou de la lumière"; that substances burn only in Priestley's dephlogisticated air (which he now proposes to call *air pur*); that in every combustion there is a "destruction ou décomposition de l'air pur," with the combustible gaining in weight exactly in proportion to the quantity of air destroyed or decom-

[57] *Ibid.*

[58] Duveen and Klickstein, *op. cit.* (note 51), p. 346.

[59] Perhaps these opening paragraphs were written after the presentation of his paper to the *rentrée publique* of 12 November 1777. The methodological ideas closely resemble those he set forth in the "Introduction et plan" for the projected second volume of the *Opuscules,* which was apparently written in 1778. They may have been introduced for the first time when he gave his second reading of his "Mémoire sur la combustion en général" on 11 December 1779.

posed; and that the substances burned turn into acids which weigh more than the starting material. Calcination of metals is only a slow combustion and obeys the same laws, except that instead of acids being formed, metallic calxes (*chaux mêtalliques*) are produced.

These phenomena are explained, Lavoisier admits, "d'une manière très-heureuse," in a very felicitous fashion, by Stahl's hypothesis. But to accept it, he remarks, one must suppose that matter of fire or phlogiston is fixed in metals and all combustible bodies. Yet this is to indulge in circular reasoning: combustibles contain the matter of fire because they burn and they burn because they contain the matter of fire.

It is Lavoisier's major point that the existence of phlogiston or fixed fire in metals and combustibles is just a supposition.[60] The physical phenomena of combustion can be just as well explained by the opposite hypothesis: that the matter of fire has another source than the metals and combustible bodies.

To drive home this point Lavoisier outlines what he means by fire: as Franklin, Boerhaave, and "une partie des philosophes de l'antiquité" believe, it is a very "subtile," very elastic fluid, surrounding our planet everywhere, penetrating all bodies and tending to reach an equilibrium state. It is the solvent (*dissolvant*) of a great many bodies, combining with them in the same way as water combines with salts. Lavoisier then summarizes his theory of vaporization, remarking that it has already been set forth in a memoir "déposé au secretariat de cette Académie." This can only be his "De la combinaison de la matière du feu avec les fluides évaporables," deposited on 5 September, which drew heavily on the experiments performed with Laplace the previous spring.

Every aeriform fluid, he explains, is the result of the combination of some body, solid or fluid, with the matter of fire. It is to this combination that airs owe their elasticity, their lightness, their tenuity and the other properties "qui les rapprochent du fluide igné." The dephlogisticated air of Priestley, for example, is a com-

[60]This identification of phlogiston with "fixed fire" was a commonplace. It had been taught by G. F. Rouelle in his lectures as part of his instrument-element theory. See Rhoda Rappaport, "Rouelle and Stahl—The Phlogistic Revolution in France," *Chymia*, 7 (1961), 76–77. Professor Rappaport also calls attention (p. 77, note 13) to the exposition of this doctrine by P. J. Macquer. See also William Smeaton, "Guyton de Morveau and the Phlogiston Theory," *op. cit.* (note 55), p. 531.

pound of matter of fire with "une autre substance [qui] entre comme base." When this base is exposed to some substance, i.e., to a combustible, with which it has more affinity than with the matter of fire, it instantly combines with the combustible, freeing the matter of fire, though in the case of calcination this process is extremely slow and difficult to observe. The process, by contrast, is more rapid and obvious in the case of burning sulphur or phosphorus. Carbon is an interesting case; when burned it forms an acid known as *air fixé* or *acide crayeux.* It too breaks down the air, releasing the matter of fire; but less of this fire-matter is given off, because part of it combines with the newly formed acid to put it into that condition of an elastic vapor in which we obtain it. For this reason, too, the volume of air in which we burn carbon is not markedly diminished, as is the case whenever solid products are formed.

There is obviously no need, Lavoisier concludes, to imagine the existence of matter of fire or phlogiston in the bodies we describe as combustible. There is probably little *feu fixé* in metals or in solid combustibles; perhaps they only contain that *feu libre* which penetrates all bodies. Dephlogisticated air or *air pur,* he continues, is therefore the true combustible body, and perhaps the only one in nature. A powerful support for this interpretation is provided by his theory of the three states of matter. Heat (*feu fixé*) is necessary to liquify solids, and still more is required to vaporize them: "Autant donc il est prouvé que les substances aériformes . . . contiennent une grande quantité de feu combiné, autant il est probable que les corps solides en contiennent peu." This, of course, is what the experiments he had performed shortly before with Laplace had suggested might be the case, or so he believed. His new theory of combustion—not, he admitted, a "théorie rigoureusement démontrée, mais seulement une hypothèse qui me semble plus probable"—is manifestly superior to Stahl's. It involves fewer contradictions and its explanations are "moins forcées." It explains the elastic and other properties of aeriform fluids (for by combining with the matter of fire they "doivent conserver une partie des propriétés ignées"); and it explains, at the same time, the phenomenon of the gain in weight. One need no longer argue that when bodies gain in weight they actually lose part of their substance.

As Macquer's reaction shows, the older chemists were by no means persuaded. He perceived without difficulty the main thrust of Lavoisier's argument—that "son principe igné" combined with air "se dégage & produit les phénomènes de la combustion." To him, Lavoisier's *base de l'air,* "substance qu'il avoue lui être entièrement inconnue," was just as elusive and mysterious as Lavoisier believed phlogiston to be. But Lavoisier never made such an avowal, nor would he have done so, for the *base de l'air* manifested itself by the increase in weight of substances with which it combined, by the decrease in volume of the air during combustion, and by the chemical properties—the acidic character—it gave to the substances with which it combined. In the next few years, in a series of brilliant papers, Lavoisier was to make these matters much better understood.

There was, however, a familiar chemical phenomenon which, not long after, was advanced to refute Lavoisier's insistence that there could be little or no phlogiston or matter of fire in combustible substances. This was the evolution of inflammable air (hydrogen) when metals are treated with dilute acids. The study of this gas by Henry Cavendish, Richard Kirwan, and Joseph Priestley led them to identify it with pure phlogiston.[61] Experiment clearly showed, so these men and such French disciples of Priestley as Lamétherie argued, that metals obviously contain an abundance of phlogiston.[62] It was one of Laplace's important contributions to the New Chemistry, as we shall see later, to dispose of this objection.

There is little trace of active collaboration between Lavoisier and Laplace during the next two years. Although fully occupied with

[61] A letter of Joseph Priestley to Lavoisier (Birmingham, 10 July 1782) written to introduce a Dr. Stokes, "a young man who is a lover of science," imparted this interesting information: "I gave Dr. Franklin an account of some experiments which I have made with *inflammable air,* which he will probably have shown you, that seem to prove that it is the same thing that has been called *phlogiston.*" This letter, evidently referring to Priestley's successful reduction of calxes using hydrogen gas, does not appear in the *Correspondance de Lavoisier.* In the summer of 1939 I was privileged to consult it, and take extensive notes from it, at the Château de la Canière. It is doubtless now in the Archives de Chabrol. An early mention in print of Priestley's experiments on inflammable air is in a paper by Richard Kirwan, "Continuation of the Experiments and Observations on the Scientific Gravities and Attractive Powers of Various Saline Substances," *Philosophical Transactions, 72* (1782), 179.

[62] Lamétherie announced his discovery of an inflammable air obtained from metals by heat alone in a letter to the *Observations sur la physique, 18* (1781), 156. The following month (September 1781) he published further details and argued that inflammable air is a constituent principle of metals (*ibid.,* pp. 234–235).

his official responsibilities in the *Régie des poudres,* Lavoisier devoted some time to considering the chemical evidence in support of his new theory of chemistry. Laplace, for his part, must have been hard at work on those classic papers of his that appeared in the 1780's, dealing with the attraction of spheroids, the shape of the planets, and the secular inequalities of the planets and the satellites.[63]

In June 1779 Laplace and Lavoisier, together with Brisson and J.-B. Leroy, reported favorably to the Academy on a new form of air pump devised by the instrument maker Nicolas Fortin.[64] Though particularly concerned with the chemical consequences of his combustion theory, especially the constitution of acids and their compounds, Lavoisier read to the Academy, on 25 November 1780, his "Mémoire sur quelques fluides que l'on peut obtenir dans l'état aériforme, à un degré de chaleur peu supérieur à la température moyenne de la terre."[65] In this memoir he reported experiments on the properties of ether and alcohol in the elastic state and expounded his views concerning change of state and the influence of temperature and pressure. Although he described an improved apparatus for the production of water vapor, as devised by "M. de Laplace et moi," there is no evidence that Laplace took active part in these experiments.

RENEWED COLLABORATION: THE NEW KNOWLEDGE OF HEAT

The summer of 1781 marked the resumption of a collaboration that was to last until the spring of 1784 and involve the two men in investigations of major importance for the future of science. It

[63]Laplace's papers of this period ("Théorie des attractions sphéroïdes et de la figure des planètes," "Sur les variations séculaires des orbites des planètes," and others) are conveniently summarized by H. Andoyer, *L'oeuvre scientifique de Laplace* (Paris, 1922). See also *Oeuvres de Laplace.*

[64]*Oeuvres de Lavoisier, 4,* 427. Fortin became a few years later Lavoisier's favored instrument-maker. See Maurice Daumas, *Les Instruments scientifiques aux XVIIe et XVIIIe siècles* (Paris, 1953), *passim.*

[65]See above, note 42. In this paper Lavoisier used for the first time the term "air vital" for Priestley's "dephlogisticated air," following the suggestion of Condorcet, who proposed the expression in *Hist. Acad. R. Sci.,* 1777 (1780), p. 22. According to Lamétherie (*Essai analytique sur l'air pur,* 2nd ed., 2 vols. [Paris, 1788], *1,* 121), Condorcet owed the suggestion to Turgot.

began modestly enough with the problem of making a barometer with a flat meniscus, following the method of a Dom Casbois of the Benedictine College of Metz.[66] A certain M. Gaux had attempted to confirm Dom Casbois' results, but wished to have the matter examined by members of the Academy of Sciences. Laplace and Lavoisier worked with Gaux on experiments which involved the accurate determination of the thermal expansion of mercury, and Lavoisier reported their results to the Academy in August 1781.[67]

Later that year we learn of more significant inquiries. The two men became convinced of the importance of making precise measurements on the thermal expansion of glass and various metals. Several months must have been devoted to devising the telescopic pyrometer to be used in these experiments, and to supervising its construction. The remarkable apparatus, which Laplace may have helped design, was described to the Academy on 22 December 1781.[68] The experiments with this equipment were performed in the garden of the Arsenal in 1781 and 1782 but were not published in Lavoisier's lifetime; other preoccupations prevented the men from computing their results, and they did not return to the subject until the Revolution, when the question of establishing metric standards gave new importance to these earlier measurements.[69]

66Fols. 98–105 of Lavoisier's laboratory *registres* in Berthelot, *op. cit.* (note 13), p. 281. The results were described long after in the memoir "De l'action du calorique sur les corps liquides," *Mémoires de Chimie,* 1ère partie, pp. 295–311 (reprinted in *Oeuvres de Lavoisier, 2,* 773–781). The experiments probably exposed Laplace, for the first time, to the phenomenon of capillarity, on which he was later to write important papers.

67*Procès-verbaux,* 1781, f. 175. The reading, begun on 8 August, was completed on 11 August.

68*Procès-verbaux,* 1781, f. 254. The pyrometer, based on what Maurice Daumas calls a "comparateur optique," is described in *Oeuvres de Lavoisier, 2,* 752–755. The illustration on p. 753 was redrawn from Biot's *Précis élémentaire de physique expérimentale,* 2 vols. (Paris, 1817). Biot's drawing (*1,* Pl. III., Fig. 41) was a reconstruction made after consulting Laplace and Mme. Lavoisier for their memories of the apparatus. Later, however, Grimaux found the original plates and reproduced them (*Oeuvres de Lavoisier, 6,* Plates I and II). A model of this pyrometer exists in the Conservatoire des Arts et Métiers. See Daumas, *Lavoisier théoricien et expérimentateur,* p. 154.

69First printed in the posthumous *Mémoires de chimie* (1ère partie, pp. 246–280), the memoir is entitled "De l'action de calorique sur les corps solides." See also *Oeuvres de Lavoisier, 2,* 339–359. A reference to these experiments occurs in a letter of F. X. Schwediauer to Torbern Bergman, written from Paris on 19 October 1782: "Dnus [i.e., Dominus] Lavoisier nunc experi-

There is a significant reference to these experiments in the famous joint "Mémoire sur la chaleur," where it is suggested that it was not practical considerations alone, but also a theoretical one, that inspired them. Here, following their table of the specific heats of different substances, they comment:

> Nous nous proposons de continuer cette table, y comprenant un plus grand nombre de substances; il serait intéressant d'avoir dans un même tableau les pesanteurs spécifiques des corps. . .les dilatabilités des corps et leurs chaleurs spécifiques; la comparison de ces quantités ferait peut-être découvrir entre elles des rapports remarquables: nous avons fait, *dans cette vue,* un grand nombre d'expériences sur les dilatations, que nous nous proposons de publier, lorsqu'elles seront entièrement terminées.[70]

To determine what relationships might exist between the densities of bodies, their thermal expansibilities, and their specific heats or heat capacities was the objective of some of these experiments. It is therefore of significance to note that the first record of Lavoisier's interest in, or knowledge about, specific heats dates from precisely this period. Indeed it was only after 1780 that Lavoisier, like other French scientists, became aware of the work of Joseph Black and of Johann Karl Wilcke on this aspect of the science of heat.

Black, indeed, had published nothing relating to his discoveries concerning heat. The first report on Black's work on heat to reach France was a *précis* of his experiments on latent heat, read to the Academy of Sciences in August 1772 and published the following month in Rozier's Journal.[71] It was this report of Black's work that

mentis occupatur, quae diversos dilatationis et contractionis diversorum metallorum gradus in diversis caloris et frigoris gradibus exacte determinabunt." See Göte Carlid and Johann Nordström, eds., *Torbern Bergman's Foreign Correspondence* (Stockholm, 1965), *1,* 369.

[70]*Oeuvres de Lavoisier, 2,* 301-302. The emphasis in the quotation is my own. But see the more practical reason—the improvement of accurate astronomical clocks—advanced by Laplace and Lavoisier in the *Mémoires de chimie,* 1ère partie, p. 248. Even here they speak of other considerations, perhaps the theoretical ones, "qu'il seroit trop long de détailler," that prompted the experiments on the expansion and contraction of solids.

[71]*Introduction aux observations sur la physique,* 2 vols. (Paris, 1777), *2,* 428-431. I cite this more accessible quarto reprint edition, although the anonymous *précis* ("Expériences du Docteur Black, sur la marche de la Chaleur dans certaines circonstances") first appeared in the excessively rare duodecimo edition in September 1772.

impelled Lavoisier to publish in October 1772 in the same journal some observations on the thermal behavior of melting ice he had made "dans le courant de Septembre de l'année dernière," i.e., in 1771.[72] If we are to understand the extent, or better the limitations, of Lavoisier's knowledge of heat before 1781, we must look closely at these two documents.

Lavoisier's discovery of the latent heat of fusion was apparently quite accidental. He had wished to compare the readings of several thermometers at a series of gradually increasing temperatures. He placed the thermometers in crushed ice, intending to melt the ice and gradually warm the water. To melt the ice rapidly he poured over it a quantity of well-water, expecting the resulting temperature to be somewhere between the temperatures of the ice and the water. To his astonishment he observed no such result. After a quarter of an hour, when the ice was still not completely melted, the thermometers all remained at the freezing point; they did not begin to rise until the last parcel of ice was melted. The explanation of this surprising effect, he remarked cryptically, can be derived from a "System on the Elements" he has already registered with the Secretary, and which he proposes to present to the Academy on a future occasion. The explanation, which he did not give, clearly involved the theory he had already developed that heat, like air and water, can exist in two states, free and combined.

Reading the Black *précis* could have done little more than confirm, or perhaps clarify, Lavoisier's conjectures. After an introduction, evidently written by the Abbé Rozier, the anonymous author briefly describes a series of experiments. They are the classic experiments Black used to illustrate the phenomenon of the latent heat of fusion of ice. Evidently all bodies absorb heat in passing from the solid to the liquid state, and fluids turn into solids when deprived "d'une manière sensible de cette chaleur cachée qui causoit leur fluidité." There are therefore two kinds or conditions of heat, "l'une sensible & l'autre cachée." And the author of the *précis* continues with a brief reference to the latent heat of vaporization. According to Doctor Black, when water reaches the boiling point it

[72] "Expérience sur le passage de l'eau en glace, communiquée à l'Académie des Sciences, par M. Lavoisier," *Introduction aux observations sur la physique, 2,* 510–511. Lavoisier's communication first appeared in the duodecimo edition in October, 1772. For an account, and interpretation, of this episode see my *Lavoisier—The Crucial Year,* pp. 92–94.

absorbs a prodigious amount of heat from its surroundings, and it is this that changes it into a "vapeur élastique." To the thermometer, water vapor seems to be no hotter than boiling water, yet when condensed in an alembic it communicates more heat to the refrigerant than the same quantity of boiling water.

It is no wonder that Lavoisier saw in these experiments confirmation of his own doctrine of the two forms of heat and his theory of vaporization. He had no way of knowing, when he rushed his own observation into print, how much earlier Black had made his discoveries or how much more earnestly the Scottish chemist had pursued them. And in the *précis* we should note two striking omissions. There is no mention of Black's discovery of *specific heats,* and there is only the barest indication that it was possible to measure latent heats as distinguished from temperature.

There is good evidence that not until 1781 did Lavoisier become fully aware of these matters. In 1779, when he wrote the revised form of his "Mémoire sur la combustion en général," he did not believe that different solids at the same temperature could have different amounts of heat in them. He was, as we noted, anxious to show that the heat liberated during combustion came from the gas rather than from the solid,[73] and that no fixed fire existed in the combustible. He was, of course, aware that different substances gain or lose heat at different rates: this "marche de la chaleur" or "progrès de la chaleur" had been investigated by Boerhaave and more recently by Buffon. In the autumn of 1777 Lavoisier and Bucquet, we have seen, made similar studies on the rate at which different substances absorb heat. Whereas Boerhaave had concluded that dense bodies gain heat less rapidly than less dense ones, Buffon found the rates to be related to the fusibilities.[74] Of the liquids they studied, Lavoisier and Bucquet found that mercury gained heat most rapidly, but that the other liquids (ether, alcohol, essential oils, among others) gained heat in proportion to their inflammability.[75] Berthelot in paraphrasing one of these experiments labels it a measurement of specific heat.[76] This was certainly not the case, although

[73] See above pp. 220-221.

[74] By Boerhaave in his *Elementa chemiae* and by Buffon in his *Introduction à l'histoire des minéraux,* in the *Supplément à l'histoire naturelle* (Paris, 1774).

[75] E. McDonald, "The Collaboration of Bucquet and Lavoisier," *Ambix,* 13 (1966), 74-84.

[76] Berthelot, *op. cit.* (note 13), p. 278.

in the doctrine of specific heats lies the explanation for these effects. Indeed it was just such an observation recorded by Boerhaave (on the different rates of heating of water and of mercury) that led Joseph Black to his theory of different heat capacities.[77] The point that Lavoisier did not grasp until later was that all bodies (not merely bodies changing to a liquid or vapor state) must contain heat that is in some measure "fixed."

For nearly a decade Lavoisier's knowledge of the behavior of heat was, therefore, both vague and limited, and he was almost entirely ignorant of the work being carried out in the Scottish universities and by Wilcke in Sweden.[78] But a series of events in 1780–1781 profoundly altered the picture. In January 1780 the *Observations sur la Physique* published a French version of Peter Dugud Leslie's *A Philosophical Inquiry into the Causes of Animal Heat* (London, 1778). After referring to the views of some Ancient and seventeenth-century physicians, Leslie discusses those of Cromwell Mortimer, Stephen Hales, Robert Douglas, William Cullen, and finally Joseph Black. Of the latter he writes:

> Une des plus ingénieuses hypothèses sur la chaleur animale, est celle du Docteur Black. Ce savant a observé, que non-seulement les animaux qui respirent sont les plus chauds de tous, mais encore qu'il y a une connexion si frappante & si intime entre l'état de la respiration & le degré de chaleur dans les animaux, que ces deux choses paroissent être dans une proportion exacte l'une avec l'autre; & il en a conclu que la chaleur animale dépend de l'état de la respiration; qu'elle est produite dans le poumon par l'action de

[77] See my sketch of Joseph Black in *Dictionary of Scientific Biography, 2* (1970), 173–183.

[78] For Wilcke, see Douglas McKie and Niels H. de V. Heathcote, *The Discovery of Specific and Latent Heats* (London, 1935), pp. 54–121. Wilcke's discovery was made early in 1772 and published in the *Handlingar* of the Royal Swedish Academy of Sciences in that year; his paper became accessible in a German translation in A. G. Kästner's *Die Königl. Schwedische Akademie der Wissenschaften, Abhandlungen auf das Jahr 1772* (Leipzig, 1776), pp. 93–116. Only the vaguest hints appeared in J. A. Deluc's *Recherches sur les modifications de l'atmosphère,* 2 vols. (Geneva, 1772), a work with which Lavoisier became familiar. There is no reason to think that he or anybody else in France had encountered the anonymous compilation, pirated from Black's unpublished lectures: *An Enquiry into the General Effects of Heat: with Observations on the Theories of Mixtures* (London, J. Nourse, 1770). On this work see Douglas McKie and Niels H. de V. Heathcote, *Specific and Latent Heats,* p. 51.

l'air sur le principe de l'inflammabilité, à peu près comme on le voit dans l'inflammation ordinaire, & que de là, elle se repand par le moyen de la circulation dans le reste du système vital.[79]

No mention is made of Adair Crawford or of the Swedish experiments. But this ignorance was soon remedied. The first volume of Torbern Bergman's *Opuscula physica et chemica,* published in Stockholm in 1779, contained a brief mention of Wilcke's experiments on latent heat. Late the following year appeared the first volume of Guyton de Morveau's French translation of the *Opuscula.*[80] It was Bergman's brief account that brought Wilcke's work to the attention of European scholars. The next year (1781) the Chevalier Landriani included in his *Opuscoli Fisico-Chimici* a dissertation on latent heat, which refers to Wilcke but credits the discovery of the phenomenon to Joseph Black.[81] In this same year the Baron de Dietrich published his French translation of Carl Wilhelm Scheele's book on air and fire, a work on which, it should

[79] Pierre Dugud Leslie, "Recherches philosophiques sur la cause de la chaleur animale," *Observations sur la physique, 15* (1780), 24-29. See esp. p. 28. For Leslie see Everett Mendelsohn, *Heat and Life. The Development of the Theory of Animal Heat* (Cambridge, Mass., 1964). Mendelsohn incorrectly gives Leslie's first name as Patrick.

[80] *Opuscules chymiques et physiques. Recueillis, revus et augmentés par lui-même. Traduits par M. de Morveau, avec des notes,* 2 vols. (Dijon, 1780-1785). The mention is in *1,* 247-248, in the article "Des eaux médicinales chaudes, artificielles," note p. 247. For the date of the appearance of the French translation see *Bergman's Foreign Correspondence, 1,* 107-113.

[81] *Opuscoli Fisico-Chimici del Cavaliere Marsilio Landriani* (Milan, 1781). The dissertation "Del calor latente" appears on pp. 81-149. Landriani seems to have been influenced by the précis of Black's work that had appeared in Rozier's Journal. He knew also of Lavoisier's short paper on the melting of ice; after describing the experiments of Black showing that a mixture of ice and warm water gave a temperature of 32° Fahrenheit, he adds "come è stato osservato anche del Sig. Lavoisier." Lavoisier's short paper in Rozier's Journal is cited in Landriani's note. But he was also familiar with the work of Wilcke through the "bellisima dissertazione" of Bergman. Similar references to Black and to Lavoisier's paper are found in Romé de L'Isle, *L'Action du feu central démontré nulle à la surface du globe,* 2nd ed. (Paris, 1781), pp. 104-105 and 114-117, but there is no mention of Wilcke. In his *Leçons élémentaire d'histoire naturelle,* 2 vols. (Paris, 1782), *1,* 44-56, Fourcroy has a long discussion of the material theory of heat and mentions Bergman, but there is no discussion of the new theory of latent heat. A full account is, however, given in the second edition of this work, now called *Elémens d'histoire naturelle et de chimie,* 4 vols. (Paris, 1786), *1,* xxvi, xxx, 119, 125.

be mentioned, Lavoisier and Berthollet reported to the Academy in August. In the translator's preface we read: "Tandis que M. Lavoisier se voyoit réduit à ne semer qu'avec précaution, dans ses différens Ouvrages, les apperçus de principes nouveaux sur la physique de la chaleur et du feu, M. Wilke les publioit à Stockholm: le Docteur Black à Edinbourg & le Docteur Irvine à Glasgow, en étoient imbus dès longtemps."[82] And Dietrich adds a note to his preface which clearly refers to the joint paper of Lavoisier and Laplace, which the latter read on their behalf at the Easter *rentrée publique* of 1777.[83]

The real stimulus for this interest in the new physics of heat was the publication in 1779 of Adair Crawford's *Experiments and Observations on Animal Heat.*[84] Crawford's book has been severely criticized by historians, especially those concerned to establish Black's priority in the discovery of latent heat.[85] Yet its influence on the Continent, as even these historians have perceived, was

[82]*Traité chimique de l'air et du feu, Par Charles-Guillaume Scheele, ouvrage traduit de l'allemand. Par le Baron de Dietrich* (Paris, 1781). For the review by Lavoisier and Berthollet see *Procès-verbaux,* 1781, f. 175 r. & v., and *Oeuvres de Lavoisier, 4,* 377-378.

[83]Dietrich's note reads: "M. Lavoisier engagea un de ses amis à lire dans une des Séances publiques de l'Académie Royale des Sciences de 1776 [*sic*], un Mémoire qui contenoit ses idées à ce sujet: elles parurent très-extraordinaires, & n'ont germé que dans la tête d'un petit nombre de Physiciens. Nous sommes malheureusement sujets à accueillir assez mal les nouveautés: elles choquent notre amour-propre, & dérangent le petit cercle d'idées que nous nous étions formées."

Philippe-Frédéric de Dietrich (1748-1793), a mineralogist and industrialist of Strasbourg, was well known to Lavoisier. On a visit to Paris, in March 1778, he repeated with Lavoisier the experiments of Volta on the inflammable air obtained from marshes. Since Lavoisier and Berthollet reviewed Dietrich's book in manuscript (they recommended its publication under the *privilège* of the Academy) it is likely that Lavoisier may have inspired Dietrich's comments in his "Avis du traducteur."

[84]*Experiments and Observations on Animal Heat, and the Inflammation of Combustible Bodies. Being an Attempt to Resolve those Phenomena into a General Law of Nature* (London, 1779). A second edition "with very large additions" appeared in 1788. For the general problem of animal heat in this period see Mendelsohn, *Heat and Life.* A brief account of Crawford's influence on Lavoisier is given in my "Laplace's Collaboration with Lavoisier," *Actes du XIIe Congrès International d'Histoire des Sciences,* Tome III B (Paris, 1971), pp. 31-40. See also Robert J. Morris, "Lavoisier and the Caloric Theory," *The British Journal for the History of Science, 6* (1972), esp. pp. 9-15.

[85]For example, McKie and Heathcote, *Specific and Latent Heats,* pp. 37-40.

tremendous. The book contains, to be sure, very little on the theory of latent heat. The first part of the book is devoted to setting forth the doctrine of specific heats (though the term itself is not used; Crawford speaks instead of heat capacities of bodies) and to recording experimental results obtained by the "method of mixtures," which he illustrates by the use of water and linseed oil. Later sections apply these experiments and theories to his doctrine of animal heat and to his theory of combustion. It is easy to see why Lavoisier should have found these ideas particularly compelling. Crawford believed that animal heat comes from the air an animal breathes, for he thought he had proved that the heat capacity of atmospheric air (or of "pure" or dephlogisticated air) was greater than that of the expired "fixed air." By the same token in the combustion of bodies—which Crawford describes in phlogistic terms—the heat and light must come from the dephlogisticated air. When the phlogiston leaves the combustible, it increases that substance's heat capacity, and by phlogisticating the air it decreases its specific heat. According to his results, experiment confirms this: metals, for example, have lower heat capacities than their corresponding calxes.

Crawford's book was not translated into French, and the first London edition of 1779, which was sold out within a few months, became exceedingly rare.[86] Yet the French scientists were familiar with his theories at least as early as the summer of 1781.[87] And if they were not able to consult the book directly, they could learn of its contents from that busy intelligencer J.-H. de Magellan. As early as August 1779 Magellan was writing to his European correspondents about Crawford's book, describing it as setting forth a "nouvelle branche de physique expérimentale," and summarizing the ideas

[86]See "Lettre de M. Magellan à l'Auteur de ce Journal, sur le Mémoire suivant," *Observations sur la physique, 17* (1781), 370. An Italian translation of Crawford's book was published in C. Amoretti and F. Soave, *Opuscoli scelti sulle scienze,* in 1780.

[87]In a report by Lavoisier and Macquer (27 June 1781) on a work of chemistry by Berthollet (*Oeuvres de Lavoisier, 4,* 379–387) appears the following sentence: "Il [Berthollet] développe, dans un discours préliminaire, les opinions modernes sur la nature des métaux, sur les phénomènes de leur calcination, de leur revivification. Il rend compte de la théorie de Stahl, de celle de M. Scheele, de celle de M. Crawfort [*sic*] et de quelques autres; et . . . il se décide pour la théorie de Stahl, à laquelle néanmoins il est obligé d'apporter des modifications considérables, et à cet égard il se rapproche beaucoup de l'opinion de Macquer." Berthollet had not yet come around to Lavoisier's theory.

contained in it.[88] By the end of the year he had begun to prepare an analysis or synopsis of it.[89] Soon thereafter he called the attention of the French scientists to it in a letter to the editor of the *Observations sur la physique,* in the course of which he wrote as follows:

> Je voudrais bien savoir, si l'Ouvrage de M. Crawford sur la chaleur animale, qui fut publié ici cet été dernier, a engagé vos Philosophes à parcourir le vaste champs qu'il ouvre dans la Physique. Le principe de la *chaleur cachée* (*latent heat*) découvert par le Docteur Black lui sert admirablement pour résoudre le grand problème animal; c'est sur ce même principe qu'il a établi son beau système de la combustion & de l'ignition; s'est d'après lui qu'on peut comprendre la cause de la chaleur produite par différens mélanges de certain fluides froids, de la chaleur excitée par l'eau dans la chaux vive, & d'autres phénomènes qui étoient inconcevables jusqu'à présent. C'est d'après l'évidence incontestable des faits que M. Crawford a établi sa doctrine & non pas sur des observations à la hâte. . .comme plusieurs de nos prétendus Philosophes modernes le font assez souvent.[90]

In 1780 Magellan published in London a long essay, actually something more than a synopsis of Crawford, entitled *Essai sur la nouvelle théorie du feu élémentaire, et de la chaleur des corps;* this was reprinted in Rozier's Journal in the issues of May and June 1781, prefaced by a letter to the editor, announcing that his memoir

[88]Magellan to Volta (5 August and 14 September 1779) in *Epistolario di Alessandro Volta, 1* (Bologna, 1949), 359 and 370. It is interesting that in these letters Magellan, who later gave the priority for the discovery of latent heat to Wilcke, here accords it to Black: "On doit les premiers pas au Dr. Black d'Edimbourg pour l'invention du *Principe* . . . le développement au Dr. Irwine de Glasgow; et l'application au Dr. Crawford." (*Op. cit.,* p. 70.) This may have influenced Landriani's opinion.

[89]Magellan to Volta (28 December 1779), *Epistolario, 1,* 388. Magellan's transmission of a copy of Crawford's book to Italy may have helped stimulate publication of Landriani's work on latent heat. Landriani had close relations with Volta.

[90]"Extrait d'une lettre de M. Magellan, de la Société Royale de Londres, sur les Montres nouvelles qui n'ont pas besoin d'être montées, sur celles de M. Mudge & sur l'Ouvrage de M. Crawford," *Observations sur la physique, 16* (1780), 62-63. On Magellan's role as an "intelligencer," transmitting scientific information between Britain and the Continent, see my *Lavoisier—The Crucial Year, passim.*

dealt with "une branche nouvelle de connoissances dans la Physique."[91]

It was Magellan's purpose to outline in more methodical fashion than Crawford had done the main discoveries of this new branch of physics.[92] He introduces and defines the new terms, using apparently for the first time the expression "chaleur spécifique";[93] he explains in detail the method of mixtures for determining specific heats; and he publishes the first table of specific heats of familiar substances, a contribution of his friend Richard Kirwan. The greater part of Magellan's *Essai* is devoted to specific heats, and though he summarizes Crawford's ideas about animal heat and his theory of combustion, much less space is devoted to them.

There can be little doubt that it was Crawford's book, or rather Magellan's exposition of it, that launched Laplace and Lavoisier on the best known aspect of their scientific collaboration, their joint work on calorimetry. Crawford is several times mentioned by name in their great "Mémoire sur la chaleur," of which the first part is devoted to specific heats and the method of mixtures. Lavoisier's

[91]*Observations sur la physique, 17* (1781), 369–386 and 411–422. Before his *Essai* was finished, Magellan had encountered Bergman's reference to Wilcke's work and wrote to James Watt to learn when Black had hit upon the notion of latent heat. Watt accordingly wrote Black asking the latter "if you choose to furnish me with a few facts relative to the time, &c., of your invention, and an account of any publications in which it has been mentioned. . . ." Magellan, who ultimately gave grudging priority to Black as a result of Watt's prodding, nevertheless thought Wilcke, because he had published his experiments, deserved the principal credit. For the letters of Watt and Black see James Patrick Muirhead, *The Mechanical Inventions of James Watt,* 3 vols. (London, 1854), *2,* 116–120.

[92]Magellan's justification for writing his *Essai,* besides the difficulty of obtaining a copy of Crawford's book, was that "he has been so much harassed writing to his friends abroad concerning these new discoveries, as published by Dr. Leslie and Dr. Crawford." See Muirhead, *op. cit.* (note 91), p. 117. It was published before the end of August 1780, when he sent a complimentary copy to Bergman.

[93]The term *chaleur spécifique* seems to have been introduced by Magellan. See J. R. Partington, *History of Chemistry, 3,* 156. But in the Swedish equivalent (*specific-varme*) Wilcke employed it soon after. His work on specific heats was published in *Kongl. Vetenskaps Academiens nya Handlingar, 2* (1781), 49–78. A French translation was published in Rozier's Journal four years later (*Observations sur la physique, 26* [1785], 256–268 and 381–389). Landriani's essay on latent heat appeared in French dress in the same year (*op. cit.,* pp. 88–100 and 197–207).

contemporaries, and even Lavoisier himself, spoke of these experiments as repetitions of those of Crawford.[94]

THE EXPERIMENTS WITH ALESSANDRO VOLTA: A SURPRISING INTERLUDE

Lavoisier and Laplace seem to have devoted the last months of 1781 and the early months of 1782 to their experiments on the expansion of solids and to exploring ways of measuring latent and specific heats.[95] During these months the famous ice calorimeter must have been conceived and designed, for it was used for the first time, in preliminary fashion, in the summer of 1782. But for several

[94]On 17 March 1783, before the results of the experiments had been reported to the Academy, Richard Kirwan wrote to Torbern Bergman from London: "Mr Lavoisier vient de Répéter les Expériences du Docteur Crawford et les a trouvé [*sic*] vraies." (*Bergman's Foreign Correspondence, 1,* 186.) This echoes a letter of Lavoisier to Schwediaur, the original of which is now apparently lost. It is referred to as follows in a letter from Schwediaur to Bergman dated the same day: "Mr Lavoisier mihi in epistola heri tradita dicit, sese experimenta circa calorem animalem methodo nova, a crawfordiana diversa et multum eâdem certiori, repetisse, *avec des résultats presque les mêmes que ceux de Mr. Crawford.* Hic ultimus novam editionem opusculi sui cum pluribus novis experimentis hac aestate edere intendit." (*Op. cit.,* p. 373.)

What is apparently an undated draft of the letter to Schwediaur is published in *Correspondance de Lavoisier, 3,* 734–735. Lavoisier writes: "Nous sommes toujours occupés M. de la place [*sic*] et moi d'expériences sur la chaleur. Nous avons trouvé un moyen de mesurer la chaleur spécifique des corps d'une manière plus sure que la méthode des mélanges. Elle peut d'ailleurs s'appliquer à des dégagemens successifs et lents de chaleur, ce que l'on ne peut mesurer par les moyens indiqués par M. Kraffort [*sic*]. Nous publierons quelque chose incessament sur cet objet. Nos résultats s'accordent asses exactement jusqu'icy avec ceux de M. Kraffort quoique notre manière d'opérer soit bien différente." I have modernized the punctuation and capitalization in the above citation, as generally elsewhere.

[95]On 16 November 1781, if we may trust Berthelot's interpretation of Lavoisier's laboratory *registres,* Lavoisier was attempting to determine the heat absorbed when ether vaporizes and on 8 December he was using the method of mixtures to determine the heat necessary to melt wax and tallow. (Berthelot, *op. cit.* [note 13], p. 282). This is a clear echo of Crawford. In his *Experiments and Observations,* 2nd ed. (1788), p. 82, we read: "Dr. Black has found that many substances, as tallow, spermaceti, bees-wax, rosin, have a fixed temperature, when they are undergoing a change of form by melting, or by congelation. Whence he infers, that during the former process they absorb, and during the latter, evolve heat." I have not determined whether this passage appears in the first edition of Crawford's book.

months of the preceding winter and spring a quite different, though not wholly unrelated, problem occupied the two men.

On 6 March 1782 Lavoisier and Laplace announced to the Academy that when substances vaporize, detectable electric charges are produced.[96] The occasion for these experiments, and the more successful ones that followed in March and April, was the presence in Paris of the Italian physicist Alessandro Volta.

Volta had set out in September 1781 on a scientific grand tour of Europe.[97] After visiting Switzerland, Germany, and Belgium, he reached Paris before Christmas, intending to spend the winter in the French capital. In mid-January he wrote his brother that he had met the principal *savants* of the city, had attended a meeting of the Academy of Sciences, and was faithfully following a course in physics given by J.-A.-C. Charles and one in chemistry by Balthazar-Georges Sage.[98] Later letters tell of his contacts with Franklin and Buffon, and the dinners he enjoyed with these men and with J.-B. Le Roy and Lavoisier.[99] Before private groups and at the Academy of Sciences, Volta showed off his newly devised condensing electroscope, a sensitive instrument for detecting extremely weak electrostatic charges, performed various experiments, and expounded his theories.[100] Among these theories was one that had occurred to

[96]*Procès-verbaux,* 1782, f. 35v. "MM. Lavoisier et Delaplace ont pris date pour une suite d'expériences déjà commencées dont il résulte que différentes substances au moment où elles passent de l'état de liquide à l'état aériforme" show signs of electrification.

[97]Volta had already a substantial reputation in Paris by the time of his arrival: his experiments on marsh gas had been repeated by the Baron de Dietrich in 1778 (Volta had earlier demonstrated them to Dietrich in Strasbourg); in the same year there was published his letter to Priestley describing his "pistolet," a glass eudiometer in which common and inflammable airs could be exploded by an electric spark (*Observations sur la physique, 12* [1778], 365–373). A second letter containing further experiments with this device appeared the following year (*op. cit., 13,* 278–303).

[98]Postscript to a letter to his brother, Arcidiacono Luigi, of 15 January 1782, in *Epistolario di Alessandro Volta, 2* (Bologna, 1951), 79. Deluc later wrote that he had the good fortune "de me lier personnellement à Paris avec M. Volta." (*Idées sur la météorologie,* 2 vols. [Paris, 1787], *1,* 3.) See below note 132.

[99]Letters of 9 and 19 February, 24 March, and 1 April 1782 in *Epistolario, 2,* 81–82, 85–87, 93–95, 96–97. Volta's last letter from Paris was written on 23 April, the eve of his departure.

[100]Volta's demonstration before the Academy took place on 13 March, as we learn from notes that Gaspard Monge took on this occasion. See René Taton, *L'Oeuvre scientifique de Monge* (Paris, 1951), p. 326. Volta wrote an account of this to his brother on 24 March (*Epistolario, 2,* 94).

him long before: atmospheric electricity, he believed, might be caused by the evaporation of water into the atmosphere. He had even, so he tells us, attempted unsuccessfully an early experiment to show that evaporation was in fact accompanied by electrification.[101] Some two years before his Paris visit, having found his newly invented *condensatore* able to accumulate very small charges, he thought of repeating his earlier experiments, "but various occupations deferred these experiments till the months of March and April of the present year 1782," when he performed them "in company with some members of the Royal Academy of Sciences."[102]

To his report on these experiments, written shortly after they were performed, Volta appended what is probably the best account of the stages in this investigation.[103] After Volta showed his experiments with the condensing electroscope to Lavoisier and Laplace ("two intelligent philosophers and members of the Royal Academy of Sci-

[101]According to Volta, his first ideas on atmospheric electricity are contained in his Latin memoir of 1769: *De vi attractiva ignis electrici.* But much earlier Benjamin Franklin had advanced a similar theory, though without conviction, in a letter to John Lining of 18 March 1755. Here Franklin mentions an experiment—identical in principle with those successfully carried out in Paris—to determine whether steam is electrified. Franklin's letter was first printed in the fourth English edition (1769) of his *Experiments and Observations;* it appears also in the third French edition of 1773.

A similar theory was advanced by Horace-Benedict de Saussure in his *Dissertatio physica de electricitate* (1766), a work I have not seen; but he refers to these early speculations in his *Essais sur l'hydrométrie* (Neuchâtel, 1783) where he writes (p. 274, note 1): "J'ai vu depuis que Mr. Francklin [*sic*] avoit eu avant moi la même pensée."

Volta was aware that Franklin, De Saussure, and apparently others had advanced the same explanation of atmospheric electricity. In recounting the Paris experiments in a letter to Count Firmian (7 May 1782) he wrote: "Negli ultimi giorni che dimorai in Parigi ho avuto occasione di fare delle sperienze in compagnia del sig. Lavoisier, e del sig. De La Place . . . nelle quali fummo più felici che il sig. Franklin e il sig. de Saussure, ed altri non erano stati." (*Epistolario, 2,* 104.)

[102]Volta's account of his measuring instrument and of the Paris experiments was published in the *Philosophical Transactions, 72* (1782), 237-280. It is followed by an English translation, doubtless the version read to the Royal Society. This translation, from which I quote, appears as a separately paginated Appendix to Part I of that volume (pp. vii-xxxiii). Here too Volta shows his awareness that others beside himself had considered the evaporation theory: "It has been questioned, whether evaporation, fermentation, &c produced any electricity." (*Op. cit.,* p. xviii.)

[103]For the historical account by Volta see *op. cit.,* pp. 274-280. The English version appears on pp. xxix-xxxiii of the Appendix.

ence"), the three men discussed the possibility of detecting the weak electrical charges that might be produced on evaporation. And Volta's account continues: "Accordingly Mr. Lavoisier ordered a large condenser with a marble plane to be made. The first experiment I attempted with this instrument, in company with Mr. de la Place, proved unsuccessful; but the weather was at that time bad, the room was narrow, and full of vapours, and the apparatus was not quite in proper order."[104] But not long after, Volta reports, successful experiments were performed by Laplace and Lavoisier "at a country-place of the latter."[105] Volta did not take part in this experiment, the success of which led Lavoisier and Laplace to make their report to the Academy on 6 March. Whether it was water or some other substance that was vaporized is nowhere mentioned.

This success, Volta continues, "incited us to repeat and diversify the experiments, by which means the discovery was completed; having obtained unequivocal signs of electricity from the evaporation of water, from a simple combustion of coals, and from the effervescence of iron filings."[106] Their experiments are described in the short memoir by Lavoisier and Laplace, initialed by Condorcet on 7 April, and published in the *Mémoires* of the Academy for 1781.[107] In their memoir the two authors acknowledge that the experiments were made possible by the "condensateur électrique imaginé par M. de Volta," and they conclude the article by testifying to Volta's participation in these last experiments.[108]

[104] *Op. cit.*, p. xxix.

[105] Here the English version is misleading; we read only that "Mr. de la Place and Mr. Lavoisier repeated these experiments in the country." Volta actually wrote (*op. cit.*, p. 275): "All incontre quelli che ripeterone l'estesse Sig. de La Place e Sig. Lavoisier *ad una compagna di quest' ultimi.*" My emphasis. This probably does not refer to Lavoisier's estate at Fréchines near Blois, but to the small property at Le Bourget, on the outskirts of Paris, which he had inherited from his mother and used earlier at least once for an experiment.

[106] *Op. cit.*, p. xxix.

[107] *Mém. Acad. R. Sci.*, 1781 (1784), pp. 292–294, and *Oeuvres de Lavoisier, 2,* 374–376.

[108] "M. de Volta a bien voulu assister à nos dernières expériences, et nous y être utile; la présence et le témoignage de cet excellent physicien ne peuvent qu'inspirer de la confiance dans nos résultats." (*Oeuvres de Lavoisier, 2,* 376.) According to Deluc, who visited Paris the following year, there was no great enthusiasm among the French scientists for Volta and his theories, except for Laplace and "un peu Mr. Lavoisier." Deluc to Volta, letter of 29 June 1783 in *Epistolario, 2,* 162–166.

It should not surprise us that Lavoisier and Laplace welcomed Volta's presence in Paris, profited by their conversations with him, and were prompted to explore the electrical phenomena association with vaporization. As we saw, the subject of vaporization had led to their first cooperative experiments in 1777. This was the aspect of the experimental results obtained with Volta which chiefly interested Lavoisier. In a memoir on the formation of vapors, published in 1784, he alluded to these experiments and the theoretical interpretation he placed upon them. This he said he proposed to develop in a later memoir, when he would explain why he was led to believe that: "les phénomènes électriques que nous observons ne sont qu'un effet de la décomposition de l'air; que l'électricité n'est qu'une espèce de combustion, dans laquelle l'air fournit la matière électrique, de même que suivant moi, il fournit la matière du feu et de la lumière dans la combustion ordinaire."[109] The promised memoir was never written; but long after, when he revised for publication in his *Mémoires de chimie* the joint paper with Laplace on the electricity produced by vaporization, he added the following concluding paragraph:

> Ces expériences pourroient porter à croire que le fluide électrique est susceptible d'entrer dans la composition des gaz et des fluides aëriformes, à peu-près de la même manière que la calorique et la lumière, mais en beaucoup plus petite quantité. Il ne seroit pas étonnant dans cette supposition que ce fluide reparut au moment où les fluides aëriformes reviennent à l'état liquide: c'est ce qui doit arriver toutes les fois qu'il y a précipitation d'eau dans l'atmosphère, c'est-à-dire, formation de nuages et de pluie. C'est sans doute à cette portion de fluide électrique devenue libre, que sont dus les phénomènes électriques qu'on observe dans les tems d'orage. L'évaporation de l'eau à la surface de la terre, enlèveroit continuellement, s'il en étoit ainsi, du fluide électrique qui seroit

[109] "Mémoire sur quelques fluides que l'on peut obtenir dans l'état aériformes," *Oeuvres de Lavoisier, 2,* 269–270. Later, early in 1786, Lavoisier attempted an experiment to test the relation between electrification and the presence of air, using a device built by Nicolas Fortin in which an electrostatic generator was operated in the receiver of a vacuum pump. Lavoisier records in his laboratory *registres* that "l'électricité tend vers zero, à mesure que le vide augmente." The same entry (23 February 1786) explains: "Je soupçonne depuis longtemps que les phénomènes électriques ne sont, comme ceux de la combustion, qu'un effet de la composition de l'air; que l'électricité n'est autre chose qu'une combustion très lente." (Berthelot, *op. cit.* [note 13], p. 304.)

ensuite rendu à son état de liberté, lorsque l'eau, dissoute dans l'atmosphère, reprendroit son état liquide.[110]

The evaporation theory of atmospheric electricity, so clearly set forth by Lavoisier in these posthumously published *Mémoires*, has long since been abandoned. Yet for a time in the early nineteenth century it carried considerable conviction, enough indeed to stimulate a heated controversy—with respect to the credit for the idea and the experiments that seemed to support it—between the partisans of Volta on one side and those of Lavoisier and Laplace on the other.[111] As late as 1888 the theory was still mentioned, though with skepticism, in a widely popular textbook of physics.[112] Even today we still have no satisfactory or complete explanation of how clouds acquire their electricity; the problem has proved to be immensely complex.

We should not imagine that the experiments with Volta were a mere distraction, unrelated to the central interest in the subject of heat. In the minds of all three men the two problems were closely linked because of that "analogy"—which Benjamin Franklin among others often stressed—between the behavior of the electrical fluid and that of the "fluid of heat." Both Lavoisier and Laplace, in their memoir, and Volta—who had arrived in Paris quite familiar with the discoveries concerning heat—were convinced that the results of their experiments had greatly strengthened this analogy. If Lavoisier and Laplace merely mention this in passing,[113] Volta on the other hand

[110]*Mémoires de chimie, 1,* 333–334.

[111]*Collezione dell' Opere del Cavaliere Conte Alessandro Volta,* 3 vols. (Florence, 1816), *1,* Part 2, 148 note. See also the sketch of Volta in *Oeuvres de François Arago, 1* (Paris, 1854), 187–240. In this *éloge,* read on 26 July 1831, Arago devotes several pages to the problem of atmospheric electricity and the evaporation theory, evidently considering it a discovery of significance. He also discusses at length the problems of priority for the discovery; but his treatment of the Paris experiments of 1782 has several inaccuracies: even the year he gives is wrong.

[112]A. Privat Deschanel, *Elementary Treatise on Natural Philosophy,* tr. J. D. Everett, 8th ed. (London, 1888), p. 650. Even more recently, in a quite different connection, R. A. Millikan (*The Electron,* 2nd ed. [1924], p. 46) saw fit to refer to the experimental results of Lavoisier and Laplace ("that the hydrogen gas evolved when a metal dissolves in an acid carries with it an electric charge") as a preliminary to his account of early attempts to measure the charge on the electron.

[113]"Il était naturel de penser, d'après ces résultats, que les corps qui se réduisent en vapeurs enlèvent de l'électricité à ceux qui les environnent, ce qui paraît d'ailleurs conforme à l'analogie observée entre l'électricité et la chaleur." (*Oeuvres de Lavoisier, 2,* 376.)

is more explicit. In the paper he read soon after to the Royal Society of London he wrote that from these experiments it appears that when bodies are transformed into volatile fluids "their capacity for holding electric fluid is augmented, as well as their capacity for holding common fire." And he goes on: "This is a striking analogy by which the science of electricity throws some light upon the theory of heat, and alternatively derives light from it: I mean on the doctrine of latent or specific heat, the first notions of which were suggested by Dr. Black and Wilke [*sic*], and which has been afterwards much elucidated by Dr. Crawford, who followed the experiments of Dr. Irwin [*sic*]."[114]

THE EXPERIMENTS ON HEAT AND CALORIMETRY

We can hardly doubt that the problem of investigating heat—in the light of the new concepts and new methods that came to them from Britain—was a subject of active discussion during the time of Volta's visit, for the calorimeter must have been nearing completion. This throws at least some light upon the strange letter that Laplace wrote Lavoisier on 7 March 1782, the day following the presentation of the joint paper on the electricity of vaporization.[115] Clearly there had been an agreement between the two men to embark together on an ambitious program of research on heat, for Laplace begins his letter by referring to "l'engagement que j'ai pris, de faire avec vous une suite d'expériences & de recherches sur la dilatation, la chaleur & l'électricité des corps." Laplace then requests, with warm expressions of appreciation, to be relieved of this commitment, so as to be free to devote himself to mathematics. Surely, if he was ever to dissolve this partnership, this must have seemed the moment to do so, for the two men would soon be deeply involved in a major effort of experimentation. One of the arguments that Laplace advances is that if he continues the collaboration he will be forced to study and master (*compulser*) all the new publications that have been appearing. Not all of them, he remarks, are written with the conciseness one could desire. In some of them "peu de vérités sont

[114] *Philosophical Transactions, 72* (1782), xxxii.

[115] *Correspondance de Lavoisier, 3,* 712–714. This letter was first published by Denis I. Duveen and Roger Hahn in the *Revue d'histoire des sciences et de leurs applications, 11* (1958), 337–342.

noyées dans de gros volumes," a remark that could admirably apply to the diffuse, but indispensable, volumes of J.-A. Deluc.[116]

Lavoisier, quite evidently, was successful in dissuading Laplace from this course. The first experiments were undertaken in July—attempts to measure the relative specific heats of mercury and water[117]—but the investigators discovered that it was necessary to operate when the weather was colder. The experiments, temporarily set aside, were resumed in November and continued through the early months of 1783.[118] The results of their calorimetry investigation, forming the substance of the famous "Mémoire sur la chaleur," were read to the Academy of Sciences at the sessions of 18 and 25 June.[119] Before its appearance in the *Mémoires,* this famous paper

[116]Deluc's *Recherches sur les modifications de l'atmosphère,* 2 vols. (Geneva, 1772), contained interesting material on precise thermometry, and on the need to reform Réaumur's thermometer. See *Recherches, 1,* Chap. 2, Nos. 422 *et seq.* It was frequently cited by Lavoisier. Both Lavoisier (*Oeuvres, 2,* 288) and Magellan (*Observations sur la physique, 17* [1781], 377) cite Deluc as having proved that mercury expands regularly, or nearly so, with increasing heat. On Deluc see the article by Robert P. Beckinsale in the *Dictionary of Scientific Biography, 4,* 27–29. Deluc's work on thermometry is briefly discussed by W. E. Knowles Middleton in his *History of the Thermometer and Its Use in Meteorology* (Baltimore, 1966), Chap. 5, p. 117, n. 112, who describes the work as "incredibly discursive" and cites J. H. Lambert as saying of it caustically, in his *Pyrometrie* (Berlin, 1779, p. 70), "Had Deluc written a work four times as small, he might have said four times as much as he did say." Cf. Henry Carrington Bolton's *Evolution of the Thermometer* (Easton, Pa., 1900), pp. 83–84.

[117]Berthelot, *Révolution chimique,* p. 285.

[118]*Ibid.,* pp. 285–286. A letter of Laplace to his friend the mathematician Lagrange (10 February 1783) shows that at this time Laplace was in close association with Lavoisier and his wife. See Lagrange, *Oeuvres* (Paris, 1892), *14,* 123. In the "Mémoire sur la chaleur," first published as a preprint in the summer of 1783, the authors describe the experiments as having been performed "pendant l'hiver dernier," i.e., during the winter of 1782–1783 (*Oeuvres de Lavoisier, 2,* 283).

[119]The *Procès-verbaux* (1783, f. 144) suggest that the memoir was read by Laplace: "M. Delaplace [*sic*] a lu un mémoire conjointement avec M. Lavoisier, sur une nouvelle méthode de mesurer la chaleur." Later we read (f. 146v) "MM Lavoisier et Delaplace ont continué la lecture de leur mémoire sur la chaleur." See also Daumas, *Lavoisier théoricien et expérimentateur,* p. 48. Guyton de Morveau wrote to Bergman that summer: "M. M. Lavoisier et de la place [*sic*] ont lû plusieurs mémoires sur la *chaleur spécifique des corps,* on les imprimera separément, ainsi vous les aures plutôt. . . ." (*Bergman's Foreign Correspondence, 1,* 133.)

(3)

MÉMOIRE
SUR LA CHALEUR.

CE Mémoire eſt le réſultat des expériences ſur la chaleur, que nous avons faites en commun, M. de Lavoiſier & moi, pendant l'hiver dernier; le froid peu conſidérable de cette ſaiſon, ne nous a pas permis d'en faire un plus grand nombre: nous nous étions d'abord propoſé d'attendre, avant que de rien publier ſur cet objet, qu'un hiver plus froid nous eût mis à portée de les répéter avec tout le ſoin poſſible, & de les multiplier davantage; mais nous nous ſommes déterminés à rendre public ce travail, quoique très-imparfait, par cette conſidération que la méthode dont nous avons fait uſage, peut être de quelque utilité dans la théorie de la chaleur, & que ſa préciſion & ſa généralité pourront la faire adopter par d'autres Phyſiciens, qui placés au nord de l'Europe, ont des hivers très-favorables à ce genre d'expériences.

Nous diviſerons ce Mémoire en quatre articles; dans le premier, nous expoſerons un moyen nouveau pour meſurer la chaleur; nous préſenterons dans le ſecond, le réſultat des principales expériences que nous avons faites par ce moyen; dans le troiſième, nous examinerons les conſéquences qui ſuivent de ces expériences; enfin dans le quatrième article, nous traiterons de la combuſtion & de la reſpiration.

ARTICLE PREMIER.

Expoſition d'un nouveau moyen pour meſurer la chaleur.

QUELLE que ſoit la cauſe qui produit la ſenſation de la chaleur, elle eſt ſuſceptible d'accroiſſement & de diminution, & ſous ce point de vue elle peut être ſoumiſe au calcul: il ne paroît pas que les Anciens aient eu l'idée de meſurer ſes

A ij

Figure 2

was published separately in August of 1783.[120] A letter transmitting a copy of this preprint to his friend Lagrange gave Laplace an opportunity to enter once more a mild protest at his captivity:

> Voici deux exemplaires d'un Mémoire sur la chaleur [he wrote on 21 August], d'après quelques expériences que nous avons faites en commun, M. de Lavoisier et moi, sur cette matière. . . . Je serais bien charmé d'avoir votre avis sur ce Mémoire, si vos occupations vous laissent assez de loisir pour le parcourir. Je ne sais en verité comment je me suis laisser entrainé à travailler sur la Physique, et vous trouverez peut-être que j'aurais beaucoup mieux fait de m'en abstenir; mais je n'ai pu me refuser aux instances de mon Confrère M. de Lavoisier, qui met dans ce travail commun toute la complaisance et toute la sagacité que je puis désirer. D'ailleurs, comme il est fort riche, il n'épargne rien pour donner aux expériences la précision qui est indispensable dans des recherches aussi délicates.[121]

THE MEMOIR ON HEAT AND CALORIMETRY

Even a cursory reading of the famous "Mémoire sur la chaleur"[122] makes quite evident the influence of Crawford's book, or perhaps Magellan's essay, and the importance Lavoisier and Laplace attached to the novel concept of specific heats.[123] A closer inspection, moreover, discloses the extent of Laplace's personal contribution to this important memoir. Of the four articles making up the memoir only the last two can have been largely the work of Lavoisier. The first article, on the general nature of heat and the new method of measur-

[120] *Mémoire sur la Chaleur. Lu à l'Académie Royale des Sciences, le 28 Juin 1783. Par Mrs. Lavoisier & de La Place de la même Académie* (Paris, De L'Imprimerie Royale, 1783), 56 pp. and two folding plates. There is a copy of this preprint in the Lavoisier Collection of the Cornell University Library. The version published in the Academy's *Mémoires* did not appear until 1784: *Mém. Acad. R. Sci.*, 1780 (1784), pp. 355–408.

[121] Lagrange, *Oeuvres, 14,* 123–124.

[122] For convenience I cite the text as printed in *Oeuvres de Lavoisier, 2,* 283–333. Except for spelling and other accidentals, it is virtually identical with the pamphlet of 1783.

[123] It is not certain that Lavoisier and Laplace were directly familiar with Crawford's book, although at one point they mention it with date and place of publication. The book is not recorded in Lavoisier's library. Various points suggest that their chief source for Crawford's ideas was Magellan's *Essai.*

ing it, is obviously strongly influenced by Laplace's habits of thought; the third is almost certainly his exclusive work; and the last—on combustion and respiration—is clearly Lavoisier's. The second article, a factual account of the experiments performed with the calorimeter, could have been drawn up by either man.

At the beginning of the first article heat is discussed in general. We are reminded that the laws of heat exchange are poorly understood, but that experiments have disclosed two important phenomena. The first is latent heat (the term, however, is not used) and is a phenomenon that Lavoisier, in a general way, had long recognized.[124] The second phenomenon is that equal masses of different substances, though the thermometer shows them to be at the same temperature, do not contain the same amount of heat: they have different "capacities" for heat. The heat that registers on the thermometer Lavoisier and Laplace propose to call *chaleur libre;* the hidden heat, for which different bodies have different capacities, they will call *chaleur spécifique.*[125]

Then follows a much-quoted section in which the authors set forth the two rival theories of heat. A number of "*physiciens*"—among whom we should certainly include Lavoisier—believe heat to be a fluid that is distributed throughout nature and penetrates all bodies. Sometimes it combines with them, and under these circumstances it cannot pass freely from one body to another, as does *chaleur libre,* nor can it therefore act upon the thermometer. Others, however, believe heat to be the invisible motions of the particles of matter.[126]

124 See above, pp. 227-229.

125 This is Magellan's term, not Crawford's. Lavoisier and Laplace did not adopt the expression *chaleur sensible* used by Magellan, probably because it seemed more appropriate for the subjective sensation of heat. Later, in a discussion of the sensation of heat, Lavoisier wrote: "On pourrait donner à cette chaleur le nom de chaleur sensible, si M. Crawford et quelques physiciens anglais modernes n'eussent donné un autre sens à cette expression." In the *Traité élémentaire,* when referring to the expression "capacité des corps pour contenir la matière de la chaleur," Lavoisier wrote that it is an "expression fort juste, introduite par les Physiciens Anglois, qui ont eu les premiers des notions exactes à cet égard" (*Oeuvres de Lavoisier, 1,* 27); compare *Mémoires de chimie* (1ère partie, p. 7): "C'est dans ce sens qu'on doit entendre ce qu'à dit le docteur Crawford, que les corps avoient différentes capacités pour contenir le calorique; cette expression . . . est parfaitement juste."

126 This was the view favored by Macquer in the second edition of his *Dictionnaire de Chymie,* 4 vols. (Paris, 1778), *2,* 163-192, article "Feu." Macquer, to be sure, gives both views and presents the theory that heat is "un état particulier, une maniere d'être, dont toute substance matérielle est

The exposition of this alternative hypothesis, of what we call today the "kinetic" or "dynamic" theory of heat, was doubtless the handiwork of Laplace and may possibly reflect his actual preference at this time. All bodies, the exposition begins, are filled with pores or tiny vacuous spaces and these greatly exceed in volume the matter of a given body.[127] These empty spaces allow the particles of bodies to *oscillate* (a term more familiar to mechanics than to chemistry) in all directions. To develop this hypothesis it is necessary to understand the general law that mathematicians have developed under the name of "*principe de la conservation des forces vives.*" Simply stated, this means that in a system of bodies acting on each other in whatever fashion, the *force vive*—which is the product of each mass by the square of its velocity—is constant. If a body is acted upon by accelerative forces, the *force vive* is equal to that which the body possessed as an initial condition (*à l'origine du mouvement*) plus the sum of the masses multiplied by the squared velocities due to the action of accelerative forces. It is easy to see how this principle can be applied to the transmission of heat as particulate motion. If two bodies at different temperatures are placed in contact, the quantities of motion they transmit from one to the other will be unequal at first, but gradually the *forces vives* of the colder body will increase, while those of the warmer body will decrease, and this exchange will

---

susceptible" simply as a conjecture. Fourcroy accepts Macquer's immaterial theory in 1782 in his *Elémens élémentaires d'histoire naturelle, 1,* 44–45, but later changes his opinion. We read in the article "Feu" in Diderot's *Encyclopédie:* "Le caractère le plus essentiel du *feu* . . . est de donner la chaleur. . . . Mais le feu est-il une matière particulière? Ou n'est-ce que la matière des corps mise en mouvement? C'est sur quoi les philosophes sont partagés." Brisson wrote in 1781: "Le plus grand nombre des physiciens regardent le *Feu* comme une matière simple, inaltérable & destinée à produire la chaleur et l'embrasement. D'autres prétendant que son essence consiste dans le mouvement seul des parties du corps qui s'embrase," but suggests that they attribute the motion to the aether or subtle matter which must be truly "la matière du feu." Fire "dans son principe" must be something other than a motion imparted to the corpuscles of matter. (*Dictionnaire raisonné de physique,* 2 vols. [Paris, 1781], *1,* 603, col. 1.)

[127]This point was set forth by Newton in his *Opticks,* and Newton's insistence upon it is stressed by Arnold Thackray in an article in *Ambix, 15* (1968), 29–53, and in his *Atoms and Powers. An Essay on Newtonian Matter-Theory and the Development of Chemistry* (Cambridge, Mass., 1970), pp. 53–57. Thackray calls this Newton's "Nut-Shell" theory.

continue until the quantities of motion that are transmitted back and forth become equal.[128]

We may stop for a moment to ask how Laplace, who of course used the principle of the conservation of *vis viva* in treating the dynamics of the solar system, came to apply this approach to the study of heat. To be sure, in a purely *qualitative* fashion, the kinetic theory of heat had been advanced by Francis Bacon and by Isaac Newton. But, as most historians of science remember, in 1738 Daniel Bernoulli included in his *Hydrodynamica* a derivation of Boyle's Law from a consideration of the motion and impact of particles of an imaginary aeriform fluid.[129] It is less often remembered that in that same chapter he implied, on the same assumptions, that the pressure of this fluid must also be proportional to the temperature, an early statement of what we know as the Law of Charles or of Charles and Gay-Lussac. The key to his derivation was Bernoulli's recognition that heat is associated with particulate motion.[130] It is widely believed that Bernoulli's work fell into total oblivion, yet directly or indirectly it seems to have influenced these passages by Laplace we have been discussing. Laplace may well have been familiar with the *Hydrodynamica,* but he did not need to be; Bernoulli's theory and his results are admirably summarized in a work

[128] In an interesting passage we read that the dynamical theory helps explain why the "impulsion" of solar rays is inappreciable, although they produce great heat. By impulsion is meant momentum, the product of mass into velocity. The mass is small though the velocity is swift, and their product "est presque nul." But the heat depends upon the *vis viva,* that is, on the square of the velocity, which explains why "la chaleur qu'elle représente est d'un ordre très-supérieur à celui de leur impulsion directe." This argument convinced Partington that "this part is certainly by Laplace." See his *History of Chemistry, 3,* 428. A further point is that Lavoisier seems to have favored an undulatory theory rather than the corpuscular theory implied in this discussion of solar rays.

[129] *Hydrodynamica, sive de viribus et motibus fluidorum commentarii* (Strasbourg, 1738), Chap. 10 (*Sectio decima*), pp. 200–202. There is an English version in *Hydrodynamics by Daniel Bernoulli & Hydraulics by Johann Bernoulli,* tr. Thomas Carmody and Helmut Kobus (New York, 1968), pp. 226–230. The translation in W. F. Magie, *Source Book of Physics* (New York and London, 1935), pp. 247–251 is unreliable. For a recent study of Bernoulli and an evaluation of his *Hydrodynamica* see the article by Hans Straub in the *Dictionary of Scientific Biography, 2* (1970), 36–46.

[130] The crucial passage (*Hydrodynamica,* p. 202) begins: "Elasticitas interim aëris nonsolum à condensatione augetur, sed & ab aucto calore, & quia constat calorem intendi ubique crescente motu particularum intestino, sequitur, etc."

with which both Laplace and Lavoisier were familiar, and which indeed was, as we have seen, a standard work on thermometry: the first edition of J.-A. Deluc's *Recherches sur les modifications de l'atmosphère* of 1772.[131] The author was known personally to Lavoisier and to Laplace. Deluc had been named a corresponding associate of the Royal Academy of Sciences in 1768, as the correspondent of C.-M. de La Condamine; when the latter died in 1774 he became the correspondent of the astronomer Lalande. Although he left Geneva permanently to reside in England, Deluc made many trips to the Continent. From a letter he wrote to his brother we know that he was in Paris in October 1781 and in touch with J.-E. Guettard, Lavoisier's old mentor, and familiar with the mineralogical collection of the Duc d'Orleans of which Guettard was the curator.[132]

Deluc, it should be recalled, was the last in a line of Swiss and Genevan scholars who—unlike other scientists and mathematicians of the Continent—displayed a marked partiality for various kinds of kinetic theory. These speculations began with Jacob and Johann Bernoulli, and with an early work of Leonhard Euler that the great mathematician subsequently disavowed. The St. Petersburg Academy, to which many Swiss mathematicians migrated for longer or shorter periods, was for a time the center of this interest. It was here that Jacob Hermann, who had studied with the elder Bernoullis at

[131] *Recherches sur les modifications de l'atmosphère, 1,* 165-171 and *2,* 368. The article on Deluc by Robert P. Beckinsale in *Dictionary of Scientific Biography, 4* (1971), 27-29, regrettably says nothing about the *Recherches.* But there is useful material in W. E. Knowles Middleton, *The History of the Barometer* (Baltimore, 1964), *passim.*

[132] The letter, addressed to "Monsieur Deluc Hospitallier, A Genève," and dated "Paris, Du 10e 8bre 1781 au Soir" is owned by Cornell University. It is worth recording that Deluc received a copy of the pamphlet version (1783) of the "Mémoire sur la chaleur" and sent it to James Watt, who in turn forwarded it to Joseph Black for whom it was ultimately destined. See Watt to Black, letter of 25 September 1783, in E. Robinson and D. McKie, *Partners in Science* (London and Cambridge, Mass., 1970), p. 128.

During his stay in Paris in 1781 and part of 1782 Deluc began work on his *Idées sur la météorologie* "pour satisfaire au désir de quelques physiciens avec qui je m'étois entretenu de mes nouvelles recherches en météorologie." Obliged to suspend his writing, after his arrival in London, he took it up again "& j'écrivis de nouveau sous la forme de lettres adressées à M. De La Place; parce que ce savant étoit celui des Académiciens de Paris avec qui je m'etois le plus entretenu de tous les objets qui le composoient." *Idées sur la météorologie, 1,* 5. This reference was called to my attention by Roger Hahn.

Basel, enunciated in his *Phoronomia* a proposition relating to the motion of air particles; in St. Petersburg, too, Daniel Bernoulli began work on his *Hydrodynamica,* and Russia's great polymath M. V. Lomonosov advanced his own theory of the role of heat in his *De vis aeris elastica* (1748–1749), assuming a rotation of particles combined with a translational motion.[133] This has been called the last eighteenth century kinetic theory, but this is hardly accurate, for a much talked-about theory, towards the end of the century, was that of George-Louis Le Sage (1724–1803). This theory was nevertheless only made known posthumously by Le Sage's disciple, Pierre Prévost.[134] At all events, it was familiar to an intimate circle of Genevan scholars as early as 1772 when Jean-André Deluc made a cryptic allusion to it in his *Recherches.* Here it is clearly implied that Le Sage believed himself to be completing the work of Daniel Bernoulli, with whom he had studied medicine at Basel.[135]

In Le Sage's theory, as in Bernoulli's, air and other elastic fluids are thought to consist of hard particles, separated by distances much greater than their diameters. Again, as with Bernoulli, these are in constant and swift translational motion, which Bernoulli has not troubled to account for. Le Sage, in contrast to Bernoulli—and it is for this that Deluc especially praises him—sought the underlying cause in an active aethereal matter pressing against the opposite sides of the particles of air.

To return to our main theme, it is important to show how Deluc sums up Daniel Bernoulli's principles concerning the general properties and motions of elastic fluids. He writes as follows:

> Les fluides sont élastiques, quand leurs particules se meuvent rapidement en tout sens. La force expansive de ces fluides peut être augmentée par deux causes, savoir, par le plus grand nombre de particules renfermées dans le même espace, & par leur plus grande vitesse. La première de ces causes produit la condensation: Si la compressibilité est regardée comme infinie, la force de ressort qui en résulte doit être en raison inverse des espaces occupés par le

[133] For the early history of the kinetic theory see G. R. Talbot and A. J. Pacey, "Some Early Kinetic Theories of Gases: Herapath and his Predecessors," *The British Journal for the History of Science, 3* (1966), 133–149.

[134] Pierre Prévost, *Deux traités de physique mécanique* (Geneva and Paris, 1818). The first treatise, Prévost says, is based on the work of Le Sage "et j'en suis simplement l'éditeur."

[135] *Recherches sur les modification de l'atmosphère, 1,* 166–167 and note (a).

même nombre de particules.... La densité d'un fluide élastique augmente donc proportionnellement au poids dont il est chargé....

Then Deluc summarizes Bernoulli's conjectures about the relation of heat to the problem at hand:

La chaleur donne une plus grande vitesse aux particules; & leur force impulsive augmente en raison doublée de cette augmentation de vitesse. Les effets de cette seconde cause sont proportionnels au nombre de particules, ou, ce qui revient au même, à la densité. Ainsi la force élastique des fluides est en raison composée simple de leur densité et doublée de la vitesse de leurs particules.[136]

To consider first the "Mémoire sur la chaleur" of Lavoisier and Laplace. There is a famous, a striking–and positivistic-sounding–passage in which the authors decline to decide between these two hypotheses of the nature of heat, for in essential matters they yield equivalent results. Indeed both lead to the common principle of conversation on which the study of heat must be based. In the simple mixture of bodies the total amount of heat is conserved, whether we think of it as a material substance that is neither created nor destroyed, or whether we believe that it is the *forces vives* that are conserved.[137] Once again, the voice of Laplace can be heard in the extremely general way in which the principle is stated and applied to all thermal effects: "Toutes les variations de chaleur, soit réelles, soit apparentes, qu'éprouve un système de corps, en changeant d'état, se reproduisent dans un ordre inverse, lorsque le système repasse à son premier état."[138] Problems involving heat, then, are problems treat-

[136] *Ibid.*, p. 171.

[137] Fourcroy misunderstood the thrust of this passage and wrote: "Enfin, MM. Lavoisier & de la Place semblent soupçonner que les deux hypothèses sur la chaleur, sont vraies & ont lieu en même-tems; c'est-à-dire, que la chaleur consiste dans l'existence d'un corps particulier, & dans les oscillations des corps excitées par sa présence," *Elémens d'histoire naturelle et de chimie,* 3ème édition, 5 vols. (Paris, 1789), *1,* 122. Lavoisier and Laplace said nothing of the sort.

[138] *Oeuvres de Lavoisier, 2,* 287–288. Laplace evidently uses the word *état* in the broadest sense of "condition." He does not mean "state" as in "change of state." Partington points out that this implies the fundamental law of thermochemistry: as much heat is absorbed in the decomposition of a chemical compound as is evolved in its formation (*History of Chemistry, 3,* 428).

ing systems of bodies that are, as we say today, *conservative* and *reversible.*

After these important introductory paragraphs, the authors proceed to define the meaning they attach to the term *chaleurs spécifiques.* By this expression they mean the ratios of the quantities of heat necessary to raise equal masses of different substances by the same number of degrees. Then follows immediately a careful description of how these specific heats can be measured by the method of mixtures. Two miscible substances are mingled together and the final temperature of the mixture is then measured. If m and m' are the masses of the two substances, a and a' their initial temperatures, and b the final temperature of the mixture, then q and q , the specific heats, are given by the relation

$$\frac{q}{q'} = \frac{m'(b - a')}{m(a - b)}.$$

If the specific heat of some substance, say water, is taken as a standard of reference and arbitrarily defined as unity, the *relative* specific heats of other substances can be given a numerical value.[139]

Lavoisier and Laplace point out various inconveniences and sources of error in this method of mixtures. Its chief deficiency, however, is its limited application: by this method it is not possible to compare substances that react chemically together or to determine the heat given off or absorbed in such reactions; and it is quite out of the question to use it to determine the heat evolved in combustion or animal respiration. Yet these, they remark, are the most interesting aspects of the problem of heat. For this reason, they have devised a general experimental method to enable all such heats to be determined with considerable precision.

The method, as we know, was to use the melting of ice as a measure of the heat; the embodiment of the new method was their famous ice calorimeter. The method, if not the actual form of the device used, was the invention of Laplace.[140] The way the method is

[139]The method of mixtures was of course adopted from Crawford or Magellan. The Kirwan table discloses the practice of referring specific heats to water as the standard.

[140]Lavoisier gives Laplace credit for the idea of the sphere of ice and the fundamental principle used for measuring heats. See *Traité élémentaire de chimie* (in *Oeuvres de Lavoisier, 1,* 29 and 285). Earlier, in referring to the measurement of heat ("Réflexions sur le phlogistique," *Oeuvres, 2,* 645–646) Lavoisier wrote: "Nous n'avons encore de moyen exact pour remplir cet objet

introduced and explained in this memoir—by means of a purely imaginary, abstract model—suggests the habits of the mathematical physicist of the team rather than of the chemist. Imagine, we are asked, a hollow sphere of ice, with walls thick enough so that the external heat of the atmosphere cannot affect the interior surface. Conceive, further, a heated body placed inside the sphere. The heat from this body will not escape outwards, but will melt successive layers of the inner surface of the ice. Imagine the hot body cooled to the temperature of ice; in the process a certain amount of ice absorbs this heat and melts to form water. Since no other cause has melted the ice, and an effect must be proportional to its cause, the weight of the water formed will be precisely proportional to the heat given off by the hot body.

With the picture of the imagined sphere of ice still before us, we are told in a general way how various thermal effects can be measured. To determine the specific heat of a solid, for example, we heat the solid to a given temperature, place it in the hollow sphere of ice, leaving it until it cools to zero,[141] then we weigh the water produced. This weight of water, divided by the mass of the body and its initial temperature, will be proportional to the specific heat. The specific heat of a fluid can be determined in like fashion if we take into account the heat capacity and weight of the vessel in which it is contained.[142]

To measure the heats produced by chemical reactions, the reactants and the vessel in which they are to be mixed are cooled to zero; these are placed inside the sphere of ice and the reaction is allowed to take place. When the whole system has cooled, the water from the melted ice is weighed. By similar methods the latent heat of fusion of solids and the heats of combustion and of animal res-

---

que celui imaginé par M. de Laplace (Voy. *Mém. de l'Acad.* 1780, page 364)." Joseph Black, who neither devised nor used an ice calorimeter, though such an invention has been attributed to him, gave Lavoisier credit for having "invented a most ingenious apparatus for the purpose of measuring all productions of heat, by the quantity of ice melted by those operations of nature in which it appears." See *Lectures on the Elements of Chemistry,* ed. J. Robison, 2 vols. (Edinburgh, 1803), *1,* 175. In a letter to James Watt (May 1784) he is more specific: "I am told it was contrived by La Place." (Muirhead, *Correspondence of James Watt,* p. 66.)

141 Lavoisier's thermometer registered zero at the melting point of ice. See below, note 147.

142 For suggestions as to how the specific heats of gases might be measured, see *Oeuvres de Lavoisier, 2,* 295–296.

piration can also be measured, although for the latter measurements some communication must be established between the interior of the reaction vessel and the external atmosphere (or source of *vital air*) so that the air can be replenished.

Their new method, the authors argue, can be applied to all phenomena where there is evolution or absorption of heat. For purposes of general comparison what is necessary is a unit of heat: for example the amount of heat necessary to raise a pound of water from the temperature of melting ice to the boiling point. All other heat measurements can be expressed in terms of this unit which, of course, is comparable to our calorie.

All this exposition is quite abstract and doubtless the work of Laplace. The hollow sphere of ice is, obviously, purely imaginary and impractical. But the principle of the ice sphere is applied in the calorimeter—the word seems to be Lavoisier's coinage—which they proceed to describe. Lavoisier had two of these made, similar in design, except that one was modified for the experiments on combustion and respiration. It is hardly necessary to say that Lavoisier, rather than Laplace, had the financial resources to foot the bill.

This historic piece of apparatus, about three feet high, consisted of cylindrical metal containers, nesting in one another with spaces between, and ending in conical outlets. Inside the outermost container was a space to be filled with ice, to insulate the inner container from the heat of the external atmosphere, as the outer layers of the thick sphere of ice were supposed to do. A wire basket, to hold the heat-producing objects, was suspended in the inner container. This basket was in contact with an inner layer of ice (corresponding to the inner surface of the imaginary sphere of ice) melted by the experimental object in the basket. The water so produced from the melting ice flows down the inner cone and out a nozzle fitted with a stopcock, to be caught in a receiver for weighing.[143]

[143]It has not been ascertained for certain which artisan was employed to build these two calorimeters. It may have been a tinsmith named Naudin. See Daumas, *Lavoisier théoricien et expérimentateur,* p. 142. Surviving examples of the ice calorimeter are now in Paris in the Musée des Techniques of the Conservatoire des Arts et Métiers. Two plates drawn by Fossier and engraved by Y. le Gouaz accompanied the original publication, and that in the *Mémoires* of the Academy, 1780 (1784). The two plates in *Oeuvres de Lavoisier, 2,* Pl. 1 and 2, were engraved by E. Wormser. Photographs of the surviving examples are shown in Maurice Daumas, "Les appareils d'expérimentation de Lavoisier," *Chymia, 3* (1950), Fig. 3 opposite p. 50.

Significantly, the first set of experiments performed with this equipment that are described involved the determination of specific heats.[144] A table giving the specific heats for twelve substances—solids, liquids, and aqueous solutions—is followed by detailed illustrations of the procedure for the case of iron-plate (*tôle*) and for nitric acid. An interesting feature of these computations is the expression of weights as decimal fractions of the French pound, the *livre poids de marc*. For example, the weight of the sheet iron—"7 livres 11 onces 2 gros 36 grains"—is given as 7^{ff}, 7070319.[145] Henceforth Lavoisier frequently used this method and advocated it strongly (in his *Traité élémentaire de chimie*) until such a time as men "réunis en société, se soient déterminés à n'adopter qu'un seul poids et qu'une seule mesure."[146] It was, therefore, a sort of way station, a step towards the French Revolution's historic adoption of the metric system, a project with which Lavoisier and Laplace were to be intimately associated.

One of the most significant quantities to be determined as accurately as possible was the amount of heat necessary to melt a pound of ice. This was found, after averaging a number of experiments, to

144An excellent study of the accuracy of the measurements made by Lavoisier and Laplace with the ice calorimeter, of contemporary criticism of the device, and of its later use and ultimate abandonment is T. H. Lodwig and W. A. Smeaton, "The Ice Calorimeter of Lavoisier and Laplace and Some of Its Critics," *Annals of Science, 31* (1974), 1-18.

145The disregard, or ignorance, of significant figures is characteristic of a period before the full grasp of error theory and experimental error. Such decimal fractions of traditional measures may have been used earlier in experiments that Lavoisier and Laplace carried out in 1781 and 1782. Their pyrometer gave them, they claimed, a precision of "un dixième de ligne." See *Procès-verbaux*, 1781, f. 254. The paper embodying these measurements was first published in the posthumous *Mémoires*, Part I, p. 246. Entitled "De l'action du calorique sur les corps solides," it can be consulted in *Oeuvres de Lavoisier, 2*, 739-759. It is possible that the data were recalculated after 1789 in decimal form. The title, with the word "calorique," suggests that changes must have been made in the text after 1781-1782. In a very early paper, "Recherches sur les moyens de déterminer la pesanteur spécifique des fluides," Lavoisier proposed to divide his areometer into decimal divisions (*Oeuvres de Lavoisier, 3*, 440-441).

146*Oeuvres de Lavoisier, 1*, 249. Here he also writes: "Frappé de ces considérations, j'ai toujours eu le projet de faire diviser la livre poids de marc en fractions décimales, et ce n'est que depuis peu que j'y suis parvenu. M. Fouché, balancier, successeur de M. Chemin . . . a rempli cet objet avec beaucoup d'intelligence et d'exactitude."

be equal to the amount of heat necessary to raise the temperature of a pound of water by 60° of Lavoisier's thermometer;[147] or—as they expressed it in more general terms—the heat required to melt a given amount of ice is equal to three fourths the quantity of heat necessary to raise the same amount of water from the temperature of melting ice to the boiling point of water. Since this expresses the basic constant "indépendamment des divisions arbitraires des poids et du thermomètre," we again detect the authors' discontent with the reigning confusion in weights and measures.

Lavoisier and Laplace had earlier remarked—and this may express Lavoisier's opinion rather than Laplace's—that the most interesting aspect of the theory of heat concerned the heat evolved in combustion and respiration.[148] Not surprisingly, therefore, they tabulate the results obtained from detonating nitre (potassium nitrate) with carbon and with sulphur, as well as the results of burning phosphorus, carbon, and ether. But the experiment everyone remembers—and which is described at greatest length in the concluding section of the memoir—is the one in which they determined the heat output of a guinea pig placed for ten hours in the calorimeter.

The "Mémoire sur la chaleur" concludes with two long sections, one almost certainly written by Laplace,[149] the other certainly

[147]The thermometers used by Lavoisier and Laplace in these experiments were probably made by Fortin, who began work for Lavoisier at about this time. The scale was graduated *à la Réaumur* into eighty divisions between the melting point of ice, at 0°, and the boiling point of water at 80°. This resembles the scale of Deluc's mercury thermometer as reproduced by W. E. Knowles Middleton, *History of the Thermometer,* p. 118. Réaumur's thermometers, using alcohol, had 80 divisions between the melting point of ice and the highest point the alcohol thermometer could reach when immersed in boiling water and still remain liquid. Lavoisier, perhaps following Deluc, simply adopted the division into 80 degrees for the region between the two critical points. But he may have been following a common practice, for most thermometers in late eighteenth-century France used 0 for the temperature of melting ice and 80 for that at which water boils. See Arthur Birembaut, "La contribution de Réaumur à la thermométrie," *Revue d'histoire des sciences et de leurs applications, 11* (1958), 138-166.

[148]"L'observation de ces phénomènes étant la partie la plus intéressante de la théorie de la chaleur . . .," *Oeuvres de Lavoisier, 2,* 291.

[149]Much later, in the first memoir of his posthumous *Mémoires de chimie,* intended to introduce and clarify the difficult "Mémoire sur la chaleur," which is reprinted as the "Second Mémoire" under a somewhat different title, Lavoisier writes at one point: "Mais comme M. de la Place a envisagé cet objet d'une manière plus mathématique, ce mémoire, fait en commun, ne me

drafted by Lavoisier for it deals with matters that particularly concerned him. Article III (Laplace's Article, if we may call it that) is an extraordinary *tour de force* and deserves a detailed treatment I cannot give it here.[150] It begins with an attempt to consider the factors that must be known if there is to be "une théorie complète de la chaleur." But these facts are not known, he remarks, and can only be determined by an almost infinite number of delicate experiments performed over a wide range of temperatures. Nevertheless a few particular problems can be examined. We are first reminded that specific heats do not give the ratio of the *absolute* quantities of heat in two bodies, but only the ratio of the quantities of heat necessary to raise the bodies a given number of degrees. Specific heats give us, strictly speaking, only "le rapport des différentielles des quantités absolues de la chaleur." Unless we make the risky assumption that the absolute quantities are proportional to their differences, we cannot conclude that we know the ratios of the absolute heats. Laplace explores in some detail the possibility of using the specific heats of substances and of their compounds to predict the heats of combustion, and so lead to determining the ratio of the absolute quantities of heat, but he concludes that only experiment can yield the knowledge of these heats of formation.

Particularly sophisticated is Laplace's treatment of the congealing of supercooled liquids in terms of states of thermo-molecular equilibrium, progressively less stable (*ferme*) as the temperature drops below the freezing point. Then follows an attempt to elucidate in similar terms the lowering of the freezing point of water by the addition of acid. Crystallization of ice is the result of intermolecular forces of attraction or affinity between the molecules of water. Heat acts as an opposing force tending to separate these molecules. If one mixes an acid with the ice, the ice will melt because the affinity of the acid molecules with those of the water is greater than the mutual attraction of the water molecules. The water will now freeze only when the acid has been sufficiently diluted so that its attraction for the water becomes equal to the force between the water molecules. This intermolecular force between water molecules is greater the lower the temperature; thus the concentration of acid at which it

---

dispense pas d'entrer dans quelques détails plus analogues à ma manière d'envisager cet objet" (*1*, 1ère partie, 8–9). The "objet" is the general conception of specific heats.

150 *Oeuvres de Lavoisier, 2*, 307–318.

ceases to melt ice will be greater the more the temperature of the mixture drops below zero. And Laplace suggests that by comparing the concentrations at which different acids lower the freezing point of water to different temperatures one can determine the respective affinities of these acids for water. By the same method the relative affinities of various solvents and solutes can be determined. I shall return later to the extension Laplace gave to this notion.

Article IV, which I attribute largely to Lavoisier, is entitled "De la combustion et de la respiration," and about equal space is devoted to these two subjects, first his theory of combustion and then his ideas about respiration and animal heat.[151] I shall discuss only the first section of this Article IV, because it illustrates one of the chief points I wish to make: that the deeper understanding of heat phenomena, stimulated by the work of the British school, and by the experiments and discussions with Laplace, put the capstone on Lavoisier's antiphlogistic theory of combustion. Just as the experiments of 1777 with Laplace encouraged Lavoisier to come out openly against the phlogiston theory, though with considerable caution, so the calorimetric experiments of 1782–1784 and their theoretical comprehension contributed important, if not decisive, elements to his famous manifesto, his "Réflexions sur le phlogistique" of 1786.

In the opening paragraphs of Article IV Lavoisier reviews the re-

[151] *Ibid.*, pp. 318-333. In his *Opuscules* (1774), Lavoisier advanced a physical, not a chemical, view of respiration. But after the discovery of the remarkable ability of oxygen to sustain respiration, the problem became one of physiological chemistry. Lavoisier's first experiments, performed in the autumn of 1776, were embodied in his "Expériences sur la respiration des animaux," *Oeuvres de Lavoisier, 2,* 174–183, where he offers two alternative explanations: *air éminement respirable* (oxygen) is either converted in the lung into *acide crayeux aériforme* (carbon dioxide), or there is an exchange in the lung, the oxygen being absorbed and carbon dioxide given off. In his "Mémoire sur la combustion en général," published in 1780, he suggests that respiration must be analogous to the combustion of carbon. When equal amounts of carbon dioxide are given off by a respiring animal or by the combustion of carbon, the heat given off should be the same. It was to verify this conjecture that the animal experiments with the calorimeter were performed. The results were troublesome, since more heat was given off by the respiring animal than from the combustion of carbon. See Mendelsohn, *Heat and Life,* pp. 151–156, and Charles A. Culotta, "Respiration and the Lavoisier Tradition," *Transactions of the American Philosophical Society,* New Series, *62* (1972), 3–40. Unfortunately a number of errors occur in the section of Culotta's paper dealing with Lavoisier's work.

cent work on combustion: the discovery of the role played by "air déphlogistiqué, air pur ou air vital," and reminds the reader that in his earlier "Mémoire sur la combustion" he had advanced the notion that the heat and light accompanying combustion was owing, at least in large part, to the decomposition of the *air pur.* Mr. Crawford, Lavoisier continues, has offered a somewhat similar explanation in his recent book, and they agree in regarding the *air pur* as the principal source of the heat developed in combustion and respiration. They differ only in so far as Lavoisier believes the heat so produced is *combined* in the *air pur* which owes its aeriform state to the expansive force of the heat so combined. Crawford, on the other hand, believes that the matter of heat is "free" in the air and separates from the air during combustion only because the *air pur* loses a large part of its heat capacity. Crawford based his interpretation on measurements of the specific heat of the *air pur.* If the experiments of Crawford are accurate—and Lavoisier is not ready to agree that they are—they show that the "free heat" in *air pur* is sufficient to account for the heat given off in combustion.

Lavoisier then describes the experiments performed with Laplace, the purpose of which—without examining whether the heat of combustion comes from the air or from the combustible—was to compare the heat given off during combustion of different substances with the corresponding changes in the *air pur* used in the experiments. An ounce of carbon was burned over mercury under a bell-jar, then the amount of *air pur* absorbed and the amount of fixed air produced were both determined. This was compared with the heat produced when an ounce of carbon is burned in the calorimeter. From this is could be computed that in the combustion of carbon the absorption of an ounce of *air pur* can melt 29.5 ounces of ice.[152] A similar experiment showed that an ounce of *air pur* when absorbed by phosphorus can melt 68.6 ounces of ice. This gave the striking result that the free heat released from *air pur* when it combines with phosphorus is 2 1/3 times greater than when it combines with carbon to form fixed air (CO_2). This supported an earlier conjecture of Lavoisier (at that time based on little or no quantitative evidence) that when *air pur* passes into a solid product all of its com-

[152]Lavoisier was not too confident about this quantitative result, for the experiment was performed only once, though under favorable conditions. He thought the important thing to be the method, and that the experiment should be repeated.

bined matter of fire is released, but when an aeriform fluid is produced, like the fixed air from burning carbon, some of it is "retained" to keep it in the elastic state. Clearly Lavoisier saw this as confirming his covering theory of vaporization, and as a strong argument that the heat and light in combustion comes largely, if not exclusively, from the *air pur.*

The importance Lavoisier attached to these experiments on heat with Laplace, and to his deeper understanding of thermal phenomena, is made clear by what is perhaps the most famous of his papers, the "Réflexions sur le phlogistique," the memoir in which, as Duveen and Klickstein have put it, Lavoisier launched his attack on phlogiston in earnest,[153] and which Douglas McKie called "one of the most notable documents in the history of chemistry."[154] It is interesting that, in spite of such encomiums, the full thrust of Lavoisier's argument in that memoir has escaped historians of chemistry.[155]

The "Réflexions" was written sometime after the publication of the "Mémoire sur la chaleur," and probably after the partnership with Laplace was amicably dissolved. It was read to the Academy in the summer of 1785 and published the next year in the *Mémoires* of the Academy.[156] Since this volume of the *Mémoires* appearing in 1786 is designated as for the year 1783, it is easy to be deceived about the chronological sequence.

The early pages of the "Réflexions" are devoted to the refutation not only of Stahl, but of two of Lavoisier's contemporaries—Antoine Baumé and Macquer—who had attempted to adjust the phlogistic

153 Duveen and Klickstein, *Bibliography of the Works of Lavoisier,* p. 68.

154 Douglas McKie, *Antoine Lavoisier, the Father of Modern Chemistry* (London, 1935), p. 220.

155 McKie, for example, wrote that the revolution "that Lavoisier had aimed at in 1773 was now, ten years later [*sic*], approaching its climax and he was at last able to engage the phlogiston theory in the open, to rout its defenders with a destructive fire of chemical facts and to seize for his own theory the position that it now properly deserved." (*Ibid.,* p. 220.) More recently Maurice Daumas summarized the memoir in like terms, stressing the purely chemical arguments, and omitting all reference to the fact that about a third of this memoir is devoted to setting forth his theory of heat and his covering theory as arguments in support of his theory of combustion. See Daumas' *Lavoisier théoricien et expérimentateur,* p. 58.

156 *Procès-verbaux,* 1785, f. 140: "M. Lavoisier a lu des 'Réflexions sur le phlogistique'." This reading took place on 28 June, but was continued at later meetings. The reading was finished on 13 July. The memoir may be conveniently consulted in *Oeuvres de Lavoisier, 2,* 623–655.

theory to accommodate such inescapable facts as the gain in weight of metals on calcination and the absorption of air in such processes. In refuting each man, Lavoisier draws upon the recent experiments on heat.[157]

Baumé's phlogiston differed from Stahl's in that he supposed that it is composed of "*feu libre*" and "*élément terreux*" that can mix in an infinite number of different proportions. He attempted to explain the weight gain during calcination by assuming that metals lose phlogiston, but that this phlogiston is replaced by "*feu libre*," or by that fire charged with some "*élément terreux.*"

Lavoisier points out that Baumé will have to accord the element of fire a very great weight, for some metals increase tremendously in weight during calcination. Moreover the "*feu libre*" must also compensate for the loss of weight due to the escape of phlogiston, for in Baumé's theory the phlogiston has weight. But no experiment has succeeded in showing that heat has weight. Indeed, far from "feu libre" being absorbed during the calcination of metals, a great deal of heat is *liberated.* This can even be measured in the case of the calcination of iron and zinc. At this point Lavoisier remarks: "Les expériences faites depuis peu en Angleterre, en France et en Suède, sur la chaleur, fournissent encore de nouvelles objections contre le système de M. Baumé."[158] His argument is this: if, as Baumé also believed, "*feu libre*" has the property of combining with metallic substances and reducing their calxes to the metallic state, bodies containing the most heat ought to be the best reducing agents. Water vapor is an example; it is combined with a large amount of heat: "de matière du feu presque dans un état de liberté, qui lui communique de l'élasticité." Yet it does not act as a reducing agent.

Macquer's theory was a more radical compromise.[159] From

157Not only the results contained in the "Mémoire sur la chaleur" but also from experiments with Laplace carried out during the winter of 1783-1784. These were not published until after Lavoisier's death. See *Oeuvres de Lavoisier, 2,* 724-738, where the paper is reproduced from the *Mémoires de chimie.*

158*Oeuvres de Lavoisier, 2,* 626. Baumé's theories are set forth in his *Chymie expérimentale et raisonnée,* 3 vols. (Paris, 1773). Partington (*History of Chemistry, 3,* 92) writes that Baumé "defended Meyer's theory of *acidum pingue* but rejected this name, calling it 'feu pur'."

159The theory is found in the second edition of Macquer's *Dictionnaire de chymie,* 4 vols. (Paris, 1778), chiefly—as Lavoisier points out—in the articles "Calcination" and "Phlogistique."

Lavoisier's point of view, this was a wholly new theory, scarcely related to Stahl's. Macquer's phlogiston is pure "matter of light" and it is contained in all combustible bodies including the metals.[160] To explain the weight increase on calcination and combustion Macquer adopts Lavoisier's explanation that it results from the combination of air, or *air pur,* with metals or combustibles. But when this takes place the "matter of light" is given off by the metal or combustible. In the course of a detailed refutation of Macquer's views, Lavoisier points to a particular difficulty in this new compromise theory:

> Si comme il le prétend, le phlogistique n'est autre chose que la pure matière de la chaleur et de la lumière, il en résulte que les métaux, dans leur état métallique, doivent contenir beaucoup plus de la matière de chaleur que les chaux métalliques; et cependant les expériences de M. Crawford, celles de M. Wilke [*sic*], celles de M. de Laplace et les miennes, prouvent le contraire.[161]

That is, the heat capacity or specific heat of metals should be greater than that of their calxes (oxides), but experiments showed that in fact the specific heats of the oxides of mercury, lead, and antimony are greater than the specific heats of the metals.[162]

At this point Lavoisier introduces a discussion of the nature of heat and its general effects with these words:

> Il est temps de ramener la chimie à une manière de raisonner plus rigoureuse, de dépouiller les faits dont cette science s'enrichit tous les jours de ce que le raisonnement et les préjugés y ajoutent; de

[160] Macquer rejected the notion of an element of fire and supposes that there is no matter of heat: heat consists only in the rapid movement of the "molécules élémentaires des corps." Light is the subtlest of all kinds of matter and, as Lavoisier puts it, Macquer "la regarde comme plus susceptible qu'aucune autre de prendre le mouvement qui constitue la chaleur." To Lavoisier this was merely a way of disguising the matter of fire by calling it the "matter of light."

[161] *Oeuvres de Lavoisier, 2,* 636.

[162] These experiments are not specifically mentioned in the "Réflexions." They were however performed in the winter of 1783–1784, when one of the chief objectives was to measure "la chaleur spécifique d'un plus grand nombre de substances, principalement des métaux et de leurs oxydes." See "Mémoire contenant les expériences faites sur la chaleur, pendant l'hiver de 1783 à 1784," *Oeuvres de Lavoisier, 2,* 724–738. The specific heats for mercury, lead, antimony, and their oxides are given in the table on p. 729.

distinguer ce qui est de fait et d'observation d'avec ce qui est systématique et hypothétique; enfin, de faire en sorte de marquer le terme auquel les connaissances chimiques sont parvenues, afin que ceux qui nous suivront puissent partir de ce point et procéder avec sûreté à l'avancement de la science. . . .[163]

It was with good reason that Lavoisier announced to the reader that before developing his ideas on calcination and combustion he proposed to set forth "quelque considérations sur la nature de la chaleur et sur les effets généraux qu'elle produit." We need not devote space to this important summary, for we know what Lavoisier's views on heat in fact were.[164] It is interesting, however, to stress the space that he devotes to the doctrine of specific heats, to note that he mentions Crawford at least twice by name and that he explicitly refers back to the "Mémoire sur la chaleur" and to the method of measuring heat "imaginé par M. de Laplace."[165]

THE COMPOSITION OF WATER

Almost equally significant for the progress of Lavoisier's new chemistry—and far more dramatic than the work on heat and calorimetry—was another research problem in which Laplace took a minor but nonetheless significant part. This was the experiment that led to the recognition that water, generally considered since Antiquity to be an element, was in fact a compound of two of the recently identified gases, oxygen and hydrogen.

Lavoisier's theory of acid-formation—that combustion of substances other than metals, their combination with *air vital* or oxygen,

[163] *Oeuvres de Lavoisier, 2,* 640.

[164] The close relationship that existed in Lavoisier's mind between the "Réflexions sur le phlogistique" and the "Mémoire sur la chaleur" is clear. When about 1792 he began the compilation that was subsequently published after his death, the *Mémoires de chimie,* the first of his earlier papers he felt he should include was the "Mémoire sur la chaleur." But the difficult and technical character of that memoir led him to preface it with a more general exposition of his ideas about heat. Much of the content of this introductory memoir is drawn directly, sometimes word for word, from the "Réflexions." This prefatory compilation is labelled "Première Mémoire" and has a misleading note which says nothing about the use made of the "Réflexions" but simply says: "Extrait des Mémoires de l'Académie des Sciences, an. 1777, et des élémens de Chimie, publiés en 1789."

[165] See above, note 140.

seemed invariably to produce some sort of acid—led Lavoisier to expect that when *air inflammable* (hydrogen) is burned with *air vital* the product should be an acid. For some reason Lavoisier's earlier collaborator J.-B. Bucquet (who had died in January 1780) had conjectured that the product of such a combustion should prove to be "fixed air" (CO_2). Lavoisier and Bucquet accordingly carried out in September 1777 an experiment to test this hypothesis, but the nature of the product eluded them, though they did discover that nothing was formed that could precipitate limewater, and therefore that Bucquet's guess was wrong.[166]

Lavoisier took up these experiments again in the winter of 1781–1782 with the assistance of a certain M. Gingembre, but when they burned inflammable air with "*air vital*" there was no indication that any acid was formed. Puzzled by these negative results, and convinced that some undetected product must have been formed ("rien ne s'anéanit dans les expériences"), Lavoisier at this point determined to repeat these experiments "avec plus d'exactitude et plus en grand." Laplace seems to have shared in this decision; together the two men designed a combustion apparatus with a double nozzle which enabled them over a considerable time to maintain and control the flow of the two gases.[167] This apparatus was completed by June 1783 and was immediately put to a quite spectacular use by Lavoisier and Laplace scarcely a week after they had presented to the Academy their memoir on heat and calorimetry.

In that month an English visitor, Charles Blagden, arrived in Paris. A distinguished Fellow of the Royal Society of London, who soon after became its Secretary, Blagden brought with him the news that his eccentric and brilliant colleague, Henry Cavendish, had obtained pure water when—in a device like Volta's "*pistolet*"—he had detonated a mixture of hydrogen and oxygen in closed vessels. According

166 For the background of this investigation I have accepted Lavoisier's account in his memoir on the decomposition of water in *Oeuvres, 2,* 334–373.

167 I infer this from the account given by Lavoisier soon after presenting his paper on the decomposition of water to a public session of the Academy of Sciences on 12 November 1783. This appeared in Rozier's Journal in December (*Observations sur la physique, 23,* [1783], 452–455). Here, after recording that Lavoisier and Laplace "se sont proposé de le constater en grand," they add: "Ils avoient préparé une espèce de lampe à air inflammable à double tuyau." (*Op. cit.,* p. 452.) Lavoisier's autograph draft, written in the first person, is in Dossier 376, Lavoisier Papers, Archives of the Academy of Sciences, Paris.

to his report, the amount of water corresponded closely to the weights of the two gases that were consumed. It was this news that impelled Laplace and Lavoisier to attempt to verify Cavendish's result by putting their new combustion apparatus to use.

The experiment was performed by the two men on 24 June in the presence of Blagden and a number of their colleagues of the Academy of Sciences, among them Vandermonde, Meusnier, and J.-B. Le Roy.[168] The product, as Blagden had asserted, was pure water. The next day the two men reported this remarkable result to the Academy.[169] They saw at once the implications of their experiment: that water was in fact a compound of the two gases. The question that had to be answered was whether the weight of the water produced was in fact exactly equal, or as equal as the errors of experiment allowed, to the weight of the two gases combined, as Blagden insisted that Cavendish had demonstrated. Lavoisier admitted that under the conditions of their experiment it was not possible to determine precisely the quantity of the two gases that reacted together, yet he believed that the weight of the water must have equalled the sum of the weights of the two gases. His reasoning, to say the least, was hardly conclusive: just as it is true in mathematics, he argued, so it is true in physics that the whole must equal the sum of its parts; and since only pure water, with no detectable residue, was produced in the experiment, "nous nous sommes crus en droit d'en conclure que le poids de cette eau était égal à celui des deux airs qui avaient servi à la former."[170] Yet despite Lavoisier's use of the first person plural,

[168]The event is described in Lavoisier's laboratory *registres* for that date. See Berthelot, *Révolution chimique,* p. 293.

[169]*Procès-verbaux,* 1783, f. 146. "MM. Lavoisier et Delaplace [*sic*] ont annoncé qu'ils avoient dernièrement répété en présence de plusieurs membres de l'Académie, la combustion de l'air combustible combiné avec l'air déphlogistiqué, ils ont opéré sur sioxante pintes environ de ces airs et la combustion a été faites dans un vaisseau fermé. Le résultat a été de l'eau très pure." For a summary of recent work on the water synthesis, see my article "Lavoisier" in the *Dictionary of Scientific Biography, 8* (1973), 66-91.

[170]*Oeuvres de Lavoisier, 2,* 338-339. In the abstract for Rozier's Journal, Lavoisier merely wrote: "L'expérience finie, on parvint à rassembler presque toute l'eau, par le moyen d'un entonnoir, & son poids se trouva près de 5 gros; ce qui répondoit à-peu-près au poids des deux airs réunis." (*Observations sur la physique, 23* [1783], 453.) But in his laboratory *registre* he recorded: "On peut évaluer à 3 gros la quantité d'eau; on aurait dû retirer 1 once 1 gros 12 grains d'eau. Ainsi il faut supposer une perte de deux tiers de l'air ou qu'il y ait perte de poids." (Berthelot, *Révolution chimique,* p. 293.)

Laplace was much more cautious. On 28 June 1783, just four days after the successful experiment, Laplace wrote a description of the event to Deluc in London. It is a straightforward account written while the episode was still fresh in his mind:

> Nous avons répété, Mr. Lavoisier et moi, devant M. Blagden et plusieurs autres personnes l'expérience de Mr. Cavendish sur la conversion en eau des airs déphlogistiqué et inflammable, par leur combustion; avec cette différence, que nous les avons fait brûler sans le secours de l'étincelle électrique, en faisant concourir deux courants, l'un de l'air pur, l'autre de l'air inflammable. Nous avons obtenu plus de 2 1/2 gros d'eau pure, ou au moins qui n'avait aucun caracterè d'acidité et qui étoit insipide au goût; mais nous ne savons par encore, si cette quantité d'eau représente le poids des airs consumés, c'est une expérience à recommencer avec toute l'attention possible, et qui me paroit de la plus grande importance.[171]

The two experimenters were unaware that almost simultaneously, and quite independently, Gaspard Monge had carried out, in the laboratory of the Ecole de Mézières during the months of June and July, similar experiments with almost identical, but much more accurate, results. When, not long afterward, Vandermonde read to the Academy a memoir of Monge describing his experiments, it became quite clear that water is quantitatively formed when the two gases are burned. As Lavoisier wrote afterwards:

> L'appareil de M. Monge est extrèmement ingénieux: il a apporté infiniment de soin à déterminer la pesanteur spécifique des deux airs; il a opéré sans perte, de sorte que son expérience est beaucoup plus concluante que la nôtre, et ne laisse rien à désirer: le résultat qu'il a obtenu a été de l'eau pure, dont le poids s'est trouvé, à très-peu de chose près, égal à celui des deux airs.[172]

[171] Muirhead, *Correspondence of James Watt,* p. 41.

[172] *Oeuvres de Lavoisier, 2,* 339. In Rozier's Journal, after describing their own experiments, Lavoisier wrote: "Peu de temps après, M. Monge a fait adresser à l'Académie le résultat d'une combustion semblable, faites à Mézières, avec un appareil tout différent, & qui peut-être est plus exact. Il a déterminé, avec un grand soin, la pesanteur des deux airs, & il a trouvé de même, qu'en brûlant des quantités d'air inflammable & d'air déphlogistiqué, on avoit de l'eau très-pure, & que son poids approchoit beaucoup d'être égal au poids des deux airs employés." (*Observations sur la physique, 23* [1783], 453.) Vandermonde read the memoir of Monge on 6 August 1783. See *Procès-verbaux,*

The experiments of Monge seem to have removed for Laplace the doubts he had expressed in his letter to Deluc. And on several occasions during the late summer of 1783 he remarked to Lavoisier that he was convinced that the inflammable air given off when iron or zinc is dissolved in vitriolic acid or "l'*acide marin*" (hydrochloric acid) comes not from the metal but from the decomposition of water. In September he set down on paper for Lavoisier the reasons for this opinion,[173] arguing as follows: metals dissolved in dilute acids form calxes, i.e., they are combined with *air vital*. Whence comes this *air vital?* Since solution can occur in closed vessels, it cannot come from the atmosphere. Nor can it come from the acid, for when vitriolic acid is deprived of its vital air it yields either *acide sulphureux* (sulphurous acid) or simply sulphur. Yet neither is produced when iron is dissolved in dilute acid, as it should be if the vital air is transferred from the acid to the metal. The only possible source is the water (which experiment has recently shown to be formed from the combustion of inflammable air with vital air). What of the inflammable air produced when these metal calxes are formed by the action of acids? It cannot come from the metal, for one ought to obtain it by treating metals with *acide nitreux* (nitric acid). Yet this is not the case; the product is *air nitreux* (nitric oxide) which can be shown to contain no *air inflammable* (hydrogen). The same *air nitreux* is formed when nitric acid acts on mercury, but the mercury can hardly have supplied *air inflammable* to the *air nitreux,* because the mercuric calx can be "revivified" or reduced by the action of heat alone, without the addition of *air inflammable.* Clearly, if the mercury contains hydrogen and loses it in being converted to the calx, it ought not to be necessary to add hydrogen to revivify the mercury.

This is a remarkable document, displaying a surprising amount of chemical knowledge on the part of Laplace. Rather than representing the first notions he entertained about the origin of the inflammable air, and which Lavoisier says he "m'avait répété bien des fois," this would seem to have been the summation of evidence the two men acquired in late August and through the month of September 1783. Lavoisier's laboratory notebooks for this period record experi-

1783, f. 164v, where we read "M. Vandermonde a lû un mémoire de M. Monge sur le résidu de la déflagration du gaz inflammable et de l'air vital."

[173]Lavoisier quotes at length from this paper in his memoir on the decomposition of water, *Oeuvres de Lavoisier, 2,* 342.

ments, in which Laplace played at least the part of observer and occasional participant, on precisely the reactions described by Laplace: the action of nitric acid on mercury (25 August, 8 September, and 21 September) and the action of vitriolic acid on iron (27 September). According to Berthelot, at least one of the entries is partly in the handwriting of Lavoisier, partly in that of Laplace.[174]

In the same period Lavoisier undertook what for him was a necessary step in proving the compound constitution of water: to find evidence for the *decomposition* of water by the action of metals. In Sweden, Torbern Bergman had recently reported in his *Dissertatio chemica de analysi ferri* (1781) that iron filings could be oxidized (converted into *éthiops martial*) by simple exposure to distilled water, and that when this happened a large amount of inflammable air was given off. Others, Lavoisier learned, had made a similar observation. During the recess of the Academy Lavoisier confirmed this effect; at least the experiment, which lasted several months, had progressed sufficiently for him to cite it in a paper on the compound nature of water, read at the public session of the Academy of Sciences on 12 November 1783.[175]

THE INFLUENCE OF LAPLACE

Not long after, on 20 December 1783, Lavoisier registered with the Academy a remarkable series of memoirs, several of which derive from, or were extensions of, the experiments we have described. In one of these, his "Mémoire sur l'union du principe oxygine avec le fer," he discusses the oxidation of iron by vitriolic acid.[176] In concentrated acid no inflammable air is given off and the oxygen is derived from the acid, but in dilute solutions it is the water that is decomposed. The explanation he offers of these phenomena he attributes to Laplace. Concentrated vitriolic acid is known to take up water to a given point of saturation. If more water is added, the solution is made up of "combined water" tightly bound to the acid and of water that is "en quelque façon" free. When, as in dilute acid

[174] Berthelot, *Révolution chimique,* p. 293.

[175] The experiment is briefly mentioned in the abstract published in Rozier's Journal in December (*op. cit.,* pp. 453–454) but the fullest description is in the paper published in 1784 (*Oeuvres de Lavoisier, 2,* 341–343).

[176] *Oeuvres de Lavoisier, 2,* 557–574. This paper was published in the *Mémoires* of the Academy of Sciences for 1782, a volume that appeared in 1785.

solutions, there is abundance of "free water," the iron acts preferentially on this "parce qu'elle [the water] n'est contre-balancée par aucune force," unlike the water that is combined with the acid.

Now, in a rather casual way, Lavoisier had made use of the concept of elective chemical affinity in a number of his early writings. But the matters that concerned him in the summer and autumn of 1783—not least the composition of water—led him, probably under the influence of Laplace, to a more detailed and intensive use of the affinity concept. The extraordinary reactivity of oxygen gas became the subject of close study. In his earlier "Considérations générales sur la nature des acides," the paper in which he advanced his oxygen theory of the formation of acids, he contented himself with the general statement that the *principe acidifiant,* or *principe oxygine,* has—like other chemical individuals—varying affinity for different substances. In his "Mémoire sur différentes combinaisons de l'acide phosphorique," read in 1780 but quite evidently reworked in 1783, Lavoisier wrote: "J'expliquerai, dans un mémoire auquel je travaille, sur les degrés d'affinités de l'air vital ou déphlogistiqué avec differentes substances, la cause de cette indissolubilité [of copper in phosphoric acid]; je prouverai qu'elle tient à ce que l'air vital ou le principe oxygine a plus d'affinité avec le phosphore qu'avec les métaux."[177] Another of the papers deposited on 20 December 1783, his "Mémoire sur la précipitation des substances métalliques les unes par les autres," treats from his new point of view the venerable question of the replacement series of the metals. Lavoisier determined the oxygen content of the different metallic oxides produced by dissolving the metals in acids, or by their mutual replacement and precipitation from solution. At the end of the memoir he summed up his conclusion: "Enfin, on conclura de tout ceci que ce n'est point en raison de leur plus ou moins grande affinité pour les acides que les métaux se précipitent les uns les autres, mais principalement en raison de leur affinité plus ou moins grande pour le principe oxygine."[178] What is particularly significant is the language in which these affinity problems are now treated. Lavoisier is willing to describe affinities in terms of forces, i.e., of forces of attraction. In one of the other papers, published in 1785, he remarks that experiments have now shown that the solution of metals is not as simple a process as people have hitherto imagined: "un grand nombre de forces

[177] *Ibid.,* p. 275.

[178] *Ibid.,* p. 545. This paper was published in 1785.

agissent chacune avec l'énergie qui leur est propre." The more we deepen our knowledge of chemistry, the more our results, simple in appearance, are seen to be complicated: "Nous ne connaissions que deux ou trois forces qui avaient lieu dans la dissolution des métaux, et il s'en trouve aujourd'hui un beaucoup plus grand nombre." He then lists seven forces—forces exerted by molecules of different species on each other, forces exerted by heat, and so on—and continues: "Connaître l'énergie de toutes ces forces, parvenir à leur donner une valeur numéraire, les calculer, est le but que doit se proposer la chimie; elle y marche à pas lents, mais il n'est pas impossible qu'elle y parvienne."[179] The echo of the work on heat with Laplace is quite apparent; and I cannot avoid the suspicion that this new language, so Newtonian in character, and the quasi-algebraic formulas in this paper, suggest the influence of Laplace who, in his own magisterial papers, worked in terms of forces of attraction and recognized the extent to which even in the vastly more simple planetary system perturbing forces greatly affect the result.[180]

Laplace continued his association with Lavoisier during the autumn of 1783 and the winter of 1783-1784, and with a degree of energy and enthusiasm that belied his apologetic remarks to his friend Lagrange. In December of 1783 he wrote to Lavoisier:

[179] "Considérations générales sur la dissolution des métaux dans les acides," *Oeuvres de Lavoisier, 2,* 509-527. See especially pp. 515 and 524-525. For a perceptive account of this paper see Charles Coulston Gillispie, *The Edge of Objectivity* (Princeton, 1960), pp. 241-246. Gillispie correctly finds Lavoisier "groping . . . toward the mathematization of chemistry" and points to the similarity of approach in this memoir and in the joint "Mémoir sur la chaleur." And he cautiously suggests that Lavoisier found in his collaboration with Laplace the idea of expressing chemical quantity in an abstract and mathematical manner. My own interpretation goes a bit farther than this.

[180] I surely do not exclude the influence of Bergman who was perhaps the first on the Continent to treat affinities as forces of attraction. His interest in attractionist theories was early; see his *Dissertatio physico-mathematica* (Uppsala, 1758). His paper applying such ideas to affinity theory (his famous *Disquisitio de attractionibus electivis*), first published in *Nova acta Regiae Societatis scientiarum Upsaliensis, 2* (1775), 161-250, was printed in revised and enlarged form in the *Opuscula physica et chemica* (*3,* 1783). Lavoisier was probably familiar with this volume, for he cites Bergman's *De diversa phlogisti quantitate in metallis* in his "Mémoire sur la précipitation des substances métalliques," and is more likely to have read the paper of Bergman in this volume of the *Opuscula* than in its separate publication in 1780 (Moström No. 158).

> Profitons du tems favorable qui se présente, pour nos expériences. Faittes bruler, je vous prie, du charbon dans une de nos machines & répétés deux fois l'expérience; faittes y respirer, trois moineaux francs, ensuite, un cochon d'inde, & répétés ces expériences le plus grand nombre de fois qu'il sera possible. Je crois que si l'accord que nous avons trouvé dans nos premières expériences, se soutient dans celles ci, nostre théorie de la respiration sera suffisament établie.
>
> Dans nostre austre machine, nous pouvons commencer par les métaux. Il sera bon de répéter l'expérience sur le mercure, ensuite sur le fer, sur le plomb, l'étain &c. Après cela nous éprouverons les chaux métalliques.
>
> Ne perdons pas de tems, je vous prie, & si vous avés besoin de moi, je serai à vous, quand vous me le ferés sçavoir. Si vous pouviez avoir fait d'icy à mercredi, les expériences sur le charbon & sur le cochon d'inde, cela me ferait un grand plasir, car je ne vous dissimule point que j'ai quelque inquiétude sur nos premières expériences.[181]

Laplace ends by remarking that he planş to set down in writing some of his ideas on the affinities of saline substances with water, "& j'aurai l'honneur de vous les communiquer afin que nous discutions ensemble cette matière intéressante."

Laplace's greater caution, and his insistence on repeating the experiments of the summer (which, we know, had already appeared in print), are no less interesting than the clue this letter gives to his role as planner of the experiments carried out that winter and spring. The work went forward intensively, with much attention paid, as we saw earlier, to heats of combustion and the determination of the specific heats of various substances, chiefly the metals and their oxides. I have already pointed out that the publication of these results was long delayed; and when they were written down, sometime around 1792, Laplace's preponderant role is underscored by his listing as senior author of that paper when it finally appeared in Lavoisier's posthumous *Mêmoires de chimie.*[182]

[181] *Correspondance de Lavoisier, 3,* 757–758. This letter was first published by Denis I. Duveen and Roger Hahn in *Revue d'histoire des sciences et de leurs applications, 11* (1958), 337–342. The letter is undated but was probably written on 21 December 1783.

[182] *Oeuvres de Lavoisier, 2,* 724–738. See above note 135.

The experiment that Laplace mentioned on the combustion of carbon was performed in due course and formed the basis of a memoir published in 1784 on the production and nature of carbonic acid.[183] Indeed one of the last bits of evidence of active collaboration between these two men of genius is to be found in Lavoisier's *registres* for 10 May 1784, where the account of an experiment on the combustion of carbon begins in Lavoisier's handwriting and concludes with that of Laplace.[184]

In this paper on the combustion of carbon, which confirmed and extended his theory of the formation of acids, Lavoisier also had his eye on the problem of affinity: "La combustion du charbon est donc un jeu des différents degrés d'affinité du principe oxygine; elle prouve que le principe oxygine a plus d'affinité avec la matière charbonneuse qu'avec la matière du feu et de la chaleur."[185] This is obvious, since the oxygen combines with the carbon, while the heat (which is a constituent part of oxygen gas, according to Lavoisier's covering theory) "se dégage avec les caractères qui lui sont propres, avec chaleur et lumière." This, as we have seen, was central to Lavoisier's understanding of combustion.

Perhaps the most interesting of the group of papers registered with the Academy on 20 December 1783 is his "Mémoire sur l'affinité du principe oxygine."[186] Since this was not published in the Academy's *Mémoires* until 1785, it obviously could have been, and probably was, revised so as to take into account the work with Laplace in the winter of 1783-1784. It contains, besides evidence drawn from these later experiments, a reference to Laplace's most remarkable and farsighted contribution to chemical theory: an outgrowth of those ideas on affinity of substances in solution which are set forth in the "Mémoire sur la chaleur" and which he mentioned to Lavoisier in his letter of December 1783.[187]

Early in this short paper Lavoisier enumerates the many difficulties that stand in the way of building up reliable tables of chemical affinity, a popular enterprise which up to this time—though he had

[183]"Mémoire sur la formation de l'acide nommé air fixe ou acide crayeux," *Oeuvres de Lavoisier, 2,* 403-422. Here Lavoisier remarks (p. 404) that the experiments "ont été faites tantôt avec M. de Laplace, tantôt avec M. Meusnier, et quelquefois avec l'un et l'autre réunis."

[184]Berthelot, *Révolution chimique,* pp. 297-298, Nos. 28-29.

[185]*Oeuvres de Lavoisier, 2,* 404.

[186]*Ibid.,* pp. 546-556.

[187]See above note 181.

used the general concept here and there in casual fashion in his early papers and notes—had not appealed to him. Such tables, he remarks, usually treat only "simple affinities," when in fact these do not exist in nature, but only double or multiple affinities. Another defect is that the influence of water is neglected; it is treated as a passive agent, whereas in reactions carried out in solution it affects the results by its attractive force, and even by its decomposition. Water exerts on the reactants what, in terms reminiscent of the celestial physics of Laplace, Lavoisier calls a "force réelle et perturbatrice." Still another source of confusion is that substances combine together with different and discrete degrees of saturation: for example, sulphur combines with different amounts of oxygen and forms two different acids. These different degrees of saturation affect the *force attractive* between the molecules of bodies. The affinity of oxygen with sulphur (when it forms sulphurous acid) is different from its affinity with sulphurous acid when it combines with it to form vitriolic acid.

More interesting still is Lavoisier's emphatic reminder that affinities are functions of the temperature. Bodies are held together by a cohesive force (the *affinité d'aggrégation*). The degree of separation of the constituent particles (*molécules constituantes*) of bodies is caused by forces opposing this affinity of aggregation. These forces result from the fine, elastic fluid of heat permeating the body, from its intensity as registered by the thermometer, and by its total amount. It is this that explains why a body can become successively solid, liquid, or aeriform.

By the same token, when different bodies come together, the effect they exert one on the other will vary with the temperature. If both are solid, no action takes place because the *affinity of aggregation* is stronger than the attractive force the particles of the two substances exert upon each other. Clearly, then, a table of affinities is without meaning unless the temperature of the reaction is specified.

To drive this point home, Lavoisier cites the behavior of mercuric oxide. At about its boiling point mercury combines with oxygen to form the red calx; but if the temperature is raised to that at which glass softens, the oxygen is given off. Clearly mercury's affinity for oxygen is different at the two temperatures. Obviously a different affinity table will be necessary for each temperature.

Lavoisier then introduces his table of the affinities of oxygen for various substances, adding some significant comments. This table, he admits, has all the defects he has just enumerated, but it may have some value until such a time as a multiplication of experiments, and the application of mathematics to chemistry, will deepen our understanding:

> Peut-être, un jour, la précision des données sera-t-elle amenée au point que le géomètre pourra calculer, dans son cabinet, les phénomènes d'une combinaison chimique quelconque, pour ainsi dire de la même manière qu'il calcule le mouvement des corps célestes. Les vues que M. de Laplace a sur cet objet, et les expériences que nous avons projetées, d'après ses idées, pour exprimer par des nombres la force des affinités des différents corps, permettent déjà de ne pas regarder cette espérance absolument comme une chimère.[188]

He then makes the revealing statement that the table that follows gives the affinity of oxygen for different substances "résultant principalement . . . des changements dans les degrés de chaleur." Since by "degré de chaleur" Lavoisier always means temperature, the table would seem to have been built up by determining the temperature at which each of the substances begins to be oxidized.

But if so, what were the "vues que M. de Laplace a sur cet objet," and what sort of experiments did they envisage in order to express by numbers the force of affinity of different bodies?

Now, of course, the idea of expressing affinities by numbers was by no means new; in fact, it was a widespread occupation of a number of Lavoisier's contemporaries.[189] Some, like William Cullen and Joseph Black in their unpublished lectures, assigned numbers that were intended to express purely *relative* dispositions to combine, as in cases of double elective affinity. Of greater significance were the attempts to *measure* some quantity assumed to be an expression of affinity. Thus K. F. Wenzel (1777) suggested that the affinities of different substances for a given solvent varied inversely as the times they took to dissolve. Similarly Kirwan, and later J. B. Richter, used

[188] *Oeuvres de Lavoisier, 2,* 550–551.

[189] See, for eighteenth-century work on affinities, Maurice Crosland, "The Development of Chemistry in the Eighteenth Century," *Studies on Voltaire and the Eighteenth Century, 24* (Geneva, 1963), 382–390, and my "Background to Dalton's Atomic Theory" in D. S. L. Cardwell, *John Dalton & the Progress of Science* (Manchester, 1968), pp. 57–91. I give (note 43 of that article) other references to the history of affinity theory.

the *combining* weights of acids and bases. Guyton de Morveau, taking a more direct physical approach, believed with Buffon that only a single law of attraction operated in chemical affinities, and that the varying affinities resulted from the several shapes of the constituent particles of different bodies. He attempted to make exact determinations by measuring the force required to lift discs of uniform size made of different metals from contact with the surface of mercury.

It is clear that none of these methods is what Laplace had in mind. And it is certainly not the case that they would have confined themselves to carrying out the kind of study of displacement reactions that had filled the chemical literature since the time of Geoffroy. The proposed method was a sophisticated physical one, but we have to look elsewhere to determine its nature. Once again we must turn back to the "Mémoire sur la chaleur" of 1783. Here we find a brief but illuminating passage in that Article III whose authorship I have felt justified in attributing to Laplace. The passage is as follows:

> L'équilibre entre la chaleur, qui tend à écarter les molécules des corps, et leurs affinités réciproques, qui tendent à les réunir, peut fournir un moyen très-précis de comparer entre elles [les] affinités.[190]

The method, then, was to be a calorimetric one, using the ice-calorimeter. Laplace's proposal seems to have been nothing less than to use the *heats of combustion,* or *heats of formation* of compounds, as a numerical measure of chemical affinity. That this is so is confirmed by a chapter of Laplace's *Exposition du système du monde,* a chapter that underwent substantial, and interesting, alterations between the first edition (1796) and the later ones. Let me refer first of all to the sixth edition (1835) which I happen to own. Here, in Chapter XVIII entitled *De l'attraction moléculaire,* Laplace writes that all chemical combinations are the result of forces, and the study of those forces should be one of the chief objects of the science of chemistry.[191] Molecular attractive force is the cause of the

[190]*Oeuvres de Lavoisier, 2,* 317. Cf. Berthelot, *Révolution chimique,* p. 295, No. 201. I do not suggest that Laplace had already envisaged the proposed method in terms of heats of combustion, for at this time he was thinking only of "affinities" of solutes and solvents, as the remainder of the passage (as well as his letter of December 1783 to Lavoisier) makes clear.

[191]*Exposition du Système du Monde; par M. le Marquis De Laplace,* sixième édition (Paris, 1835), pp. 323-364.

aggregation of like molecules (*molécules homogènes*) and the cause of affinities between molecules of different kinds. In sum, all these chemical problems are reduced to microphysical ones: the underlying causes of chemical and physical change are intermolecular or interparticulate attractive forces. These forces depend upon the shape and relative positions of the elementary particles, on the laws of their attractive forces, as well as on the repulsive forces of electricity and heat, and perhaps also on other forces yet unknown.

In this late edition of the *Exposition,* Laplace concludes on a pessimistic note. The various attractive forces, he writes (in words that recall those of Buffon and Guyton de Morveau), should depend upon the variety of shapes of the *molécules intégrantes* and their respective positions. If so, this would bring chemical phenomena under the one great law of physics and chemistry, that of universal attraction. But the impossibility of knowing the shapes of the molecules and their distances from one another make this sort of explanation so vague as to be useless for the advancement of the sciences.

In the edition of 1796, however, this pessimistic passage is followed by some extremely interesting sentences, eliminated from the last editions:

> Au milieu de ces incertitudes, le parti le plus sage est de s'attacher à déterminer par de nombreuses expériences, les lois des affinités; et pour y parvenir, le moyen qui parait le plus simple, est de comparer ces forces avec la force répulsive de la chaleur, que l'on peut comparer elle même avec la pesanteur. Quelques expériences déjà faites par ce moyen, donnent lieu d'espérer qu'un jour, ces lois seront parfaitement connues; alors, en y appliquant le calcul, on pourra élever la physique des corps terrestres, au degré de perfection, que la découverte de la pesanteur universelle a donné à la physique céleste.[192]

Such experiments seem never to have been seriously undertaken by Laplace and Lavoisier, and after the spring of 1784 there is no evidence of further collaboration in the laboratory. Yet there is no reason to believe, as Maurice Crosland wrote not long ago, that towards the end of Lavoisier's career—as evidenced by what he says in the *Traité élémentaire de chimie* of 1789—he began to have grave

[192] *Exposition du Système du Monde,* 2 vols. (Paris, 1796) *2,* 197.

doubts about the possibility of constructing tables of affinity. His final position, Crosland would have it, was "one of retreat."[193]

I do not think that this was the case (in 1796, at least, this was not true of Laplace). To be sure, in the *Traité*, Lavoisier shows himself to be fully aware, as indeed he had been a few years before, of the complexities and difficulties attending the study of affinities. But in 1789 his confidence in the long-range possibilities remained undiminished: the subject of chemical attractions and affinities is, he wrote in this famous book, the branch of chemistry best capable of being turned into a systematic body of knowledge, and it will come to hold the same place with respect to the rest of chemistry that "the transcendental geometry" bears to the elementary parts of mathematics.[194] Perhaps this rather sibyline remark makes sense if we take it that in 1789, as earlier in 1783–1784, affinity theory, by means of this thermochemical approach, must ultimately become a branch of a higher and more mathematical chemistry. If so, this is indeed prophetic, in a general way at least, of the transformation chemistry was to undergo in the nineteenth century and our own day. By means of chemical thermodynamics the physical chemist can indeed, without leaving his "cabinet," predict the course of many chemical reactions.

There is even later evidence than the *Traité* that not long before his tragic execution, Lavoisier had not lost faith in the eventual value of affinity investigations. This is suggested by his manuscript notes for a proposed "Cours de chimie experimentale," notes first discovered and described by Maurice Daumas, and written in December 1792.[195] In this project, which he never lived to undertake, and

[193]Crosland, "Development of Chemistry in the Eighteenth Century," *op. cit.* (note 189), p. 388. See also his *The Society of Arcueil* (Cambridge, Mass., 1967), p. 267, where he writes that in his collaboration with Lavoisier "Laplace had been particularly interested in the problems of chemical affinity and had even led Lavoisier to hope that it would be possible to measure chemical affinity exactly." But Crosland adds: "Yet Lavoisier soon realized that he had been led to hope for the impossible." It should be pointed out that Laplace is known to have influenced Berthollet and may have aroused the latter's interest in affinity theory. Berthollet quotes Laplace on affinity in his *Essai de statique chimique*, 2 vols. (Paris, 1803), *1*, 532. See Crosland, *Society of Arcueil*, p. 245.

[194]*Oeuvres de Lavoisier*, *1*, 6.

[195]Maurice Daumas, "Les conceptions de Lavoisier sur les affinités chimiques et la constitution de la matière," *Thalès*, *6* (1949–1950), 69–80. See

which was intended to be frankly speculative and to reveal "le fonds de ma pensée," Lavoisier planned to discuss, along lines that we may assume resembled what Laplace later wrote, his particulate or molecular theory of matter. He mentions a proposed "Sixième Partie" to be devoted to affinities or elective attractions. Because Lavoisier in these notes mentioned Guyton de Morveau, Daumas has concluded that Lavoisier, had he written this part, would have summarized the facts and ideas that Guyton had set forth in the *Encyclopédie méthodique.* At the very least, we can be sure that this was only part of what Lavoisier had in mind.

ACKNOWLEDGMENT

A portion of this study was carried out under a research grant from the National Science Foundation. I wish to express my appreciation for the assistance afforded by the grant, by my research assistants, Dr. Carleton Perrin and Dr. J. R. Gough, and by the referees of this paper.

---

also his *Lavoisier thèoricien et expérimentateur,* pp. 110–112. For the eighteenth-century preoccupation with affinity tables and theory useful references are A. M. Duncan, "Some Theoretical Aspects of Eighteenth-Century Tables of Affinity," *Annals of Science, 18* (1962), 177–194 and 217–232, and my "Background of Dalton's Atomic Theory," *op. cit.* (note 189).

Mayer's Concept of "Force": The "Axis" of a New Science of Physics

BY P. M. HEIMANN*

Mayer's conception of nature has received inadequate attention from historians. His works tend to be regarded as speculative, confused, and obscure, yet containing strong elements of positive science which justifies his inclusion among the "pioneers" of energy conservation.[1] The analysis of Mayer's ideas from the perspective of the "simultaneous discovery" of energy conservation presupposes that though the intentions of the different pioneers may have been different, their ideas can be understood in terms of concepts common to them all.[2] Mayer's first published paper, "On the Forces of Inorganic Nature" in 1842,[3] which is traditionally regarded as his first publication on the conservation of energy, opened with the statement that "the purpose of the following pages is to seek the answer to the question: what are we to understand by 'forces' and how are these interrelated?"[4] Mayer's claim, as he informed his friend the physician Wilhelm Griesinger later that year, was to have represented "the connection of many phenomena much more clearly than has been seen hitherto," and to have given "a clear and good idea of what a force

*Department of History, Furness College, University of Lancaster, Bailrigg, Lancaster, LA1 4YG, England.

[1] D. S. L. Cardwell, *From Watt to Clausius* (London, 1971), pp. 229 ff.

[2] This is the approach adopted by Thomas S. Kuhn in his important study, "Energy Conservation as an Example of Simultaneous Discovery," in *Critical Problems in the History of Science,* ed. M. Clagett (Madison, 1959), pp. 321–356. The crucial feature of Kuhn's analysis is his claim to have pointed to factors that are *specific* to the period 1830 to 1850 and hence to have explained simultaneous discovery. Kuhn's approach is discussed, with reference to his stress on the interconversion of forces as one such specific factor, in P. M. Heimann, "Conversion of Forces and the Conservation of Energy," *Centaurus, 18* (1974), 147–161.

[3] J. R. Mayer, "Bemerkungen über die Kräfte der unbelebten Natur," *Ann. d. Chem. u. Pharm., 42* (1842), 233–240. I have used the reprint of this and other papers of Mayer's in J. J. Weyrauch, ed., *Die Mechanik der Wärme in gesammelte Schriften von Robert Mayer* (Stuttgart, 1893). Mayer's correspondence and unpublished papers are collected in Weyrauch, *Kleinere Schriften und Briefe von Robert Mayer* (Stuttgart, 1893). These two volumes are cited below as Weyrauch, *1* and Weyrauch, *2,* respectively. All translations from these volumes are my own.

[4] Weyrauch, *1,* 23.

is."[5] The kernel of Mayer's thought is his concept of force and his theory of the relations between forces, and its understanding is essential to our comprehension of Mayer's fundamental intentions. It is this kernel that is obscured or distorted by the search for correspondences and connections between Mayer's ideas and those of other "pioneers" of energy conservation.

The distortion of Mayer's intentions goes back to nineteenth century controversies over the scientific merits and priority of Mayer's papers and his claim to be judged as an originator of the law of the conservation of energy. Central to this claim was the discussion of the value and validity of Mayer's calculation in 1842 of the mechanical equivalent of heat. Mayer based his claim to scientific originality on this calculation, and indeed it would distort Mayer's view that "numbers are the sought-for foundation of an exact natural science"[6] to fail to note the significance of his calculation. Mayer drew attention to his statement of the "law of the equivalence of heat and motion" and his derivation of "its numerical expression, the mechanical equivalent of heat,"[7] by noting that "the celebrated English physicist Joule was led to the thesis that the phenomena of heat and motion rest essentially on the same principle, or, as he expresses it in a similar way to myself, that heat and motion can be transformed into one another."[8] John Tyndall, in defending Mayer's claim to be a pioneer of energy conservation, respected Mayer's theory of nature and drew attention to the analogy between Mayer's doctrine of the indestructibility and transformability of forces and Joule's notion of the indestructibility and convertibility of natural powers.[9] The priority controversy led to a blurring of the conceptual basis of Mayer's calculation, for Mayer rejected what was central to Joule's work, the mechanical theory of heat; analogies between Mayer's and Joule's papers have limited validity. By contrast, Joule's champion P. G. Tait poured relentless scorn on Mayer's theory of nature. He called it "entirely subversive of common sense and logic in an experimental science," being "without experimental bases."[10] Mayer's intentions

[5] Weyrauch, *2,* 181.

[6] J. R. Mayer, "Bemerkungen über das mechanische Aequivalent der Wärme" [1851], in Weyrauch, *1,* 237.

[7] Weyrauch, *1,* 272.

[8] *Ibid.*

[9] John Tyndall, "Remarks on the Dynamical Theory of Heat," *Philosophical Magazine, 25* (1863), 380.

were obscured by polemics whose purpose was as limited as their conceptual finesse and historical acumen. Moreover, those who have attempted to view Mayer as one of the pioneers of energy conservation have judged his importance almost solely on the basis of that work that seemed relevant to energy conservation.

The present paper attempts to escape from this historiography by focusing on the primary intentions of Mayer's natural philosophy. The genesis of Mayer's ideas lay in physiology, which is essential to recognize if we wish to understand how the fundamental feature of his thought came to be a clarification of the concept of "force." The apparent obscurity of Mayer's conception of nature has led to considerable confusion over Mayer's philosophical indebtedness. Although claims that Mayer was a Kantian owing to his appeal to the law of causality,[11] or an adherent of Schelling's philosophy,[12] or that he "generalized" Leibniz' concept of force[13] may be fanciful, the confusion surrounding his natural philosophy is conducive to claims of this kind. The present paper, which aims at a clarification of Mayer's concept of force, will conclude with a brief discussion of the validity of such claims.

Mayer's account of his observation, which he made while performing a venesection in Java in 1840, that venous blood in the tropics was unusually bright red—so bright that he thought he had struck an artery—is a familiar story. Arguing that this observation implied that a person living in the tropics needed less oxygen than one living in a cooler climate for the maintenance of body heat, he speculated on the transformation of food material into body heat and on the relation between this transformation and the performance of bodily exertion.[14] From this speculation he derived his concept of the equivalence of heat and mechanical motion and his theory of the transformability and indestructibility of forces. Although Mayer in

[10] P. G. Tait, "On the History of Thermodynamics," *ibid., 28* (1864), 292.

[11] Charles C. Gillispie, *The Edge of Objectivity* (Princeton, 1960), pp. 375, 385.

[12] Erwin Hiebert, in his comments on Kuhn's paper on energy conservation, in Clagett, *op. cit.* (note 2), p. 394.

[13] Alwin Mittasch, *Julius Robert Mayers Kausalbegriff* (Berlin, 1940), pp. 40 ff.

[14] J. R. Mayer, "Die organische Bewegung in ihren Zusammenhange mit dem Stoffwechsel" [1845], in Weyrauch, *1,* 105 ff.

his first two papers—the second of which was published in Justus Liebig's *Annalen der Chemie und Pharmacie* in 1842—did not mention this physiological observation and its implications, he alluded to his derivation of the concept of the transformation of forces from physiological investigations in a letter to his friend the physicist Carl Baur on 24 July 1841.[15] The first clear reference to this derivation is in a letter to Griesinger on 14 June 1844, in which Mayer emphasized that his theory was not concocted at the writing table but arose from observations of the "changed physical condition" in the tropics, and to his reflections on "the conditions of the blood [which] directed my thoughts primarily to the production of animal heat by the respiratory process."[16] Mayer gave a full account of the biological origins of his theories in his 1845 paper, "Organic Motion in Its Relation to Metabolism." His reason for not having done so earlier was in part because his 1842 paper was a short note intended as a priority claim[17] and because, as he told Griesinger, "if one wants to achieve clarity on physiological questions, a knowledge of physical matters is essential if one does not choose to consider the question metaphysically, to which I have an infinite distaste."[18] Mayer thus believed that it was essential to provide a general physical rather than physiological framework for his concept of the indestructibility and transformability of forces.

Mayer's interpretation of the brighter color of venous blood in the tropics in terms of animal heat reflected contemporary thinking. Indeed, his account of his observation in 1845 shows that he drew upon Liebig's *Animal Chemistry* (1842) for much of his discussion of animal heat, which suggests that his interpretation of his observation made in 1840 may well owe something to retrospective reconstruction. Liebig asserted that "the mutual action between the elements of the food and the oxygen conveyed by the circulation of the blood to every part of the body is the source of animal heat."[19] The experimental basis of his argument that respiration was the only source of animal heat was not original but was founded on Lavoisier's

[15]Weyrauch, *2,* 110.

[16]*Ibid.,* pp. 212 f.

[17]Weyrauch, *1,* 246 f.

[18]Weyrauch, *2,* 213.

[19]Justus Liebig, *Animal Chemistry or Organic Chemistry in its Application to Physiology and Pathology,* trans. W. Gregory (1842; Johnson reprint: New York and London, 1964), p. 17.

conclusion that respiration was a chemical process comparable to the combustion of carbon.[20] Later experimental work had left the combustion theory of animal heat in an ambiguous state; C. Despretz and P. L. Dulong implied on experimental grounds that the chemical theory of animal heat was inadequate.[21] Liebig's attempts to meet such criticisms were unsatisfactory, though two of his arguments—that anomalies occurred as a result of the physical situation of the experimental animals[22] and that inaccurate data were used for the heats of combustion[23]—were employed by Mayer in an attempt to support Liebig.[24]

Mayer's use of Liebig's arguments here shows that when he wrote his account of physiological problems in 1845 he made use of Liebig's extended discussion of animal heat. But there are more significant affinities between their writings. Liebig had supported his firm statement of the respiratory theory of animal heat by a series of examples, each purporting to illustrate the relation between the quantity of food consumed, the supply of oxygen, and the amount of heat given off by the animal. He claimed that "our clothing is merely an equivalent for a certain amount of food," so that "the more warmly we are clothed the less urgent becomes the appetite for food, because the loss of heat by cooling, and consequently the amount of heat to be supplied by the food, is diminished." Thus, he argued that if men living in temperate zones were to go naked they would need to consume enormous quantities of food, and that the loss of appetite experienced by Europeans resident in the tropics was owing to their diminished need for food.[25] Liebig's assumption was that an animal needed a smaller amount of oxygen in warmer climates, for there was a relation between the amount of oxygen supplied to the animal and the difference between the temperature of the environment and that of the animal. The smaller the temperature

[20] For an account of work on animal heat at this time, see F. L. Holmes' Introduction to the Johnson reprint of *Animal Chemistry*.

[21] C. Despretz, "Recherches expérimentales sur les causes de la chaleur animale," *Ann. chim. phys., 26* (1824), 337-364; P. L. Dulong, "Mémoire sur la chaleur animale," *ibid., 1* (1841), 440-455.

[22] Liebig, *op. cit.* (note 19), p. 36. For comments, see Holmes' Introduction, p. xxxiii.

[23] J. Liebig, "Ueber die thierische Wärme," *Ann. d. Chem. u. Pharm., 53* (1845), 63-77.

[24] See Mayer's comments in his 1845 paper, Weyrauch, *1,* 81ff.

[25] Liebig, *op. cit.* (note 19), pp. 21 ff.

difference, the less oxygen the animal needed, and hence a smaller quantity of nutrients would be oxygenated.[26] Mayer used the same argument in his account of his 1840 observation of the bright red color of venous blood in the tropics. He explained the red color by relating the difference between the temperature of the environment and that of the animal to the amount of oxygen supplied to the animal. He saw the color difference between arterial and venous blood as "an expression of the magnitude of the oxygen consumption," so that the unusually red color of venous blood in the tropics implied that a smaller amount of oxygen was supplied to the body.[27] Because of the smaller temperature difference between the animal and the environment in the tropics, a smaller quantity of nutrients was oxygenated and a smaller quantity of oxygen was supplied to the blood: "nature had the task of decreasing the chemical process [the oxygenation of food] to a corresponding degree."[28] Mayer's interpretation of his observation used terms that show his adherence to Liebig's theory of animal heat. Although there seems no good reason to doubt Mayer in claiming that his 1840 observation was the starting-point of his speculations, it is clear that the 1845 account of this observation was explicitly couched in Liebig's terms.

As in the case of Mayer's theory, Liebig's theory of animal heat was part of a general theory of the transformation and indestructibility of natural agents. Although there would thus seem to be obvious affinities between the two theories, the affinities are superficial. Liebig's theory of animal heat was part of a general chemistry of physiological processes, in which the chemical "force" supplied from respiration was the source of animal heat, which in turn was responsible for the work done by bodily exertion. Moreover, Liebig argued that a special vital force was essential to the explanation of bodily phenomena such as growth, motion, and resistance to disease.[29] The vital force was one of several "forces" that were equivalent in the sense that one force could generate another. The existence of the vital force was known from its distinct effects: "it is a peculiar force, because it exhibits manifestations which are found in no other

[26] *Ibid.*, p. 19.

[27] Weyrauch, *1*, 106.

[28] *Ibid.*, p. 107.

[29] See T. O. Lipman, "The Response to Liebig's Vitalism," *Bull. Hist. Med.*, *40* (1966), 511–524, and *idem*, "Vitalism and Reductionism in Liebig's Physiological Thought," *Isis*, *58* (1967), 167–185.

known force."[30] Just as the phenomena of gravity were ascribed to a gravitational force, vital phenomena were to be ascribed to a vital force. The laws of vitality and gravitation were to be studied by investigating the effects of these forces, and scientific analysis could not extend to the study of the nature of life any more than it could to that of the nature of gravity.[31] Moreover, "no force, no power can come of nothing,"[32] so that forces could be transformed into one another but could not be annihilated.[33]

Mayer vehemently rejected the concept of a vital force; he considered that the appeal to a vital force was an appeal to a miracle and a return to mysticism,[34] and he included Liebig among those he attacked for their vital force explanations.[35] Having accepted Liebig's chemical theory of animal heat, Mayer argued that physiological processes could be completely explained in chemical terms.[36] He argued that Liebig's application of the law of the transformation, equivalence, and indestructibility of forces to the vital force did not render the concept of vital force intelligible. He regarded the use of vital force to explain the special characteristics of organisms as gratuitous and without foundation. In his 1845 paper Mayer stated that only the following "forces" could be held to exist: "fall force" (the force which caused the fall of bodies), motion, heat, magnetism, electricity, and chemical force.[37] The operations of nature were determined by the transformation of these forces into one another, and Mayer specifically excluded the vital force from the operations of nature. He also believed that Liebig's notion of forces as causes was confused. Mayer argued that since Newton had considered gravity as a *causa mathematica* and that since the "fall force" was the *causa physica,* Liebig's notion of the gravitational force as the physical cause of the fall of bodies contradicted Newton.[38] Mayer's theory

[30]Liebig, *op. cit.* (note 19), p. 221.

[31]*Ibid.,* p. 7.

[32]*Ibid.,* p. 28.

[33]*Ibid.,* p. 196.

[34]See his comments in his 1845 paper, Weyrauch, *1,* 81, 84.

[35]*Ibid.,* pp. 132–138, in passages which Mayer omitted from the 1867 reprint of his papers (*Die Mechanik der Wärme*) so as not to detract from the significance of Liebig's achievement by old polemics: see a letter of Mayer's to H. Schaaffhausen, 20 August 1867, Weyrauch, *2,* 412.

[36]Weyrauch, *1,* 88 ff.

[37]*Ibid.,* p. 71.

[38]*Ibid.,* p. 59.

of the indestructibility of forces was grounded on the equality of causes and effects, and his critique of Liebig's view of the causal principle was uncompromising and highlights the difference between their conceptions of nature. Mayer wryly noted to Griesinger in a letter in December 1842 that Liebig had written to him that the nature of force and of cause and effect was a confused question, as if Liebig "knew himself to be long ago elevated above the general confusion"; Mayer was satisfied that Liebig was not so elevated.[39] Despite superficial similarities, Mayer's conception of nature was essentially different from Liebig's; the basis of this difference in worldview lay in Mayer's concept of force, an analysis of which provides the key to an understanding of Mayer's thought.

Mayer's first paper on the nature of force, "On the Quantitative and Qualitative Determinations of Forces,"[40] which he wrote in 1841, provides an insight into his primary intentions. His declaration that "all bodies are subject to changes . . . [which] cannot happen without a cause . . . [which] we call force," that "we can derive all phenomena from a basic force [*Urkraft*]," and that "forces, like matter are quantitatively invariable" is the key to his natural philosophy. His intention was to create a new science of physics concerned with "the nature of the existence of forces." He compared physics with chemistry, which was concerned with the existence of matter. The basic principle of both chemistry and physics was that "the quantity of [their] entities [was] invariable and only the quality of these entities [was] variable."[41] Although the forces of nature might change their form, they could not change their magnitude. Mayer's meaning is clarified by letters he wrote to Baur in this period, in which he emphasized that forces are determined by principles analogous to the basic principles of chemistry. Just as "the chemist has to deal with a given quantity of matter, the physicist [deals] with a given quantity of force," so that both chemistry and physics "must be based on the same principles."[42] The basic principle of chemistry was the "indestructibility of substance" during chemical changes: "when H and O are destroyed (become qualitatively zero) and HO

[39]Weyrauch, *2,* 190.

[40]J. R. Mayer, "Ueber die quantitative und qualitative Bestimmung der Kräfte," Weyrauch, *2,* 100–107.

[41]*Ibid.,* pp. 100 ff.

[42]Weyrauch, *2,* 121.

[water] ensues . . . the chemist is not permitted to accept that H and O really become zero." Forces must be discussed in the same way; they are "as indestructible as a substance, for in combining with one another they are destroyed in their old form (to become qualitatively zero),"[43] though they remain quantitatively indestructible. Mayer's distinction between quantitative and qualitative differences was clarified by a chemical analogy: "Chemistry teaches us to recognize the qualitative changes that matter undergoes," for "in chemical processes only the *form* and not the *magnitude* of the given matter is changed."[44] In an analogous way the forces—the subject matter of physics—could change their qualitative form, but not their quantitative magnitude.

Mayer attempted to demonstrate his argument that forces were quantitatively invariable by examining the inelastic collision of two bodies. Although his argument suffers because he defined the "force" or "motion" of the bodies quantitatively as mass multiplied by velocity and because he asserted that there was a loss of "force" in collision, his basic conceptual point remains clear enough: from the "presupposition of the invariability of the quantity of force,"[45] the "neutralized" force apparently "lost" in collision was not really lost; it was heat. There was no quantitative loss, but only a qualitative transformation of force. Mayer concluded that "motion, heat and . . . electricity are phenomena which can be explained by a single force . . . and can be transformed into one another in accordance with definite laws. Motion is transformed into heat by being neutralized by an opposite motion."[46]

Despite the weakness of Mayer's attempt at a quantitative argument, his 1841 paper contains a major insight: mechanical losses could be quantitatively equated with the evolution of heat. Moreover, from the first Mayer realized that he was creating a new science that would develop the implications of the interconnection of heat and mechanical processes. His intention was further clarified in his second paper, "On the Forces of Inorganic Nature," which he published in 1842. From this paper it is clear that the analogy he had

[43] *Ibid.*, pp. 110 ff.

[44] See his 1845 paper, Weyrauch, *1*, 48. Mayer's stress on the chemical analogy is noted by George Rosen, "The Conservation of Energy and the Study of Metabolism," in *The Historical Development of Physiological Thought*, ed. C. M. Brooks and P. F. Cranefield (New York, 1959), pp. 243-263.

[45] Weyrauch, *2*, 103.

[46] *Ibid.*, p. 105.

drawn between chemistry and physics—the former as the science of matter, the latter as the science of force—was not merely a heuristic metaphor, but expressed a fundamental feature of his conception of nature: nature constituted a duality of matter and force. Mayer remarked that whereas matter was readily characterized by properties such as weight and volume, the nature of force remained "unknown, impenetrable, hypothetical." From the outset he declared it his intention to explicate the framework of a science of forces. His paper was "an investigation to make the concept of force as exact as that of matter," so as to create a science of physics that would be as exact as that of chemistry.[47] Hence, as he told Baur, "my first endeavor now is to secure for the science of forces the axis around which the science of matter rotates."[48]

It is significant that one of the crucial steps in the argument—one which was contained in two passages which Mayer omitted from the 1867 reprint of the 1842 paper—involved an appeal to chemical principles. "Chemistry," he noted, "whose problem is to elaborate the existing causal relations between material entities, teaches us that a material cause has a material effect."[49] In a letter to Griesinger later in 1842, he indicated that the principle of the relation between material causes and material effects in chemistry expressed the indestructibility of substances in chemical processes that was known from chemical experiments. "The law which unconditionally governs all ponderable entities (matter)," he stated, "is that no given material entity is ever reduced to nothing and none arises out of nothing; material substances change into one another and assume different manifestations."[50] Because of the analogy between the laws of matter and those applicable to the operations of force, the proof of the law of the indestructibility of forces could be seen to depend on the relation between causes and effects. The indestructibility of forces was grounded in the equivalence of causes and effects: "forces are causes," he declared in his 1842 paper, so "with them we may make full use of the principle: *causa aequat effectum* . . . this first property of all causes we call their indestructibility."[51]

This statement of the indestructibility of force in terms of the

[47]Weyrauch, *1,* 23.
[48]Weyrauch, *2,* 121.
[49]Weyrauch, *1,* 31.
[50]Weyrauch, *2,* 176.
[51]Weyruach, *1,* 23.

equality of causes and effects expresses the kernel of Mayer's concept of force. The relationship between his concept of causality and the quasi-substantial concept of force may be illustrated by reference to Leibniz' statement of the causal principle, for Mayer's view of the law of causality resembles Leibniz' doctrine that "there is always a perfect equivalence between the full cause and the whole effect . . . each entire effect is equivalent to the cause."[52] In discussing Leibniz' statement of the equality of causes and effects, Emile Meyerson pointed out that the equality was "none other than the principle of identity applied to objects in time."[53] Mayer's statement of the indestructibility of forces in terms of the law of causality exemplifies Meyerson's thesis that the quantitative equality of causes and effects implies their ontological status as substantial entities. Mayer asserted that "to force as a cause corresponds force as an effect,"[54] and that hence "a force is no less indestructible than a substance";[55] he concluded that force possessed the property of "substantiality."[56] His claim that the indestructibility of force was grounded on the equality of causes and effects was thus fundamental to his view of the ontological status of force. Given the parallels between Mayer's causal principle and Leibniz', it may be noted that Leibniz' close association of the notion of cause and the principle of sufficient reason is perhaps echoed in Mayer's appeal to the "general laws of human thought . . . the principle of logical reason [*Satz vom logischen Grunde*]."[57] Mayer's appeal was in support of his claim that the indestructibility of forces was dependent on the relation between causes and effects; in 1845 he grounded the indestructibility of forces in "the laws of thought" and in "experience," by which he meant the principle of causality and the law of the indestructibility of matter, respectively.[58]

Mayer's statement of the analogy between the laws applicable to matter and those applicable to force and his statement of the "substantiality" of force did not entail the material status of force. The

[52] Leibniz' reply to Catelan (1687), quoted in Pierre Costabel, *Leibniz et la Dynamique* (Paris, 1960), p. 33.

[53] Emile Meyerson, *Identity and Reality* (London, 1930), p. 43.

[54] Weyrauch, *1*, 31.

[55] Weyrauch, *2*, 115.

[56] Weyrauch, *1*, 73. See also Meyerson, *op. cit.* (note 53), chapter 5.

[57] Weyrauch, *2*, 177.

[58] Weyrauch, *1*, 59. See also Gerd Buchdahl, *Metaphysics and the Philosophy of Science* (Oxford, 1969), pp. 40 ff.

central feature of his 1842 paper was his attempt to explicate a concept of force as a substantial, yet nonmaterial, entity. Whereas matter possessed the properties of weight and impenetrability, force did not, and Mayer concluded that forces were nonmaterial, or "imponderable," entities. Employing his distinction between quantitative and qualitative changes in nature, Mayer defined "causes" as "(quantitatively) *indestructible* and (qualitatively) *transformable entities.*" This definition covered both material and "imponderable" entities, and Mayer asserted that no transitions could occur between the two classes of entities. Matter was ontologically distinct from forces, which were "*indestructible, transformable, imponderable* entities."[59] The concept of force was characterized by the "*union* of indestructibility and transformability."[60] The transformation of forces implied that "in innumerable cases we see motion cease without having caused another motion"; nevertheless, "a force once in motion cannot be annihilated, but can only change into another form."[61] The transformability of forces was thus associated with their indestructibility. Moreover, forces which were transformed one into the other were "two different manifestations of one and the same entity";[62] it was this commitment to the notion of the different forces of nature as phenomenal manifestations of a single force, an *Urkraft* as he had termed it in 1841, that led Mayer to argue that to view one force as the cause of another did not entail that any one phenomenal force could be reduced to any other.

The nonreducibility of phenomenal forces to one another bore on Mayer's conception of the transformation of mechanical motion into heat. He emphasized in 1842 that though "we prefer the assumption that heat is caused by motion to the assumption of a cause without an effect and an effect without a cause," we should not infer from this causal relation any similarity between heat and motion other than that they both belonged to the category of force: "for it to become heat," he stated, "motion must cease to be motion."[63] From this statement it is clear that Mayer did not affirm the mechanical theory of heat, and he did not presuppose it in his calculation of the mechanical equivalent of heat. His denial that heat consisted of mo-

[59]Weyrauch, *1*, 24.
[60]*Ibid.*, p. 25.
[61]*Ibid.*, p. 26.
[62]*Ibid.*, p. 24.
[63]*Ibid.*, p. 28.

tion followed directly from his understanding of the nature of force. Motion and heat had the same ontological status, each a different manifestation of a single *Urkraft*. The causal relation between forces only implied their indestructibility and transformability, not the reducibility of one to the other.

Mayer elaborated on this point in a letter to Griesinger in November 1842, in which he affirmed that "what heat, what electricity etc. are in their essence I do not know, as little as I know of the inner essence of a material substance." As a result of his investigations he felt that "I know this, that I see the connection of many phenomena much more clearly than has been seen hitherto, and that I can give a clear and good idea of what a force is."[64] Mayer was arguing that the intelligibility of his concept of force was grounded in his account of the connection between phenomena and hence of the connection between forces, but his claim that he made no attempt to specify the "essence" of force is ambiguous. Mayer did not speculate on the nature of the *Urkraft* or on the nature of matter, as is indicated by his disavowal of knowledge of the "inner essence" of matter, and his claim that he did not know the "essence" of the different forces reflects his concern to give an account of the connection between forces rather than to discuss the nature of, say, heat. However, he was concerned to specify the ontological status of force: in the mature statement of his theory in 1845, he stated that force possessed the property of "substantiality," adding that "there is no immaterial matter"[65] and emphasizing that the substantiality of forces did not grant them the status of material entities. There is clear disharmony between his claim that he refused to speculate on or had knowledge of the essence of force and was only concerned to define the relations between forces and his statement that force was a nonmaterial, yet substantial, entity. There is disharmony between his discussion of the substantive and of the relational aspects of force.[66]

Mayer stressed not only the ontological status of force, but also the quantitative expression of force. In 1842 he argued that force was to be measured by the Leibnizian *vis viva,* noting, as had Leibniz,

[64]Weyrauch, *2,* 180 ff.

[65]Weyrauch, *1,* 73.

[66]In stressing Mayer's emphasis on the relational aspects of force, Ernst Cassirer failed to do justice to the complexity of Mayer's ideas. See Cassirer's interesting discussion of this point in *The Philosophy of Symbolic Forms* (New Haven, 1957), *3,* 461 ff.

that "the law of the conservation of *vis viva* is based on the general law of the indestructibility of causes."[67] The law of the conservation of *vis viva,* which applied to mechanical motions, was a special case of Mayer's law of the indestructibility of forces. In 1842 he associated *vis viva* with the measure of "fall force," the force which caused the fall of bodies, and in 1845 he extended that measure to the force of "motion."[68] These two forces, as he pointed out in 1841, were associated with changes in the spatial relations of material entities,[69] and hence the Leibnizian *vis viva* had obvious application to them. From the first, Mayer had distinguished these "mechanical forces," as he later termed them,[70] from the forces associated with electrical, thermal, and chemical phenomena.[71] Given Mayer's emphatic statement of the view that one kind of force could not be regarded as constituting the essence of another and that the causal relation between forces only implied their transformability, the way in which the Leibnizian *vis viva* was related to forces such as heat remained obscure. Nevertheless, in calculating the mechanical equivalent of heat he made no attempt to specify the way the nature of heat was related to its quantitative expression.[72] It was in this sense that Mayer's theory of the indestructibility of force required no statement of the "essence" of force.

Despite his claim to have clarified the concept of force by an explication of the phenomenal manifestations of and connections between forces, Mayer's papers and correspondence show his continued concern to define the nature of force. In a letter to Baur of July 1842 he argued that "from the actual state of experience we are able to consider only five forces in inorganic nature as being objective: these are fall force, motion, heat, electricity and chemical difference of matter. These five forces stand in such connection that one can transform into another."[73] He amplified this view in a letter to Griesinger the following December, in which he explained that the

[67]Weyrauch, *1,* 25. Kuhn is mistaken in supposing that Mayer employs $1/2mv^2$ as the measure of force (*op. cit.* [note 2], p. 349).

[68]Weyrauch, *1,* 62.

[69]Weyrauch, *2,* 101.

[70]Weyrauch, *1,* 71.

[71]Weyrauch, *2,* 101.

[72]For interesting comments on the failure of eighteenth century scientists to perceive this relation, see Erwin Hiebert, *Historical Roots of the Principle of Conservation of Energy* (Madison, 1962), pp. 2, 92–94, 103.

[73]Weyrauch, *2,* 134 ff.

transformability of these five forces into one another was a consequence of the fact that they were "one and the same entity in different manifestations."[74] He published his five fold division of forces in 1845, noting again that they were "in truth only a single force."[75] He explained to Griesinger that forces could be defined simply in terms of their capacity to produce effects: forces were "causes which produce motion."[76] He elaborated on this point in his 1845 paper, declaring that "an entity which through its expenditure brings motion we call force."[77] These statements illustrate the ambiguity and dual nature of Mayer's discussion of force: forces were defined as possessing "substantiality" and as manifestations of the *Urkraft,* yet they were also defined in terms of their phenomenal effects and quantitative equivalence. The basis of Mayer's claim to have clarified the concept of force can thus be seen in his demonstration of the transformability of forces from the indestructibility of causes and in his explication of the phenomenal manifestations of force.

The obscurity surrounding Mayer's conception of nature has led to some confusion as to what he meant by saying that forces were causes and why he emphasized the role of forces in maintaining the operations of nature. In this connection, there has been some speculation as to Mayer's probable philosophical sources. The preceding analysis permits an assessment of the conceptual similarities between Mayer's natural philosophy and philosophical traditions and of the possible influence of such traditions on his theory of nature.

The preceding analysis shows that the principle that causes are equivalent to their effects was fundamental to Mayer's natural philosophy. His theory "in no way rests on an unusual and arbitrary definition of the causality condition," he assured Griesinger in 1844.[78] Kant's importance in the German cultural sphere at this time and the central place of the causal principle in Kant's philosophy might seem to suggest Mayer's adherence to Kant's philosophy. But the analysis of Mayer's concept of causality above shows that although Mayer may have been indebted in some sense to Kant for his causal principle, his view of causality was different from Kant's. For

[74] *Ibid.,* p. 201.
[75] Weyrauch, *1,* 48.
[76] Weyrauch, *2,* 201.
[77] Weyrauch, *1,* 47.
[78] Weyrauch, *2,* 226.

Kant the principle of causality established the law-likeness of nature: nature is dependent upon the causal principle "as the original ground of its necessary conformity to law."[79] There is no suggestion in Mayer's writings that causality is to be considered as a condition of the law-likeness of nature; rather, Mayer regarded the causal principle as asserting the indestructibility of causes and the equivalence of causes and effects.[80]

The conceptual analogies between Mayer's view of the causal principle and Leibniz' doctrine of the equality of causes and effects, Mayer's adoption of the Leibnizian concept of *vis viva,* and the possible programmatic analogy between Mayer's attempt to create a new science of physics and Leibniz' attempt to create a "new *science of dynamics*"[81] hardly justify our viewing Mayer as a Leibnizian. Leibniz emphasized that force was something different from size, figure, and motion, and that it corresponded to something real in bodies.[82] Leibniz' science of dynamics was the culmination of his critique of the Cartesian conception of body as essentially extension: "the concept of *forces* or *powers*," he wrote, "for whose explanation I have set up a distinct science of *dynamics*" was to bring "the strongest light to bear upon our understanding of the concept of *substance.*"[83] Hence, for Leibniz a true science of bodies would be concerned with the nature of force. His science of dynamics was a consequence of his theory of substance, in which his metaphysics differs so demonstrably from that of Mayer's natural philosophy. For Leibniz the nature of material substance was to be understood in terms of the concept of force, not in terms of impenetrability and extension, which he considered to be an "incomplete" and "analyzable and relative concept"; whereas extension was an "attribute resulting from many substances existing continuously at the same

[79] I. Kant, *Critique of Pure Reason,* trans. N. Kemp Smith (London, 1929), p. 173.

[80] See Gerd Buchdahl, "The Conception of Lawlikeness in Kant's Philosophy of Science," *Synthese, 23* (1971), 24–46.

The Kantian approach is fundamental to Helmholtz' philosophy of nature: for a full analysis, see P. M. Heimann, "Helmholtz and Kant: the Metaphysical Foundations of *Über die Erhaltung der Kraft,*" *Studies in History and Philosophy of Science, 5* (1974), 205–238. For a discussion of these two different senses of the principle of causality, see Meyerson, *op. cit.* (note 53), chapter 1.

[81] Gottfried Wilhelm Leibniz, *Philosophical Papers and Letters,* ed. L. E. Loemker (Dordrecht, 1969), p. 435.

[82] *Ibid.,* p. 315.

[83] *Ibid.,* p. 433.

time,"[84] force was "absolutely real."[85] The primitive active force was substantially present in all corporeal entities and was manifested phenomenally as the derivative active force—the "force by which bodies actually act and are acted upon by each other"—which appeared either as "dead force" or as "living force [*vis viva*]."[86] The conservation of *vis viva* expressed on the phenomenal level the unfolding activity of the monads which constituted the essence of material substances. Leibniz' concept of force as a defining characteristic of material substance stands in sharp contrast to Mayer's matter-force duality. For Mayer, force possessed the property of "substantiality," but was ontologically distinct from matter.

Mayer's emphasis on the role of forces has led to the suggestion that *Naturphilosophie* influenced his thought. As this philosophical movement is often held to have had an influence on science,[87] some discussion of its possible relation to Mayer's work is in order. The fundamental aim of Schelling's philosophy was to discover certain a priori principles that were inaccessible to empirical cognition. His physics was concerned with the inner essence of nature in contrast to empirical physics which was concerned with phenomenal principles. For Schelling, empirical physics was the science of the "surface" of nature, whereas the aim of *Naturphilosophie* was to comprehend a level of reality that by definition was known only a priori.[88] Schelling's conception of physics stands in sharp contrast to that of Mayer, who sought to render the concept of force as empirically meaningful as the concept of matter. Schelling conceived of nature as the product of two opposing tendencies, productive activity and the limitation of that activity; physical objects were the result of the interaction of these tendencies.[89] The tendencies were manifested on the empirical, phenomenal level as two primitive forces, known as

[84] *Ibid.*, pp. 516, 520. [85] *Ibid.*, p. 445.

[86] *Ibid.*, pp. 436 ff. For an analysis of the relations between Leibniz' dynamics and metaphysics, see Martial Gueroult, *Leibniz, Dynamique et Métaphysique* (Paris, 1967); Costabel, *op. cit.* (note 52); Carolyn Iltis, "Leibniz and the *Vis Viva* Controversy," *Isis, 62* (1971), 21-35.

[87] For a useful general discussion, see Barry Gower, "Speculation in Physics: The History and Practice of *Naturphilosophie*," *Stud. Hist. Phil. Sci., 3* (1973), 301-356.

[88] F. W. J. Schelling, "Einleitung zu dem Entwurf eines Systems der Naturphilosophie," in Schelling, *Sämmtliche Werke*, ed. K. F. A. Schelling (Stuttgart, 1857), *3*, 275.

[89] *Ibid.*, p. 288.

Grundkräfte, so that the principles that characterize the inner essence of reality, though only known a priori, had empirical counterparts. Schelling's theory of the polarity of forces as an expression of the tension between productivity and its limitation has no parallel in Mayer's writings. The general thrust of the writings of *Naturphilosophen* such as Ritter was to account for phenomenal changes by an appeal to inner essences, and their notions of the unity of nature and the polarity of forces bear only a remote analogy to Mayer's concept of the transformation of forces.

Mayer felt that his work had forced "the twaddle of the *Naturphilosophen* to stand in wretched nakedness in the pillory,"[90] if for no other reason than because of the quantitative content of his work. "A single number," he exclaimed, "has more real and lasting value than a costly library of hypotheses."[91] Statements of this kind do not disprove the possibility that Mayer was influenced by the *Naturphilosophen.* It seems that he was familiar with their speculations, and it is possible that the influence of *Naturphilosophie* was so pervasive in this period that it was implicit in all scientific activity concerned with the unity and conversion of "forces"; indeed, Mayer's reference to an *Urkraft* and to the unity of forces suggests an indebtedness to the ideas of the *Naturphilosophen.* But while such an influence is possible, it must be emphasized that in other respects Mayer's conception of nature was fundamentally at variance with the ideas of the *Naturphilosophen.* His matter-force duality and stress on the concept of force as an empirical quantity suggest that there would be little reason to attribute to *Naturphilosophie* any substantive role in shaping Mayer's natural philosophy. His intentions were fundamentally opposed to those of the *Naturphilosophen,* and claims for the influence of their ideas on Mayer's work must therefore be regarded with some suspicion.

Mayer provided one clue in connection with his indebtedness to philosophical traditions. He hinted that he had been influenced by the notion of attractive and repulsive forces as postulated by the "Kantian school of natural philosophy."[92] Mayer's intentions, however, show little correspondence with Kant's central concern in the *Metaphysische Anfangsgründe der Naturwissenschaft* (1786), which was to discuss the ways in which the a priori, transcendental cate-

[90] Weyrauch, *2,* 181.
[91] *Ibid.,* p. 226.
[92] *Ibid.,* p. 378.

matter. Kant's metaphysics of nature was to be in "unison with the gories of the *Critique* were to be applied to the empirical concept of mathematical doctrine of motion"[93] and to show that the explication of the concept of matter demonstrates the possibility of Newtonian science. Mayer's reference to the Kantian "school" is difficult to interpret. It is unlikely that he had in mind J. F. Fries, whose *Mathematische Naturphilosophie* (1822) was an attempt to develop the Kantian approach in opposition to the *Naturphilosophie* of Schelling. Fries' exposition of Kant's metaphysics and stress on a firm link between mathematics and metaphysics has no counterpart in Mayer's natural philosophy.[94] It would seem likely that Mayer's reference to the Kantian theory of forces reflected no more than an awareness—possibly at second-hand—of Kant's explication of matter in terms of attractive and repulsive forces in the chapter on "Dynamics" in the *Metaphysische Anfangsgründe*. Kant had argued there that the dynamical concept of matter was to be explicated in terms of an attractive force limiting the repulsive force, so that the quantity of matter present in a given space was determined by the limiting or balance of the repulsive force by the attractive force. Mayer's reference to the Kantian emphasis on attractive and repulsive forces—even though he only made this remark in an "Autobiographical Note" written as late as 1863[95]—suggests that he was indebted to the Kantian notion of the limiting or balance of forces for his concept of the interconversion and equivalence of forces, but there is no substantive affinity between the Kantian discussion of attractive and repulsive forces and Mayer's conception of nature.

In attempting to clarify the confusion over Mayer's indebtedness to philosophical traditions, this analysis has underlined the conceptual individuality of Mayer's natural philosophy. Nevertheless, the possibility of the influence of the Kantian notion of the balance of forces or of the unity of forces of *Naturphilosophie* cannot be discounted, however limited this influence and tenuous the filiation of ideas and despite the major differences between these philosophical traditions and Mayer's work. It would be a mistake to conclude that Mayer's intellectual development can be understood out of all con-

[93] I. Kant, *Metaphysical Foundations of Natural Science*, trans. J. W. Ellington (Indianapolis and New York, 1970), p. 17.

[94] For a discussion of Fries, see Heimann, *op. cit.* (note 80), pp. 230 f.

[95] For comments on the dating of this note, see Weyrauch, *2*, 376.

text; for textual analysis, though necessary, is not a sufficient condition for historical understanding.

ACKNOWLEDGMENTS

I wish to thank Gerd Buchdahl, Karl Figlio, and Russell McCormmach for comments on a draft of this paper.

Debates over the Theory of Solution: A Study of Dissent in Physical Chemistry in the English-Speaking World in the Late Nineteenth and Early Twentieth Centuries

BY R. G. A. DOLBY*

"I used to think theology
Was rather rough on doubt,
But chemistry with ions beats
Theology all out.

You'd better join the church before
Your course is well begun,
Because you'll need to exercise
The art of faith, my son."

Ellwood Hendricks[1]

"The history of the development of the ionic dissociation hypothesis appears to me to be one of gravest warning, which we should heed before it be too late; the grave ethical value of the lessons to be learned from it should not be overlooked."

H. E. Armstrong[2]

1. INTRODUCTION

Modern physical chemistry is often dated from 1887. In that year J. H. van't Hoff's theory of solution and S. A. Arrhenius' theory of electrolytic dissociation were made the core of a new chemical specialty promulgated by the school formed around Wilhelm Ostwald. The new approach to chemistry encountered resistance, particularly in Britain; it came to dominate British ideas of solution only after more than a decade of strenuous opposition. Arrhenius' theory of electrolytic dissociation was attacked particularly strongly. In the twentieth century, the debate continued in a subdued form in

*Unit for the History, Philosophy and Social Relations of Science, University of Kent at Canterbury, Canterbury, Kent CT2 7NR, England.

[1] Quoted by James Kendall, *Journal of Chemical Education, 2* (1925), 376.

[2] H. E. Armstrong, *Science Progress, 3* (1909), 656.

Britain, and in America, too, the new theories remained subject to an undercurrent of criticism. Only after the death of the main opponents of the Ostwald school, the emergence of new theories, and changes in the major interests of physical chemists, did the debate finally end.

There is considerable interest in following such a debate, both as an important episode in the history of science and as a case study exploring the methodology of scientific change. A debate is a valuable source of insight into the criteria affecting choices between alternative theories and is useful for revealing the role of dissent in scientific change.

In the course of the present historical study we will see how dissent functioned in one problem area of mature science; we will see that it was regarded as an important part of the process of reaching rational consensus, not an embarrassing and unintended symptom of the breakdown of normal processes. Indeed, there were institutionalized settings for debate. The British Association for the Advancement of Science, and later the Faraday Society, organized meetings for discussions between rival parties; the discussions and debates arising out of papers read to the Chemical Society were reported in a special journal;[3] journals such as the *Philosophical Magazine* accepted polemical notes and papers, and *Nature* published letters in which correspondents developed the debate. There were also informal rules for debate, as is indicated by the remarks the scientists made when they considered that their opponents had become unconstructively polemical, that is, had transgressed the rules.[4] Although the debate over the theory of solutions was at times exceptionally heated, neither side felt that they were upsetting the

[3] *Proceedings of the Chemical Society,* which commenced publication in 1885.

[4] The following sentence by P. S. U. Pickering, replying to criticisms made by James Walker, is suggestive of what was and was not acceptable in the debate. "Although Dr. Walker took upon himself the task of refuting the objections which I raised against the dissociation theory, he appears to me to have avoided the very objections for which I most sought explanations, though, perhaps, the absence of these explanations will probably be regretted by his friends less than *the absence of that courtesy which those who are seeking after the truth might naturally expect from their fellows, and the absence of which in the present instance was all the more to be regretted as I had not associated Dr. Walker's name with any of the views which I had called into question.*" *Philosophical Magazine* [5], *32* (1892), 40; my italics.

continuity of scientific development, but rather that it was by debate that science would reach the truth most effectively. When some of the textbooks of the Ostwald school expounded the theories of van't Hoff and Arrhenius without referring to the arguments of their critics, the latter considered this to be a perversion of the rational path of science.[5]

In the concluding section, I shall discuss the concept of dissent. I shall stress its usefulness in other historical studies of science and discuss the suggestion that scientific change takes many forms in which dissent is of fundamental importance, so that debate is not restricted to cases of scientific revolutions. This paper is to be considered, therefore, as a contribution to the descriptive methodology of scientific change.

Historiographic Note

In writing the history of the debate over the theory of solutions, two major problems of presentation emerged: the importance of representing sympathetically several incompatible viewpoints on the same issues, and the need to report simultaneous but tenuously connected parts of the debate.

The events in this history occurred in the social realm, in which people's behavior is affected by what they *think* is happening. By reporting the comments made at the time of the debate and later retrospects, I have tried to show that the participants frequently differed widely in their understanding of the course of the debate and of the nature of the final outcome. Since the position of each character tended to change slowly while the center of public attention moved around rapidly during the debate, I could best have demonstrated the essential coherence of each viewpoint by following it through the whole period. To work through the debate as many times as there are characters, however, would overtax the patience of the reader. I have compromised by concentrating on particular characters at particular phases of the debate. Normally, I have indicated the personality and preoccupations of each character when he first makes an important appearance; the reader may refer to these passages when considering subsequent briefer discussions of the later development of his ideas and his later involvement in the debate.

[5] See in particular the discussion of the issues raised by Armstrong's career, in section 7 below.

Often the debate broke down into separate simultaneous but interacting strands, each of which would be best understood by being followed over an extended period. I have somewhat reduced the complications of bringing them together into a single thread of narrative by ignoring parallel debates in non-English speaking countries, except when they had an immediate effect on the central characters of the British and American debates. The debate in the English speaking world was largely self-contained (though dependent on the prior work of the Ostwald school on the continent), so that the restriction is acceptable. With only a few exceptions (such as the controversy over hydration between Arrhenius and D. I. Mendeleef and his Russian supporters), the debate in other countries was less dramatic and more piecemeal, dominated by unrelated priority disputes and individual misunderstandings, and is therefore less useful historical material.

2. HISTORICAL BACKGROUND TO THE DEBATE

It is possible to trace a continuing line of discussion on the nature of solutions throughout the history and prehistory of chemistry.[6] However, until the last third of the nineteenth century, relatively few chemists dealt with the nature of solutions as a problem in its own right and as a problem around which questions about the phenomena of solution could be organized. In his general discussions of chemical affinity, C. L. Berthollet had dealt with reactions in solution and solubility. But because he believed that chemical compounds could have indefinite proportions, he did not make a sharp distinction between compounds and solutions. In the following decades, Berthollet's ideas stimulated much discussion but few experiments and no decisive advances.[7] The increasing importance of a

[6]See for example, P. Walden, "Die Lösungstheorien in ihrer geschichtlichen Aufeinanderfolge," *Sammlung chemischer und chemisch-technischer Vorträge, 15* (1910), 277–454. This paper provides a thorough though not exhaustive retrospective survey of ideas of solution, particularly in the eighteenth and nineteenth centuries. See also J. R. Partington, *A History of Chemistry* (London, 1964), *4*, chap. XX.

[7]See F. L. Holmes, "From Elective Affinities to Chemical Equilibria," *Chymia, 8* (1962), 105–145, for a historical account of the tradition. Holmes suggests (pp. 128–129) that the impasse that chemists reached in their discussion of affinity theories in the decades after Berthollet was due to their lack of knowledge of the state of substances in solution (particularly of equilibrium mixtures of dissolved substances), and that, in the early nineteenth century, the number of chemists was still small enough for the less attractive topics to be neglected.

chemistry of definite proportions in the wake of Daltonian atomic theory made questions related to the nature of solutions, which have indefinite proportions, seem peripheral. Chemists studied diverse phenomena of solution to varying extent during this period, but they showed little interest in attempting to draw the material together to give a deeper understanding of the nature of solutions.

Gradually, during the middle part of the nineteenth century, a new question about the nature of solutions came sharply into focus as physical and chemical forces became more strongly contrasted. Chemists asked if solutions, particularly aqueous solutions, were to be understood as containing definite chemical species produced by the combination of the dissolved substance and the water, or if they were to be explained in purely physical terms. One can extract from the literature fairly strong statements of either opinion, supported by different selections of evidence. For example, in 1846 one writer argued that the contraction of total volume that occurs when substances dissolve in water is of a magnitude corresponding to such an immense external pressure that it must be evidence of chemical combination.[8] In 1867 another writer, L. Dossios, considered that a satisfactory treatment of solutions could be derived from a kinetic theory that assumed that the kinetic energy of a molecule was greater than the attraction between two neighboring molecules but less than the total attraction of all the molecules on one another. Dossios argued from analogy with homogeneous bodies: even though the molecules in solution are not all identical, the kind of force that produces cohesion in homogeneous bodies also provides a satisfactory explanation for cohesion in solutions. Saturation occurs when the molecules leaving solution equal in number those reuniting with it. The solubility will increase with temperature, that is, with the increase in molecular movement.[9]

By the 1870's, thermochemistry provided one of the most important sources of information on the state of substances in solution. As is to be expected, when substances pass from the solid to the liquid state, many solids dissolve in water with the absorption of heat. Some others do not. For several chemists, M. Berthelot among them, it was most plausible to attribute the heat liberation to chemical combination of the dissolved substance with the water. Berthelot presented an influential full statement of the hydrate theory of the

[8] J. J. Griffin, *Philosophical Magazine* [3], *29* (1846), 289–310, 444–467.

[9] L. Dossios, *Vierteljahrsschrift der Zürichischen Naturforschenden Gesellschaft, 13* (1867), 1–21. Dossios' treatment was taken up by W. W. J. Nicol in 1883.

1870's in his 1879 *Essai de méchanique chimique fondée sur la thermochimie.* There he considered that solution of salts occurs with the formation in the solution of definite compounds between the salt and the water, analogous to or identical with the hydrates of constant composition known in the crystalline state.[10] Thermochemical evidence looked especially strong for the dilution of strong acids. For example, the full dilution of a sequence of solutions of nitric acid containing from zero to one hundred equivalents of water per equivalent of acid shows that substantial amounts of heat are liberated on dilution until the solution contains six or seven equivalents of water. Berthelot's study of the shape of the heat of dilution curve led him to believe that there may be a definite hydrate with two equivalents of water, and another with five or six. Similar evidence suggested to him that there were several hydrates of sulphuric acid in solution. Berthelot went on to assemble a variety of evidence to show more generally the existence of hydrates in solution and to give some idea of their possible composition.

Berthelot's conception of the importance of hydrates in solution was representative of the positions of other chemists who had a chemical conception of solution. In the 1880's, the most eminent defender of the hydrate theory of solutions was Mendeleef. He had been writing on his conception of solution since 1865[11] but attracted particular attention on this subject outside Russia only after the publication in 1886 of his theory on the determination of hydrates present in solution. The hydrate theory of solution was the most plausible method of explaining the physical changes resembling the manifestations of chemical combination that accompany the formation of a solution. But it had to be combined with a physical theory of the liquid state of solution generally and of the more important processes which occur in solution.

During the second half of the nineteenth century, physicists and chemists combined the kinetic theory of heat with the kinetic theory of gases, making possible a thorough understanding of the quantitative aspects of the gaseous state and of reversible processes involving gases. Their first applications of the theories to the liquid state were speculative in nature, for they were slow to achieve an adequate understanding of reversible processes in solution and to link solution

[10] M. Berthelot, *Essai de méchanique chimique fondée sur la thermochimie* (Paris, 1879), *2,* 162.

[11] P. Walden, *op. cit.* (note 6), pp. 386–390, 408–413.

to the gaseous state. Dossios' work was representative of the first phase of theory development, prior to the application of thermodynamics.

In his third paper on thermodynamics, "On the Equilibria of Heterogeneous Substances," J. W. Gibbs set out a systematic basis for the treatment of chemical phenomena by thermodynamic methods.[12] His general treatment adequately dealt with equilibria in solution, in particular equilibria between gases and solutions, where Henry's Law could be used. However, Gibbs wrote his paper in a compressed, abstract, mathematically elegant style which required mathematical sophistication of the reader and thus limited the distribution of the paper even though its results were of great interest to chemists. It was some time before many interested chemists were in a position to appreciate and exploit Gibbs's work.[13] Other physicists and chemists independently developed less general treatments,[14] but it was not until the late 1880's that the thermodynamics of chemical phenomena became a fashionable area.

In the meantime, a small number of chemists pioneered a physical treatment of solutions that, although less general than Gibbs's in its conception, was far simpler mathematically and was linked to a rich variety of experimental phenomena. It became the basis for a new and prolific school of research, which in turn transformed the peripheral subject area of physical chemistry into a thriving research specialty.[15]

Van't Hoff had risen to eminence as one of the two chemists who in 1874 had advanced the idea of the tetrahedral carbon atom as the basis of a theory of the arrangements of atoms in space capable of explaining the nature of optically active organic compounds. In the

[12] J. W. Gibbs, *Transactions of the Connecticut Academy, 3* (1875–1878), 108–248, 343–524. Reprinted in J. W. Gibbs, *Scientific Papers* (London, 1906), *1*, 55–353.

[13] L. P. Wheeler, in *Josiah Willard Gibbs* (New Haven, 1951), discusses the distribution of Gibbs's paper and, in an appendix, gives a list of names of those scientists to whom Gibbs sent offprints.

[14] For example, H. von Helmholtz, *Sitzungsberichte der Königlich Preussischen Akademie der Wissenschaften zu Berlin, 1* (1882), 22–39, 825–836; *2* (1883), 647–665. J. H. van't Hoff, *Études de dynamique chimique* (Amsterdam, 1884).

[15] A fuller discussion of these developments is given in R. G. A. Dolby, "Social Factors in the Origins of a New Science: The Case of Physical Chemistry" (forthcoming).

1880's, van't Hoff turned to problems of physical chemistry, publishing his influential *Études de dynamique chimique* in 1884. Concerned with chemical affinity, van't Hoff in this work took up the problem of measuring the magnitude of chemical affinity. One possible approach to the problem was, following E. Mitscherlich, to study the attraction of salts for water. Mitscherlich's method of comparing the water vapor pressure of the salt with that of pure water at the same temperature suggested that an implausibly weak force was involved. But a chance discussion with the botanist H. de Vries, his colleague at Amsterdam, about the work of H. Pfeffer on osmotic phenomena immediately suggested to van't Hoff that osmotic pressure would provide a measure of the affinity of a salt solution for water. If a salt solution is separated from pure water by a semipermeable membrane, which allows the passage of water but not of the salt, the water tends to flow into the salt solution, unless its passage is resisted by an externally applied pressure. The pressure required to stop the flow is described as the osmotic pressure. Consideration of osmotic pressure indicated that the affinity of the salt solution for water was far higher than was suggested by Mitscherlich's data. By considering the thermodynamics of the attainment of equilibrium between a solution and a pure solvent connected both through a semipermeable membrane and by the unrestricted passage of vapor, van't Hoff was able in *Études de dynamique chimique* to express the equilibrium in the standard mathematical form that he used for equilibria, $d(\log c)/dT = q/2T^2$.

After the publication of *Études de dynamique chimique,* van't Hoff studied the many idealized reversible processes that the concept of a semipermeable membrane allowed. He quickly found that osmotic pressure corresponds to gas pressure and that the law relating the pressure, volume, and temperature of an ideal gas, $PV = RT$, holds for dilute solutions also. Even the constant, R, was the same for solutions as for gases. Van't Hoff concluded that Avogadro's principle holds for dilute solutions as well as for gases and developed a theory of solutions from these results. It was difficult to make precise direct experimental studies of osmotic pressure, but by extending his thermodynamic reasoning van't Hoff was able to explain the results of F. M. Raoult, who in 1882 had established that the depression of the freezing point of a dilute solution, as the lowering of the vapor pressure of the solvent, is proportional to the molecular concentration of the dissolved substance.

Van't Hoff presented his theory of solution in 1885,[16] expressing his general result as $PV = iRT$, where P is the osmotic pressure, and i is an empirical factor which is characteristic of a given solution. For many substances, i was close to unity in dilute solution, but for electrolytes it was generally much higher, often nearer two or three. This defect of the theory Arrhenius turned into an advantage in a paper published in 1887.[17]

Physical scientists had had difficulty in explaining satisfactorily the nature of electrolytic substances ever since the discovery of the chemical action of electricity at the turn of the nineteenth century. In 1857, R. Clausius had suggested that there must be a small permanent dissociation of electrolytes into electrically conducting ions. The permanent dissociation would explain why current can pass immediately on the application of voltage without first having to produce dissociation. In the 1850's, J. W. Hittorf's studies of ionic transport during electrolysis had shown that the proportion of the current carried by each ion in a salt was usually different. This made it unlikely that during electrolysis the ions were combined in firm molecules. In the 1870's F. Kohlrausch developed the modern methods of avoiding the problems of polarization in measurements of the conductivity of electrolyte solutions. Study of dilute solutions led him in 1874 to his law of the independent migration of ions, according to which the conductivity of a dilute solution is the sum of two constants, one depending on the cation and the other on the anion.

Electrochemical phenomena had long been a source of speculation about the nature of chemical force. The electrochemical theory of J. J. Berzelius, in particular, had been an influential early conception of chemical combination. But Berzelius' essentially qualitative ideas had not proved adequate for later developments in chemical theory or in electrochemistry. In his Faraday Lecture of 1881, the physicist H. von Helmholtz had presented to chemists an electrical conception of chemical combination which was compatible

[16] J. H. van't Hoff, *Archives des Sciences Exactes et Naturelles, 20* (1886 [for 1885]), 239–302; *Kongliga Svenska Vetenskaps-Akademiens Handligar, 21,* no. 17 (1886), 3–58. These two papers are almost exactly the same except for their final sections. Van't Hoff later described the development of his views in "Wie die Theorie der Lösungen entstand," *Berichte der Deutschen Chemischen Gesellschaft, 27* (1894), 6–20.

[17] S. Arrhenius, *Zeitschrift für physikalische Chemie, 1* (1887), 631–648.

with knowledge of electricity. It was based on Faraday's study of electrolysis and the "modern chemical theory of quantivalence." From Faraday's law and the chemical atomic theory Helmholtz concluded that the ions carry a definite unit of electric charge in electrolysis. (He pointed out that this was in conflict with Berzelius' dualism in which the atoms in binary compounds did not completely saturate each other's electrical forces, so that more complex aggregates of atoms could be built up.) By a chain of reasoning Helmholtz estimated the order of magnitude of electrical forces acting between ions. It seemed sufficient to explain how the voltages used in electrolysis could disrupt chemical compounds. Helmholtz suggested that at least for typical non-aggregated compounds the forces of affinity and the constancy of chemical valency could be explained in terms of the electrical attraction between charged atoms. Although Helmholtz claimed to have avoided speculative assumptions in his treatment, he admitted that "I am not sufficiently acquainted with chemistry to be confident that I have given the right interpretation, the interpretation which Faraday himself would have given if he had been acquainted with the law of chemical quantivalence."[18]

By the 1880's, some physicists studying electrolytic phenomena considered that their primary task was to provide a mechanism for the electrical phenomena and attributed related chemical phenomena to the interaction of electrically charged particles. For example, Kohlrausch, G. Wiedemann, and others during the early 1880's developed Clausius' idea of dissociated ions in their speculations on the effects of an electric current in a conducting solution. They supposed that some of the electrolyte molecules are split into oppositely charged ions, which move freely and independently through the solution at speeds unaffected by the chemical environment and influenced only by such general conditions as viscosity. Under such conditions a small hydrogen ion would move more rapidly through the solution than larger ions. Their approach contrasted with that of chemists who considered chemical interaction between solvent and solute to be an essential feature of solution.

At this stage of the development Arrhenius devoted his doctoral thesis to the problem.[19] He made the bold and chemically implausible suggestion that a significant proportion of the molecules of all

[18] H. von Helmholtz, *Journal of the Chemical Society, 39* (1881), 303.

[19] Arrhenius' thesis was published in *Bihang till Kongliga Svenska Vetenskaps-Akademiens Handligar, 8,* nos. 13, 14 (1884).

electrolytes are permanently in an active form that conducts electricity. The proportion of molecules in the active form increases with dilution until, at extreme dilution, all the molecules are conductors. He suggested that this variation in the proportion of active molecules is more important than any changes in the resistance of the solution to the passage of ions in measured variations of electrolytic conductivity. Furthermore, it is these conducting molecules that are chemically active, so that the relative strength of acids, for example, can be explained by the concentration of active molecules.

As he later admitted,[20] Arrhenius did not have a great variety of experimental data to support his ideas. Even though he had refrained from stating his belief that the active molecules are charged radicals of the kind that, as physicists speculated, might be produced during electrolysis, the ideas expressed in his dissertation disagreed completely with the prevailing conceptions of the chemical nature of salts and related substances. Arrhenius' ideas were largely ignored, but he did get an immediate positive reaction when he sent a copy of his thesis to Ostwald in Riga. Ostwald was able to use Arrhenius' work to correlate his own measurements of the chemical affinities of acids with measurements of their conductivities. Because of Ostwald's interest, Arrhenius gained the respect of local colleagues and was able to pursue an academic career. Arrhenius spent several years studying in a number of European laboratories.

In 1887, Arrhenius read van't Hoff's formulation of the theory of solutions. He immediately saw that it provided further evidence for the permanent dissociation of electrolytes. If the sodium chloride molecule, for example, is permanently dissociated into two ions, then a salt solution will have twice the molecular concentration suggested by the formula NaCl. This explains why the factor, i, in van't Hoff's equation is so close to two for sodium chloride solution. For any electrolyte solution, the factor, i, should be the same as the number of ions produced by the dissociation of a molecule.

Arrhenius published his theory in Sweden and in Britain, but its main impact came with its publication in the *Zeitschrift für physikalische Chemie.*[21] In an earlier number of the same volume, van't Hoff

[20] S. Arrhenius, *Journal of the Chemical Society, 105* (1914), 1418.

[21] Arrhenius said (*Journal of the American Chemical Society, 34* (1912), 361) that an account of the hypothesis was included in a report of the British Association Committee for the investigation of the conductivity of electrolytes. However, Partington (*op. cit.* [note 6], p. 678) says that the report is no

had republished his theory of solutions, mentioning Arrhenius' theory of ionic dissociation as an explanation of the size of i for electrolyte solutions.[22]

Physical scientists very quickly recognized a number of important implications of the two mutually supporting theories. When the law of mass action was applied to the equilibrium between normal and dissociated substances, it was found that the calculations did not agree with experiment for the slight variation of conductivity with dilution of most salts, but gave a very good agreement for the much greater variation with weak acids. Several people noted this, but it was Ostwald who published first what is now called "Ostwald's dilution law."[23] In 1888–1889 W. Nernst made another application. By treating the ions of a solution in terms of their motion under an osmotic pressure gradient, Nernst was able to calculate the contact potential between a metal and a solution of its ions, and also between two solutions.[24]

The most important figure in the dissemination of the new treatment of solutions was Wilhelm Ostwald. When, in 1887, he was appointed to the second chair of chemistry at Leipzig, he founded a research school which became the nucleus of a research specialty of physical chemistry. The exploitation of the new treatment of solutions was a central theme of the new specialty. A new journal, *Zeitschrift für physikalische Chemie,* which he founded in the same year, and his many influential textbooks publicized the new area of scientific growth. The graduates of Ostwald's laboratory spread Ostwald's conception of the new discipline in Germany and then in other countries in which they could establish themselves. They became professors specializing in physical chemistry, wrote textbooks

---

longer in the British Association files. The publication in *Zeitschrift für physikalische Chemie, 1* (1887), 631-648, embodied two communications read before the Swedish Academy of Sciences and published in *Översigt af Kongliga Svenska Vetenskaps-Akademiens Forhandligar,* 1887.

[22] J. H. van't Hoff, *Zeitschrift für physikalische Chemie, 1* (1887), 481-508.

[23] Arrhenius discusses the work that was being done on this law independently of Ostwald in *Journal of the American Chemical Society, 34* (1912), 361-362. Ostwald's route to the discovery is given by F. G. Donnan in his Ostwald Memorial Lecture, *Journal of the Chemical Society* (1933), p. 325.

[24] W. Nernst, *Zeitschrift für physikalische Chemie, 2* (1888), 613; *4* (1889), 129. A discussion of this and related work is given by J. R. Partington in *Journal of the Chemical Society* (1953), p. 2853.

modelled on Ostwald's, and founded specialist journals similar to the *Zeitschrift für physikalische Chemie.*

It was in America that the new discipline took root most rapidly. At the end of the nineteenth century, the American university system was growing rapidly. Large numbers of Americans were going overseas to qualify themselves for university teaching. Students studying sciences such as chemistry went to Germany, because it provided the best scientific training. Besides being drawn by the reputation of German chemical research, they went to obtain the German degree of doctor of philosophy, which was a suitable qualification for a university teacher and relatively cheap to obtain. The Americans who had gained a German degree by research on a laboratory subject were soon teaching the same way in America; after the founding of Johns Hopkins University as a graduate university in 1878, many other American universities reformed their graduate schools or set up new ones. The students in these American graduate schools in turn went on to teach in other universities, spreading rapidly the interests of the first generations of German trained graduates in any subject suitable for laboratory research by graduate students. In chemistry, it was organic chemistry which dominated Germany in the later part of the century and which benefitted most from the American expansion. But Ostwald's physical chemistry soon became sufficiently fashionable to attract the interest of many young American chemists who were not enamored of organic chemistry, particularly since at that time the fortunes of inorganic chemistry were at a low ebb.

In Britain, the growth of the university system and of chemistry within it was far less rapid. However, the graduates of the Ostwald laboratory were again of great importance in the establishment of British physical chemistry in accordance with the Ostwald program.

3. THE BRITISH DISCUSSION OF ELECTROLYSIS AND SOLUTION 1880–1887

In contrast to the situation in America, where physical chemistry can be represented as expanding almost unresisted into the growing university system, the introduction of the new ideas of the Ostwald school into Britain was strongly affected by earlier chemical opinion then prevalent in Britain. Two relevant general topics were especially important: theories of the nature of solutions and the process of dissolution, and discussion of the role of electricity in chemical action and combination, especially as illuminated by the study of electrolyte solutions.

Solution

The Daltonian atomic theory had accentuated the distinction between stoichiometric compounds and solutions. Chemists had been especially successful at developing an understanding of the former, but problems remained in the treatment of the latter. By the early 1880's, the prevailing opinion held that, when a salt (or any solute) dissolves in water, the solvent first forms hydrates which are then dispersed throughout the liquid. In 1878, for example, W. A. Tilden had maintained that "a solution contains *a mixture of several hydrates, the constitution of which depends partly on the temperature of the liquid, and partly on the proportion of water present.*"[25] A similar view was expressed by the "hydrate theorists" of the 1880's, and after the publicity associated with Mendeleef's development of the hydrate theory after 1886, a number of scientists who had earlier held this view were encouraged to claim priority.[26] But the ideas of the many precursors of the hydrate theory of the 1880's were vague and not related to very much evidence.

In 1883 W. W. J. Nicol mounted an attack on the "hydrate theory" of solution. He presented the theory in an extended quote from Berthelot's *Mécanique Chimique,* which he considered to be the most concise statement of the generally received hydrate theory that he could find.[27] Nicol argued that none of the water molecules in a solution are chemically combined with the solute in a manner analogous to water of crystallization, but that the process of dissolution is to be understood in terms of general attractive forces. A solution is formed when the attraction of the molecules of water for a molecule of the salt exceeds the attraction of the molecules of the salt for one another (at least under conditions where the changes in the water-water attraction have no effect). Nicol amassed experi-

[25] W. A. Tilden, lecture to Bristol Naturalists' Society, February 1878, published in the Society's *Proceedings* and quoted by Tilden in *The Progress of Scientific Chemistry in Our Own Times,* 2nd ed. (London, 1913), p. 279.

[26] Tilden's retrospect of 1913 illustrates the trend. See also the claim by T. Sterry Hunt for an essay of 1855, made in *Chemical News, 58* (1888), 151-153.

[27] W. W. J. Nicol, *Philosophical Magazine* [5], *15* (1883), 91-92.

mental evidence in support of his position in many papers in the 1880's, engaging also in a polemical exchange with P. S. U. Pickering, who is a major figure in the present history.

Tilden and Nicol were representative of a growing group of British chemists in the mid-1880's who shared an interest in the experimental study of solutions. The meetings of the British Association for the Advancement of Science reflected the rise of interest in the nature of solutions. In the years immediately before 1884 no papers on the subject had been presented; but in 1884 four papers (involving six authors) were read. The following year two committees (involving five members) reported on solutions and continued to report for several years. In 1887 a larger committee (initially with seven members) first reported on the bibliography of solution. The reports of these committees, which did not contain theoretical discussions, reflected the considerable British interest in the nature of solutions predating the theories of Arrhenius and van't Hoff.

H. E. Armstrong and the Role of Electricity in Chemical Action

The development of ideas on electricity and chemical action before 1887 can best be presented by a look at H. E. Armstrong and his activities.[28] Armstrong was the key figure in the opposition to the new physical chemistry in England. He was an extreme individualist, a man who would never yield to the social pressures of a scientific community or follow scientific trends. Armstrong arrived at his position before he knew of the theories of Arrhenius and van't Hoff, and he developed them with a consistency that ignored the tides of fashion that he felt to be moving around him. After fifty years he wrote what he called his "considered message at the end of seventy years of constant study" in a polemical letter criticizing the dissociation theory and restating his nearly unchanged position.[29]

Armstrong had been trained by E. Frankland at the Royal College of Chemistry in London and by H. Kolbe at the University of Leipzig. As his first research he had made a chemical analysis of polluted water for Frankland, and he retained an interest in water throughout his career. In Leipzig he had studied orthodox organic chemistry. Armstrong's personality came to closely resemble Kolbe's—per-

[28]For a full-length biography of Armstrong, see J. Vargas Eyre, *Henry Edward Armstrong* (London, 1958).

[29]H. E. Armstrong, "Ionomania in Extremis," *Chemistry and Industry, 14* (1936), 916–917.

haps as a result of deliberate imitation. Armstrong also inherited from Kolbe a belief in residual affinity as a feature of chemical combination.

Armstrong had many strands to his career, though the most important were undoubtedly his efforts in chemistry and in education. In education he was against the tendency towards specialization; his wide range of chemical interests, illustrated by the variety of his many contributions to the discussions at meetings of the London Chemical Society, confirms his opposition to specialization.[30] His research followed three main lines: the chemistry of naphthalenes, the chemistry of camphor, and the mechanism of chemical change. His ideas of chemical change led him into conflict with the Ostwald school of physical chemistry.

Many of the issues on which Armstrong developed such strong opinions he had initially taken up by chance. One interest, for example, grew out of circumstances related to an early, part-time position in chemistry at the London Institution.[31] The chair had once been held by W. R. Grove, inventor of the Grove cell, and Armstrong found some of Grove's apparatus in the laboratory. Later he bought a bundle of *Transactions of the Royal Society,* which included Grove's memoirs. The study of these gave him a continuing interest in electrolysis.

Armstrong first stated his ideas on chemical action in 1885, when he was thirty-seven. By this time he had built up an influential position in British chemistry, particularly in his autocratic secretaryship of the Chemical Society (1875–1893). The immediate stimulus to his public speculation was a report by H. B. Baker to the Chemical Society in March 1885. Baker had extended the work of H. B. Dixon on the reactivity of very pure chemical substances and had found that pure carbon and phosphorus are both incombustible in pure oxygen. He believed moisture to be essential for the reactions. In the discussion that followed, Armstrong suggested that all chemical reaction is "reversed electrolysis." He thought that Baker's report was providing evidence that the same kind of aggregation is required for combustion as is involved in electrolysis.[32] Armstrong elaborated his ideas as president of the Chemical Science Section of the British As-

[30] See the record of *Proceedings of the Chemical Society.*

[31] Armstrong was appointed to the chair of chemistry at the London Institution in 1870.

[32] H. E. Armstrong, *Proceedings of the Chemical Society, 1* (1885), 40.

sociation Meeting at Aberdeen that summer.[33] He attributed special importance to the concept of molecular aggregates. Employing the notion of residual affinities, he insisted that it is rare for atoms to satisfy one another's affinities completely in simple substances. Atoms, he said, will readily aggregate, especially by the mutual attraction of negative radicles, and thus provide the special conditions under which chemical reactions can most readily take place. In solutions, aggregation is influenced by the interaction between solvent and dissolved substance. Armstrong remarked that this kind of aggregation seems to be involved in electrolysis. He noted that, while simple binary liquids like hydrochloric acid and water are not electrolytes, a mixture of them is. Furthermore, of the binary metallic compounds it is the least volatile (and therefore the most associated) that show electrolytic properties.

Armstrong's view of chemical action, although highly speculative, was compatible with the prevailing outlook in chemistry, and Armstrong could employ it in the explanation of chemical phenomena. But, as Armstrong recognized, it was not in accord with an increasingly popular conception of the nature of electrolysis being developed by a number of physicists. In his 1881 Faraday Lecture to the Chemical Society, Helmholtz had expounded a physical view of the relation between electrical and chemical phenomena; Armstrong developed his discussion from a criticism of Helmholtz' arguments.[34] Helmholtz had challenged chemists: "I shall consider my work of today well rewarded if I have succeeded in kindling anew the interest of chemists in the electrochemical part of their science."[35] Armstrong took up the gauntlet. He was keen to speculate on the rela-

[33] H. E. Armstrong, *Report of the British Association* for 1885 (1886), pp. 945–964, especially pp. 952ff.

[34] Armstrong considered Helmholtz' lecture very influential, even though he disagreed with Helmholtz' reasoning. In 1896, he was to comment that "probably Helmholtz' Faraday lecture was the one Faraday lecture which was distinctly an original contribution, which we can be sure exercised a very important influence on the scientific world. A very large share of the attention which has been drawn to this subject of late years, which van't Hoff, Arrhenius, and others have developed to such an extraordinary extent, has arisen out of the Faraday lecture by Helmholtz. Not only here but in Germany also it attracted very great attention, and was of very much consequence." *Proceedings of the Chemical Society, 12* (for 1896), 28–29.

[35] H. von Helmholtz, *op. cit.* (note 18), p. 304.

tionship of chemical and electrical phenomena, but wished to argue from the considerations that *chemists* held to be most important.

While his criticism of Helmholtz was restricted to some of the assumptions at the basis of Helmholtz' argument, Armstrong attacked directly the ideas being developed by other German physicists. Armstrong's conception of electrolysis gave an essential role to aggregation, with the solvent playing a vital part in aqueous solution. The physicists, on the other hand, treated electrolysis as a process in which current is carried by the dissolved substance, and they explained it by the partial dissociation of the dissolved substance into parts that are even simpler than the chemical molecules. Armstrong was sure that their ideas were very bad chemistry. He thought it nonsense to suppose that the crucial stage in electrolysis was dissociation rather than association and to treat the water in aqueous solutions simply as a resisting medium. Armstrong felt that the chemically active nature of water showed that it must be involved in the electrolytically active components of the solution. He pointed out that the concentration of solutions that show maximum conductivity (as measured by Kohlrausch) is usually close to that of solutions that develop the maximum heat of solution (as measured by J. Thomsen).[36] The approximate agreement, he claimed, must surely be due to the maximum formation of electrolytically active aggregates. As he was later to argue at greater length, it is possible that it is not the dissolved substance but the water which is the actual electrolyte. He saw further support for his belief that chemical action and electrolysis were to be understood in the same (chemically inspired) terms in the work of Ostwald and Arrhenius relating the conductivity of acids to their chemical activity. As Arrhenius had not then explicitly stated that the active components of electrolyte solutions were dissociated solute ions, Armstrong did not appreciate that Arrhenius' result was intended to support an even more extreme theory of ionic dissociation.

In 1885 Armstrong sought the support of a physicist friend, Oliver J. Lodge. In 1884 Lodge had contributed a long paper, "On the Seat of the Electromotive Forces in the Voltaic Cell," to the British Association.[37] Lodge then turned to electrolytic phenomena

[36] H. E. Armstrong, *Report of the British Association* for 1885 (1886), p. 957.

[37] O. J. Lodge, *Report of the British Association* for 1884 (1885), pp. 464–529; published also in *Philosophical Magazine* [5], *19* (1885), 153–190, 254–

at Armstrong's request, not because research in this area was one of his chief interests. As Lodge explained later,[38] Armstrong was rather hostile to the physical explanations of electrolysis in terms of dissociated ions and to their favorable reception by chemists. He wanted them scrutinized by a physicist. At the Aberdeen meeting in 1885, Lodge discussed the matter in a long paper, "On Electrolysis," but did not come to any firm conclusions.[39] A committee was then appointed to study the matter further. The committee on electrolysis in its physical and chemical bearings gave its first report to the British Association in 1886. Armstrong and Lodge were joint secretaries.

Before the work of the committee had appeared, Armstrong published an elaboration of his views.[40] He distinguished as electrically active: metals; simple electrolytes such as fused silver iodide; pseudo-dielectrics, which behave as dielectrics when pure, but as electrolytes when mixed with other members of their own class (such mixtures are "composite electrolytes"); dielectrics. The discussion was primarily of composite electrolytes. The reasoning can be illustrated by the uncomplicated case of hydrochloric acid and water. Each of the pure substances is a non-electrolyte—Armstrong refused to accept Kohlrausch's work suggesting a small conductivity for pure water—while a mixture of the two is an electrolyte. It seemed incomprehensible to Armstrong that anybody could seriously argue that the

280, 340-365. Later additions and subsequent discussion were published in *Philosophical Magazine,* including a contribution from Ostwald, *Philosophical Magazine* [5], *22* (1886), 70-71.

[38] O. J. Lodge, *Past Years. An Autobiography* (London, 1931), pp. 190-191. Lodge's recollection is slightly misleading; at that stage Armstrong's opposition was directed mainly against the physicists who were developing Clausius' theory of dissociation. Arrhenius became the main opponent a little later.

[39] O. J. Lodge, *Report of the British Association* for 1885 (1886), pp. 723-772. Lodge's correspondence with Armstrong makes it clear that he was not able to put much time into preparing the paper. He accumulated materials and theories and then told Armstrong that he could not manage the "monotheism" of Armstrong's unitary chemical theory of electrolysis. He considered that the matter needed looking at from *all* points of view. Armstrong Papers (Imperial College Archives): letters from O. J. Lodge, first series, nos. 423-428 (May to December 1885).

[40] H. E. Armstrong, "Electrolytic Conduction in Relation to Molecular Composition, Valency and the Nature of Chemical Change: Being an Attempt to Apply a Theory of Residual Valency," *Proceedings of the Royal Society, 40* (1886), 268-291.

dissociated atoms of hydrochloric acid are the active species, while the water is merely the mechanical means of separating the ions. Since only the mixture is conducting, it seemed more likely to him that the active material is a kind of molecular aggregate of acid and water. An analogy could be drawn between electrolyte solutions and metal alloys, for both show some of the features of chemical combination yet lack definite proportions. The analogy provides a counter example to which Armstrong frequently returned in his later papers. Armstrong noted that a very small percentage of lead can make gold quite brittle and urged that lead can have this effect only if each atom of lead affects larger molecular aggregates of gold. He expected that the formation of a simple compound with contrasting properties would lead to simple additive changes in physical properties. The position Armstrong maintained throughout the later debates was now established. He always stressed the importance of association and insisted that water is too reactive to be ignored in solutions. He drew support for his views from chemical analogies between solutions and other substances.

I want now to return to the main forum of debate on electrolysis, the British Association committee on electrolysis. In the *Report of the British Association* for 1886, Lodge published an important collection of discussions and abstracts of papers and letters, including the latest work by continental scientists. Arrhenius' ideas up to January 1887 were well represented. Lodge included his correspondence with Arrhenius and a long critical abstract he had made of Arrhenius' dissertation of 1883. Lodge was especially appreciative of the second, theoretical part of the dissertation in which Arrhenius had suggested that at all times (and not just when a current is flowing) a substantial proportion of the molecules of an electrolyte are active, both electrolytically and chemically. Arrhenius believed, but had not dared to say until 1887, that the active molecules were in fact dissociated in the manner Clausius had suggested.[41] In commenting on Arrhenius' "Chemical Theory of Electrolytes" Lodge had written: "But it is a bigger thing than this: it is really an attempt at an electrolytic theory of chemistry."[42] He was appreciative of Arrhenius' theory but not committed to it.

[41] A fuller discussion of Arrhenius' ideas and their development is included in R. G. A. Dolby, *op. cit.* (note 15).

[42] O. J. Lodge, *Report of the British Association* for 1886 (1887), p. 362.

Most of the letters and abstracted papers of overseas authors in the 1886 *Report* were part of a three way discussion between Arrhenius, Kohlrausch, and E. Bouty on the nature of electrolytic conductivity, especially under the simplifying circumstances of extreme dilution. The argument interested physicists rather than chemists. Two problems stimulated the debate over the conclusions to be drawn from experiment: the determination of the conductivity of pure water, which became a major problem as investigations of dilute solutions increased, and the self-induction resulting from the use of alternating current in conductivity measurements. Kohlrausch and Arrhenius disagreed as to whether or not a connection could be assumed between the electrical conductivity and the internal friction of a solution. Arrhenius argued that the most important explanatory factor is the proportion of an electrolyte in the active form, and Bouty claimed that for most electrolytes (those we call "normal") the mobility of each ion is about the same at extreme dilution. There were also British contributions to the committee report, consisting mainly of experimental studies checking the assumptions and assertions of the continental scientists.

In the *Report of the British Association* for 1887 the committee continued the discussion. Again Lodge was the editor of the printed report. The most interesting items in the report are his comparison of the views of Arrhenius and Armstrong, together with a reply by Armstrong.[43] Although Lodge had entered the discussion of electrolysis at Armstrong's request, his treatment was not very critical of Arrhenius. In Lodge's opinion, Arrhenius had provided a perfectly orthodox view of the nature of electrolysis which was of special interest in its application to chemistry. Lodge clearly identified Arrhenius' "active" molecules with the dissociated molecules of Clausius.[44] Lodge's criticisms of Armstrong's views were more severe. Writing as a physicist, he avoided commenting on the chemical features of Armstrong's ideas, but pointed out their physical difficulties. For example, he stressed that any chemical theory that assumes that the electromotive force produces the changes necessary for the passage of electric current should be able to explain why a very small electromotive force is sufficient.

[43]*Report of the British Association* for 1887 (1888). Lodge, pp. 351–353; Armstrong, pp. 354–357.

[44]O. J. Lodge, *Report of the British Association* for 1887 (1888), p. 351.

In his reply, Armstrong suggested that the experimental results of Arrhenius' work should be separated from the theoretical conclusions.[45] He insisted that Arrhenius was unjustified in stressing dissociation, when association is so important in chemical reactions. He argued for the importance of third substances in chemical reactions, suggesting that we still have much to learn about simple chemical changes. At this early stage in the discussion, Armstrong was still prepared for compromise, though recognizing the difficulties of reconciling chemical and physical considerations. "In conclusion, I would add that I urge these pleas on behalf of my hypothesis with the greatest diffidence, feeling that I am unfortunately unable to fully appreciate the force of the mathematical and physical arguments."[46] After stressing the uncertainties of the current state of understanding of intramolecular structure, he went on: "It is impossible at present to quantify peculiarities and relationships which are patent to the chemist, but these must be taken into account; and for this reason it is all-important that chemists and physicists should cooperate."[47] Armstrong and Arrhenius continued to exchange criticisms after 1887.

The Early Ideas on Solution of P. S. U. Pickering

One of the most prominent, though not authoritative, figures in the debates of the 1880's and 1890's was P. S. U. Pickering.[48] Pickering was a man of private means who showed great independence of outlook. One of his first published papers had been a criticism of a view expressed by his tutor at Oxford.[49] Pickering did his experimental studies by himself, without any laboratory assistant

[45] As previously explained, Armstrong had earlier argued that Arrhenius' and Ostwald's data supported his theory.

[46] H. E. Armstrong, *Report of the British Association* for 1887 (1888), p. 357.

[47] *Ibid.*

[48] The most useful biographical source on Pickering is T. M. Lowry and E. J. Russell, *The Scientific Work of the Late Spencer Pickering F.R.S.* (London, 1927). The correspondence between Arrhenius and Ostwald reprinted in *Aus dem wissenschaftlichen Briefwechsel Wilhelm Ostwalds,* ed. Hans-Günther Körber (Berlin, 1969), pt. 2, suggests that Pickering was the English opponent of the new theory who was the most trouble to the Ostwald school.

[49] E. J. Russell, *Journal of the Society of Chemistry and Industry, 39* (1920), 448R. The tutor was W. W. Fisher; the paper appeared in *Journal of the Chemical Society, 33* (1876), 409.

or attendant; he published only a very few joint papers, mostly based on the work of his women students at Bedford College, London.[50] He developed a number of highly original methods, some of which might have been more valuable had he discussed them with another scientist in their development. Sometimes he based them on highly doubtful assumptions. He had the odd habit of giving his experimental results with more figures than he considered reliable. The apparent overprecision of his figures misled many of his opponents into thinking that he was ignorant of the errors in his work and naive of the uses of mathematics in chemistry.[51]

Pickering was readily drawn into controversy, and his discussions of the nature of solutions are not the only example of his polemical exchanges. Pickering's ideas about solution developed out of his studies of sulphates and of sulphuric acid, in which a great variety of chemical and physical methods were used to draw chemical conclusions. Independently of Armstrong[52] and of E. J. Mills[53] he had developed a theory employing the notion of residual valences to explain such molecular aggregate compounds as hydrated salts and double salts.[54] In 1886, he published a series of thermochemical papers, building on the work of Thomsen and Berthelot. Among his studies were experiments investigating the heat changes in dilutions of solutions. The irregularities in the curves obtained by plotting heat of dissolution against concentration for hydrated salts at various temperatures led him to conclude that different hydrates were being formed in the solutions at various temperatures.[55]

In the same year he published a paper on water of crystallization which discussed the difficulties in the general notion of hydration and water of crystallization.[56] When Tilden criticized the experimental basis of this paper,[57] Pickering accepted some of the criti-

[50] *Ibid.*

[51] See the discussion by T. M. Lowry in T. M. Lowry and E. J. Russell, *op. cit.* (note 48), pp. 8–9.

[52] H. E. Armstrong, *Proceedings of the Chemical Society, 1* (1885), 40; *Report of the British Association* for 1885 (1886), pp. 945–964.

[53] E. J. Mills, *Philosophical Magazine* [4], *28* (1865), 364.

[54] P. S. U. Pickering, "Atomic Valency," read to the Chemical Society in December 1885. *Proceedings of the Chemical Society, 1* (1885), 122–125; published in full as a pamphlet in 1886.

[55] P. S. U. Pickering, *Journal of the Chemical Society, 49* (1886), 260–311.

[56] P. S. U. Pickering, *Journal of the Chemical Society, 49* (1886), 411–432.

[57] W. A. Tilden, *Proceedings of the Chemical Society, 2* (1886), 198–199.

cism and repeated the work.[58] His comment in accepting the force of the criticism revealed the problems of his method of working in relative isolation. "A worker is no doubt apt to be misled by having had for a long time too close and, perhaps, a one-sided view of his own work, but something must also be subtracted from the critic's opinion, from the fact that he has not followed the work in the whole of its progress, especially so when he himself has performed work of a similar nature, but under totally different conditions, and with different and comparatively imperfect instruments."[59] Even when his isolated reasoning had not led him astray, Pickering's work was often so different from that of his fellow scientists that it was not clearly understood or appreciated. The intuitive insights that the experimental worker gains from intimate experience of his materials can only be communicated to fellow workers with very similar interests.

In 1886, in the course of his discussion of hydration and water of crystallization, Pickering came into conflict with W. W. J. Nicol. Nicol had argued[60] that water of crystallization does not exist in solution at all. The two men carried their debate through a number of meetings of the Chemical Society and published rival experimental papers in *Philosophical Magazine.*[61]

A convenient summary of Pickering's position before 1887 is contained in a paper read to the British Association in 1886.[62] He described the "hydrate theory of solution" as the view that there were chemical compounds of definite proportions between solvent and solute; against this he argued that solutions contain molecular compounds of indefinite proportions, held together by residual valences. When a solid dissolved, the molecular aggregates of the solid state were broken down into simpler forms (which absorbed heat) and replaced by aggregates with the liquid (which evolved heat).

Another aspect of Pickering's work which was soon to lead him

[58] P. S. U. Pickering, *Journal of the Chemical Society, 51* (1887), 290-356.

[59] *Ibid.*, p. 291.

[60] W. W. J. Nicol, *Philosophical Magazine* [5], *16* (1883), 121-131; *18* (1884), 179-193.

[61] For examples of their polemical exchanges, see Nicol, *Proceedings of the Chemical Society, 2* (1886), 220-222; *3* (1887), 40-42 (the papers and the subsequent discussion are reported); *Chemical News, 54* (1886), 191; *Journal of the Chemical Society, 51* (1887), 389-396; and Pickering, *Journal of the Chemical Society, 51* (1887), 75-77.

[62] P. S. U. Pickering, *Chemical News, 54* (1886), 215-217.

into direct opposition to the theories of the Ostwald school of physical chemistry was his explanation in 1887 of heats of neutralization.[63] He employed his idea of residual valences, arguing that because the affinities of acids and bases are only partially saturated, they readily combine with one another to form molecular aggregates.

In 1887 and the years following, the main direction of Pickering's experimental work was to be reoriented by Mendeleef's hydrate theory. But before considering the further development of Pickering's career, the early impact of the work of the Ostwald school should be discussed.

3. EARLY STAGES OF THE CONFRONTATION, 1887-1888

The Penetration of the Ideas of the Ostwald School into Britain and Early Reactions of its British Supporters

The theories of van't Hoff and Arrhenius and their development within the Ostwald school were quickly regarded, by supporters and opponents alike, as a unified theory. I shall frequently have occasion to refer to the unified theory below and will use the suitable, neutral label "the Ostwald school theory." There is adequate justification for using Ostwald's name rather than that of Arrhenius or van't Hoff, because, as one commentator wrote, "Prof. Ostwald is one of the warmest supporters of the physical theory, and has done more, perhaps, than any other, to make it what it now is."[64] Scientists at the time referred to the theory as "the new theory of solution" or "the physical theory of solution," or by some longer locution. Their briefer labels, however, often had polemical connotations.

The theory of the Ostwald school gained its first exposure in England through the direct intervention of some of the continental enthusiasts, and through the expositions of a very limited number of early British sympathizers. Arrhenius' theory was given very early publicity. Arrhenius had been in contact with the British Association electrolysis committee, and his 1887 theory was first made public in a letter to the committee which was circulated among the members.[65] M. M. P. Muir, a Cambridge chemist, best known as a textbook writer and historian of chemistry, had been giving sympa-

[63] P. S. U. Pickering, *Journal of the Chemical Society, 51* (1887), 593-601.
[64] J. W. Rodger, *Nature, 45* (1891), 193.
[65] See note 21.

thetic expositions of Ostwald's work since 1879. In 1887, Ostwald had given van't Hoff's theory of solutions in the second volume of his influential *Lehrbuch der allgemeinen Chemie,* and in 1889, Muir had made extensive use of Ostwald's *Lehrbuch* in the second edition of his *Principles of Chemistry.* Muir's influence, however, was mainly through his textbooks and the only early supporter of the Ostwald school theory among leading British research chemists was William Ramsay.[66]

Ramsay had done a little experimental work on solutions since the mid-1880's, and when Raoult's method of determining molecular weights in solution became the subject of active British discussion at the meeting of the Chemical Society in May 1888, Ramsay was among those presenting papers employing the method. Ramsay was equally enthusiastic about van't Hoff's theory, which built on Raoult's method. On June 9, 1888, he read a translation of van't Hoff's 1887 paper to the Physical Society.[67] At about the same time, Ramsay began corresponding with Ostwald. He soon caught Ostwald's enthusiasm for Arrhenius' theory and for the possibilities of the new approach to the study of solutions. Later he became a close friend of Ostwald.

Ramsay had to overcome some confusions before he clearly understood all the implications of the Ostwald school theory, in particular the electrolyte dissociation theory. An early source of difficulty for a great number of chemists was the distinction between the dissociated ions of an electrolyte and the same substance in the free state—between an ion of sodium and a free sodium atom, for example. Arrhenius, in his 1887 memoir, had stressed that in solution the dissociated atoms have a high electric charge and so cannot easily be separated from one another. But his notation did not at first indicate the difference. Thus, in a note to the British Association,[68] Arrhenius wrote about water being very slightly "dissociated into H and OH," without indicating immediately that the separate parts have high electrical charges. Muir in his 1889 discussion of

[66] There are two book length biographies of Ramsay: W. A. Tilden, *Sir William Ramsay: Memorials of his Life and Work* (London, 1918); M. W. Travers, *A Life of Sir William Ramsay* (London, 1956).

[67] Ramsay's translation was published in *Philosophical Magazine* [5], *26* (1888), 81-105.

[68] S. A. Arrhenius, *Report of the British Association* for 1888 (1889), p. 353.

Arrhenius' theory[69] repeated Ostwald's summary of the distinction (but pointed out that as we do not yet know the significance of an electric charge on an ion, Arrhenius' distinction does not *explain* the difference between electrolyte ions and the products of gaseous dissociation). Ramsay was initially far more confused, as M. W. Travers shows in his biography.[70] For example, when Ostwald wrote to him describing an experiment to demonstrate the electrostatic production of ions, Ramsay sent on the letter to Lodge (as Secretary to the British Association electrolysis committee), referring to it as an argument for the existence of free atoms of potassium and chlorine (rather than free ions). Ostwald's letter was published in *The Electrician,*[71] a journal that had been selected that year as a suitable medium for communications among the electrolysis committee between the annual meetings of the British Association. In the letters that were published in subsequent issues, a number of confusions became evident. Lodge, for example, criticized the experiment when he failed to replicate it. Ostwald replied that Lodge should have read the original paper on which the letter to Ramsay was based; he would then have seen that, far from being the *crucial* experiment that Ramsay had thought it to be, it was an "ideal experiment." However, a comparable experiment had subsequently been carried out.[72]

Armstrong's Reaction to the Ostwald School Theory, 1888–1889

In 1889 the British Association published Arrhenius' "Reply to Professor Armstrong's Criticisms Regarding the Dissociation Theory of Electrolytes."[73] The reply, written soon after van't Hoff and Arrhenius had developed their theories, made a point by point criticism of Armstrong's ideas and also of the work of Armstrong's student, H. Crompton. Armstrong appended two notes, commenting on the reply, and giving a general discussion of the current state of

[69]M. M. P. Muir, *A Treatise on the Principles of Chemistry,* 2nd ed. (Cambridge, Eng., 1889), p. 462.

[70]M. W. Travers, *A Life of Sir William Ramsay* (London, 1956), pp. 89–93.

[71]W. Ostwald, *The Electrician, 22* (1889), 493–494.

[72]The exchange took place in *The Electrician, 22* (1888–1889), 493–494, 676, 691–692; *23* (1889), 30, 44; ending with an abstract of a paper in which an actual experiment was detailed in *23* (1889), 300–301, 323.

[73]S. A. Arrhenius, *Report of the British Association* for 1888 (1889), pp. 352–355.

ideas on electrolysis.[74] Arrhenius' note and the discussion by Armstrong show the sorts of misunderstandings that may be associated with the early stages in the confrontation and development of theories, and which are most naturally described as "incommensurable." While Arrhenius was developing a theory that built only on the quantitative phenomena of dilute solutions, Armstrong was evaluating theories that made the most sense in terms of the general understanding chemists had of chemical reactivity. Not dilute solutions, but solutions with comparable numbers of solvent and solute molecules appeared to be the most natural as a basis for theoretical study. By Armstrong's standards, the behavior of dilute solutions was irregular and complex, the Ostwaldians making far too much out of idealized simplifications of solution behavior. Arrhenius clearly believed that the role of water in electrolysis was minimal, and that, when alcohol was substituted for water and solvent, the main reason for the decrease of the conductivity of the electrolyte was the higher resistance of alcohol to the passage of ions. Armstrong immediately found such a view of the role of the solvent implausible. Indeed the ionists increasingly abandoned the view of the solvent as inert. At this early stage of the debate, the position that each side developed was—by later standards—frequently based on unreliable experiments and buttressed by implausible or speculative claims. It was consequently easy for negative attitudes to form and harden over issues that to later scientists did not seem to merit so much fuss.

In the discussion that Armstrong, as chemical secretary of the electrolysis committee, added to the 1888 report after the summer meeting, he criticized both Arrhenius' theory and, indirectly, van't Hoff's theory of solution. Armstrong was able to refer to J. J. Thomson's *Applications of Dynamics to Physics and Chemistry* (1888), one of the first English responses to the theories of van't Hoff and Arrhenius. Thomson accepted van't Hoff's conclusion that the work of Pfeffer and Raoult showed that the molecules of a dissolved substance exert the same pressure as they would in the gaseous state for equal volume and equal temperature.[75] But he did not accept that knowledge of the structure of the molecule of the solute could be

[74] H. E. Armstrong, *Report of the British Association* for 1888 (1889), pp. 355-356, 356-360.

[75] J. J. Thomson, *Application of Dynamics to Physics and Chemistry* (London, 1888), p. 175.

derived from the van't Hoff relationship. Thomson's treatment suggested that any physical influence that would change the mean Lagrangian function of the water on the two sides of a semipermeable membrane (such as the evolution of heat on dilution) would lead to the observed relationship of osmotic pressure and concentration.[76] More studies were needed to investigate different solvents and to make more use of the relationship with absolute temperature. Thomson was more critical of Arrhenius' dissociation theory, although he did not mention it by name. He wrote:

> Indeed the theory has recently been stated that in dilute aqueous solutions the dissolved acid or salt is in most cases dissociated and that to a very considerable extent; thus it has been stated that in dilute solutions of HCl as much as 90 per cent of the acid is dissociated. The reasons given for this conclusion do not seem to me to be very convincing, and the experimental results on which they are based seem to admit of a different interpretation. The supporters of this theory urge that for the salt to produce the effect which in some cases it does, it is necessary to suppose that the molecules of the salt exert a greater pressure than they would if they occupied the same volume at the same temperature when in the gaseous condition. This reasoning is founded on the assumption that all the effects due to the dissolved salt may be completely explained merely by supposing the volume occupied by the solvent to be filled with molecules of the salt in the gaseous condition. Now though we may admit that the salt does produce the effects that would be produced by this hypothetical distribution of gaseous molecules, still it does not follow that these are the only effects produced by the salt. The salt may change the properties of the solvent and the effects attributed to the dissociation of the molecules may in reality be due to this change.[77]

Naturally, Armstrong was pleased that a physicist as eminent as J. J. Thomson should come to this conclusion; as part of his critique of the dissociation hypothesis he quoted part of the above passage. Armstrong went on to say that dissociation of gaseous hydrogen chloride only occurs to a tiny extent at temperatures of 1300–1500°. "That a gas of such stability should be almost entirely dis-

[76] *Ibid.*, pp. 189–190.
[77] *Ibid.*, pp. 212–213.

sociated by mere dissolution in water is to me incredible."[78] He summarized his general opinion of the advocates of the dissociation theory in a final paragraph.

> Arrhenius, Ostwald and others regard both electrical conductivity and chemical activity as similarly conditioned by the degree of dissociation—in their opinion, very active substances, such as sulphuric acid, are to a large extent dissociated in solution; inert substances, such as acetic acid, are but to a slight extent dissociated in solution. But the adherents of this school all overlook the fact that there are two distinct theories of chemical interchange: the older theory that the interacting molecules initially combine and that the resulting complex then splits up—which may be termed the integration theory; and the more modern dissociation theory. I am led to regard the former as the more comprehensive and generally applicable, especially as comparatively so few compounds are electrolytes, and I venture to think that physicists also would incline to my belief if they would assume a somewhat different mental attitude towards the facts, and would seek to fully unravel the entire series of changes involved in chemical interactions.[79]

Armstrong's remarks to the electrolysis committee on van't Hoff's theory were brief and limited to attacking its status as a buttress of the Arrhenius theory. He had, however, set out his view of Raoult's method of determining molecular weights more fully in a paper and in discussions at the Chemical Society in 1888 and 1889. As secretary of the society, he reported his own comments thoroughly. Although he appreciated the law expressing the analogy between the osmotic pressure of the solutions and the pressure of an ideal gas, labelling it as a "masterly generalisation,"[80] he doubted that very much could be inferred from it as to actual molecular structure. In particular, he doubted that Raoult's method reliably gave actual molecular weights, or that it gave acceptable evidence that electrolytes are really dissociated. In his reasoning he invoked the case of the atomic heats of the elements. By using Dulong and Petit's law, we can infer something about the *atomic* heat of the elements, even though our measurements are actually made on *molecular* sub-

[78] H. E. Armstrong, *Report of the British Association* for 1888 (1889), p. 356.

[79] *Ibid.*, p. 357.

[80] *Ibid.*, p. 356.

stances. Similarly, even though solutions might be made of large molecular aggregates, variations in the behavior of these aggregates might, by Raoult's method, tell us something about the chemical units of which they are composed. If this was so, we could not legitimately infer that electrolytes, for example, are dissociated into separate atoms, or that colloids have very high molecular weights.[81]

It should be noted that Raoult's work on molecular weights put Armstrong's theory into a crisis situation. For the simplest interpretation of the quantitative studies of freezing point depression was that the molecular weights of most substances in solution are very close to their fundamental formulae. Armstrong blurred the issue; he could not really explain why there should be such a sharp quantitative relationship between freezing point depression and fundamental formulae. It was an argument for electrolyte dissociation that he could not easily evade.

Mendeleef's Hydrate Theory and its Critics 1887–1889

An important development of 1887 from the point of view of the British opponents of the Ostwald school theory was a method suggested by Mendeleef to study hydration in solution. Basing it on Berthelot's similar but less precisely expressed ideas,[82] Mendeleef presupposed that all the water in a solution is combined with the dissolved substance and that there will be changes in the solution properties at different concentrations as one hydrate is replaced by another. In his paper[83] Mendeleef found that, although a plot of the density of a solution against the percentage composition gave a relatively continuous curve, the differential of the density gave a series of straight lines when plotted against composition. Each of these straight lines was represented as due to varying proportions of two hydrates in equilibrium. As the percentage concentration was increased, the points of discontinuity indicated where a higher hydrate had disappeared, leaving a single hydrate containing a lower proportion of water. As concentration increased still further, an increasing

[81] H. E. Armstrong, *Proceedings of the Chemical Society, 5* (1889), 42–43, 98, 109–113.

[82] P. S. U. Pickering, "Solutions," in *Watts Dictionary of Chemistry,* ed. M. M. P. Muir and H. F. Morley, revised ed. (London 1894), *4,* 492.

[83] The English version of Mendeleef's paper appeared in the *Journal of the Chemical Society, 51* (1887), 778–782.

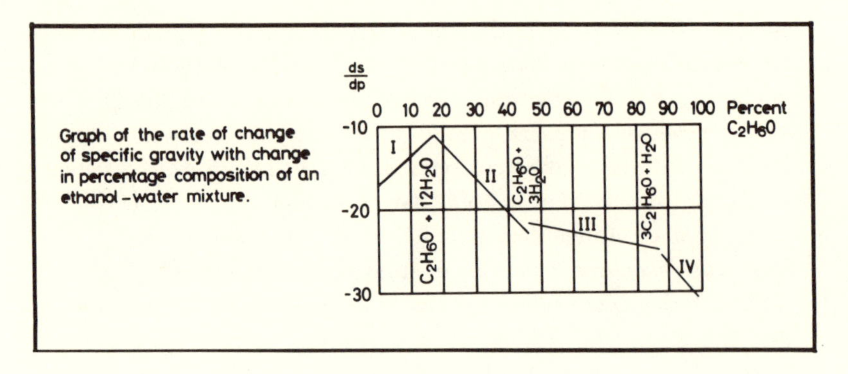

amount of the next lower hydrate appeared in the solution.[84] (See graph.) The method was in principle applicable to any property successive hydrates displayed differently; their succession could be revealed by manipulating the basic curve of solution property against concentration in a variety of ways.

Mendeleef's theory was taken up enthusiastically by some British chemists but was also criticized by Arrhenius and other members of the Ostwald school at early stages in its development. It had to develop in the face of fierce opposition.

Perhaps the first English chemist to take up the method was Armstrong's student Crompton.[85] Armstrong added a note to Crompton's paper,[86] recapitulating his own position and commenting enthusiastically on the support that Crompton's use of the method had given his own theoretical but qualitative arguments. Armstrong welcomed the method as a general method for demonstrating the nature and significance of hydration in solution, but he made no experimental use of Mendeleef's method after this time. It was Pickering who developed it into a major research program. He had remarked on the importance of the method when Crompton's paper was first read,[87] and he was soon applying it in his experimental work.

In an early paper exploring the method, Pickering applied it to his studies of the heat of dilution of solutions.[88] Armstrong had become much more critical of the method, commenting after the reading of Pickering's paper that many chemists would probably hesitate to

[84] The graph is from D. I. Mendeleef, *Journal of the Chemical Society, 51* (1887), 780.

[85] H. Crompton, *Journal of the Chemical Society, 53* (1888), 116–125.

[86] H. E. Armstrong, *Journal of the Chemical Society, 53* (1888), 125–133.

[87] P. S. U. Pickering, *Proceedings of the Chemical Society, 3* (1887), 128.

[88] P. S. U. Pickering, *Proceedings of the Chemical Society, 4* (1888), 35–37.

accept Mendeleef's explanation until evidence of the existence of definite hydrates had been obtained by investigating a greater range of properties for the nature of their dependence on the composition of the solution. If measurements on many properties indicated the same changes in hydration, Mendeleef's interpretation would come closer to being conclusive. Armstrong also pointed out, however, that the method was unlikely to be definitive, as there could be more than two hydrates present in a solution at a given concentration.[89] Very soon Pickering, too, was dissatisfied with Mendeleef's use of the method, particularly as the data Mendeleef had used did not lead to his graphical results when replotted. Pickering also came to agree with Armstrong that there might be more than two hydrates of a solute present in a solution of a given concentration.[90] Nevertheless, Pickering was convinced that the irregularities in the curves of variation of solution property with dilution *did* indicate the formation of successive hydrates. Pickering was to develop the method experimentally in a series of papers in the early 1890's, but his work continually was criticized, particularly by members of the Ostwald school.

As early as 1889 Arrhenius published a paper in *Philosophical Magazine* that was critical of Mendeleef's idea and its English supporters.[91] He attempted to show that the method was fallacious, depending for its conclusions on the magnification of experimental errors. He argued that the phenomena of dilute solution, which Mendeleef had wished to explain in terms of hydration, were quite satisfactorily accounted for by the Ostwald school theory. There could only be a very small amount of hydration in very dilute solutions, and the close agreement of the Ostwald school theory with experimental data implied that what hydration there was could not vary significantly with concentration.

Since Pickering's work had at that stage only appeared in abstract,[92] he complained that Arrhenius was premature in his criticism, and that when the paper was published in full, it would be seen that it was Arrhenius who was mistaken.[93] The first exchanges between Pickering and Arrhenius were characterized by the mixture of

[89] H. E. Armstrong, *Proceedings of the Chemical Society, 4* (1888), 37.

[90] See for example P. S. U. Pickering, *op. cit.* (note 82), p. 493.

[91] S. A. Arrhenius, *Philosophical Magazine* [5], *28* (1889), 30–38.

[92] The abstract was in *Chemical News, 60* (1889), 278.

[93] P. S. U. Pickering, *Philosophical Magazine* [5], *28* (1889), 148. The full paper appeared eight months after Arrhenius' criticisms. *Journal of the Chemical Society, 57* (1890), 64–184.

acrimony and argument at cross-purposes that is typical of early stages of scientific debate. Each side thought that the other used a misleading selection of experimental results and processed its results with a method that presupposed the conclusion being sought. The hydrationists preferred to study concentrated solutions, while the Ostwald school restricted themselves to dilute solutions. But the differences were most marked in the methods of treating the experimental data. Arrhenius was seeking simple quantitative regularities and so was encouraged to smooth the results heavily so as to indicate only the broadest patterns in the data. Pickering, on the other hand, was using quantitative study only to back up qualitative arguments about the complexity of solutions. In contrast, Arrhenius suggested that since Mendeleef, Crompton, and Pickering had not agreed on the solution compositions at which they found discontinuities in their curves (and so had concluded that different hydrates were present in solution), they were measuring only their own experimental errors. The oddest feature of Pickering's method was that, following Crompton, in order to get a set of straight lines, he used the *second* differential of the curve obtained by plotting solution property against percentage composition. And as the initial graphs were not regular enough, he obtained his second differentials by smoothing the curve of the first differential. Arrhenius insisted that any irregularities that remained in the second differential were merely the result of insufficient smoothing at the intermediate stage.[94] Pickering was heated in his reply.

> Professor Arrhenius attacks me on the subject of this smoothing of the curves, remarking that "if Mr. Pickering had 'smoothed' his curves properly, he would evidently have removed these angular points or sudden changes of curvature." The question hinges on the interpretation of the word "properly." Professor Arrhenius seems to think that the "proper" amount of smoothing to be made is such that all sudden changes of curvature should be obliterated; and this too in an investigation the sole object of which is to ascertain whether there are such sudden changes or not. I must beg to differ with him. The "proper" amount of smoothing I take to be such as will allow but little more error in the experimental points than the known errors of the determinations, or than that which seems to be the probable error according to the irregularities of consecutive points in the figure. If with such smoothing we are led

[94] S. A. Arrhenius, *Philosophical Magazine* [5], *28* (1889), 37-38.

to conclusions which are obviously false, or which are at variance with the results obtained from independent sources, then and only then must we admit some further source of error, and increase the smoothness of our drawings.[95]

Pickering remained quite unrepentant during the 1890's in his use of the method, although he came to prefer the direct detection of irregularities in the curve of solution properties by fitting a bent rule to the graph of the original experimental data.[96]

In 1889, Pickering felt that he had achieved a major triumph for his method. He discovered that on cooling his solutions he was occasionally able to crystallize out the hydrate that his curve-fitting techniques had suggested should be present. The first success was $H_2SO_4,4H_2O$, a previously unknown hydrate.[97] Thus, in 1890, Pickering felt prepared to go into the offensive against the Ostwald school. In papers contributed to the *Journal of the Chemical Society,* to *Philosophical Magazine,* and at the meeting of the British Association in Leeds, he expounded his hydrate theory and attacked what he saw as the weak points of the Ostwald school theory.[98]

4. THE LEEDS MEETING OF THE BRITISH ASSOCIATION IN 1890

The Leeds meeting was important in many strands of the debate on the theory of solution and deserves a full discussion. Under the stimulus of the disagreements within the electrolysis committee of the British Association, a joint session of the physics and chemistry sections was organized at the 1890 meeting. It was labelled a discus-

[95] P. S. U. Pickering, *Philosophical Magazine* [5], *29* (1890), 429.

[96] The use of a bent rule, or lathe, was elaborated in later papers in the face of much criticism. A full and sympathetic discussion of the issue is given by T. M. Lowry in T. M. Lowry and J. Russell, *op. cit.* (note 48), pp. 57-80.

[97] The first announcement was made in *Chemical News, 60* (1889), 68.

[98] See in particular, P. S. U. Pickering, *Journal of the Chemical Society, 57* (1890), 331-369, in which he explained the implications of the crystallization of previously unknown hydrates of sulphuric acid and began the attack on the Ostwald theory. In *Philosophical Magazine* [5], *29* (1890), 427-434, Pickering replied to Arrhenius' criticisms. In *Philosophical Magazine* [5], *29* (1890), 490-501, he attacked an obsolete version of Raoult's method of deriving molecular weights from measurements of freezing point depressions and then criticized the aspects of the Ostwald theory that he mistakenly thought were based on this work. At the British Association meeting Pickering presented his paper, "The Present Position of the Hydrate Theory of Solution," *Report of the British Association* for 1890 (1891), 311-322.

sion "on the theory of solution and its connection with osmotic pressure." The leading experts from Europe were invited, and although Arrhenius could not come (a note from him was read by Ostwald's British student, James Walker) Ostwald and van't Hoff attended. Van't Hoff warned Ostwald before the meeting that he had heard that they would be attacked; he should bring his dictionary.[99] Most of the discussion took place in the sessions on Saturday September 6th, Monday September 8th, and the informal meetings in between.

Saturday: G. F. FitzGerald

On Saturday the electrolysis committee reported. Among the papers read was a discussion of electrolytic theories by the Irish physicist, G. F. FitzGerald.[100] FitzGerald was to lead the opposition of British physicists to the Ostwald school theory. He had a greater reputation among his colleagues than that which history has left him. In part this was because he was at his most impressive in critical discussion; it was then that he earned great respect for his opinion. A number of the physicists who were not actively involved in the debate over the theory of solutions came to distrust those features of the new theories of physical chemistry that he had commented on. FitzGerald was also a friend of Ramsay and an acquaintance of Armstrong, and his opinion was taken seriously by the chemists.[101] He had been a member of the electrolysis committee since its creation in 1887 and was its chairman in 1890. At the Saturday session his paper was introduced as being "in preparation for a discussion on the extreme dissociation theory of solution supported by these recent investigations [the works of Ostwald, van't Hoff, and Arrhenius], as opposed to the more customary view held by chemists, and having reference also to Dr. Armstrong's views of residual affinity. . . ."[102] FitzGerald favored a modification of the classical Grotthus theory of electrolysis. If, when polarized by an electromotive force, the molecules draw one another apart at a rate proportional to the polarization, the Grotthus theory can explain why Ohm's law holds for

[99] Letter from van't Hoff to Ostwald, 19 September 1890, in Ostwald's correspondence, *op. cit.* (note 48), pp. 226–227.

[100] G. F. FitzGerald, *Report of the British Association* for 1890 (1891), pp. 142–144.

[101] See, for example, the introduction to *The Scientific Writings of the Late George Francis FitzGerald,* ed. J. Larmor (Dublin, 1902).

[102] *Report of the British Association* for 1890 (1891), p. 142.

very small currents in electrolysis.[103] The limits of modification were clear, however, for "there seem to be very serious difficulties in supposing that uncombined atoms are for any time free in the liquid. . . ."[104] FitzGerald considered it plausible that, as Armstrong had suggested, double decomposition takes place by association, but the matter should be investigated further. He went on: "There are some other phenomena that have been explained upon the supposition that free atoms are gadding about in a liquid. Such are the lowering of the boiling [sic] and freezing points of solutions of salts, and their effect on osmotic pressure."[105] He then went on to sketch how these effects might be a product of the nature rather than the number of the molecules of an electrolyte. If the electrolyte molecules were more easily polarized, that is, turned in an effective direction more often than other molecules, they might be able to produce twice the osmotic pressure. And changes in boiling and freezing points were to be explained by exceptional effects of electrolytes on the normally constant molecular affinity of salts for water.

On Saturday also, a paper was presented by W. N. Shaw, "Report on the Present State of our Knowledge in Electrolysis and Electrochemistry."[106] The paper was a half digested review of the literature in which no clear conclusions were drawn as to the claims of the two main sides.

Informal Discussion between Saturday and Monday

It was in the informal discussion, especially on Saturday and Sunday, that the debate over the theory of solutions became liveliest. As Ostwald told the absent Arrhenius, FitzGerald brought up a whole series of objections on Saturday night and on Sunday; Ostwald and van't Hoff talked to him all day.[107] The best account of the informal discussion was given by the Leeds professor of chemistry, Arthur Smithells, in a contribution to Ramsay's biography.

> The Leeds Meeting of the British Association in 1890 is memorable as marking the first Ionic Invasion of England in the persons of

[103] G. F. FitzGerald, *Report of the British Association* for 1890 (1891), p. 143.

[104] *Ibid.*, p. 143.

[105] *Ibid.*

[106] W. N. Shaw, *Report of the British Association* for 1890 (1891), pp. 185–223.

[107] Letter from Ostwald to Arrhenius, 23 September 1890, in Ostwald's correspondence, *2, op. cit.* (note 48), p. 71.

van't Hoff and Ostwald. It was, of course, in the early days of the ionic theory of solution and I remember Ostwald remarking that the united ages of himself, van't Hoff and Arrhenius were less than a hundred years.

Ramsay and Ostwald met for the first time as fellow-guests in my house, which became accordingly a sort of cyclonic centre of the polemical storm that raged during the whole week. No meeting within my experience has more fully illustrated the fact that the most interesting and stimulating proceedings of the British Association are those which occur outside the section rooms. The discussion was, as I have said, incessant. I remember conducting a party to Fountains Abbey on the Saturday and hearing nothing but talk of the ionic theory amid the beauties of Studley Royal. The climax, however, was reached the next day—Sunday. The discussion began at luncheon when FitzGerald raised the question of the molecular integrity of the salt in the soup and walked around the table with a diagram to confound van't Hoff and Ostwald. After luncheon the party adjourned to the garden and was gradually increased by the arrival of strolling philosophers until it assumed quite large proportions. I regret that at this distance I cannot recall the names, but believe that it included, in addition to Ramsay and those named, Lodge, Armstrong, Pickering, Otto Petterson, and there were some others.

The discussion continued throughout the afternoon with alternating vehemence and hilarity. I have a particular recollection of FitzGerald walking restlessly about with his hand clasped on his brow and declaring in his rich Irish brogue, "I *can't* see where the energy comes from." Ramsay, as you can imagine, was no silent spectator. Being a convinced ionist, he was eager in helping out the expositions of Ostwald, whose English at that time was imperfect and explosive, and his wit and humour played over the whole proceedings. I wish I could do more justice to him and to the occasion. I believe it effected a good deal towards forming friendships, promoting goodwill and removing misunderstandings, and certainly it was the beginning of relations of great mutual sympathy and regard between Ramsay and Ostwald, which lasted till they were divided by their respective national sympathies at the unhappy outbreak of war.[108]

[108] As quoted by W. A. Tilden, *op. cit.* (note 66), pp. 117–118.

Monday

The formal discussion on Monday started with a paper by Pickering.[109] Pickering was too much of an individualist to give a very good impression to the ionists.[110] Much of his paper was directed to an exposition of his hydrate theory, which he considered superior to the Ostwald school theory. Among his criticisms of the electrolyte dissociation theory were a number of points that, although often expressed in a way that showed ignorance of the development of chemical thermodynamics in the 1880's, reflected serious difficulties. For example, no one then considered the possibility that salts might be ionized even in the solid state, and early investigators thought it a matter of chance that the quantity of energy required to dissociate the solid salt into separate atoms was so close to that obtained by combining atoms with high electric charges. No energy was admitted to be available from combination of the electrolyte and solvent. Pickering pointed with scorn to some of the chance equalities of energy changes which the dissociation theory presupposed in the process of dissolution. He noted that the high heat of solvation of anhydrous salts would also be surprising unless hydration in solution is recognized as the explanation. However, his arguments did not influence the ionists, because in their eyes Pickering was obviously confused about the nature of energy changes. It takes a very sympathetic study of Pickering's critical remarks to winnow out the telling criticism from the confusions and misunderstandings.

In the subsequent discussion FitzGerald made the longest contribu-

[109] P. S. U. Pickering, *Report of the British Association* for 1890 (1891), pp. 311-322.

[110] See, for example, the comments by M. W. Travers, *op. cit.* (note 66), p. 92. Pickering is presented as an unsuitable figure to lead the anti-ionist camp. Certainly, Pickering had misunderstood many features of the theories of van't Hoff and Arrhenius. See also, T. M. Lowry and J. Russell, *op. cit.* (note 48), pp. 45-46. The ionists' dismissal of Pickering was not complete, however. In the discussion which followed Pickering's paper, a note from Arrhenius was read which criticized Pickering's recent work. But as Ostwald said in his autobiography, "Arrhenius had calculated a great number of Pickering's own measurements in the freezing point depression of sulphuric acid and had found remarkably good correspondence, evidence that Pickering's measurements were incomparably better than his theory." *Lebenslinien* (Leipzig, 1927), *2,* 135; my translation.

tion.[111] Particularly because of Ramsay's respect for his friend's incisive critical mind, FitzGerald was the anti-ionist whose opinion was taken most seriously by the Ostwald camp. In the informal discussions they had convinced FitzGerald that the Ostwaldian theory *could* account for the energy changes associated with the production of charged dissociated ions in solution, but he still considered that the truth was more complicated than they recognized. Among his other comments, FitzGerald criticized van't Hoff's suggestion that the osmotic pressure of dilute solutions is produced by the kinetic pressure of solute particles unable to penetrate the semipermeable membrane. For FitzGerald the debate settled little except preliminary misunderstandings. Later he was to criticize the Ostwald school theory more sharply; he had apparently remained unaffected by the ionists' arguments. His opposition to their theory continued unabated until his early death (at the age of fifty) in 1901.

Armstrong, who was undoubtedly the leading chemist in opposition to the Ostwald school theory, did not develop his position significantly at the Leeds meeting. In the published discussion he concentrated on just a few of the difficulties of the theory that he had previously set out. But one feature of his argument gained critical notice from Ostwald. In his publications both before and after the 1890 meeting—and probably in conversation as well—Armstrong frequently appealed to the chemists' intuitive feelings or chemical common sense. For example, in an earlier discussion of the contrast that, the dissociationists claimed, existed between the highly stable hydrogen chloride gas and the highly dissociated hydrogen chloride solution, he had said: "Such a conclusion may enable certain mathematical problems to be solved in an apparently satisfactory manner, but it is hardly one which a *chemist's common sense* would lead him to accept forthwith, in the absence of any explanation accounting for so extraordinary a difference between a dissolved substance at a low temperature and the same substance in a gaseous state at a high temperature."[112] At the Leeds meeting, Ostwald dismissed this kind of argument with the remark that Armstrong should distinguish between chemical facts and chemical feelings. The facts, Ostwald claimed, were in support of the theory of his school, and the feelings would change quite easily. "Chemists will speak in a year or two as

[111] The discussion is recorded in *Report of the British Association* for 1890 (1891), pp. 323–338.

[112] H. E. Armstrong, *Proceedings of the Chemical Society, 5* (1889), 113. My italics.

quietly of the free ions as they now speak of the uncombined mixture of hydrochloric acid and ammonia in the gaseous state [of ammonium chloride].”[113]

Neither side triumphed at the Leeds meeting on theories of solution. The formal meeting concluded with J. H. Gladstone's remark that there had been a rapprochement and increased mutual understanding between the two sides.[114] Ostwald's comments on the meeting indicate that he, at least, thought that it had been relatively successful for the ionists.

> I do not think I am wronging our hosts in supposing that the invitation had been given first of all with the friendly intention of persuading us that we were in error and of sending us back home again after a good lesson. And during the first few days our adversaries alone held the floor, so that one might have thought up to a certain point that we were already scientifically dead. But when, after long and lively personal discussions, the representatives of the modern ideas finally had a chance to speak, even at the public sessions, the appearance of things was not slow in changing, and we were able to separate from our hosts in amiable fashion and not without triumph.[115]

Even if Ostwald thought that the debate had been relatively successful from the ionist point of view, the British opposition continued to argue against the new theories, and particularly Arrhenius' theory of electrolyte dissociation through the 1890's.

5. THE DEBATES OF THE 1890'S

Further Development of the Hydrate Theory

In the early 1890's Pickering continued the polemic. His primary target in his criticisms of the Ostwald school theory was Arrhenius' theory of electrolyte dissociation, but he also attacked van't Hoff's view of osmotic pressure. Although van't Hoff had established a quantitative relationship, Pickering considered it to be of limited value to him since he believed that van't Hoff's conception of the underlying molecular processes was defective. He welcomed a paper

[113] W. Ostwald, *Report of the British Association* for 1890 (1891), p. 334.

[114] J. H. Gladstone, *Report of the British Association* for 1890 (1891), p. 338.

[115] Quoted in translation. G. Bruni, “The Scientific Work of J. H. van't Hoff,” *Annual Reports of the Smithsonian Institution* (1913), p. 779. See also E. Cohen, *Jacobus Henricus van't Hoff* (Leipzig, 1912), p. 282.

by R. H. Adie, read at the Chemical Society in 1891, in which Adie reported irregularities of the osmotic pressures of solutions.[116] Pickering stated that the irregularities told strongly against the physical theory of solution and developed his criticisms further: "Mr. Adie has touched on one of the most fundamental objections to the present physical theory of solution—the existence of osmotic pressure—for this is due to the impermeability of a membrane to the dissolved substance; and how can it be contended that if the molecules of the latter were single, and still more if they were dissociated into ions, they cannot get through holes which the water molecules have no difficulty in threading?"[117]

Pickering's most important work in the 1890's was his development of the Mendeleef method of studying hydrates in solution. Colleagues attacked him for his faith in the method, and particularly for his use of a flexible lathe to process his results, and his adherence to the method more than anything else served to isolate him from the other opponents of the Ostwald school. A discussion of his method at the Chemical Society on 4 June 1891[118] revealed the character of the criticism. Other chemists were worried by difficulties in working out an adequate mathematical theory of the method and they could not decide whether or not the discontinuities obtained were due to errors in experiment and to the smoothing of curves. The major obstacle they encountered was that they could not reproduce Pickering's results. As one critic, Rücker, pointed out, "as in the case of all other experimental methods, the final test of validity is whether concordant results are obtained by different observers. . . ."[119]

In the same discussion Armstrong indicated that even for those who accepted the importance of hydration, Pickering's work was objectionable.

> It appeared to him . . . that Mr. Pickering's conclusions were in many respects open to question from a chemist's point of view; he thought, in fact, that Mr. Pickering both proved too much and was illogical. Prepared as the speaker was to believe in the existence of hydrates in solution, he could not imagine that so large a number as was suggested would arise, or that the 102 breaks in the sulphuric acid curves, for example, could possibly be interpreted

[116] R. H. Adie, *Proceedings of the Chemical Society,* 7 (1891), 25–26.

[117] P. S. U. Pickering, as reported in *Proceedings of the Chemical Society,* 7 (1891), 26–27.

[118] See *Proceedings of the Chemical Society,* 7 (1891), 105–109.

[119] *Ibid.,* p. 107.

as evidence of as many distinct hydrates. There was no independent evidence to support such a conclusion.

Then he thought Mr. Pickering was illogical, because he interpreted all the breaks as indicative of hydrates, notwithstanding that he asserted—doubtless, with justice—that both water and sulphuric acid in the pure state consisted of complex molecules: surely in this case, as change would set in at either end of the curve, it must be impossible to say which of the breaks are to be interpreted as indicative of change in the composition of the complex molecules of acid and water respectively, which are due to the formation of hydrates consisting of simple water and acid molecules and which are due to the formation of hydrates, say, of simple water and complex acid molecules."[120]

Pickering had ready replies to most of the objections raised in the discussion, and although there is no evidence that he convinced a significant number of the scientists he addressed, he remained active in his advocacy of the method until he gave up his research on the subject in about 1896.

In a series of papers, mostly in *Philosophical Magazine,* Pickering widened the experimental data on which his version of the hydrate theory was based, proposed explanations in terms of an increasing range of phenomena, and engaged in a running battle with younger members of the Ostwald school. Following the early attacks by Arrhenius on the British hydration theorists, in 1891 James Walker took up the Ostwaldian cause.[121] Walker set out a collection of criticisms similar to those that Arrhenius had employed, showing Pickering's misunderstandings of the Ostwaldian theory and picking out the more objectionable features of Pickering's own arguments.[122] When Pickering replied in 1892,[123] he suggested that his criticisms of the Ostwald school theory in earlier years no longer applied in full, as the official ionist position now conceded that the dissociated ions are probably hydrated, and that the heat of their hydration may help to stabilize the ionized solution. However, in case the old view lingered on, Pickering presented some thermochemical studies intended to show the implausibility of any theory of solution that treats the ions as moving so freely through the solution that their behavior

[120] *Ibid.*, p. 106.

[121] For an account of Walker's career, see *Journal of the Chemical Society* (1935), pp. 1347–1354.

[122] J. Walker, *Philosophical Magazine* [5], *32* (1891), 355–365.

[123] P. S. U. Pickering, *Philosophical Magazine* [5], *34* (1892), 35–46.

shows a quantitative analogy with gas molecules moving through empty space. He also replied to Walker's criticisms, although he apparently felt that the Ostwaldian had overstepped the bounds of the etiquette of scientific debate.

An even more heated exchange took place between Pickering and H. C. Jones; it began in *Berichte der Deutschen Chemischen Gesellschaft* and transferred to *Philosophical Magazine.* Pickering had attempted to carry the debate to the Ostwaldians' home territory, arguing that the hydrate theory was better able to explain the freezing point depressions of dilute solutions than was the Ostwald school theory. Jones replied, and the series of polemical papers followed.[124] Jones was later to become the most enthusiastic American popularizer of the Ostwaldian school; his far from impartial textbooks and historical retrospects have had great influence. Like Arrhenius and Walker, Jones claimed in 1893 that Pickering's methods of calculation depended on the magnification of random experimental error, and he presented rival data in support of the ionic dissociation theory. Pickering pointed out that Jones's work disagreed significantly with his own and suggested some sources of error which Jones—in Pickering's view—had not avoided. When Pickering applied his own method of graphical analysis to Jones's data, Jones reacted strongly. "Mr. Pickering has applied his method of curve drawing to my results from sodium chloride, which differ from his to the extent of more than 50 per cent, and with it claims to have found the same 'breaks' as in his own results. He has thus shown the true value of his method, which seems to be largely independent of the experimental data."[125]

Pickering was not as foolish about experimental accuracy as Jones seems to have thought. He explained the difference between Jones's results and his own as probably being due to error in instruments which he had eliminated in his own work by the use of several thermometers.[126] Such a systematic error would not significantly affect the arguments his calculations were intended to support. And with

[124] P. S. U. Pickering, *Berichte der Deutschen Chemischen Gesellschaft, 25* (1892), 1314; *26* (1893), 1221, 1977; *Journal of the Chemical Society, 65* (1894), 293-312; *Philosophical Magazine* [5], *37* (1894), 162-164; H. C. Jones, *Berichte der Deutschen Chemischen Gesellschaft, 26* (1893), 551, 1635; *Philosophical Magazine* [5], *36* (1893), 465-497; *Proceedings of the Chemical Society, 10* (1895), 101, discussion 101-104.

[125] H. C. Jones, *Philosophical Magazine* [5], *36* (1893), 484-485.

[126] P. S. U. Pickering, *Philosophical Magazine* [5], *37* (1894), 163.

backing from E. H. Davies he pointed out to Jones at a meeting of the Chemical Society that he had not introduced new errors in the treatment of Jones's data. He had not subtracted one experimental figure from another, thus making the error in the result greater than that in either of the initial figures, as Jones had claimed, but had subtracted from Jones's data a precisely calculated quantity, designed to magnify the variation already present in the data.[127]

After this exchange Pickering was still prepared to refer to his treatment of Jones's result as further support for the hydrate theory. But Jones, in a later retrospect, considered that nothing of value remained after his own and other criticisms of the Mendeleef hydrate theory. "More accurate work has shown that most, if not all of the irregularities in the Mendeleef plot of solution property against percentage concentration are due to experimental error; and that there is not the slightest evidence for the theory of Mendeleef."[128]

Jones's conclusion is rather uncharitable, as by 1899 he was developing his own solvate theory, which he presented as a natural development of the Ostwaldian theory. While it is certainly the case that Pickering denied the existence of dissociated ions, even if they were solvated, his arguments were not restricted to the Mendeleef method, and many were of value to *any* theory of solvation. There was a much greater degree of continuity between the old hydration theories and Jones's later work than Jones found it expedient to recognize.

Pickering summarized his work on the hydration theory in his article on "Solutions" in *Watts Dictionary of Chemistry* in 1894.[129] The editors of this edition of the dictionary had decided to invite Arrhenius and Pickering to contribute rival entries.[130] Pickering's position remained essentially unchanged after this article.

Other Developments in the Early 1890's

During the early 1890's Ostwald was producing illustrations and arguments in support of the Ostwaldian theory. He gave quantitative

[127] P. S. U. Pickering, *Proceedings of the Chemical Society, 10* (1895), 101–102; E. H. Davies, *ibid.*, p. 103.

[128] H. C. Jones, *A New Era in Chemistry* (London, 1913), p. 166.

[129] P. S. U. Pickering, *op. cit.* (note 82), pp. 492–496.

[130] The article by Arrhenius was on solutions in general; Pickering contributed a summary of the general and particular arguments in support of the hydrate hypothesis.

explanations in terms of the ionic dissociation hypothesis of such properties as the magnetic rotation and the color of electrolyte solutions. Some of his expositions appeared in English. When Ostwald's paper "Chemical Action at a Distance" appeared in *Philosophical Magazine,*[131] it attracted critical discussion. Like the earlier exchange over the example of electrostatic production of electrolysis, Ostwald's latest presentation was construed by the British as an *argument* for the Ostwaldian theory, although Ostwald only claimed to be presenting yet another type of phenomenon that could be anticipated and explained satisfactorily by the new theory. The behavior of silver dipped in dilute sulphuric acid provides an example of chemical action at a distance. Although no appreciable action normally takes place, the silver will dissolve if it is connected to a platinum wire dipped in a solution containing an oxidizing agent such as acidified potassium bichromate, and if the two solutions are connected through a semipermeable membrane. Ostwald described a number of such experiments, most of which were already known, and showed how readily they were explained in terms of the Arrhenius theory. Indeed, Ostwald claimed:

> The description of some of the experiments, which are communicated here, was completely worked out at my writing table, *before I had seen anything of the phenomena in question.* After making the experiments on the following day, it was found that nothing in the description required to be altered. I do not mention this from feelings of pride, but in order to make clear the extraordinary ease and security with which the relations in question can be considered on the principles of Arrhenius' theory of free ions. Such facts speak more forcibly than any polemics for the value of this theory.[132]

Pickering, however, was not impressed. He argued[133] that the same phenomena could as readily be explained on the more traditional principles of the Grotthus theory. Because one always requires a liquid-liquid contact in chemical action-at-a-distance experiments, one can always describe chains of solute molecules as stretching from one electrode to the other, the only observable effects being at the

131 W. Ostwald, *Philosophical Magazine* [5], *32* (1891), 478–480.
132 *Ibid.,* p. 156.
133 P. S. U. Pickering, *Philosophical Magazine* [5], *32* (1891), 478–480.

electrodes. "All these experiments seem to be on a par with one described some time ago by Professor Ostwald, consisting in the production of a small amount of electrolysis by a current of electrostatic origin: experiments which are perfectly consistent with the old electrochemical theory, dressed up in the garb of the dissociation theory and then presented to us as proof positive of this theory."[134]

Similar objections were made by J. Brown in a paper read at the British Association meeting that year (1891) and published in *Philosophical Magazine* the following year. Brown also discussed further experiments by Ostwald which had been described in *Zeitschrift für physikalische Chemie,* and he drew the conclusion that Ostwald had yet to point to any crucial experiments. Both of the rival theories could explain the known experimental results.[135]

Ostwald's tendency to present his arguments as if there was no serious rival to the Ostwald school theory for the treatment of solutions was a constant annoyance to many British scientists.[136] His textbooks in particular drew criticism on this point. For example, in a review of Ostwald's *Solutions,* the reviewer commented: "As a concise account of the new theory of solution Professor Ostwald's work is most valuable; but it is somewhat to be regretted that he did not give some indication of the older theories, of which not a word is said. Indeed, a student taking up the book and having no previous knowledge of the subject would be led to suppose that the theory here put forward is the universally accepted one, whereas it is really regarded by the majority of chemists as quite untenable."[137] Another reviewer of *Solutions* made similar comments and provoked an exchange of letters with Ostwald.[138] The reviewer, J. W. Rodger, referred to the theory of the Ostwald school as the "physical" theory and contrasted it with the hydrate or "chemical" theory; he criticized Ostwald for concentrating so heavily on physical factors, when chemical considerations suggest that solutions are far more

[134]*Ibid.,* pp. 479–480.

[135]J. Brown, *Philosophical Magazine* [5], *33* (1892), 82–89.

[136]That this tendency was to some extent deliberate is revealed in a letter from Ostwald to Arrhenius, 2 January 1892. Ostwald reported with obvious pleasure that the section on solutions of his *Lehrbuch der allgemeinen Chemie,* which had just been translated into English, contained no mention of Pickering. See Ostwald's correspondence, *op. cit.* (note 48), p. 103.

[137]James L. Howard, *Philosophical Magazine* [5], *33* (1892), 146.

[138]J. W. Rodger, *Nature, 45* (1891-1892), 193–195, 342–343, 487. W. Ostwald, *ibid.,* pp. 293–294, 415, 606.

complex than the physical picture assumes.[139] Ostwald, in his reply, objected to the contrasting of "physical" and "chemical" processes and theories of solution. "It has never been maintained, either by me or by any other representative of the newer theory of solutions, that no interaction takes place between the solvent and the dissolved substance; on the contrary, I have for years greatly encouraged research work directed towards making clear the nature of such interactions."[140] Ostwald went on to give his version of the relationship of the two theories of solution.

> I beg Mr. J. W. R. to recall the history of the rivalry between the two "theories." Van't Hoff and his successors developed the law of solutions entirely without polemical strife, because, since the fundamental ideas of van't Hoff's theory were entirely new, there was nothing at all in its territory to combat, as till then there was nothing there. The attacks upon van't Hoff were begun by an investigator who had until then directed his attention exclusively to the phenomena which I have above characterised as individual, and who was evidently unprepared to deal with such colligative properties. The defence had to consist in an unceasing clearing up of misconception. Now, the greatest of these misconceptions is, that both "theories" are rivals. The existence and form of the laws founded by van't Hoff and his successors stand at present beyond question. . . . But what has until now been known as the hydrate theory has not been in a position to give any information whatever in regard to these laws; none of them have been discovered with its aid, and since it has for its subject not the colligative but the individual properties of solutions this will not be otherwise in future.[141]

The rest of Ostwald's letter and much of the later exchange were concerned with the use of the word "theory." The debate degenerated into a terminological disagreement, Ostwald arguing that theories were merely collections of laws, and Rodger that theories should also satisfactorily *explain* the laws they relate. Accusations of inconsistency of usage further obscured the issue.

Ostwald's textbooks soon were the guide for a large number of

[139] J. W. Rodger, *ibid.*, pp. 193–195.
[140] W. Ostwald, *ibid.*, p. 293.
[141] *Ibid.*, p. 294.

textbooks of physical chemistry in English. Most of these English textbooks were written by members of the Ostwald school and followed closely his presentation. They were criticized by many later anti-ionists for giving students a far from impartial view of the status of the rival theories of solution.[142] Some English textbooks, however, did attempt to give a fair perspective on the two theories from a British point of view. One such book was written by W. C. D. Whetham, later William Cecil Dampier. In the early 1890's Whetham had published experimental work that he considered gave experimental support to the electrolyte dissociation theory.[143] In his textbook *Solution and Electrolysis* (1895) he wrote enthusiastically about the theory of ionic dissociation, less so about the value of treating dissolved substances in solution as if they were in a quasi-gaseous state. He attempted to be fair to Armstrong and Pickering, discussing their ideas briefly and suggesting how Pickering's more effective arguments for the hydration of dissolved substances could be absorbed into the Ostwaldian theory. "We can, in fact, regard a considerable mass of the solution, containing, perhaps, several molecules and dissociated ions of salt, and hundred molecules of solvent, as chemically one large molecule, the parts of which are nevertheless to some extent physically independent of each other."[144] Whetham was to develop this compromise in the following years.

The beginnings of a search for compromise were apparent, but the debate was far from over. In his presidential address to the Chemical Society in 1895 Armstrong showed himself to be as intransigent as ever. After a discussion of the development of his own arguments, he set out his current evaluation of the dissociation hypothesis. "Whatever view may ultimately be taken of the hypothesis—whether it can be retained as a permanent addition to our theories or not—its introduction has been eminently fruitful of results, and an already too voluminous literature of the subject has grown with surprising rapidity. Yet it appears to me that it has been accepted by a particular

[142]See, for example, J. J. Howard, *Philosophical Magazine* [5], *33* (1892), 144-147; A. Smithells, *Nature, 62* (1900), 76-77; L. Kahlenberg, *Journal of the American Chemical Society, 24* (1902), 485-486; H. E. Armstrong, *Science Progress, 3* (1909), 65-66.

[143]See, for example, W. C. D. Whetham, *Philosophical Magazine* [5], *38* (1894), 392-396.

[144]W. C. D. Whetham, *Solution and Electrolysis* (Cambridge, Eng., 1895), p. 212.

school—at the head of which stands Ostwald, and who regard and treat all unbelievers as heretics worthy of the stake—not as a mere working hypothesis, but as an absolute creed, without any sufficient attempt being made to discuss its general probability."[145] Armstrong went on: "Personally, I am still entirely unconvinced of the validity of the hypothesis, although no one can be more willing to admit that in so far as *weak* solutions are concerned, a 'law' has been discovered which is broadly true in *mathematical form,* however open to question the fundamental premises may be on which it is based. I am satisfied that the phenomena of chemical change are, as a rule, far more complex in character than is assumed by the advocates of the hypothesis."[146]

The Exchange in *Nature* 1896-1897

In 1896 the debate continued with new life, drawn particularly from a prolonged exchange of letters in *Nature.* The exchange took place against the background of FitzGerald's "Helmholtz Memorial Lecture" at the Chemical Society in 1896.[147] FitzGerald's lecture has been varyingly assessed as "profound and brilliant"[148] and as "rambling and irrelevant."[149] In general it was favorably received by the opponents of the Ostwald school theory. In his lecture FitzGerald moved from an evaluation of parts of Helmholtz' work to discussion of some of the dangers of applying thermodynamics to chemistry without sufficient care. Every application of the second law of thermodynamics should be to a reversible reaction, and a complete cycle should be examined so that all thermodynamic changes are recognized. As an example of the errors that may result from unjustified applications, FitzGerald mentioned a recent publication that presented a proof that osmotic pressure is proportional to absolute pressure, unjustifiably assuming that all the heat supplied was doing osmotic work.[150] A lesson was to be learned for mathematical chemistry: "It is as risky for a chemist to apply mathematics as for a mathematician to lecture to chemists: we should work in co-operation."[151] FitzGerald attacked Ostwald's argument that when

[145] H. E. Armstrong, *Journal of the Chemical Society, 67* (1895), 1124.
[146] *Ibid.,* pp. 1124-1125.
[147] G. F. FitzGerald, *Journal of the Chemical Society, 69* (1896), 885-912.
[148] J. Lister, *Proceedings of the Chemical Society, 12* (1897), 26.
[149] J. R. Partington, *op. cit.* (note 6), p. 679.
[150] G. F. FitzGerald, *op. cit.* (note 147), p. 898.
[151] *Ibid.,* p. 899.

an electrolyte is subject to electrostatic induction, the superficial induced charges are due to a layer of electrified ions on its surfaces. "If there were no forces other than electrical ones, these ions would fly off the surface like dust."[152] He then moved on to a discussion of the theory that solutions and gases are analogous.

> So much advance has been made by assuming that bodies in solution behave in some important respects like the same body in the gaseous state, that there has been a serious danger of assuming the physical conditions are at all like. The dynamical condition of molecules in solutions is essentially and utterly different from that of a molecule in a gas. The essential condition for applying any known dynamical theory of gases to calculate their behaviour is that the time during which two molecules are within the sphere of one another's action is small compared with the time during which they are apart, and that consequently the chances of three or more molecules being in simultaneous collisions is very small. . . . Now this essential condition for the application of the dynamics of a gas to molecules in solution is very far indeed from being fulfilled. A molecule is never outside the sphere of action of its neighbors.[153]

Indeed, FitzGerald's calculations suggested that a molecule in solution was within the sphere of influence of between a hundred and a million of its neighbors. "It is, no doubt, a most remarkable thing that osmotic pressure should be even roughly the same as what would be produced by the molecules of the body in solution if in the gaseous state, but to imply that the dynamical theory of the two is at all the same, or that the dynamical theory of a gas is in any sense an *explanation* of the law of osmotic pressures is not at all in accordance with what is generally meant by the word 'explanation'."[154] FitzGerald did not just object to the van't Hoff theory of solutions; he went on to attack the idea that electrolytes show enhanced osmotic effects because they are dissociated. He sketched an alternative explanation, which was more qualitative and also more in line with Armstrong's chemical arguments. The Ostwald school were satisfied with superficial mathematical regularities when they should have been seeking mechanical explanations.

The immediate stimulus for the correspondence in *Nature* was a

[152] *Ibid.*, p. 902.
[153] *Ibid.*
[154] *Ibid.*, p. 903.

suggestion by the physicist J. H. Poynting as to how osmotic pressure might be accounted for, not as an additional pressure produced by the dissolved molecules, but from some kind of association between the solvent and solute. Poynting hoped to avoid the dissociation hypothesis in his explanation of the higher osmotic pressure of electrolytic substances.[155] The British climate of scientific opinion, influenced by men like FitzGerald and Armstrong, made the dissociation hypothesis something to be avoided if possible. Poynting's idea, in part based on an earlier paper of 1881,[156] was that the mobility of molecules of a solvent is affected by internal or external pressure; the effect of a dissolved substance is analogous to negative pressure—the solute molecules slow down the solvent in their neighborhood. Since it may be argued that electrolytes affect the solvent more than non-electrolytes, the greater effect of their molecules allows for their additional osmotic pressure. Poynting's proposed mechanism stimulated W. C. D. Whetham to write a letter to *Nature,*[157] and thus establish that journal as the forum for a continued exchange. Developing the theme of the concluding part of his textbook published the previous year,[158] Whetham argued for compromise. He suggested that Poynting's mechanism of osmosis was quite compatible with the dissociation theory. He showed for example that, if each ion unites with a certain amount of the solvent, then the dissociation of a molecule into its ions will have a greater effect on the solvent and so raise the osmotic pressure. "Thus Professor Poynting's conditions would be satisfied, and at the same time the advantages of the dissociation theory would be retained."[159] This argument Poynting accepted.[160] After a discussion with Ramsay he announced that he now accepted that ions move independently of one another in solution. Provided that the ions associated with the solvent, their indirect effect on osmotic pressure could be understood in terms of the mechanism he had postulated.

Armstrong then joined the discussion with a polemical letter filled with uninhibited rhetoric. He criticized the deviousness with which the Ostwaldians evaded criticism and described the new breed of

[155] J. H. Poynting, *Philosophical Magazine* [5], *42* (1896), 289-300.

[156] J. H. Poynting, *Philosophical Magazine* [5], *12* (1881), 32-48, 232.

[157] W. C. D. Whetham, *Nature, 54* (1896), 571-572.

[158] W. C. D. Whetham, *op. cit.* (note 144), p. 44.

[159] W. C. D. Whetham, *Nature, 54* (1896), 572.

[160] J. H. Poynting, *Nature, 55* (1896), 33.

chemists turned out by the school as nominally chemists, but as chemists without "chemical feeling." It seemed to him that in chemistry two parties were emerging that had nothing in common. He concluded his letter by commenting on the corruptive tendency of the Ostwald school in science education to present dogmatically as fact hypothetical ideas of atomic dissociation.[161] Armstrong's letter stimulated three other scientists to reply. O. J. Lodge showed that he had become a moderate advocate of the ionist cause; he was also concerned that Armstrong's intemperate language might have given the false impression that he wished to keep chemistry free of all trespassers.[162] Whetham, employing moderate language, tried to bring the debate back from polemics to serious discussion by replying to the detailed points in Armstrong's objections.[163] Whetham realized that it was the manner rather than the substance of Armstrong's letter that was intended to have the greatest impact, and he admitted that to read the letter as serious argument "considerable mental agility is needed to follow *all* the metaphors which Professor Armstrong crowds into a single sentence. . . ."[164] E. F. Herroun, another anti-ionist, was the third scientist to respond to Armstrong.[165] Herroun had devoted most of his work to the measurement of the electromotive force of cells, a topic on the borderlines of physics and chemistry. Most of his papers were published by the London Physical Society. Herroun believed that the arguments of the anti-ionists had been put very strongly, and that the ionists were employing unfair techniques of scientific debate. "It seems to me a duty of teachers to protest against the growing tendency there seems to be of putting forward the crude hypotheses of the ionist school, as though they had some claim to acceptance as well established scientific laws, about which no reasonable doubt exists. So far from this being the case, the arguments commonly advanced in support of the theory seem to consist mainly of the misapplication of physical laws to a few carefully selected cases, aided by plausible but misleading assertions."[166] Herroun devoted the bulk of his letter to questions that he felt exposed errors in the ionists' position, especially as it appeared

[161] H. E. Armstrong, *Nature, 55* (1896), 78–79.
[162] O. J. Lodge, *Nature, 55* (1896), 150–151.
[163] W. C. D. Whetham, *Nature, 55* (1896), 151–152.
[164] *Ibid.,* p. 151.
[165] E. F. Herroun, *Nature, 55* (1896), 152.
[166] *Ibid.,* p. 152.

in Ostwald's *Outlines of General Chemistry*, the "Bible of the ionists." He asked that the ionists answer these questions and that the uncommitted—"those who are at present only partly dissociated"—reflect upon them until satisfactory explanations appear.

Early in 1897 Pickering again contributed to the arguments of the anti-ionists.[167] He claimed in a letter to *Nature* that the theories of osmotic pressure and of ionic dissociation are to be treated only as numerical regularities rather than as acceptable theories. "For a theory to be acceptable it should, at the very least, be reasonably probable, and should not violate any fundamental and well-established facts; it should stand the test of any apparently crucial experiments brought forward to settle between it and its rivals, and, I think I may add, it should give some explanation, not simply of the behaviour of matter in the condition in question, but also of why matter ever assumes such a condition."[168]

The Ostwaldian theory, he argued, did not meet these conditions, while the hydrate theory did. The Ostwald school had yet to explain why thermally stable compounds fly apart when exposed to water molecules which are represented as being inert. The question of the thermochemical relationships involved in dissociation had yet to be resolved. As both electrolysis by electrostatic discharge and "chemical actions at a distance" were adequately explained by older theories, the Ostwaldians had yet to produce a significant crucial experiment. As support for his side of the debate Pickering proposed two crucial experiments which he claimed the Ostwald school theory could not handle. Referring to his arguments that solvation makes the particles of the dissolved species too large to pass through a semipermeable membrane, he claimed to have experimental evidence that while pure water and pure propyl alcohol can penetrate a semipermeable vessel, only the solvent of a dilute solution of one in the other is able to pass through.[169] Pickering's second crucial experiment was based on the claim that the freezing points of solutions of sulphuric acid and water in acetic acid suggest

[167]P. S. U. Pickering, *Nature, 55* (1897), 223-224.

[168]*Ibid.*, p. 223.

[169]It was not until 1905 that Pickering's argument was dealt with. The effect was shown to be transient, disappearing at equilibrium and so not justifying Pickering's conclusion. See A. Findlay and F. C. Short, *Journal of the Chemical Society, 87* (1906), 819-822; F. C. Short, *Philosophical Magazine* [6], *10* (1905), 1.

that the sulphuric acid and the water are associated rather than that the acid is dissociated.

No committed ionist replied to Pickering. Whetham wrote that he was sceptical of the gaseous theory of solutions but in favor of the dissociation theory.[170] Although he could think of no answer to the first crucial experiment, he suggested that Pickering's second experiment was not crucial, as the ionists did not need to assume that sulphuric acid is dissociated when dissolved in acetic acid. He suggested, furthermore, that his own compromise proposal that the ions are highly hydrated could explain Pickering's observations completely. Whetham then suggested that a different list of observations from those that Pickering had discussed provided the strongest evidence for the dissociation theory. The velocity with which an ion travels through a dilute solution in electrolysis is independent of the nature of the other ions present. The conductivity of a dilute solution is proportional to its concentration rather than to the square of the concentration (as would be required by a theory of electrolysis like Grotthus' which explained ion transport by a mechanism of collision between solute molecules). Basing his calculations on the assumption that the ions move independently and at different velocities, Whetham derived the potential difference between two solutions in contact that differ only in concentration. In the British Association meeting of 1897, Whetham outlined his compromise again. He concluded that, although ions travel independently and are free from one another for most of their existence, "it must be particularly noticed that this freedom from each other does not at all prevent the ions from forming chemical combinations with the solvent molecules. Neither does it throw any light on the fundamental nature of solutions."[171] Poynting's mechanism for osmotic pressure had provided a satisfactory alternative to the van't Hoff analogy with ideal gases.

The later part of the exchange in *Nature* in 1897 consisted largely of a discussion between leading theoretical physicists. In his endeavor to give his position respectability Armstrong had earlier quoted supporting remarks by leading physicists: in his letter of the previous November he had referred to Rayleigh's agreement with FitzGerald

[170] W. C. D. Whetham, *Nature, 55* (1897), 606–607.

[171] W. C. D. Whetham, *Report of the British Association* for 1897 (1898), pp. 244–245.

on the dangers of pushing formal analogies too far and supposing a real dynamical similarity between gases and solutions.[172] Rayleigh replied to affirm his support for van't Hoff's work.[173] He agreed that van't Hoff's supporters had often been careless in the phraseology they used and that they may thus have turned some distinguished physicists and chemists against the theory. To support the extension of Avogadro's law to dilute solutions, Rayleigh gave an alternative derivation. In the following issue a week later, Kelvin presented a counter argument. He stated that in the absence of knowledge of the solvent-solvent, solute-solute intermolecular forces, there was no satisfactory theoretical argument for applying Avogadro's law to solutions. Kelvin concluded by referring with approval to FitzGerald's "Helmholtz Memorial Lecture."[174] Two months later Gibbs argued in a letter to *Nature*[175] that it was not necessary to know very much about intermolecular forces to show that Avogadro's law applied to dilute solutions. He showed that his theoretical treatment in his 1875–1878 paper, "On the Equilibrium of Heterogeneous Substances,"[176] was able to deal with the problem. Another contribution to the discussion in *Nature* was a report of a lecture by the physicist J. Larmor, originally given to the Cambridge Philosophical Society.[177] Larmor referred to the theoretical treatments of osmotic pressure implicit in the work of Helmholtz and Gibbs which made van't Hoff's law indisputable. He also suggested what kind of processes in dilute solutions would make the result comprehensible.

The *Nature* debate ended with a final exchange between Pickering and Whetham.[178] Pickering made it clear that he was not prepared to accept Whetham's compromise. His abhorrence of dissociated ions remained even when their advocates admitted that they might be hydrated. "When a theory can only explain observed facts by driving us to assumptions of the existence of such compounds as H_xH_2O and

[172] H. E. Armstrong, *Nature, 55* (1896), 78.

[173] Rayleigh, *Nature, 55* (1897), 253–254.

[174] Kelvin, *Nature, 55* (1897), 272–273.

[175] J. W. Gibbs, *Nature, 55* (1897), 461–462.

[176] J. W. Gibbs, *Transactions of the Connecticut Academy, 3* (1874–1878), 108–248, 343–524.

[177] J. Larmor, *Nature, 55* (1897), 545–546.

[178] P. S. U. Pickering, *Nature, 56* (1897), 29; W. C. D. Whetham, *ibid.*, pp. 29–30.

$SO_{4y}H_2O$, I venture to think that that theory must be somewhat shaky."[179] Whetham concluded that only differences of opinion remained. "Such a view of the dissociation theory [with solvated ions] seems to me to offer many advantages. It may be contrary to some *opinion,* but I do not think that any *facts* have yet been pointed out which refute it. Till they are, it may possibly be of use as a working hypothesis in the investigation of that complicated structure which we call a solution."[180] Admitting that ions are hydrated was the main concession the ionists made to Pickering's camp. Pickering refused to accept even hydrated ions and any further development of the subject could only have seemed degenerative to him.

The 1896–1897 discussion in *Nature* marks the end of the first decade of opposition to the Ostwald school theory. Although the chief opponents of the theory became increasingly entrenched in their positions, the main issues were becoming clearer. Among physicists the near consensus was emerging that van't Hoff's law was theoretically satisfactory for dilute solutions but that the underlying mechanism of osmosis was still far from being elucidated. Chemists like Whetham who were searching for compromise agreed that many erroneous claims had been made by the early ionists, and that in particular a full understanding of the nature and stability of ions required recognition of the active role of the solvent. Unlike the American H. C. Jones, the British usually recognized the importance of at least some of Pickering's and Armstrong's arguments. But the debate was still far from over. In spite of the confrontations, neither side had come to a full understanding of the other. Each side still developed its own strong points and in referring to the other stressed only the opponent's weaker points. With some change in the leading personalities and locations of the debate about the turn of the century, the argument was to continue for more than another decade. The main opposition to the growing numbers of Ostwaldian physical chemists came from L. Kahlenberg and from Armstrong and their respective students. The work of these two men can be described separately. Since the crusading phase of Kahlenberg's opposition did not continue as long as that of Armstrong, I shall first consider the former and discuss the American situation.

[179] *Ibid.,* p. 29. [180] *Ibid.,* p. 30.

6. KAHLENBERG'S INFLUENCE ON CRITICAL DISCUSSION OF THE THEORY OF SOLUTION IN AMERICA AND BRITAIN

The Development of Kahlenberg's Opposition

The American chemist Louis Kahlenberg developed his opposition to the theories of Arrhenius and van't Hoff in a different manner from the English. After taking his first degrees at the University of Wisconsin, he went to Germany, as was the American custom, to further his chemical education. He studied at Leipzig in 1894, and in 1895 he obtained his doctorate with Ostwald. He also spent some time with other German chemists, returning to the University of Wisconsin in 1896.[181] Kahlenberg later wrote that during his student days and for several years afterwards he had been enthusiastic over the theories of Arrhenius and van't Hoff. It was while seeking to establish new experimental facts to support the theories further that he continually observed contradictory phenomena.[182]

The development of Kahlenberg's opposition can be followed in the succession of papers and reviews he wrote around the turn of the century. In his early experimental papers he unquestioningly employed the dissociation theory where appropriate. The first doubts appeared in a study of non-aqueous solutions. The success of the theory for aqueous solutions may make it seem natural to apply the theory to non-aqueous solutions without further question, he said, but one ought to ask whether or not in non-aqueous conducting solutions the dissolved substance is dissociated. "In view of the opposition that the theory had to meet at the time of its promulgation, and in view of the fact that even at present a number of scientists are opposed to it, we may well hesitate to apply this theory (which is based primarily on a study of aqueous solutions) to non-aqueous solutions without a firm experimental basis for doing so."[183] At the end of the same paper, in which the dissociation theory turned out to be inadequate to explain the experimental

[181] There is a biography of Kahlenberg by N. F. Hall, *Transactions of the Wisconsin Academy of Science, Arts and Letters, 39* (1949), 83–96; *40* (1950), 173–183; it contains a useful bibliography of the work of Kahlenberg and his students (though it must be used with caution as there are many errors and omissions).

[182] L. Kahlenberg, *Transactions of the Faraday Society, 1* (1905), 49.

[183] L. Kahlenberg and A. T. Lincoln, *Journal of Physical Chemistry, 3* (1899), 13–14.

results, Kahlenberg wondered whether or not "after its glorious success in explaining the properties of aqueous solutions of acids, bases and salts, the dissociation theory will need the help of its old rival, the hydrate theory (perhaps in somewhat modified form), to explain the facts in the case of non-aqueous solutions."[184] Later the same year, Kahlenberg expressed his doubts more strongly.

> From the facts here presented, it seems clear that solutions may conduct electricity fairly well and yet the dissolved substance may possess a normal molecular weight, i.e., it may be not dissociated. But this point once established, the question arises, cannot conduction then in all solutions be explained without assuming electrolyte dissociation? To be sure then the high osmotic pressure and corresponding low molecular weights found in so many cases, in aqueous solutions of electrolytes for instance, would have to be accounted for otherwise than by the dissociation hypothesis.[185]

Kahlenberg's early critical studies were based on work on the dissociative power of solvents and on concentration cells in non-aqueous solutions. Kahlenberg was also working on physiological chemistry (his first position after returning from Germany was in the pharmacy department at Wisconsin) and he was finding more evidence of the limitations of the dissociation theory, even in aqueous solutions. The Ostwald school attributed many of the properties of acids to the presence of the hydrogen ion. For example, they related the sourness of the taste of acids to the concentration of hydrogen ions. In contrast, Kahlenberg found that many acid salts tasted much more sour than suggested by the theory.[186] In a parallel study of the toxic effect of acid sodium salts on sensitive plants his conclusion was again critical of the dissociation theory.[187]

[184] *Ibid.*, p. 34.

[185] L. Kahlenberg, *Journal of Physical Chemistry, 3* (1899), 398.

[186] L. Kahlenberg, *Journal of Physical Chemistry, 4* (1900), 33-37. Kahlenberg's paper generated a brief exchange with T. W. Richards over the explanation of the sour taste. T. W. Richards, *Journal of Physical Chemistry, 4* (1900), 207-211. L. Kahlenberg, *Journal of Physical Chemistry, 4* (1900), 533-537. Kahlenberg also responded to critical comments in reviews by A. A. Noyes in *Review of American Chemical Research, 6* (1900), 73, 147, 157; *Zeitschrift für physikalische Chemie, 36* (1901), 613-615. See L. Kahlenberg, *Journal of Physical Chemistry, 5* (1901), 380fn.

[187] See, for example, L. Kahlenberg and R. M. Austin, *Journal of Physical Chemistry, 4* (1900), 553-569.

Around the turn of the century Kahlenberg also began to write unfavorable reviews of uncritical expositions of the Ostwald school theory. For example, in 1900 in his review of a textbook by H. C. Jones he wrote that although the book viewed the dissociation theory most favorably, during the last year or so facts had begun to accumulate, especially about non-aqueous solutions, that the theory could not explain. "It seems at present that even before the theory of electrolytic dissociation will have found its way into regular chemical textbooks, it will have undergone radical modifications from its present form, or will perhaps have been superseded by more adequate explanations."[188]

Two years later, Kahlenberg used stronger language: "A brief treatise on general physical chemistry which devotes such an undue proportion of its space to the exposition of the theory of electrolytic dissociation and its applications (without even attempting to indicate the shortcomings of this hypothesis) as this book does, can at the present stage of the development of the science, hardly claim to present the subject in a fair, well-balanced form."[189]

By 1901 Kahlenberg was engaged in a systematic program designed to show the untenability of the Ostwald school theory. He published a number of general papers, summarizing his arguments, and with his students he published studies extending the experimental basis of his case. The first of the general papers appeared early in 1901.[190] The paper was introduced by Kahlenberg's version of the history of the Ostwaldian theory. After its promulgation in 1887,

> the theory at once met with great opposition, notably in England, and it was by no means received with open arms on the continent. But the hypothesis inspired experimental investigation, and the results of this phenomenal activity (which at first centred in Ostwald's laboratory at Leipzig, but spread rapidly to other parts

[188] L. Kahlenberg, review of H. C. Jones, *The Theory of Electrolytic Dissociation and Some of its Applications* (New York, 1900), in *Journal of the American Chemical Society, 22* (1900), 228-229.

[189] L. Kahlenberg, review of J. L. R. Morgan, *The Elements of Physical Chemistry,* 2nd ed. (New York, 1902), in *Journal of the American Chemical Society, 24* (1902), 486.

[190] L. Kahlenberg, "The Theory of Electrolytic Dissociation as Viewed in the Light of Facts Recently Ascertained," *Journal of Physical Chemistry, 5* (1901), 339-392.

of Germany, to various other countries of Europe and to America) soon silenced the opposition in Germany and gradually diminished it also in England. It must not be supposed, however, that this silence meant that all were convinced. The silence seemed to result on the one hand because of a recognition of the futility of the debate with the knowledge of existing facts, and because of a recognition of, if not admiration for, the enthusiasm displayed by the adherents of the theory. . . .[191]

Kahlenberg attributed the success of the theory to its ability to relate the factor *i* in van't Hoff's theory of solutions to measurements of electrical conductivity for the limited range of aqueous solutions on which measurements had been made. He explained that "in view of the few experimental data at hand in 1887, and the fact that many of them had not been determined with accuracy, the poor agreement, of a goodly number of values at least, was readily overlooked in view of the generalities that the theory sought to bring, generalities which were soon incorporated without proper qualifications into text-books."[192]

The Ostwald school had tended to neglect non-aqueous solutions (most of the early studies of non-aqueous solutions were made by French-speaking scientists), believing that non-aqueous solutions yielded "normal" molecular weights and were generally nonconductors. Kahlenberg reviewed his earlier work to show that such solutions did not behave as the Ostwald school believed. "In the face of these facts the theory of electrolytic dissociation is untenable in the case of non-aqueous solutions."[193]

Kahlenberg devoted the greater part of his 1901 paper to experimental studies of aqueous solutions; he concluded from his studies that there were many cases of dilute solutions for which the dissociation theory failed to account except by using a large number of ad hoc stratagems, and that the theory failed to account quantitatively (and sometimes even qualitatively) for more concentrated solutions. Kahlenberg went on to list all other counter arguments he could muster. He referred to his physiological experiments on sour taste and toxicity. He reviewed the difficulties others had expressed about the thermochemical features of the theory. He mentioned the well-known problem of reconciling the dissociation theory with the law of mass action, particularly for strong electrolytes. He criticized

[191] *Ibid.*, p. 340. [192] *Ibid.* [193] *Ibid.*, p. 344.

the Ostwald school in this connection: "It is really unfortunate that in discussing problems of equilibrium into which strong electrolytes enter (their solubility for instance) the adherents of the dissociation theory should go right ahead with their mathematical equations and deductions as though they were in full accord with the law of mass action."[194]

Kahlenberg was particularly harsh on the dissociation theory when he discussed its application to analytical chemistry. He agreed that some physical scientists might find the language of the dissociation theory helpful, but he insisted that it was untenable to assert that the new terminology, combined with an attempt to apply the law of mass action to electrolytes, constitutes a foundation for analytical chemistry. "The very fact that chemistry has not received much benefit from Professor Ostwald's little book on 'The Scientific Foundations of Analytical Chemistry', in the way of improving existing analytical methods and discovering new ones speaks for itself."[195]

Next Kahlenberg rejected some of the more popular explanations that are expressed in terms of the dissociation theory. He argued that the additive properties of solutions need not imply dissociation, since properties such as molecular specific heat could be related additively to atomic heats in cases where there was no question of dissociation. He found the readiness of electrolytes to undergo rapid reaction to be an inadequate argument for their dissociation; non-conducting substances and mixtures can also react rapidly—as in many well-known explosives. Even the actual phenomena of electrolysis did not require Arrhenius' theory; Kahlenberg considered that Clausius' hypothesis was quite sufficient to explain all the facts.

The abandonment of the dissociation theory would require some other explanation of the high value of *i* in van't Hoff's equation for dilute solutions of electrolytes. Kahlenberg sought to overcome the difficulty by arguing that the van't Hoff law was of only limited validity, so that one had to expect deviations from it. In particular

[194] *Ibid.*, p. 381. The scientists engaged in the debate discussed here never paid much attention to the failure of strong electrolytes to behave in accordance with the law of mass action; the problem was probably too far removed from chemical intuition. It was, however, this issue which proved crucial in the further development of the dissociation theory. See H. Wolfenden, "The Anomaly of Strong Electrolytes," *Ambix, 19* (1972), 175–196.

[195] L. Kahlenberg, *Journal of Physical Chemistry, 5* (1901), 382.

Kahlenberg attacked the analogy between gases and dilute solutions. He seemed to think that to ridicule the idea that the solute in dilute solutions behaves like a gas in the volume of the solution was to undermine the whole of van't Hoff's theory of solutions.[196] Kahlenberg admitted that, as with ideal gases, the gas equation is supposed to hold strictly only for infinitely dilute solutions. But, he insisted, one has a right to expect from a modern theory of solutions that with increasing concentration a solution should behave at least qualitatively as a gas does with increase of pressure. He argued that for practical purposes a normal solution is rather dilute and showed that often there is not even qualitative agreement with gas behavior over a given range of concentration.[197]

In two years Kahlenberg's initial disquiet about the dissociation theory had grown into an attack on the theory of dissociation in general and also on van't Hoff's theory of solutions. "It is solely because of the rapid growth of the erroneous idea that the deductions drawn from the indiscriminate application of the simple gas equation to solutions and from the notion that all well-known facts harmonize with the theory of electrolytic dissociation, that I have felt compelled to call attention to the real status of the experimental facts underlying these deductions. It is hoped that this will stimulate to renewed experimental activity, for surely our theory of solutions leaves much to be desired."[198]

Early Reactions to Kahlenberg

Kahlenberg's first general paper and those that followed attracted a certain amount of sympathetic attention as well as the scathing reaction of dedicated Ostwaldians. A. A. Noyes, for example, reviewed the paper scornfully: "The author's view seems to be that because the theory does not alone explain all the relevant phenomena of all kinds of solutions, it is unjustifiable or inadvisable to employ it at all. It is scarcely necessary for the reviewer to add that the theory has received far too many approximately quantitative verifications to be overthrown by such facts as are cited by the author."[199]

[196]The significance of van't Hoff's analogy between gases and dilute solutions is discussed further in the section on the nature of osmotic pressure.

[197]L. Kahlenberg, *Journal of Physical Chemistry, 5* (1901), 389.

[198]*Ibid.,* p. 391.

[199]A. A. Noyes, *Review of American Chemical Literature,* 7 (1901), 157.

A similar but more moderate response appeared in *Nature*, in a letter by the British chemist H. M. Dawson. After commenting on points of detail, Dawson wrote:

> Although, therefore, the publication contains a large number of valuable empirical data, yet it cannot be allowed for one moment that the ionic theory has been shown to be untenable. It is far from the wish of the writer to minimise the difficulties which do admittedly confront the theory of ionic dissociation. It must not, however, be supposed that the theory has received its final and complete form; the possibilities of its rational expansion and development to explain existing irregularities are far from being exhausted. A warning note may be sounded against a too ready assumption that new experimental data prove the untenability of the theory without very careful consideration of what exactly is, and is not, stipulated by the theory.[200]

The American Electrochemical Society

A second general paper by Kahlenberg appeared in 1902 in the *Transactions* of the American Electrochemical Society.[201] He added little but rhetoric to his earlier arguments.

The American Electrochemical Society requires consideration in its own right in the present historical study: it reported the discussions at its meetings, providing historians with valuable information about the interactions of American electrochemists. Kahlenberg later described the climate of opinion in the early meetings. He characterized the first decade of activity of the society as dominated by discussion of the fundamental electrochemical theories and concepts. He considered this due to the impact of Arrhenius' electrolytic dissociation theory, which, although promulgated in 1887, only gained a foothold after Ostwald propagandized it and applied it to ordinary chemical reactions in aqueous solutions. "So it is not surprising," he wrote, "that the American Electrochemical Society, too, in its early years, should have busied itself to a considerable degree with the discussion of the pros and cons of this new electrochemical theory. On the whole our membership, following perhaps the lead of the English scientists, was rather loath to adopt

200 H. M. Dawson, *Nature, 65* (1902), 415.

201 L. Kahlenberg, "Current Electrochemical Theories," *Transactions of the American Electrochemical Society, 1* (1902), 119–124, discussion 124–125.

Arrhenius' theory. Be that as it may, the members were certainly by no means completely stampeded by it; they were, in general, content to make use of it as a working hypothesis as far as this could be done."[202]

Kahlenberg's account agrees with the record in the *Transactions* of the society. The air of caution about the dissociation theory at the society meetings reflected the hostility to the theory of several of the organizers of the society. Kahlenberg was one of the six vice-presidents elected at the first meeting. Another opponent of the theory was C. J. Reed, an electrochemical engineer who was considered the "prime mover in the organisation of the society"[203] and who was the first secretary of the society.[204] The first president of the society, J. W. Richards, was unenthusiastic.[205]

The most useful record of opinion in the society about the electrolytic dissociation theory was the general discussion on the topic held in September 1903. One of the introductory papers was by W. D. Bancroft, a graduate of the Ostwald laboratory and founder of the *Journal of Physical Chemistry*. After briefly describing the dissociation theory and its applications, he gave the following description of its status at the time:

> The electrolytic dissociation theory has carried the chemical world by storm. It is not too much to say that those who have never accepted the theory no longer exert an effective opposition. Every day, however, it becomes clearer that the early adherents of the theory are now working along two divergent lines. One group of men has been interested in increasing the number of facts to which the theory applies. These men have been very successful. . . . Other men, prominent among whom is Kahlenberg, have felt that the interesting things about a theory are its limitations. They have also been successful, and we now know a great deal about the shortcomings of the electrolytic dissociation theory. . . . The final result is that the electrolytic dissociation theory applies only

[202] L. Kahlenberg, *Transactions of the American Electrochemical Society, 51* (1928 [for 1927]), 42.

[203] *Transactions of the American Electrochemical Society, 1* (1902), 8.

[204] See, for example, C. J. Reed, *Transactions of the American Electrochemical Society, 3* (1903), 278; *4* (1903), 177–182; *5* (1904), 142.

[205] See, for example, J. W. Richards, *Transactions of the American Electrochemical Society, 5* (1904), 40–41.

to infinitely dilute solutions, which is much the same as excluding the whole of chemistry.[206]

Many of the members who contributed to the main discussion, including a number of Leipzig graduates, agreed that although the theory had been very useful as a working hypothesis, there were many objections to it, some of which were serious. The Canadian W. Lash Miller, who had studied under Ostwald, expressed a slightly different attitude: after reporting some experimental results which were apparently contrary to the theory, he gained laughter and applause when he admitted that one reason for not abandoning the dissociation theory was that it was very handy in giving lectures; so he looked for an explanation compatible with the theory.[207]

Within the American Electrochemical Society, therefore, there were a number of scientists who, like Kahlenberg, perceived serious problems for the dissociation theory. However, in the absence of a better theory they were willing to continue to use it as a basis for research, even though its use was beginning to lead to artificial complications just as theories of cycles and epicycles had become increasingly complicated in the history of astronomy.[208] Some of the participants expected the present theory to be succeeded by a revolutionary alternative.[209]

Kahlenberg saw the dissociation theory as being in an irremediable state of crisis and wished to reject it completely. He had come to think that it was corrupting the path of physical chemistry.

Further Experimental Work by Kahlenberg; the Debate between Fernekes and Smith

As an alternative to the Ostwald school theory Kahlenberg favored explanatory mechanisms which give an active chemical role to the solvent. He developed these ideas in a paper published in 1903.[210]

[206]W. D. Bancroft, *Transactions of the American Electrochemical Society, 4* (1903), 175–176.

[207]W. Lash Miller, *Transactions of the American Electrochemical Society, 4* (1903), 185.

[208]The analogy with astronomy was raised by H. S. Carhart, *Transactions of the American Electrochemical Society, 4* (1903), 186–187. Much of the language of the discussion invites the application of T. S. Kuhn's term "crisis."

[209]*Ibid.*, p. 187.

[210]L, Kahlenberg, *Journal of the American Chemical Society, 25* (1903), 380–392.

He measured the rate of interaction between magnesium and various aqueous solutions. He found considerable variation in the rate of evolution of hydrogen and claimed that that result supported the idea that the solution is the product of chemical interaction between solvent and solute. The dissociation theory, he argued, is unable to explain his results, and his alternative chemical conception will be a valuable aid to further research, particularly as it does not focus attention on very dilute solutions.

Kahlenberg's ideas were employed in an extended exchange between one of his students, G. Fernekes, and an advocate of the dissociation theory, G. M. Smith. Fernekes had made an experimental study of the reactivity of sodium and potassium amalgams in various aqueous solutions.[211] In another experimental context, Arrhenius had once suggested that ions of sodium or potassium in solution would hinder the entrance of more of these metals from an amalgam into the solution.[212] Consequently a solution of potassium chloride, for example, should slow the reaction of potassium amalgam, but not of sodium amalgam. In his experimental work Fernekes found that the potassium salt slows the reaction of both potassium and sodium amalgams with the solution. He presented this result as evidence against the dissociation theory and in favor of Kahlenberg's idea that the process of dissolution involves a chemical reaction between solvent and the dissolving substance. The nearly equal retardation of the reaction of both amalgams by sodium and potassium chlorides could be explained if the reaction rate depended on the availability of free water.

Smith replied,[213] arguing that Fernekes' result *is* compatible with the ionic theory. He explained that when sodium amalgam is placed in potassium chloride solution, potassium exchanges with the sodium in the amalgam (where it can be detected experimentally) until the amalgam and the solution contain both elements. As a result of such an exchange, both types of cations retard the reaction of either metal with the water. Fernekes' response was to introduce the case of barium amalgam which reacts three times as strongly with pure water as with potassium chloride solution, although no potassium can be detected in the amalgam.[214] Smith replied by

[211] G. Fernekes, *Journal of Physical Chemistry,* 7 (1903), 611-639.

[212] S. A. Arrhenius, *Zeitschrift für physikalische Chemie, 11* (1893), 805.

[213] G. McP. Smith, *Journal of Physical Chemistry, 8* (1904), 208-213.

[214] G. Fernekes, *Journal of Physical Chemistry, 8* (1904), 566-570.

describing further experiments in which an ion exchange took place in the amalgam. He concluded: "While all the phenomena of solutions may not be capable of explanation by the ionic theory, it is nevertheless very rash to claim that it has been shown 'conclusively' that this theory has outlived its usefulness."[215]

The debate between Fernekes and Smith might be said to illustrate the point that, when each side is supported by enthusiasts, there is virtually no end to the moves that can be made to avoid accepting the other's case as conclusive.

In another experimental study of 1904, Kahlenberg argued that electrode potentials depend not merely on the metal involved, but also on the solvent and any solute in the solution.[216] Bancroft agreed in the subsequent discussion that Kahlenberg's evidence cast doubt on the assumptions underlying the use of the van't Hoff-Raoult formula.[217]

The Discussion of Kahlenberg's Work in Britain, 1904–1905

At the end of 1904 Kahlenberg read a general paper to the recently formed Faraday Society in Britain.[218] In it Kahlenberg most clearly summarized his arguments against the theories of dissociation and solution. The paper was widely discussed: most chemists who remarked on it felt that Kahlenberg had gone too far in his rejection of the basic theories of physical chemistry. Many shared the view that, although supporters of the dissociation theory had made many claims that Kahlenberg was right to criticize, these claims were not essential to the theory. Both the extreme dissociationists and Kahlenberg were wrong to emphasize these claims.

> The objections raised [by Kahlenberg] will, I think, serve as a salutary corrective to the tendency which to my mind pervades contemporary chemical thought to exaggerate the scope of the ionic theory and to use facts which neither contradict nor support it as arguments in its favor.[219]
>
> The author [Kahlenberg] seemed to attack the dissociation

[215] G. McP. Smith, *Journal of Physical Chemistry, 9* (1905), 35.

[216] L. Kahlenberg, *Transactions of the American Electrochemical Society, 6* (1904), 53–65; discussion pp. 65–66.

[217] W. D. Bancroft, *ibid.*, p. 65.

[218] L. Kahlenberg, *Transactions of the Faraday Society, 1* (1905), 42–53; discussion pp. 53–64.

[219] H. J. Sand, *ibid.*, p. 58.

theory as if everything that had been written in favor of it formed a chapter of an orthodox Bible. The previous speakers had already shown that Dr Kahlenberg criticised many views which were not really held by the majority of modern electrochemists, and as regards the physiological action of electrolytes, Nernst differed from Arrhenius.[220]

As in the earlier discussion at the American Electrochemical Society, Kahlenberg and his audience diverged on the methodological question of whether or not a theory with as many problems as the theory of electrolyte dissociation was better than no theory at all. R. Abegg argued in a communication: "If one wishes to overthrow any successful theory, it is not sufficient for the progress of science to point out its difficulties–in that case not one theory in science would be valid!–but it is necessary to find another theory likely to explain not merely the difficulties of the old one but also the many phenomena and observations already embraced by it, a procedure a classical example of which you see in the history of the theories of light."[221] Kahlenberg replied that he was not hankering for more theories. "The pathway of the progress of science is strewn with defunct theories, and not always has a new theory replaced an old one. I would not depreciate the value of a working hypothesis, but I would also not minimise the power it has to lead the biased investigator astray."[222]

The paper Kahlenberg presented to the Faraday Society attracted the critical attention of H. C. Jones, an enthusiastic American supporter of the Ostwaldian theory. Kahlenberg had quoted work done by Jones and his students as evidence against the Arrhenius theory. Jones was annoyed, for he had presented the results as a natural development of the Ostwaldian program. He concluded his reply to Kahlenberg by quoting an earlier comment: "These substances, instead of presenting any exception to the theory of electrolytic dissociation fall directly in line with the theory, as so many similar cases have done. Indeed, it is interesting to see how many exceptions to the theory of electrolytic dissociation disappear as experimental methods become more refined and experimental work more accurately carried out. The amount of evidence for the

[220] H. Borns, *ibid.*, p. 60.
[221] R. Abegg, *ibid.*, p. 57.
[222] L. Kahlenberg, *ibid.*, p. 64.

general correctness of this most fruitful generalisation is at present so large that any apparent exception will be accepted only after it has been very thoroughly substantiated by repeated experiments."[223] Kahlenberg insisted that it was Jones's data not his interpretation that supported the case against the electrolyte dissociation theory, and he cited criticisms that pointed to the arbitrariness of Jones's method of relating his results to the Ostwaldian theory.[224]

Kahlenberg on Osmotic Pressure

In his paper read to the Faraday Society, Kahlenberg had mentioned preliminary results he had obtained in a study of osmotic pressure. This work was published in 1906.[225] His main point of attack was the analogy between gases and dilute solutions. Kahlenberg argued that semipermeable membranes act by selective solubility and that the thermodynamical idea of a passive semipermeable membrane was a fiction. Membranes can only become more nearly semipermeable if their selective action through differential solubility becomes more pronounced. Kahlenberg presented a model of the action of such a membrane: he described experiments in which three layers of fluid, chloroform, water, and ether, are contained in a vessel in such a way that chloroform is underlying water and water underlying ether. As ether is soluble in water it can pass through into the chloroform layer to form a solution of chloroform and ether. But chloroform is almost insoluble in water and so must remain in the lowest layer. Thus the water acts as a semipermeable layer. In the course of the experiment the ether layer disappears, most of it passing through into the lowest layer. Kahlenberg claimed that the model was adequate to explain the qualitative and quantitative results of studies with semipermeable membranes. Just as no liquid is completely insoluble in another, Kahlenberg argued, no membrane is completely semipermeable. Consequently the experimenter will not obtain equilibrium until the separated liquids are of the same composition. It was not customary to stir

[223] H. C. Jones, *Philosophical Magazine* [6], *10* (1905), 157. The comment was originally made in *American Chemical Journal, 27* (1902), 22.

[224] L. Kahlenberg, *Philosophical Magazine* [6], *10* (1905), 662-664. The critic of Jones's method was J. J. van Laar, *Chemisch Weekblad, 2* (1905), 1-16.

[225] L. Kahlenberg, *Journal of Physical Chemistry, 10* (1906), 141-209.

the liquids in measurements of osmotic pressure, which meant that it took some time to reach a point of apparent equilibrium across a semipermeable membrane. In this time, leakage due to incompleteness of semipermeability could substantially affect the measurement. Kahlenberg proposed that, as in solubility measurements, the separated liquids should always be stirred. Then one would obtain maximum osmotic pressure more quickly and diminish the effects of leakage. Kahlenberg presented the new sets of measurements obtained in this way as evidence of the inadequacy of van't Hoff's law, even as an idealization. "To speak of the osmotic pressure of any isolated solution without specifying what membrane separates it from what other liquid is nonsense, in the light of the facts here presented. And further, to assume that solutes are polymerized or dissociated in dilute solutions because the osmotic pressures developed by the latter in given cases happen to deviate from values computed from the gas laws is evidently equally unjustifiable practice."[226]

There was considerable interest in the mechanism of osmosis at the time, and Kahlenberg's paper attracted some attention. An early forum for discussion was the correspondence columns of *Nature,* which had published an abstract of Kahlenberg's paper.[227] The Earl of Berkeley and E. G. J. Hartley objected to Kahlenberg's experimental methods, and Whetham disagreed with Kahlenberg over the theoretical significance of his work.[228] Whetham told Armstrong, who also joined the exchange, that he had an impartial attitude toward the dissociation theory:[229] "I hold no brief for the [electrolytic dissociation] theory, as Professor Armstrong seems to imagine, and if it ever ceases to be the best hypothesis in the field, I shall willingly abandon it. Consistency always seems to me to be the meanest of virtues, and in matters scientific it may become the most deadly of vices."[230]

But even with such a Popperian attitude, Whetham considered that Kahlenberg's paper carried little weight against the theory of solu-

226*Ibid.,* pp. 207–208.

227L. Kahlenberg, *Nature, 74* (1906), 19.

228The Earl of Berkeley and E. G. J. Hartley, *Nature, 74* (1906), 54–55, 245. W. C. D. Whetham, *ibid.,* pp. 54, 102–103, 195–296. L. Kahlenberg, *ibid.,* p. 222.

229H. E. Armstrong, *Nature, 74* (1906), 79.

230W. C. D. Whetham, *Nature, 74* (1906), 103.

tions or the electrolyte dissociation theory. Whetham distinguished between the ideal concept of osmotic pressure employed in thermodynamics and the experimentally determined osmotic pressure. Whatever Kahlenberg was able to show experimentally could not affect the thermodynamic proof. "Defining osmotic pressure as the hydrostatic pressure needed to keep a solution in equilibrium with its solvent across an ideally perfect semipermeable membrane, we obtain a conception, possibly of less chemical and physiological importance, which nevertheless enables us to develop a thermodynamic theory of solution; and this theory has been verified experimentally in cases where we have reason to suppose that the actual conditions approach the ideal."[231]

Kahlenberg initially accepted Whetham's distinction between the two concepts of osmotic pressure, commenting only that the ideal thermodynamic concept has no counterpart in reality and ought to be relabelled in a way that does not involve the term "osmotic."[232] But by 1909 he reacted more strongly. In a reply to criticisms by the Dutch physical chemists E. Cohen and J. W. Commelin[233] he wrote:

> It is a well-known dodge of the thermodynamicists to claim that they are not concerned with the mechanism of osmosis and that which produces osmotic pressure, for they simply require to know the magnitude of the latter in order to proceed with their computations. Right here lies the strength, but also the great weakness of thermodynamic methods. *What we want to know above all things is what causes osmosis and osmotic pressure in order to put us into a position to discuss more intelligently the results of osmotic experiments, and to determine to what extent such results are actually useful in forming a true basis for a tenable theory of solutions.*[234]

Kahlenberg was convinced that the thermodynamic approach to the theory of solutions could be of little value since it claimed to be merely an idealization and did not represent the actual osmotic processes. The approach was also misleading for the physical

[231] W. C. D. Whetham, *Nature, 74* (1906), 295.

[232] L. Kahlenberg, *Transactions of the Faraday Society, 3* (1907), 26.

[233] E. Cohen, *Chemisch Weekblad, 3* (1906), 290; E. Cohen and J. W. Commelin, *Zeitschrift für physikalische Chemie, 64* (1908), 1–52.

[234] L. Kahlenberg, *Journal of Physical Chemistry, 13* (1909), 97–98. Italics in original.

chemist, he argued, as thermodynamicists *are* concerned with measurements of the magnitude of osmotic pressure, and they tend to pay disproportionate attention to those experiments that show approximate agreement with the gas laws.[235]

In 1907 Kahlenberg submitted a paper to the general discussion of osmotic pressure held by the Faraday Society. He added little to his earlier views, and his work did not attract very much discussion. A. Findlay and Whetham commented on the distinction between thermodynamic and experimental concepts of osmotic pressure. Findlay commented on the role each had played in the past and accused Kahlenberg of being the source of the confusion surrounding the concepts. Both Findlay and Whetham tried to make clearer the limitations and the significance of the thermodynamic concept.[236]

Kahlenberg's Later Position

Kahlenberg's views on the Ostwaldian theory did not go through any further significant development. His final position was summarized in an address given in 1909, "The Past and Future of the Study of Solutions."[237] He suggested that the Ostwald school theory should now be rejected even as a working hypothesis, but that the historian should remember at least its enthusiasm. "The pages of the history of chemistry that record this experimental work on dilute solutions will ever maintain their brilliant luster, for they reflect the enthusiastic efforts of scores of active young hands and minds that were urged on by a most inspiring leader, an able teacher and experimenter and a most loveable man—Wilhelm Ostwald. Without him the theories of van't Hoff and Arrhenius would scarcely have gained a foothold."[238] He summarized his previous arguments against the Ostwald school theory and sketched an alternative chemical theory of solution. The physical theory had distracted attention from older chemical theories, which should be returned to. "Before the advent of the physical theories of solutions considerable work was done in ascertaining the chemical relationships that must exist between solvent and solute in order that solution may take place; but during

[235] *Ibid.*, p. 98.

[236] A. Findlay, *Transactions of the Faraday Society, 3* (1907), 32. W. C. D. Whetham, *ibid.*, p. 36.

[237] L. Kahlenberg, *Science, 31* (1910), 41–52.

[238] *Ibid.*, p. 43.

the last two decades this work has been practically discontinued, which is particularly unfortunate. It clearly indicates, however, how our so-called modern conceptions of solutions, which have been pressed upon the scientific public by a species of propagandism that is, and it is hoped will remain, quite unrivalled in the history of chemistry, have really stood in the way of progress."[239] The new path for the study of solutions should be to develop chemical insights into solution and should be experimental rather than mathematical: "In the study of solutions, just as in the study of chemical compounds in the narrower sense of the word, we are continually confronted with discontinuities. Now discontinuities can not be handled by mathematicians at present. . . ."[240]

Occasional remarks in his writings from the later part of his career reveal that Kahlenberg continued his undiminished hostility to the Ostwald school theory; to a decreasing proportion of his students he still assigned experimental work designed to show its limitations. According to a biographer,[241] it so discouraged him that his criticisms of the dissociation theory had not been accepted, that he became dogmatically empirical in his approach, ignoring most developments in physical chemistry. "His distrust of theory drove him into a dogmatic empiricism which severely limited his scope, and led him on occasion to condone in himself and in graduate students mere ignorance in the name of healthy scepticism."[242] His own evaluation of his lack of success is reflected in remarks made in a discussion in 1927. After referring to his studies showing the defects of the Nernst-Thomson rule, which relates the dissociative power of a solvent to its dielectric constant, he said: "These facts are not mentioned in the so-called standard books on physical chemistry, for they do not fit in with the theoretical views there promulgated."[243]

Methodological Discussion of the Issues Raised by Kahlenberg

It is appropriate to review some of the general methodological issues raised by Kahlenberg's career. We have seen that, like Arm-

[239] *Ibid.*, p. 49.

[240] *Ibid.*, p. 51.

[241] N. F. Hall, *Transactions of the Wisconsin Academy, 39* (1949), 89–90.

[242] *Ibid.*, p. 90.

[243] L. Kahlenberg, *Transactions of the American Electrochemical Society, 51* (1927), 557.

strong and Pickering, Kahlenberg was of a polemical temperament. He distrusted mathematical and idealized physical concepts and techniques as they did. However, because he was a rebel from within the Ostwald camp, he presented his case against the Ostwald school theory in a form that differed methodologically from that of the English opposition. His claim was not that the theory was a novel approach which attempted to bypass by illegitimate means the established chemical theories, but rather that the initially plausible theory had outlived its usefulness. "Like other theories founded upon too narrow a basis of induction, it has gradually been outgrown—the facts are too much for it."[244] Unfortunately, as Kahlenberg saw it, the defenders of the Ostwald school theory continued on their old path, while their theory became increasingly incongruous. "To be sure, these investigators and also some others still try to 'harmonize' their results with the theory of electrolytic dissociation, but their efforts at this remind one strongly of the attempts of the ardent advocates of the old phlogistic theory, when they sought to harmonize the fact that bodies are heavier after they are burned, with the hypothesis they wished to uphold at all hazards."[245] Kahlenberg wrote of one such attempt to save the theory: "It is in fact nothing more than to arbitrarily make the behaviour of all solutions conform to a Procrustean bed."[246]

Once he had proved to his own satisfaction that both the theory of electrolyte dissociation and the gaseous theory of solution were not literally true, Kahlenberg treated them as working hypotheses, founded upon partial analogies. Moreover, he argued, they had now become a hindrance in research. He believed it was better to have no theory at all than to keep a misleading one. This attitude set him apart from most of his colleagues.

Much of his work was devoted to criticizing the analogies used by the Ostwald school theorists and to replacing them by ones more natural to a chemist. He insisted that the interpenetration of substances in solution is produced by chemical affinity. In the continuum noted by Thomas Graham from adhesion to solution to chemical action, it is concentrated solutions which show the greatest comparability with chemical action. "It thus becomes evident that in

[244] L. Kahlenberg, *Journal of Physical Chemistry, 5* (1901), 391.

[245] L. Kahlenberg, *Transactions of the American Electrochemical Society, 1* (1902), 121.

[246] *Ibid.*, p. 122.

investigating solutions we must begin with the most concentrated and end with the most dilute; the latter will appear simply as a limiting case."[247] In such a way Kahlenberg built up arguments to show that it was most productive to study concentrated rather than dilute solutions.

Again, Kahlenberg found the analogies of the Ostwaldian theory to be misleading when they were used to explain the mechanisms of the key processes of solution. He insisted that the process of dissolution, for example, is best viewed as the mutual interaction of solvent and solute. To explain the dissolution of metallic magnesium in aqueous solutions the ionic theories provide a far less reliable guide than a theory in which solutions are "a chemical combination of solvent and solute according to variable proportions."[248] Kahlenberg considered van't Hoff's analogy between gases and dilute solutions especially misleading. It was based only on a limited selection of aqueous solutions at infinite dilution; at higher concentrations, the behavior of solutions deviates from ideal gas behavior in quite a different manner than does the behavior of real gases. As far as the process of osmosis was concerned, Kahlenberg felt that the thermodynamicists' approach exhibited a brutal disregard for the facts. The thermodynamic idealization of osmotic pressure ignored the actual processes except in the rare cases of quantitative agreement between practical and theoretical results.

Kahlenberg attacked van't Hoff's theory of solution primarily to eliminate it as a source of support for Arrhenius' theory. Van't Hoff's theory was not nearly exact enough to be "corrected" by introducing the additional parameter of variable dissociation suggested by Arrhenius. Kahlenberg considered that whatever was of value in Arrhenius' speculations was already contained in Clausius' earlier theory. Once the quantitative basis for Arrhenius' theory had been undermined, chemists could return to the view that electrolyte solutions were like other solutions, and they could seek the mechanism of the passage of electricity in some other way. Kahlenberg believed that electrolytic conduction is probably similar to other types of conduction. Electrolysis requires electrodes as well as a conducting liquid, and it is possible that all the chemical effects

[247] L. Kahlenberg, *Transactions of the Faraday Society, 1* (1905), 50.

[248] L. Kahlenberg, *Journal of the American Chemical Society, 25* (1903), 390.

take place at the surface of the electrode and that concentration changes in the solution are a secondary effect. Typical electrolytes contain both a metal, which conducts, and a nonmetal, which insulates; often, as with the chlorides of tin, compounds with a higher proportion of nonmetal such as stannic chloride insulate, while compounds with a lower proportion such as stannous chloride are electrolytes. Thus electrolytic conduction appeared to Kahlenberg to be part of a natural gradation of properties between metallic conduction and insulation.[249]

Kahlenberg was an experimental scientist who tried to establish the real nature of a limited range of phenomena in which, he believed, mathematical relationships were hidden, if present at all. As the appropriate method of investigation in such circumstances he chose to stay close to the experimental phenomena, using analogies and guiding hypotheses only where they allowed him to draw the results together into an explanation of the basic processes without directing him into unproductive lines of investigation. Thus his opposition to the Ostwald theory came from his belief that it was based on an idealization of the facts rather than the facts themselves and that the analogies it contained conflicted with intuition and began to hamper research as chemists studied a wider range of types of solutions.

7. THE CONTINUATION OF THE DEBATE IN BRITAIN

H. C. Crompton

As professor of chemistry at the Central Technical College, London, Armstrong attracted a small but active group of graduate students. A number of these were set to work on projects designed to support Armstrong's crusade against the Ostwald school theory. One of the first of Armstrong's students was H. C. Crompton. Crompton made an independent theoretical contribution to the debate and hence deserves brief consideration. Unlike the other students, who did routine experimental work for Armstrong, Crompton attempted to develop theoretical and physical arguments for aspects of Armstrong's position at the level of the arguments of the opposing physical chemists. I have already noted Crompton's early application

[249] L. Kahlenberg, *Transactions of the American Electrochemical Society, 13* (1908), 265-272.

of Mendeleef's method of detecting hydration in solution. In 1897, after an inconclusive study of latent heats of solution, Crompton attempted to calculate theoretically the effect of association between the solvent and the solute on van't Hoff's law of osmotic pressure.[250] Associated liquids can be identified by deviations from Trouton's formula, and associated solutes from modifications of the expected properties of the solutions. Crompton argued that, as a general rule, unassociated liquids dissolve unassociated compounds, and associated liquids dissolve associated compounds. The Ostwald school had been misled by the special behavior of water: water, as the electrolytes that dissolve in it, is associated, but its degree of association is reduced by the addition of electrolytes. Crompton developed the subject by quantitative means and reached the conclusion that

> the hypothesis of electrolytic dissociation is entirely unnecessary for the explanation of those exceptions to van't Hoff's law observed in the case of freezing point reductions. Put briefly, the case is this—the exceptions are no exceptions at all, it is the law that is wrong. No account is taken of association in the liquid state, either for the solvent or dissolved substance, as, in fact, Dr. Armstrong and others in this country have so repeatedly pointed out, and it is this that necessitated the introduction by van't Hoff of the by now celebrated coefficient, *i*, for aqueous solutions, upon which the hypothesis of electrolytic dissociation was founded.[251]

Crompton continued with qualitative arguments, repeating a number of Armstrong's arguments.

In the discussion that followed Crompton's presentation of his theoretical results to the Chemical Society,[252] it became clear that Crompton's treatment, even if not seriously in error, failed to provide an alternative to the ionists' account of the facts.

Later that year Crompton again attempted to give theoretical support to Armstrong's view. His attention had been drawn "to the fact that Planck . . . has long since proved that association could have no effect on the osmotic pressure of liquids."[253] Crompton argued that

[250] H. Crompton, "The Theory of Osmotic Pressure and the Hypothesis of Electrolytic Dissociation" *Journal of the Chemical Society, 71* (1897), 925–946.

[251] *Ibid.*, pp. 941–942.

[252] *Proceedings of the Chemical Society, 13* (1897), 112–115.

[253] H. Crompton, *Proceedings of the Chemical Society, 13* (1897), 225. The reference to Planck is to a discussion between Planck and Wiedemann, *Zeitschrift für physikalische Chemie, 2* (1888), 241, 343.

Planck's reasoning was circular: Planck had assumed that there is no change in the association of the liquid on vaporization, and he had neglected the volume change involved in association. However, Crompton's arguments were not well received; it was fairly clear that he was not at all effective at such a level of theoretical argument. Indeed, he abandoned theoretical work, returning to experimental studies of latent and specific heats of gases and liquids, in which he showed the effects of molecular association without developing theoretical interpretations from the experimental results. Crompton's failure in theoretical work is informative about Armstrong's approach: the approach was inadequate at a rigorous, theoretical level of argument beyond chemical analogy and the chemists' intuitive sense of plausibility.

Further Development of Armstrong's Views

In 1900 the physics and chemistry sections of the British Association held a joint discussion on ions.[254] This was an inconclusive affair, in which the familiar protagonists once more outlined their positions. Some also ruminated on the implications for ionization in liquids of the recent discovery of the electron and the growing understanding of ionization in gases. But the new developments raised far too many problems for easy assimilation.[255]

In his later publications[256] Armstrong wrote that he had made an important development of his position after the 1900 British Association discussion. He came to think that he had been paying too much attention to the dissolved substance, and that it must be the structure of water itself which was of critical importance, the dissolved substance merely modifying this structure. In the case of electrolytes, the modification was to produce a greater proportion of electrolytically active components. This idea, a natural development of his arguments since the early 1880's, was presented in his article on chemistry in the 1902 edition of the *Encyclopedia Britannica* (after a reasonably impartial presentation of the electrolyte dissociation theory).[257] Armstrong based his theory of the active role of water on the idea that oxygen could associate with more than two atoms; he thus made possible the building up of the molecular ag-

[254] Although reported by title only in the *Report of the British Association* for 1900, a brief account of the meeting was given in *Nature, 62* (1900), 564.

[255] *Ibid.*

[256] H. E. Armstrong, *Proceedings of the Royal Society, A81* (1908), 80.

[257] H. E. Armstrong, *Encyclopaedia Britannica,* 10th ed. (Edinburgh, 1902), second of the new volumes, pp. 736–741.

gregates in which he had believed for so long. He suggested that some aggregates might have their radicles so arranged as to be electrolytically active, while others did not; in no case were the radicles free. He elaborated the theory to its greatest extent in 1908.[258]

T. M. Lowry

Another of Armstrong's students to become involved in the discussion of the theory of solutions around the beginning of the twentieth century was T. M. Lowry. Lowry moved to physical chemistry from organic chemistry and because of his limited mathematical knowledge worked at experimental topics. Like Armstrong, Lowry regularly attended and addressed scientific meetings. But unlike Armstrong he was not a controversialist and sought to resolve controversies by finding descriptions that revealed common ground between opposing theories.[259] His tendency towards reconciliation is revealed in one of his early contributions to the theory of solutions. In a development of Whetham's compromise between ionic dissociation and hydration, Lowry suggested that not merely were ions hydrated, but that the ions owed their existence to the added hydration accompanying dissociation. Lowry later gave the following account of the origin of his idea and of its place in the debate.

> The writer can claim to look upon this controversy with some measure of personal interest since, while still at school he was first attracted to the study of chemistry by the fascination of the theory of electrolytic dissociation, and in particular of its dramatic correlation of the conductivity of an electrolyte with the freezing point of the solution. Soon afterwards, as a student of Professor Armstrong–then, as now, a consistent critic of the "new" theory of solutions,–he was faced with the necessity of finding for his own use a scheme which should be compatible with the two rival points of view. This he found in the idea that *both* views were correct, that their incompatibility was imaginary and not real, and that the hydration of the ions not only provided a way of reconciling the two theories of solution, but also supplied a motive for the electrolytic dissociation of a salt which (as Armstrong had

[258]H. E. Armstrong, "Hydrolysis: Hydrolation and Hydronation as Determinants of the Properties of Aqueous Solutions," *Proceedings of the Royal Society, A81* (1908), 80–95.

[259]C. B. Allsop and W. A. Waters, *British Chemists,* eds. A. Findlay and W. H. Mills (London, 1947), p. 407.

pointed out) was conspicuously absent from the original "naked" theory of electrolytic dissociation.[260]

When Lowry presented his paper to the Faraday Society,[261] he learned in the subsequent discussion that parallel developments had already taken place on the continent.[262] Nevertheless, Lowry's work and his search for compromise continued to play an important part in the papers and discussions of the Faraday Society.

The Nature of Osmotic Pressure 1887–1907

By the middle of the first decade of the twentieth century, the Faraday Society had become the most important British forum for the discussions and controversies over the theory of solutions. "General discussions" were to become an enduring feature of the society, and the first two of these recorded in the *Transactions* were on "Osmotic Pressure" and "Hydrates in Solution"—both topics that would have invited further critical scrutiny of the Ostwald school theory.

Until about 1907 the discussions of the nature of osmotic pressure had been marked by considerable confusion. The Faraday Society meeting may be considered the occasion at which the British reached a common understanding of what had and had not been established.

A major source of the continuing misunderstandings was van't Hoff's original paper of 1887.[263] His statements had allowed his readers to attribute to him contradictory accounts of the mechanism of osmotic pressure. In one passage he wrote that given a vessel, *A*, containing a solution separated from its solvent by a semipermeable membrane, "it is known that the attraction of the solution for water will cause water to enter into *A*. . . ."[264] A page later, van't

[260] T. M. Lowry, in T. M. Lowry and J. Russell, *The Scientific Work of the Late Spencer Pickering F.R.S.* (London, 1927), p. 32.

[261] T. M. Lowry, *Transactions of the Faraday Society, 1* (1905), 197–206; discussion 206–214.

[262] The reference was to Werner, *Zeitschrift für anorganische Chemie, 3* (1893), 294.

[263] J. H. van't Hoff, *Zeitschrift für physikalische Chemie, 1* (1887), 481–508. Later references are to the translation in *Alembic Club Reprints,* No. 19 (Edinburgh, 1961).

[264] *Ibid.,* p. 6. One Ostwaldian physical chemist who took this sentence as indicating van't Hoff's own understanding of the source of osmotic pressure was A. Findlay. See, for example, *Osmotic Pressure* (London, 1913), p. 69 and footnote.

Hoff wrote of the analogy between gases and solutions: "We wish to emphasise in this connection that we are not here dealing with a fanciful analogy, but with one which is fundamental; for the mechanism which according to our present conceptions produces gaseous pressure, and in solutions osmotic pressure, is essentially the same. In the first case it is due to the impacts of the gas-molecules on the containing walls, in the second to the impacts of the dissolved molecules on the semipermeable membrane. The molecules of the solvent present on both sides of the membrane, since they pass freely through it, need not be taken into consideration."[265]

Van't Hoff's memoir contains another source of confusion for its readers, for a literal interpretation of the passage just quoted gives the wrong sign to the osmotic pressure. Van't Hoff seemed to imply that additional pressure would have to be applied to the solvent side rather than the solution side of the membrane to produce equilibrium.[266] Eventually chemists agreed that the problem is trivial. Gases behave just as solutions do, as was shown experimentally when a container with a palladium septum (which is only permeable to hydrogen) was filled with hydrogen on one side and a mixture of gases on the other. The hydrogen flowed through the septum to reach the same partial pressure on each side. At equilibrium, the other gases present on one side created an excess hydrostatic pressure on that side of the septum.[267] Analogously, the solvent attempted to flow through a semipermeable membrane into the solution. Unless a hydrostatic pressure is applied to the solution, equilibrium can only be reached at infinite dilution. Thus, although the magnitude of the osmotic pressure is proportional to the concentration of the solute, it is actually the external pressure required to stop the entry of the solvent into the solution. Any chemical action between solvent and solute increases the osmotic pressure, but that effect can be neglected at the very low concentrations to which van't Hoff's analogy of gases and liquids applied. Neither of the two mechanisms that van't Hoff mentioned were essential to his

[265] J. H. van't Hoff, *op. cit.* (note 263), p. 7. Kahlenberg considered this account of the mechanism of osmosis sufficiently authoritative to believe that an attack on the analogy between gases and dilute solutions was an attack on the whole theory of solutions.

[266] This point is made, for example, by J. G. Rhodin, *Transactions of the Faraday Society, 3* (1907), 81-85.

[267] W. Ramsay, *Philosophical Magazine* [5], *38* (1894), 206-218.

theory. Indeed, when giving a theoretical proof that Boyle's law applies to dilute solutions, van't Hoff treated the two mechanisms as alternatives with equal status. "If we consider osmotic pressure to be of kinetic origin, that is as arising from the impacts of dissolved molecules, we have to prove proportionality between the number of impacts in unit time and the number of impinging molecules in unit volume. . . . On the other hand if we see in osmotic pressure the effect of an attraction for water, its magnitude is obviously proportional to the number of attracting molecules in unit volume, with the proviso (which is fulfilled in sufficiently dilute solutions) that the dissolved molecules are without action on one another, and that each contributes on its own account a constant amount to this attractive action."[268]

Another matter that had further obscured the discussion of osmotic pressure was an early confusion over the appropriate units to employ in the calculations. In his original theory van't Hoff had assumed that the relevant volume in the equation $PV = iRT$, when applied to solutions, is the volume of the *solution.* But chemists found that they obtain better agreement when they use the volume of the *solvent* present in the solution—that is, if they calculated concentrations in gram-molecules of solute per 1000 gm of water rather than per 1000 cc of solution.[269] They then suggested theoretical reasons why this should be a better basis for calculation. Armstrong was among those who used the problem as evidence of the incompetence of the ionists.[270]

The main English speaking opponents of the Ostwaldian theory employed conflicting conceptions of the nature of osmotic pressure and of the action of the semipermeable membrane in their arguments. It has already been pointed out that in 1891 Pickering had claimed that the very existence of osmotic pressure is an argument against the electrolyte dissociation theory. His belief that the membrane allowed only the smallest molecules to pass allowed him to infer that solute molecules must be highly hydrated in solution. Kahlenberg believed that the semipermeable membrane acted by

[268]J. H. van't Hoff, *op. cit.* (note 263), p. 10. See also van't Hoff's comment in the *Report of the British Association* for 1890 (1891), p. 336, where he insisted that the kinetic interpretation of osmotic pressure was more intended to popularize than to prove the laws in question.

[269]H. N. Morse, *American Chemical Journal, 38* (1907), 175.

[270]See, for example, H. E. Armstrong, *Science Progress, 3* (1909), 651.

selective solubility, and that demonstration of this mechanism was an argument against van't Hoff's theory of solutions. Finally, Armstrong attributed osmotic effects to the action of the solute on the degree of association of the solvent, interpreting unexpectedly high osmotic pressure as evidence of a great effect on the depolymerization of water, not as evidence of dissociation of electrolytes. Thus *his* views on the origin of osmotic pressure also constituted an argument against the Ostwald school theory. As van't Hoff's original memoir allowed the Ostwaldians to claim different mechanisms of osmotic pressure as orthodox, chemists were not likely to reach an immediate consensus on the matter. By the time of the debate on osmotic pressure at the Faraday Society, scientists were generally aware of a great number of more or less acceptable possible mechanisms for osmotic pressure. None of these they could at that stage accept as basis for a convincing account of other features of electrolytes and solutions.[271]

By 1907 chemists had also understood—especially from the exchange between Kahlenberg and Whetham—that the thermodynamic treatment of solutions did not require any assumptions about a causal mechanism or the existence of a real example of a perfect semipermeable membrane.

We may see the discussion at the Faraday Society meeting of January 1907 as a record of reorientation: the main focus of the debate over the theory of solutions passed from the direct evaluation of van't Hoff's theory to new issues in which the opposing parties were drawn up in a new way. Some of the participants still expressed opposition to the Ostwald school theory, for they were not all of the same intellectual lineage, but such attitudes were no longer the main point of interest.

The Faraday Society Discussion of Hydrates in Solution, 1907

The second general discussion held by the Faraday Society in 1907 was that on hydrates in solution. It took place on 25 June with Pickering as chairman. Because Pickering took part in the discussion, the confrontation of the Ostwald school theory and the hydrate theory continued, but that was no longer the main object of the discussion. British chemists who were sympathetic to the theory of

[271] A good idea of the openness of the issue is given a little later by A. Findlay in the last chapter of his monograph, *Osmotic Pressure* (London, 1913).

electrolytic dissociation now almost invariably acknowledged the importance of the hydration of ions. The main paper of the discussion, presented by W. R. Bousefield and T. M. Lowry,[272] dealt with the idea that it is the increase in hydration that stabilizes dissociated ions in solution. When Pickering resurrected his thermochemical arguments against the ionization theory, Bousefield dismissed his objection: "The criticism of the Chairman missed the point of the paper, which was not to prove the ionization theory, but, accepting the ionization theory, to show how one of its difficulties was removed."[273] Pickering also objected that the method by which Bousefield and Lowry had calculated the heats of formation of individual ions did not allow them to use their results as an argument for the independent existence of the ions in solution. They had only shown, he argued, that the heat of neutralization of acids and bases is constant, and they had used that result in the initial calculations. Pickering assumed that he was opposing an argument for the ionization theory and had failed to find any. But if one considers the calculation merely as an elaboration of the conceptual apparatus of the ionization theory, Pickering's criticism of circularity in the reasoning is not justified. In fact, Lowry claimed in his reply that he and Bousefield had shown that the additivity of the heats of formation holds for a wider range of electrolytes than strong acids and bases, so that they *had* gained further support for Arrhenius' theory.[274]

The participants at the meeting gave serious attention to Armstrong's idea that the properties of solutions should all be understood in terms of changes in *association.* G. Senter read a paper in which he criticized Armstrong's theory for its lack of quantitative explanations and for its failure in two cases to provide even correct qualitative explanations.[275] The participants of the meeting based the discussion on their concern to assimilate the known phenomena of hydration without having to abandon the successful explanations of the Ostwald school theory. British physical chemists of the early twentieth century generally knew more mathematics than those of two decades earlier, and they were more concerned to find quanti-

[272]W. R. Bousefield and T. M. Lowry, "The Thermochemistry of Electrolytes in Relation to the Hydrate Theory of Ionisation," *Transactions of the Faraday Society, 3* (1907), 123–139.

[273]*Ibid.*, p. 161.

[274]*Ibid.*, p. 162.

[275]G. Senter, *Transactions of the Faraday Society, 3* (1907), 146–152.

tative explanations for their experimental observations. With such a requirement, the Ostwald school theory had no serious rival; the only satisfactory way of treating hydration effects was to start from that theory. Armstrong's warning against the distorting effects of undue emphasis on the quantitative aspects of theory carried decreasing weight with physical chemists. Although the chemical phenomena on which Armstrong had based his analogies were studied further, chemists were not interested in returning to his theoretical position. By the second decade of the twentieth century, the dissociation theory was in a state of rapid development anyway. Physical chemists made a succession of attempts to calculate the effects of ionic interaction, and they recognized the advantage of treating strong electrolytes as completely dissociated. Although Armstrong continued his opposition, there was no longer any real debate but rather a feeling that Armstrong was out of touch with the main issues of the field.

Methodological Discussion of the Issues Raised by Pickering's Career

Pickering's remarks in the discussion at the Faraday Society in 1907 were his last direct contribution to the debate on the theory of solutions. They show how out of touch he had become with the subject. His experimental work on solutions had ceased in the mid 1890's; an accident forced a change of career and Pickering devoted himself to fruitgrowing and horticultural research.[276] There is no sign that Pickering recanted his views on the theory of solution. His contribution to the 1907 Faraday Society is that of a dedicated opponent of the Ostwald school theory. His last theoretical paper (on residual affinity)[277] did not contain an explicit discussion of the Ostwald school theory, but in the course of an explanation of heats of neutralization in terms of affinity he claimed that one thermochemical effect that his explanation could account for could not be accounted for by the dissociation hypothesis.[278]

Pickering's career reminds us that there are several ways in which a scientific idea can come to dominate an area of science. Its proponents may succeed in converting its opponents to the new idea but

[276] The factors involved in the change are discussed in Pickering's biography, T. M. Lowry and J. Russell, *op. cit.* (note 48).

[277] P. S. U. Pickering, *Proceedings of the Royal Society, A93* (1917), 533–549.

[278] *Ibid.*, pp. 547–548.

they may also come to dominate the field if their opponents withdraw without changing their views. The appearance of greater unanimity is thus restored to the field by default.

The history of Pickering's opposition to the Ostwald school theory reflects the difference in methodological attitudes between the Ostwald school and its British opposition on the use of quantitative methods in chemistry. In many ways, the Ostwald school was positivistically inclined: for its members, even approximate mathematical laws were a triumph for science. They held that science should build upon the quantitative laws that were first established in limited domains. But for Pickering, as for Armstrong, the wide-ranging experience of the chemist provided a far better basis for a full understanding of the phenomena of solution than the hypotheses erected on approximate regularities holding over limited domains. He formulated this methodological issue quite clearly in 1890:

> No one can doubt the mathematical correctness of the conclusions which Arrhenius, van't Hoff, Ostwald and others draw from the premises with which they start in their arguments respecting osmotic pressure, nor can we doubt the value of connecting numerous actions with one and the same cause, or that there are a large number of instances in which the observed facts are in substantial agreement with their conclusions. But we may, I think, legitimately doubt whether the premises of the argument are sound, whether the conclusions harmonize as well as they should with the experimental data, whether the theory is more than a mathematical exercise, or more than a convenient working hypothesis of a rough character, instead of being, as its supporters maintain, an hypothesis established so firmly that we may build upon it a physical theory of solution.[279]

A few months later Ramsay responded to Pickering's criticism of the dissociation theory.

> Professor Ramsay agreed with Mr. Pickering that many difficulties still remained to be solved before accepting the theories of Arrhenius and van't Hoff in their entirety. He thought, however, that a theory whatever its nature, could not be regarded as an absolute explanation of phenomena, but merely as a mental picture, whereby phenomena familiar to our senses could be con-

[279] P. S. U. Pickering, *Philosophical Magazine* [5], *29* (1890), 490.

ceived of as analogous to those which do not directly appeal to our senses. It might well be possible that the analogy was a defective one; we "explain" many phenomena by the "atomic theory" but have a very limited conception of an "atom"; and the analogy between dilute solutions and gases must also be accepted as merely provisional—as a means of connecting together a great number of phenomena which would otherwise remain isolated facts. By all means let us draw attention to seeming discrepancies; to explain them may involve modification of the theory; but till we have a better one, let us accept one which correlates a large number of phenomena which have not otherwise been united under any other scheme.[280]

In retrospect, it can be understood why Pickering remained unconvinced by the arguments of the Ostwald school, and it can also be seen why his ideas persuaded so few people. His arguments were not as muddled as they were sometimes said to be. But the telling arguments were always presented side by side with misunderstandings and personal idiosyncracies which enabled his opponents to sidestep the difficult questions and answer only those that suited them. His opponents were probably not conscious of this feature of their response, for it is much easier to comment only on those features of a criticism that one is prepared to answer. It is really only in sympathetic historical retrospect that Pickering's work can be fully appreciated.

H. E. Armstrong's Later Research Program

By 1906 Armstrong had set up a systematic research program on the theory of solutions, involving many of his students. The program had three main lines, yielding three series of papers.[281] In his research papers Armstrong was less concerned to show that the dissociation theory was unsatisfactory (though he liked it as little as ever) than to prove that his alternative ideas could effectively guide research. In referring to the dissociation hypothesis in one paper in 1906 he wrote: "It is neither desirable to dwell on the inherent improbability of the conception nor to enter into any discussion of the hypothesis, beyond saying that it is difficult to discover any argument of which it is the unavoidable consequence among the reasons

[280] W. Ramsay, *Proceedings of the Chemical Society, 6* (1890), 172.

[281] A bibliography of these papers is contained in J. V. Eyre, *Henry Edward Armstrong* (London, 1958).

put forward in support of its acceptance, as these are inconclusive when not based on uncertain premises; my object is to consider an alternative explanation."[282]

One of the most fruitful topics Armstrong found for the development of his alternative conception was hydrolysis. The dissociationists had suggested that hydrolysis is due to the presence of the free hydrogen ion and will occur at a rate dependent on its concentration. Some of the papers in the studies of the Armstrong school on enzymes were related to the work on hydrolysis. The hydrolytic actions of enzymes are very specific. Their action can only be explained, Armstrong argued, by assuming that the key stage of hydrolysis is an association of the substance hydrolyzed, the enzyme, and water. Hydrolysis certainly cannot be explained by assuming that the same active substance (the hydrogen ion) is liberated by all hydrolytic agents. In a long series of papers, "Processes Operative in Solutions," the arguments about the specificity of hydrolysis reactions were extended to inorganic substances. For example, some acids have an increased hydrolytic activity in the presence of their metallic salts, although by the dissociation theory this should depress the formation of the hydrogen ion.

In his series of papers, "The Origin of Osmotic Effects," Armstrong and his students developed a similar case for osmotic phenomena. The work extended the idea developed earlier that the main osmotic effect is produced by the solvent, and the solute merely affects its degree of polymerization.

The Armstrong school published papers until about 1915. Armstrong remained as polemical as ever, but his active research in science diminished after his chemistry department was progressively shut down in 1911–1913.

Armstrong's Last Polemical Writings and a Methodological Discussion of the Issues Raised by his Career

Armstrong's later polemical remarks are of special interest to the present historical study because, as he perceived that the debate was not going as he wished, he described its course in sociological terms, using in particular the metaphors of fashion and religious dogma. He is thus a valuable source of insight into the interpenetration of the methodology and the sociology of scientific practice. Of course,

[282] H. E. Armstrong, *Proceedings of the Royal Society, A78* (1906), 264.

Armstrong was far from impartial in his representation of the conflict. He was frustrated by his opponents' failure to follow his conception of rational scientific discourse. This led him to note a social dimension in the historical choice between rival scientific ideas. Although he believed that science should resolve disagreements by open and fair argumentation, he described how the Leipzig school resisted criticism very much as Kuhn describes the maintenance of orthodoxy in normal science.

Armstrong compared the methods by which the new school had come to prominence with the nonrational social processes he considered to be at work in women's fashion and the transmission of religious dogma. His fiery polemics put the case more strongly than could any paraphrase.[283] The metaphor of fashion was used to indicate the degeneration of the discipline.

> After all, we scientific workers (or should it not rather be said we workers in science? Because, although evil communications corrupt good manners, the work of science has not, as a necessary consequence, the establishment in the worker of a scientific habit of mind), like women, are the victims of fashion: at one time we wear dissociated ions, at another electrons, and we are always loath to don rational clothing; some fixed belief we must have manufactured for us: we are high or low church, of this or that degree of nonconformity, according to the school in which we are brought up—but the agnostic is always rare among us and of late the critic has been taboo.[284]

In another context, Armstrong discussed the importance of fashion in the scientific education system. "Of late years, in science, as in ordinary life, fashion has ruled the day and there has been a tendency to adopt extremes. 'Authority', exercised through textbooks and fostered by our examination system, dictated the fashion. Students of chemistry all the world over have been led to profess their belief in the doctrine of electrolytic ionic dissociation, much for the

[283]The most interesting polemical documents of Armstrong's later career are "The Thirst of Salted Water," *Science Progress, 3* (1909), 638-656, a polemical continuation of the ideas set out in simple form in "A Dream of Fair Hydrone," *Science Progress, 3* (1909), 484-499; and "Ionomania in Extremis," *Chemistry and Industry, 14* (1936), 916-917.

[284]H. E. Armstrong, *Science Progress, 3* (1909), 643.

same reason that they have turned up their trousers, not because the practice is rational but because it is conventional. . . ."[285]

Armstrong also described the state of the field by the metaphor of religious dogma. In a polemical letter published late in his life, he made clear his attitude to all forms of dogma. "More than seventy years ago, I not only cast religious dogma aside but also asserted my spiritual freedom when I walked out of church because I could not listen to the nonsense talked from the pulpit. . . . Most so-called physical chemistry is just religious dogma—faith with no works."[286] The decline of rationality in chemistry Armstrong attributed to the replacement of rational discussion by dogma.

> It is difficult to avoid the conclusion that we have offended against all the old canons of practice by which former workers were guided. They disputed, often vigorously and violently—they held the strongest opinions; but as a rule they were careful to balance arguments and to allow arguments to be balanced.
>
> The modern method is not even to present the case of the opponent—the student is not allowed the choice of alternatives, he is rarely, if ever, informed that there are alternatives. Prof. J- simply asserts: "This is the truth; believe it you must and shall." Of such kind has been the Leipzig message from the beginning. The spirit of intolerance is abroad among us. If we are not almost back to the days of the Inquisition, we are at least as dogmatic as are the adherents of any religious persuasion. All this in the name of science and of scientific method, of the discipline upon which we are placing so much hope of future enlightenment of society.[287]

He used the same metaphor of religious dogma to espress one of his main worries—corruption of young scientists by Ostwaldian physical chemists. "All the major channels of communication and most of the minor are secured by the high priests of the cult: they command the almost universal obedience of student youth; and now their technical jargon confronts us everywhere."[288] Any benefits in the productiveness of the Ostwaldian dogma could not outweigh the cost to the discipline. "It will be held by some, perhaps even by many, that even

[285] H. E. Armstrong, "Graham Memorial Lecture," *Science Progress, 6* (1912), 606.

[286] H. E. Armstrong, *Chemistry and Industry, 14* (1936), 917.

[287] H. E. Armstrong, *Science Progress, 3* (1909), 655.

[288] *Ibid.*, p. 656.

if my indictment be true, it matters little nevertheless that a visionary scheme has been advanced—or even that it should have been forced into use for a time. It has inspired workers. But at what expense has victory been gained—if indeed there be true victory of any kind? What nature of example is it that we have set? To what extent are almost all sources of information available to the youthful mind polluted for years to come?"[289] Armstrong was certainly stressing a widely recognized problem. I have already noted a number of scientists who criticized the Ostwald school for omitting reference to their critics in textbook expositions of their theories. But when scientific education is closely tied to the research frontiers, it is perhaps inevitable that the special needs and practices of such teaching dictate the selection of material. The Ostwald school theory was well suited to the needs of the lecturer, and the mainly critical ideas of Armstrong were less suitable for textbook presentation than the positive ideas of the Ostwald school.[290]

Armstrong disapproved of the distinguishing feature of the work of the Ostwald school; namely, the use of mathematical and especially thermodynamic reasoning. He insisted that such standards were inadequate for chemistry.

> As a chemist and a friend of the poor molecules, I feel that the aspersion of immorality should not be allowed to rest upon them

[289]*Ibid.*, p. 655.

[290]This was made especially clear by A. Smithells in a review of J. W. Walker, *Introduction to Physical Chemistry* (London, 1899). After noting that Walker had failed to give the full case against the Ostwald school theory of solution, Smithells went on to point out the difficulty of stating the opposition's case explicitly:

> The theory of ionic dissociation has been applied to explain and co-ordinate a very large number of chemical facts, and has thrown light on matter that was previously dark. The contention of the objectors appears to be mainly that this light is illusory. The present writer is far from claiming judicial functions in the matter; but he ventures to think that the opposition to the dissociation theory would be more respected, both here and on the Continent, if it were of a more positive character, and if a more tangible alternative theory could be presented than the one which is assailed. The history of science shows plainly enough that a comprehensive theory with some weak points will hold its ground until a not less comprehensive theory with fewer weak points makes its appearance. It is probably on this ground that Professor Walker takes his stand in freely imparting the doctrines of electrolytic dissociation to elementary students of physical chemistry.

A. Smithells, *Nature, 62* (1900), 77.

> forever unless the evidence be really condemnatory beyond question. In any case, it is important that we should discover the true nature of the crime committed in solution; to cloak the inquiry by restricting it to thermodynamic reasoning—a favorite manoeuvre of the mathematically minded—is akin to using court influence in abrogation of full and complete investigation; such a course may satisfy the physicist but is repulsive to the chemist, who, although able, perhaps to imagine the existence of a frictionless piston, yet desires, in the first place to get nearer to a knowledge of what happens to the real tangible piston of practice.[291]

By 1909 Armstrong had read a comment by Helmholtz suggesting that chemists following van't Hoff and Ostwald and working on experiments on solutions cannot adequately appreciate thermodynamic laws, which can be grasped in their abstract form only by rigidly trained mathematicians.[292] Armstrong referred to this comment with approval and accepted Helmholtz' implication that Ostwald and his supporters were not physicists. Armstrong went on to say: "The fact which Helmholtz did not sufficiently appreciate was that the men who were taking the liberties he deprecated were not chemists, at least in feeling—that they were men who had thrown chemistry to the winds and were proceeding on hypothetical let-it-be-granted principles. The physico-chemical school, in fact, has never been a school of chemists."[293] By 1936, Armstrong admitted that he was dealing with chemists (though of a special kind); the same criticism took a new form.

> The fact is, there has been a split of chemistry into two schools since the intrusion of the Arrhenic faith, rather it should be said, the addition of a new class of worker into our profession—people without knowledge of the laboratory arts and with sufficient mathematics at their command to be led astray by curvilinear agreements; without the ability to criticise, still less of giving any chemical interpretation.
>
> The fact is, the physical chemists never use their eyes and are most lamentably lacking in chemical culture. It is essential to cast

[291] H. E. Armstrong, *Nature, 74* (1906), 77.

[292] Helmholtz made the comment in a letter in 1891. It was quoted by L. Königsberger in *Hermann von Helmholtz,* 3 vols. (Braunschweig, 1902–1903), which had recently appeared in a slightly abbreviated English translation (Oxford, 1906). See p. 340.

[293] H. E. Armstrong, *Science Progress, 3* (1909), 648.

out from our midst, root and branch, this physical element and return to our laboratories.[294]

The critics of the Ostwald school did not share a unified position. It is of interest, therefore, to quote Armstrong's retrospective view of the relation between his own ideas and those of other opponents of the Ostwald school. Of Pickering he wrote: "My work has so often been referred to in conjunction with that of Pickering on the determination of the composition of hydrates in solution that it is desirable to point out that the questions considered by us were often of a very different character. I always believed in the existence of hydrates but I was in search of something more—of a process, in fact, to account for the reciprocal character of the effect which solute and solvent exercised; one which at the same time would make it possible also to explain the effects produced by non-electrolytes."[295] He had a more positive view of his relationship with FitzGerald. FitzGerald had regularly deferred to Armstrong's chemical judgment, and Armstrong was encouraged by the support of such an eminent physicist and quoted him to lend respectability to his own position. "But what has weighed more than almost any other consideration with me has been the absolute and uncompromising attitude of objection to the hypothesis taken by FitzGerald at British Association meetings and especially in his Helmholtz Memorial Lecture. He alone appeared to me always to understand the situation, and to appreciate the difficulties."[296]

Our study of Armstrong's ideas and arguments is now complete. His position changed very little throughout his long career, though its relationship to the prevailing attitudes of the field was radically transformed. His ideas were presented in positive and negative forms in alternate decades, and his own theory never progressed beyond an increasingly unfashionable kind of qualitative speculation. Thus, his influence on theory was primarily as a stimulant to sharpening the ideas of his opponents. However, he did represent a viable alternative approach to *experimental* physical chemistry. Many of his students were faithful to that approach throughout their careers. There continued to be room for laboratory-oriented non-mathematical physical chemists, and those Armstrong influenced are hard to

[294] H. E. Armstrong, *Chemistry and Industry, 14* (1936), 917.

[295] H. E. Armstrong, *Science Progress, 3* (1909), 652.

[296] *Ibid.*, p. 651.

identify by their publications alone, unless they indulged in chemical speculation or in historical review. But because there is a direct line of intellectual influence from Armstrong to succeeding generations, it is impossible to say that the Ostwald school reached universal dominance in physical chemistry.

8. THE RELATIONSHIP OF THE DEBATE TO LATER DEVELOPMENTS IN PHYSICAL CHEMISTRY

In the early decades of the twentieth century the new specialty of physical chemistry flourished. The theory of solutions, which had been so important in the institutionalization of the field during the 1880's and 1890's, was displaced by new problem areas from its position as the most fashionable interest of the discipline. In part the shift occurred because physical chemists began to exploit twentieth-century developments in physics, and in part it was a natural outcome of the debate. As a result of the intensive study of electrolytes and solutions triggered by the debate, physical chemists had exploited the most productive topics and had either settled or postponed the divisive issues. As the topic of solutions became less fashionable, the unresolved problems lost their urgency. The theory of electrolyte solutions continued to develop, but the changes were due as much to the influence of related developments in physics as to the impetus from research in the problem area itself. Conceptions of atomic and molecular structure were undergoing rapid change in the first decades of the century, leading to new ideas about the relationship between chemical and electrical forces. The concept of a charged ion increasingly became central to chemical theory. In such a context, the development of the theory of electrolyte solutions could only be seen as a natural extension of Arrhenius' original theory, for the original debate had been most strongly polarized on the question of the existence of charged dissociated ions. Although later theories of electrolysis made full use of ions, some of Arrhenius' other original postulates were abandoned. For example, Arrhenius had explained the properties of solutions in terms of an equilibrium between ions and undissociated solute molecules. The later theories treated strong electrolytes as being *completely* dissociated, the interaction between an ion and its "atmosphere" of oppositely charged ions being the primary factor in the deviation of ions from completely independent behavior.[297] Further modifications of the theory took some account of solute-solvent interaction.

Van't Hoff's theory of solution became a subordinate part of the general problem area of the application of thermodynamical methods to chemistry. More powerful mathematical methods (stemming in particular from the initially unappreciated work of J. W. Gibbs) displaced the simple reasoning of van't Hoff and other early popularizers of the field. The analogy between ideal gases and dilute solutions and the early thermodynamic treatment of osmotic processes were inevitably evaluated in terms of the more sophisticated theory. As such van't Hoff's work appeared as a rather useful special case of the general relationships, and the ideas that had helped him to develop his theory and which were not reflected in the fuller treatment, were ignored. Except for those who, like Armstrong, could not follow or appreciate abstract thermodynamical reasoning, physical chemists generally agreed that understanding in the field had progressed from, but had not undermined, van't Hoff's work.

It is easy to conclude, therefore, that the main development of physical science has vindicated the claims of the Ostwald school theory. But although the issues of the debate are no longer alive, our obvious preference for one side is largely to be explained by the historical accident of unexpected discoveries in other fields and by the way physical chemistry became institutionalized, rather than by some logical criterion of which theory best survived exposure to experimental test. The increased importance of ions in chemical theory was a consequence of the new understanding of the atom; few physical chemists played a pioneering role in these developments, perhaps because Ostwald so strongly opposed atomism. Also the rapid rise of the new discipline of physical chemistry had as much to do with the favorable conditions for its growth in Germany and America as it had with the productiveness of its central ideas. The growth of the discipline reflects the larger number of chemists with mathematical and physical training. Moreover, because of their training, such scientists were far more interested in building on quantitative regularities over limited ranges of phenomena than in employing qualitative chemical speculations that linked a rich variety of chemical phenomena but were not amenable to quantitative

[297]For the later history of the theory of strong electrolytes, see H. Wolfenden, *op. cit.* (note 194).

treatment. They would have preferred the Ostwald school theory and its direct successors.

9. CONCLUSION

A General View of the Nature and Significance of the Debates

The manner in which the debates over the theory of solution arose suggests one particular mechanism for generating dissent in science.

Many of the influences encouraging conformity in scientific work are most active during the period of training and the scientist's early career. If there is any geographical variation in conceptions of the problem field of a science, the scientists who have been trained at a particular institution will tend to acquire the local variant. The main influences on such local views do not come uniformly from the work of the whole international scientific community, but mainly from local work or, for peripheral regions, from work done at the centers of scientific excellence that are recognized locally. Under conditions in which international communication is imperfect distinct patterns of scientific influence can emerge or be perpetuated. Thus, if Frenchmen cite Frenchmen most often, while Scandinavians, Germans, and Austrians cite Germans, there is a tendency towards regional stratification, so that the main ideas of a science are carried forward in parallel but quasi-independent channels of influence. Such parallel streams often correspond closely to one another (this is one source of simultaneous independent discovery by people in different countries), but the conceptual, linguistic, political, or geographical barriers to communication also allow significant divergences. So do individual genius, quirk of personality, or local circumstances. For example, distinctive scientific approaches may emerge under favorable institutional conditions, without immediately being diffused internationally. When different local scientific variations do come into contact, the exposure of scientifically significant differences may generate dissent. An analogous situation may occur when scientists working in different but related problem areas lack adequate means of communication.

The debates over the theory of solutions are an example of disagreement between national groups of scientists who developed different solutions for the same problems. In countries in which science was not strongly dominated by German science, chemists developed chemical theories of the nature of solutions. In Germany, organic chemists dominated, and only a few chemists, mostly Bunsen's associates and students, worked on topics in physical chemistry. Arrhenius, van't Hoff, and Ostwald developed the key ideas on solution in countries that took their scientific orientation from German science, but in which chemistry was not as completely dominated by the most fashionable themes of organic chemistry as in Germany. The new ideas on solution were more closely linked to conceptions of theoretical physics than to the chemical theories of solution chemists favored in Britain, France, and Russia (Mendeleef). When Ostwald moved from Riga to Leipzig in 1887, he was able to build up a new school of physical chemistry rapidly because there was so little *active* opposition from German chemists, and because his approach was in harmony with German thermodynamic theory. Ostwald's laboratory dominated the new field and was able to attract a community of students from all nations. Ostwald aggressively sought to establish his new approach (with the help of Arrhenius and other supporters)[298] and by his manner increased the intensity of the confrontation with the English. The initial reactions of research scientists in England were critical. The first to be won over, Ramsay, may have been converted so readily because he was more interested in quantitative results than in understanding the processes involved in solution.[299] After chemists recognized the extent of the divergence of views between the two sides, the Ostwald school were invited to the British Association meeting in 1890. The people on the two sides gained quite different impressions of the outcome of this meeting. Ostwald said several times that he thought that his side had done well. The English, however, continued to hold to their earlier arguments.

The debate over the nature of solutions was intense and prolonged. One factor contributing to its intensity may have been the crucial importance of theory in physical chemistry at that time. It was not difficult to gather data on the physical properties of solutions (though it was harder to do it well). What was in short supply

[298]See, for example, the comment by Arrhenius, *Journal of the American Chemical Society, 34* (1912), 363.

[299]The discussion by Travers, *op. cit.* (note 70), suggests that Ramsay's initial understanding was rather superficial; his inclination to regard scientific theories primarily as provisional tools for representing quantitative regularities is illustrated by the comment cited in note 280 above.

was an adequate theoretical treatment of the data. There was no theory that provided an adequate explanation of all the data available. Rival theoretical approaches treated different ranges of data in different ways with varying precision, and each of the two sides naturally paid greater attention to the kinds of data it could treat by its methods. While Ostwald's approach gave a quantitative treatment of infinitely dilute solutions, the English chemists sought a mainly qualitative treatment of solutions of all strengths. The disagreement between the Ostwald school and the English chemists over theoretical orthodoxy raised so many other points of disagreement that the arguments of one side often seemed circular to its opponents.

By the late 1890's the two sides understood each other better. This did not prevent further debate; for example, the exchange in *Nature* in 1896–1897. Except for Armstrong's polemics, the exchanges were more constructive. However, the differences between the two sides were not to be settled by rational exchange; the outcome was determined by indirect factors. By the 1900's, it was clear that the original opposition to the Ostwald school had failed to develop alternative theories very far. The Ostwald school program of extending quantitative treatment from infinite dilution to more concentrated solutions and then to non-aqueous solutions had some success, but also limitations, which were exposed by Kahlenberg. There were fewer difficulties about the experimental aspects of the Ostwald program. But Armstrong, too, was able to guide experimental research in topics devised under the stimulus of his prejudice against ions and his belief that chemical change involved association rather than dissociation. Thus the experimental work of each side could proceed, even though it was largely unappreciated by the other. As time went on, the debate over the theory of solutions became less central to the field without having produced a generally accepted resolution. Although the theories of van't Hoff and Arrhenius had been the spearhead of the new specialty of physical chemistry disseminated by the Ostwald school, the new science was not restricted to the study of solutions, and other topics such as the study of chemical affinity and reaction rates gained greater prominence as productive methods were applied to them. New problem areas emerged, particularly with the study of radioactivity and atomic structure, and later with the rise of quantum theory. As later theoretical treatments assumed that strong electrolytes were completely ionized in solution, theoretical discussion moved away from the issues of the older debate. The final exchanges in the debate, therefore, had little urgency for anyone except for people like Armstrong and Kahlenberg, who had committed a significant proportion of their careers and reputations to the debate. Others could afford to ignore it if they wished.

The rational procedures employed in the debate can be said to have failed, for they did not lead to a general consensus about its outcome. Those who were not directly involved tended to accept simplified versions of what happened, if they thought about the issue at all. Since the training of young physical chemists had come to be dominated by teachers and textbook authors descended directly from the Ostwald school, most later physical chemists accepted judgments about the outcome of the debate from that perspective. As far as the views of the participants of the debate are concerned, the following four judgments, as well as intermediate positions, may be clearly identified from the available retrospects. Each is illustrated by a quotation.

1. The Ostwaldian theory prevailed in the face of intense opposition.

> It is not at all surprising that the chemists who had been thinking of reactions in terms of atoms during a long lifetime should have hesitated to welcome the new conception with open arms. Yet the evidence produced for this theory by Arrhenius was so varied and quantitative, that the new theory soon had many adherents. It quickly acquired the support of Ostwald and van't Hoff, and this gave it the stamp of authority.
>
> The objections which were at first offered to the theory were of two kinds; those based on a lack of familiarity with the theory itself, and with the phenomena with which it was meant to deal, and these are of no interest to us, or to any one else.
>
> Then there came the objections which were based upon an intelligent desire to get at the truth. When chemists began to think of chemical phenomena in terms of the new theory, they encountered real difficulties, partly on account of the newness of the theory itself. The theory was called upon to prove itself, as it should be able to do. This kind of thoughtful, conservative criticism is always most useful in science. It is an antidote for extreme radicalism, which is hurtful in science as in everything else. The result has been that during the past quarter of a century, about every rational objection has been offered to the theory of electrolytic dissociation that could be thought of. Facts have been cited,

which, taken at their face value, seemed distinctly at variance with the theory. When these supposed facts have been tested by careful experimental work, they have in practically every case been found to be in error. The theory has met the unusually large number of objections unflinchingly; and it stands today as one of the cornerstones of the modern developments in chemistry. All things considered, it is certainly one of the most important generalizations that has been reached in chemistry, certainly since the discovery of the law of conservation of mass and the law of conservation of energy.[300]

2. The opponents of the Ostwald school theory were triumphant in that their major claims and criticisms were vindicated.

But in 1887 Arrhenius put forward the view of electrolytic *dissociation* which was in complete contrast to this view [that of H. E. Armstrong on the importance of the solvent and of *association*], in that the influence of the solvent was entirely ignored, it being supposed that mere dissolution caused a stable salt to tumble apart in complete defiance of the laws of chemical affinity and chemical attraction. It is therefore hardly to be wondered that Henry Armstrong, having already realised the vital importance of the solvent in electrolysis, refused to be carried away by the hypothesis in which its essential functions were completely overlooked. His opposition to the crude expositions of the German physical chemists in their early enthusiasm for the new theory was based on considered judgement, arrived at before the theory was promulgated, and was too well founded to be swept away by the current of popular approbation.[301]

3. A compromise was reached in which the best points of each side were retained and the defects eliminated. The difference between the two theories became no more than that of complementary working hypotheses, each guiding a different kind of approach.

[300] Jones, *op. cit.* (note 128), pp. 121–123. Jones's perspective has been influential among later historians of chemistry. For example, A. J. Ihde writes: "Perhaps the most comprehensive account of the rise of physical chemistry is Harry C. Jones' *A New Era in Chemistry.*" *The Development of Modern Chemistry* (New York, 1964), p. 809. A similar historical evaluation to Jones's is given by Arrhenius, *op. cit.* (note 298).

[301] Eyre, *op. cit.* (note 28), pp. 220–221.

> It is surely now time that all the irrelevant and intemperate things that have been said and written by the supporters of the osmotic pressure and electrolytic dissociation theories on the one hand, and by those of the hydrate theory on the other, should be forgotten. Far from being irreconcilable, the theories are complementary, and workers may, each according to his proclivity, pursue a useful course in following either.[302]

4. The Ostwaldian theory continued to dominate, but it inspired chemists with an increasing sense of crisis as its limitations became more conspicuous.

> Both [the ionic theory and the phlogiston theory] postulated the existence of an entity which no one could succeed in isolating: phlogiston in one case and the dissolved ion in the other. Both neglected facts which they could not explain, though these facts might well be vital. Both wore an attractive air of simplicity which vanished on closer inspection. And both had to be twisted and contorted into all kinds of queer shapes to meet the needs of the moment. . . . The ionic theory appears to occupy very much the position which was held by the phlogiston theory immediately before the work of Lavoisier: many chemists are enthusiastic about it, and the majority are content to follow their lead, but the air is heavy with portents and revolutionary changes may be at hand.[303]

We see, then, that a degree of consensus was restored to the field not because chemists reached a rational resolution of their scientific disagreements, but because the main issues of the field were replaced by new central concerns. The remaining differences in opinion about the issues of the debate were not crucial to the research practice of the younger physical chemists. Was the debate of little consequence then? It is difficult to assess the overall impact of any internal development in a science that is also substantially affected by external factors. However, it is clear that the debate did have a significant impact. In general, the impact of a scientific debate cannot be judged solely by the overt outcome, by whether or not everyone agrees that one side has won or that a compromise has been reached. The typical direct effect of a debate is that the opinions of the opposing sides converge, either because those holding extreme opinions change their

[302] J. Walker, *Report of the British Association* for 1911 (1912), p. 356.
[303] E. J. Holmyard, *The Great Chemists* (London, 1928), p. 121.

minds or because they come to be ignored in the science. But the indirect effects of debate can also be substantial. At the least, the debate may produce a shift in the line between generally agreed upon facts and controversial conjectures, and the polarization of viewpoints brought about by the debate may assist in the perpetuation of the heterogeneity of the intellectual lineage of the field. Critical discussion may also, however, expose and put under pressure assumptions of underlying technique, inference, methodology, and philosophical method, so that they have to be more clearly stated and thought out. In the absence of a challenge, the supporters of a theory may only exploit its strongest points and not look very hard at its limits. But even in a debate in which "rational" processes appear to have failed, theorists may be stimulated to tackle problem areas that emerged not through the internal development of their respective theories, but in the course of the debate. If a theory can handle such areas better than its rival, that may add to its plausibility and recruiting power, and thus increase its subsequent influence in the field. A scientific field develops most rapidly in its focal areas, the themes or problem areas which attract most interest and attention. The temporary effect of a debate is to make the points of contention into focal issues. Some of the lines of research which are so stimulated may be successful, that is, they may lead to the formulation and solution of a productive sequence of research problems. If, however, even after continued research, those who are not heavily committed to one of the sides in a dispute decide that science in its current state is unlikely to settle the points at issue, then the debate may have a negative effect as research interest swings away from its issues. The belief that it is "time to get on with something worthwhile now," especially when expressed by younger members of the field, may encourage a new direction of development of the field. Thus, even the most inconclusive of debates may be of historical importance.

Several aspects of the indirect effects of debate that I have discussed are illustrated by the influence on physical chemistry of the debate over the theory of solution. However, it is difficult to demonstrate the precise significance of these effects through historical documentation. It is clearly the case, for example, that the debate forced attention upon a range of controversial issues, revealing which of them were suitable for further research and which were not yet amenable to scientific study. Thus, the debate helped to redirect lines of investigation within the general problem area. But it must be conceded that the part played by the debate was not crucial to the most prominent subsequent developments in the science. In an ideal version of internally generated scientific change, experimental investigation and theoretical representation go hand in hand, each stimulating and drawing upon the other. But in practice—and physical chemistry around the turn of the century is no exception—there are many theoretical developments that are insufficiently linked to experimental work, and there are many experimental studies that remain theoretically problematic because there is no single, well-attested way of making their unexpected results fully accord with theory. The debate over the theory of solution did influence theoretical studies and experimental studies, but it did not produce the ideal situation in which key developments of theory and experiment are closely linked. For example, the debate brought to the fore the study of the interaction between solvent and solute, which had not been a prominent part of the original Ostwaldian theory. Lines of investigation can be traced from the nineteenth century critics of the Ostwald school theory to constructive studies in the twentieth century. Thus, an effect of the debate was to keep solvation as an important problem, even though it was not a simple consequence of the thermodynamic treatment of the Ostwald school or an application of a successful experimental technique. In the first half of the twentieth century, there were continuing attempts to invoke solvation as an explanation of residual discrepancies between theory and experiment in the study of other phenomena of solution. These were not fully satisfactory, as the extent of hydration of the dissolved species varied with the solution property it was related to. The attempts continued, because many chemists accepted the idea that there is considerable solvent-solute interaction in aqueous solutions. The issue added little to the achievements of physical chemistry, but it added to the number of open problems in the field that investigators could take up with some hope of success.

A second problem that became prominent through the debate was that of non-aqueous solutions. Kahlenberg pointed out the early difficulties chemists encountered when they applied the Ostwald school program to non-aqueous solutions. The study of non-aqueous solutions remained an important area for empirical study, but theory lagged far behind experiment in the early part of the century. Even such theoretical suggestions as could be made (such as the relation between dissociation in a solvent and its dielectric constant) had a very limited applicability and gave only limited insight.

The Usefulness of the Concept of Dissent in the Historical Study of Scientific Change

The present study is intended to provide not merely a historical account of a particular scientific debate but also sharper definitions and richer descriptive categories to be applied to the study of other debates. The historical development of science has been rich in controversy, which is well worth examination. If historical attention is directed primarily to the development of individuals or of specific ideas, social interactions tend to be less studied and less understood; by focussing on social processes such as debate, the historian can gain a more balanced understanding of the nature of scientific activity. Studying cases of scientific debate can reveal the underlying assumptions of scientific practice. When debate is triggered by or feeds on implicit differences of fundamental assumptions or values, these can be made explicit as they come to the participants' attention. For example, the significance for the practice of science of differing extra-scientific commitments can be more readily examined as these commitments are revealed in controversy.[304] Useful insights into the social processes of science can sometimes be gained from the scientists themselves in a debate. Social mechanisms of scientific change are not normally considered by the participants of a scientific debate, but they may be if the scientists find that their aims are frustrated during a controversy.[305] The historical study of dissent and debate is especially interesting in the development of general theories of scientific change, in ways that the present study has attempted to illustrate, and which are discussed in more general terms in the remainder of this concluding section.

Even the most general treatments of the nature of scientific change can benefit from the study of dissent. One view of scientific change holds that science proceeds steadily in an atmosphere of general agreement. Another view holds that science generates the best account of natural phenomena by developing and successively eliminating alternatives. The latter view, which separates the creative

[304] This was an important theme emerging in a meeting of historians and sociologists of science sponsored by the Science Studies Unit at the University of Edinburgh in September, 1974.

[305] The comments of H. E. Armstrong in the present study illustrate this point.

and critical phases, especially if new ideas require a major investment of creative effort before they can be exposed to critical assessment, will gain depth by studies of the social processes of criticism and dissent. Since both of these general kinds of views have been held by historians,[306] it would be of interest to compare them in connection with a study of the prevalence and character of dissent in science. In recent discussions in the philosophy of science stemming from the exchange between T. S. Kuhn and K. R. Popper,[307] very different roles have been given to debate. A historical investigation of its character and prevalence might produce the insights required to clarify its role in scientific change. Popper suggests that dissent is a widespread phenomenon in scientific change, that scientists actively seek to falsify and to overthrow other scientists' theories. The attempt to falsify theories must be in part a social process for the obvious psychological reason that it is a lot easier to try to overthrow the other man's theory than it is to criticize one's own while there is still some hope that it might be developed further. Kuhn, on the other hand, sees a large proportion of scientific activity conducted within the harmonious consensus of "normal" science. In the first edition of his *The Structure of Scientific Revolutions,* Kuhn suggested that debates in mature science that involved more than individual acrimony—for example, those over priority—were limited to the revolutionary episodes of a whole science in crisis. But although, as Kuhn stressed, the methods of training in science encourage relatively uncritical conformity in normal science, there are other parallel factors that encourage dissent. Kuhn's later discussions[308] suggest that his analysis should be applied to smaller groups of scientists. His arguments still apply to the nature of the consensus within paradigm-sharing communities, but because of the small size

[306]The first view has often been held by scientists who took an extreme empiricist view of science. The latter view is clearly illustrated in E. G. Boring, "The Psychology of Controversy," *History, Psychology and Science: Selected Papers* (New York, 1963), 67–84, and in M. Polanyi, "Passion and Controversy in Science," *Bulletin of Atomic Scientists, 13* (1957), 114–119.

[307]See in particular, *Criticism and the Growth of Knowledge,* ed. I. Lakatos and A. Musgrave (Cambridge, 1970).

[308]See in particular, T. S. Kuhn, postscript to the second edition of *The Structure of Scientific Revolutions* (Chicago, 1970); "Logic of Discovery or Psychology of Research?" and "Reflections on my Critics" in *Criticism and the Growth of Knowledge,* pp. 1–23, 231–278; and "Second Thoughts on Paradigms," in *The Structure of Scientific Theories,* ed. F. Suppe (Urbana, 1974), pp. 459–482.

of these communities they are now more dependent on external resources and stimuli in their work. There is therefore greater room for minor conflicts between social groups, particularly if they are geographically separated or tackle different but overlapping ranges of problems. The subject of the present study may be regarded as an example of controversies at an intermediate level between major revolutions and conflicts between individuals. Further studies of controversies at the intermediate level should be informative about the social and intellectual mechanisms underlying scientific change.

The historical study of scientific controversies can illuminate another methodological interest in the role of dissent in scientific change. Two extreme patterns of dissent may be distinguished. The first is a rational pattern in which consensus is achieved by rational exchange over an isolated disagreement. This version of debate is acceptable to most traditional philosophies of science. It is the ideal of H. E. Armstrong in the present study. At the other extreme is the account of incommensurable viewpoints developed by P. K. Feyerabend and Kuhn.[309] This account applies best to theories or paradigms which are so different in their fundamental structure and in the methodological values presupposed that every concept employed in them is affected. Every observational situation is construed differently from the two theoretical orientations, because the descriptive and explanatory terms employed are not linked in the same way to the rest of their respective system, even though some words may be common to both systems. If two scientific viewpoints are regarded as incommensurable in this way, it becomes difficult, if not impossible, to make logical comparisons between particular statements or sets of statements when they are extracted from their original theoretical context. Even comparisons of whole systems become difficult, except in the most intuitive way, especially if the range of phenomena covered and the kind of knowledge claimed by each are not identical. Hence, logical relations of inclusion and contradiction must be replaced by such metaphors as gestalt switch or translating between languages of radically different cultures. If the major confrontations of ideas in scientific change are best described as incommensurable in this way, it might be thought that debate can only be a *symptom* of scientific change, revealing the deep divisions between the two sides. Rational

[309] For a recent statement of Feyerabend's position on incommensurability see his book, *Against Method* (London, 1975). For Kuhn see "Reflections on my Critics" in *Criticism and the Growth of Knowledge.*

processes would be so ineffective in them that arguments merely harden attitudes rather than modify them.

The clearest examples of incommensurability in science come from cases in which the two sides confront each other in the philosopher's mind. When the philosopher considers incommensurability in terms of static confrontation, he tends to crystallize the thought of each side into a more precise and rigorously logical structure than was employed, or at least agreed upon, by the practitioners themselves. The philosopher then discusses incommensurability in terms of the difficulty of relating the logical structures in a noncontroversial way. But the total thought of an actual science while it is developing is a combination of explicit semi-formal structures of argument set in a context of informal and flexible assumption and conjecture, which is being articulated in the course of further work. When the formal parts of a scientific approach are compared, they do not appear incommensurable; rather, they seem to agree, to contradict, or to relate to different issues. It is only when the *total* problem-solving orientations are reconstructed that the nature and range of incommensurability becomes apparent. We can learn more about science if, instead of making static confrontations of crystallizations of systems of belief, we study the confrontation of opposing approaches in actual historical situations. The historical study will lead us to analyze the processes involved in confrontation and the manner in which debate changes the situation. As the present study illustrates, even though the *initial* confrontation appears incommensurable, the two sides soon came to understand one another more fully, though they continued to disagree. Although methods of rational persuasion had very limited effect, the two sides came to be separated less by misunderstanding than by contrasting commitments. What incommensurability remained stemmed from the tendency of each side to see the gaps and latent tensions in the informal aspects of the opposing position as basic defects or falsifications, and the corresponding features of its own position as research puzzles already implicitly solved and to be spelled out fully in further research.

The examination of historical cases of debate can aid us in assessing the extent to which dissent plays a part in scientific change. It can also show to what extent the different scientific beliefs are selected and transformed by rational discussion, by the indirect effects of confrontation, or by mechanisms in which contrasting beliefs are developed independently, their relative fortunes not being affected by confrontation at all.

The Rise of Physics Laboratories in Britain

BY ROMUALDAS SVIEDRYS*

1. INTRODUCTION

In this paper I will discuss some factors in the origin of teaching and research physics laboratories in nineteenth-century British colleges and universities. I will pay close attention to the laboratories of the Universities of Glasgow, Edinburgh, Oxford, and especially Cambridge, for these provided the directors and often the models for laboratories elsewhere. I will examine in less detail a number of other laboratories. By the end of the century there were more than twenty-five academic physics laboratories; their rapid development in the last third of the nineteenth century in Britain paralleled that in several other countries. The invention, recognition, and spread of academic physics laboratories represent an important stage in the institutionalization of the physics discipline in the nineteenth century.

In the first part of the paper I will discuss certain antecedents of academic physics laboratories; these include chemical, engineering, and private physical laboratories. The relations between physics and engineering were close in the middle of the nineteenth century; it is therefore necessary that I examine certain industrial problems—especially those arising from electrical telegraphy and electrical standards—that physicists worked on. I need to examine too the place of physics laboratories in engineering training, and the later, gradual separation of physics and engineering, which occurred as each acquired its own specialized professional curriculum and associated institutional forms. At the close of the paper I will discuss the respects in which and the reasons why the Cavendish Laboratory of the University of Cambridge played the leading role among the entire group of academic physics laboratories in nineteenth-century Britain.

2. ANTECEDENTS OF FORMAL ACADEMIC PHYSICS LABORATORIES

The Rise of Academic Engineering Education

Innovations in British engineering education occurred outside the established institutions of higher learning. In 1807 the East India

*Polytechnic Institute of New York, 333 Jay Street, Brooklyn, New York 11201.

Company created at Addiscombe a college of engineering to train its own engineers, providing the first formal engineering education in Britain. Graduates of Addiscombe harnessed Indian rivers and constructed irrigation works, canals, and railroads.[1] The college closed following the Indian mutiny of 1857, when the British dissolved the East India Company. The second British venture in formal engineering education was the School of Naval Architecture at Portsmouth, established in 1811. Directed by the Cambridge graduate James Inman,[2] it produced forty-two graduates before it was closed in 1832 owing to bureaucratic rivalries. Another British venture in engineering education was the College of Civil Engineering, Putney, which was established in 1839 during the railroad boom. It was supported by the nobility who sought to stake a claim on a profitable industry. Few students came to the college with sufficient background; the college was soon in financial difficulties and closed in 1859. Another engineering college was established at Queenswood in 1846, closing in 1863 following the death of its director. In 1864 the Royal School of Naval Architecture was founded at South Kensington, with students chosen by competitive examination and staffed by distinguished Cambridge graduates.

In the first half of the nineteenth century, the established British colleges and universities began to offer professional training in engineering. The major impetus for this was the development of railways in Britain in the 1830's and 1840's. The demand for engineers increased, and the introduction of steel into railroad stations, locomotive sheds, tunnels, and railroad bridges created new engineering problems. The old apprenticeship system gave way to other means of training engineers. Table 1 shows the first instruction in engineering and the first chairs of engineering in British higher educational institutions.

In general, professors of engineering had either industrial and railroad connections or academic training and scientific interests. Examples of the latter were James Thomson and William J. M. Rankine at Glasgow and John Tyndall, who supervised the engineering de-

[1] W. H. G. Armytage, *A Social History of Engineering* (New York, 1961), p. 123.

[2] James Inman (1776–1859) was a senior wrangler in 1800. In addition to directing the School of Naval Architecture, he was also principal and professor of mathematics at the Royal Naval College at Portsmouth.

TABLE 1

Engineering Chairs at Established British Colleges and Universities*

Institution	Year Founded
University College, London	1828 Instruction
King's College, London	1831 Instruction
University of Durham	1837 Instruction
Cambridge University	1837 Chair
University of Glasgow	1840 Chair
University College, London	1841 Chair
Trinity College, Dublin	1842 Chair
Queen's College, Belfast	1843 Chair
Royal School of Mines, London	1853 Instruction and Chairs

*For sources of the information in Table 1, see Romualdas Sviedrys, "The Rise of Physical Science at Victorian Cambridge," *Historical Studies in the Physical Sciences, 2* (1970), 127–145, on pp. 131–135.

partment of Queenswood College and later went into physics. Boundaries between physics and engineering were still fluid and easily and often crossed. Rankine and James Thomson, for example, produced many papers on experimental and theoretical problems stimulated by their engineering interests. Some of the best texts on engineering were written by Cambridge wranglers such as Robert Willis and Oxford trained men such as Henry Moseley. During the 1860's and 1870's, engineers found physical laboratories to be adequate training grounds for their careers.

Chemistry Laboratories

Chemistry laboratories, which existed in most British institutions of higher education by 1845,[3] provided precedents for private and academic physics laboratories. Indeed, chemists often did research in their laboratories on subjects that later became part of physics, but which they still viewed—conceptually or experimentally—as part of chemistry. Faraday's laboratory at the Royal Institution, although initially a chemistry laboratory, became the first physics laboratory

[3]In 1820 Thomas Thomson began practical chemical training at Glasgow University, which in 1829 established a chemical laboratory. At the University of Edinburgh, practical chemistry teaching began in 1823 under Charles Hope. University College, London, introduced chemical laboratory practice in 1829.

TABLE 2

Private and Nonacademic Physics Laboratories*

Founder and Location of Laboratory	Year Founded
Michael Faraday, Royal Institution, London	1830's
Charles Wheatstone, King's College, London	1835
James Prescott Joule, Manchester	1838
James David Forbes, Edinburgh	1840
George Gabriel Stokes, Cambridge University	1849
Thomas Andrews, Queen's College, Belfast	1850
William Thomson, University of Glasgow	1850
William Froude	1853
Augustus Matthiessen, London	1857
John Peter Gassiot, London	1857
James Clerk Maxwell, King's College, London	1862
David Kirkaldy, London	1865

*For works on individual institutions, laboratories, and directors discussed in this article and summarized in Tables 2–4, see Romualdas Sviedrys, *James Clerk Maxwell and the Cavendish Laboratory: A Case Study of Science and Society in Victorian England* (diss., The Johns Hopkins University, 1970).

in Britain. The transition paralleled Faraday's work on electricity, which gradually changed from a subject in chemistry to one in physics during the 1830's. Other experimentalists made the same transition: James Cummings, professor of chemistry at Cambridge, became better known for his research in electricity than in chemistry proper; Thomas Andrews established a chemical laboratory in 1850 at Queen's College, Belfast, which he soon transformed into a physics laboratory.

Early Private Physics Laboratories

Private physics laboratories in Britain (see Table 2) were small and informal; seldom more than two students would be found working in one at the same time. In the laboratories of James Prescott Joule, William Froude, Augustus Matthiessen, and David Kirkaldy, no students worked at all; in George Gabriel Stokes's laboratory, students were admitted, but only for demonstrations. Academic laboratory work was not a degree requirement; it usually was a voluntary arrangement designed to give a professor free assistance and students practical experience with physical apparatus. When a private academic laboratory became recognized by the university and its expenses incorporated into the official budget, a financial burden was removed from the physics professor. He received new facilities and maybe a salary for a laboratory assistant to help with the increased number of students. The emergence of the official academic from the private physical laboratory reflected an awareness of the similar trend in institutions of higher learning on the Continent.

3. WILLIAM THOMSON'S GLASGOW LABORATORY AND THE DEVELOPMENT OF THE ELECTRICAL TELEGRAPH

The rapidly developing, science-based, electrical communications industry was another major stimulus to the growth of the physical laboratory in Britain. Beginning in the mid-1840's, private investors spent large sums on the construction of an adequate telegraph network at home, while the government did the same abroad. The technical and scientific problems arising from the development of the telegraph required the cooperation of scientists and the use of their laboratories. The growth of electrical telegraphy also demanded instruction in theoretical electricity and practical training in electrical precision measurements and hence encouraged the establishment of special laboratories for these purposes.

The first practical electrical telegraph in Britain dated from 1837; it was a product of the cooperation between the scientist Charles Wheatstone and the entrepreneur William F. Cooke. On his way home from India, Cooke first saw a telegraph demonstrated in science lectures at the University of Heidelberg. He tried to construct a better instrument in London, and, having failed, he sought the advice of Wheatstone, who, as professor of natural philosophy at King's College, London, was using the basement of the college to perform experiments in metallic conduction. They soon established a partnership and took out a patent on a perfected telegraph,[4] which they

[4]The relationship of Cooke and Wheatstone has been discussed by Geoffrey Hubbard, *Cooke and Wheatstone and the Invention of the Telegraph* (London, 1965). On the early history of the telegraph see J. J. Fahie, *A History of the Electric Telegraph to the Year 1837* (London, 1884). For developments during the crucial years from 1845 to 1853 see the work of one of the pioneers, John W. Brett, *On the Origin and Progress of the Oceanic Electric Telegraph* (London, 1858).

demonstrated as a railroad traffic monitor with a permanent thirteen-mile long telegraph line. In 1846 they established the Electric Telegraph Company. Wheatstone devoted most of his time to the telegraph and practically abandoned his academic teaching.

Progress in telegraphy was rapid. By 1850 short submarine cables were laid from England to Ireland and Holland, and an attempt was made to lay an underwater cable from Dover to Calais. The construction of an Atlantic submarine cable was seriously discussed from 1854 on. The New York, Newfoundland and London Telegraph Company was organized in the same year.[5] In 1858 the first Atlantic cable was laid.

William Thomson, professor of natural philosophy at the University of Glasgow, played a major role in the scientific work on the Atlantic cable. Thomson's interest in electricity dated from 1846, when he began an investigation of the analogies between elasticity and electricity. His work, which displayed the mathematical approach characteristic of those trained at the University of Cambridge, raised questions that required experimental answers. In 1850 he established a laboratory in an unused wine cellar near his lecture room, where he began investigating the electrical properties of matter with the help of a few volunteers from his natural philosophy class. In 1853 Thomson verified the oscillatory nature of the electric discharge from the Leyden jar, after which, following Faraday's explanation of signal retardation in long telegraph wires, he investigated insulated wires, a problem which provided his students with research problems for the next few years.[6]

From the results of his investigation Thomson predicted that the Atlantic cable would show a significant retardation of signals. He proved to be right. Besides, due to faults in its insulation a gradual attenuation in the strength of signals set in. Six weeks after the laying of the cable, a final break in communications occurred. The cable failed due to haste, poor planning, and the neglect of research results from Thomson's laboratory. Thomson maintained that to diminish

[5] On the general history of the Atlantic cable see Bern Dibner, *The Atlantic Cable* (Norwalk, Conn., 1959). William Thomson's role is also discussed in *Silvanus P. Thompson, The Life of William Thomson Baron Kelvin of Largs* (London, 1910), pp. 325-396 and 481-508.

[6] See William Thomson in a letter to his brother James, 7 December 1855, in Thompson, p. 311.

signal retardation the diameter of the copper wires had to be increased to reduce the electrical resistance and capacity of the cable. His proposed solution went counter to the views of the company's official electrician, E.O.W. Whitehouse, a retired medical practitioner. It also made the construction of the cable more expensive and required strict testing of materials. Although Thomson was on the board of directors of the company, he was unable to overrule Whitehouse. Working in his laboratory with his students, Thomson found that samples of copper supplied to the cable manufacturers varied greatly in their electrical resistance and that the conductivity of copper decreased significantly with even small amounts of impurities.[7] Thomson established a special testing laboratory at the cable factory to help copper wire manufacturers make low-resistance cable. The example spread within the industry. Thomson was told that "the conductivity question is occupying the best of all of us—every hank of wire is being carefully tested to a given standard. The attention of the most eminent smelters is being directed to the matter, and they have been invited to the Gutta Percha Works to see for themselves the variations in conducting power of the several parcels sent in."[8] Thomson established another testing laboratory on board the cable-laying ship when, in 1858, he took charge of the electrical operations during the cable-laying voyage.[9]

The failure of the first Atlantic cable led to inquiries. In 1859, the Board of Trade and the Atlantic Telegraph Company appointed a special committee to make a definitive study. Its eight members, carefully selected from among the most eminent scientists and engineers,[10] critically examined the state of the art. Engineers, manufacturers, and scientists gave testimony, and witnesses and members of the committee wrote research papers to answer experimentally key points raised in the inquiry. Their report, which was published in 1861, immediately became an important manual for the tele-

[7] William Thomson, "On the Electric Conductivity of Commercial Copper of Various Kinds," *Proceedings of the Royal Society, 8* (1857), 550–555.

[8] Thompson, p. 351.

[9] *Ibid.*, p. 357.

[10] Four members were appointed by the Board of Trade (Douglas Galton, William Fairbairn, Charles Wheatstone, and George P. Bidder, who became President of the Institution of Civil Engineers the following year), and four by the Atlantic Telegraph Company (Cromwell F. Varley, Latimer Clark, Edwin Clark, and George Saward, secretary of the Atlantic Telegraph Company).

graphic engineer.[11] It attacked trial and error methods and affirmed those of exact science,[12] reinforcing the research trend already prominent in the telegraph industry. It was recognized that submarine telegraphy required not only new skills and new machinery, but above all physicists trained in precision electrical measurements and mathematical theory. The failure of the first Atlantic cable marked the demise in telegraphy of untrained men like Whitehouse and the rise of physicists like Thomson.

To understand more fully the industrial importance of physicists, it is necessary to look at the growth and needs of telegraphy. Between 1854 and 1867, Britain's telegraphic network doubled in size. The cost of a message fell by one half and the volume of traffic increased fourfold. Those who earned their living from the manufacture and use of electric conductors, insulators, batteries, and telegraphic equipment greatly increased, creating a demand for instruction in telegraphy and, indirectly, in electricity. The Royal Engineers, who as a result of the Crimean War and the Indian Mutiny began to appreciate the value of the telegraph, established in 1857 a Military Telegraph School at Chatham,[13] which instructed hundreds of noncommissioned officers and enlisted men in the basics of electricity, chemistry, field telegraphy, and the applications of electricity to military technology. Industrial testing rooms performed a similar service for industrial workers, though they offered mostly practical training in electrical measurements and little instruction in the theory of electricity. At the pinnacle of the industry emerged a "whole new race of engineers calling themselves telegraph engineers and electricians,"[14] whose efficiency depended in great measure on their knowledge of the physics of electricity. Since neither the school of telegraphy at Chatham nor the testing rooms prepared men to advance the scientific basis of the electrical industry, associations

[11] *Report of the Joint Committee appointed by the Lords of the Committee of Privy Council for Trade, and the Atlantic Company, to inquire into the Construction of Submarine Telegraph Cables, together with the Minutes of Evidence and Appendix* (London, 1861).

[12] *Ibid.*, p. xxxvi.

[13] Concerning the Chatham School of Royal Engineers, see Whitworth Porter, *History of the Corps of Royal Engineers, 2* (London, 1889), especially pp. 169–254.

[14] *Report from the Select Committee on Scientific Instruction, together with the Proceedings of the Committee, Minutes of Evidence, and Appendix* (London, 1868), p. 118 (2208).

were formed between industrialists and consulting scientists. One such consultant was William Thomson, who used his academic laboratory to help alleviate the pressing need for trained telegraphic engineers.

Just as physicists served the telegraph industry, so did—and most clearly in the case of Thomson—the industry serve physicists. During the late 1850's and early 1860's Thomson's laboratory was the only academic site for theoretical and practical instruction in electricity. Thomson had obtained a share in the telegraph industry early in the 1850's by entering into partnership with James White.[15] Through his patents Thomson soon became the major shareholder in the firm, which in turn provided research problems for Thomson's laboratory students and jobs for his laboratory graduates for more than four decades. By 1859 as many as twenty of his students were engaged in research on telegraphic instruments, galvanometers, resistance standards, and the Daniell cell. In 1862 Thomson added a room to the cellar that housed his laboratory, to which he assigned half a dozen of his better students to work on improving electrometers. During the academic session of 1864-1865, when Thomson served on an advisory committee of scientists that supervised the design and construction of the second Atlantic cable, research work in his laboratory was particularly intensive. The success of the second cable in 1866, whose electrical operations Thomson directed, marked a turning point of his career at Glasgow. That same year the government conferred knighthood on him for his scientific, technological, and entrepreneurial contributions. It was an unusual distinction for a university professor, and it enhanced his stature in the eyes of the university officials who were at the time drawing up plans to transfer the university to a new site. Realizing that Thomson had placed the University of Glasgow in the forefront of British science, they acknowledged their debt by providing him with excellent experi-

[15]On James White and his partnership with Thomson, see "Lord Kelvin's Laboratory in the University of Glasgow," *Nature, 55* (1897), 486-492. Additional information in Thompson, pp. 717, 754-755, 769, 994, and 1155. After White's death in 1884, the business was taken over by William Thomson, his nephew and personal assistant James Thomson Bottomley, and David Reid. By that time the company employed some two hundred workers in addition to the staff of engineers. The laying of the Atlantic cable in 1866 established Thomson as an electrical industrialist; most of the electrical equipment and instruments used for the cable were built in the workshops of James White.

mental facilities in their new building and thus conferring official status on his laboratory.

In his earlier cellar laboratory Thomson admitted only those students who could do independent research. He assigned them problems that bore either on his own work or on that of the firm of White; as a result students had the satisfaction of advancing knowledge and its applications. The work was informal: "There was no special apparatus for students' use in the laboratory, . . . no laboratory course, no special hours for students to attend, no assistants to advise or explain, no marks given for laboratory work, no workshop, and even no fee to be paid."[16] Students learned while engaged in research, which often began with the construction of their own research equipment. After 1870, when Thomson moved his laboratory to six specially adapted and well-lighted rooms, things became more organized. Assistants and laboratory fees appeared, followed by more systematic laboratory training. During Thomson's frequent absences from Glasgow on consulting duties, his nephew and personal assistant, James Thomson Bottomley,[17] acted as his deputy in the lecture room and the laboratory.

In Thomson's laboratory the distinction between a student and an employee of White's became increasingly blurred as Thomson began to use his students on engineering assignments abroad. Between 1866 and 1874, when submarine telegraph construction was at its peak, his laboratory repeatedly experienced epidemic outbreaks among his laboratory students "of desire to become telegraph engineers."[18] Beginning in the late 1880's, his laboratory had to compete hard with three rival electrical engineering laboratories in London, directed by John Hopkinson at King's College, London, John A. Fleming at University College, London, and William E. Ayrton, one of his former

[16] William E. Ayrton, "Kelvin in the Sixties," *Popular Science Monthly, 72* (1908), 262–263.

[17] James Thomson Bottomley (1845–1926) came to Glasgow in 1870 to act as private assistant to his uncle. He had studied at Queen's College, Belfast, but graduated with honors from Trinity College, Dublin, where he obtained laboratory training from and assisted Thomas Andrews. He gained further experience as demonstrator in physics in the laboratory of William G. Adams at King's College, London. In 1875 he was appointed first Arnott and Thomson demonstrator at Glasgow University and began giving lectures there. He held this post until 1899, resigning from it at the same time that his uncle withdrew from active academic life.

[18] Letter to Jessie Crum, 29 March 1872, in Thompson, p. 622.

students who was now at City and Guilds Central Technical College, London. By then, telegraphy had been replaced by the generation and transmission of electrical power as the mainstay of the electrical industry, topics which Thomson had less interest in.

4. ACADEMIC RECOGNITION OF PHYSICS LABORATORIES

The eight years following 1866, the year of official recognition of Thomson's laboratory in Glasgow, saw the recognition of physics laboratories at nine other major academic institutions in Britain (see Table 3). Unlike Thomson, some of the directors had only modest means, but others, such as Robert B. Clifton at Oxford and James Clerk Maxwell at Cambridge, received separate laboratory buildings costing around fifteen thousand pounds each. The new physics laboratories taught practical techniques in all branches of physics and had no direct links with the electrical industry.

The laboratory that Peter Guthrie Tait[19] established in 1868 at the University of Edinburgh may be taken as an example of the emerging laboratories. Before acquiring the Edinburgh chair of natural philosophy in 1860, Tait had been professor of mathematics at Queen's College, Belfast, where he had worked in the laboratory of Thomas Andrews. In the course of writing a joint textbook on natural philosophy with William Thomson, Tait paid frequent visits to Glasgow and became well acquainted with Thomson's laboratory there. After four years of struggle, and helped by the success of the Atlantic cable and the grim lessons for British industry of the 1867 Paris Exhibition, Tait acquired funds and space for a laboratory at Edinburgh.[20] In his opening lecture of the 1868–1869 academic year, he talked of his desire to emulate Thomson's laboratory:

[19]For information on Tait see his biography by one of his students, Cargill Gilston Knott, *Life and Scientific Work of Peter Guthrie Tait, Supplementing the Two Volumes of Scientific Papers Published in 1898 and 1900* (Cambridge, 1911).

[20]In a letter to Thomas Andrews, 20 December 1867, Tait wrote: "I am about to get a laboratory for practical students. The money has been voted. Henderson [professor of pathology] has been induced to give up his classroom (which is sitting right over my apparatus room), and during the holidays it will be put in order for work. . . ." (Knott, p. 20.)

TABLE 3

The First Formally Recognized Physics Laboratories at British Institutions of Higher Learning

Institution	Year Founded	Founder-Director
University of Glasgow	1866	William Thomson
University College, London	1866	George C. Foster
University of Edinburgh	1868	Peter G. Tait
King's College, London	1868	William G. Adams
Owens College, Manchester	1870	Balfour Stewart
University of Oxford	1870	Robert B. Clifton
Royal School of Mines (later Royal College of Science)	1872	Frederick Guthrie
Royal School of Science, Dublin	1873	William Barrett
Queen's College, Belfast	1873	Joseph D. Everett
Cambridge University	1874	James Clerk Maxwell

> A room has been fitted up as a practical laboratory, where a student may not only repeat and examine from any point of view the ordinary lecture experiments, . . . but where he may also attempt original work, and possibly even in his student days make some real addition to scientific knowledge. That this is no delusive expectation is proved by the fact that in Glasgow, under circumstances as to accommodations and convenience far more unfavourable than I can offer, Sir William Thomson's students have for years been doing excellent work, and have furnished their distinguished teacher with the experimental bases of more than one very remarkable investigation. What has been done under great difficulties in the dingy old building in Glasgow, ought to be possible in so much more suitable a place as this.[21]

In his laboratory Tait could only accommodate the five or six students from his advanced course on the applications of higher mathematics to physical science. He worked intensively with them to train them in the methods of research. He required each student to take an elementary course in "the application of the various physical instruments," beginning by "practicing them in measuring time, . . . then measuring very carefully by the micrometer and the sphe-

[21] Knott, pp. 70–71.

rometer, length, curvature; and by other appropriate instruments, temperature, electric current, electric potential, and so on."[22] Once students mastered the measurements, they went on to research problems; Tait allowed them great freedom in choosing problems, so that work in his laboratory was not dominated, as it was in Thomson's, by electricity.

Tait's hopes were not fully realized; his laboratory did not eclipse or even challenge Thomson's, which he himself acknowledged some six years later.[23] Despite having had a similar mathematical training under William Hopkins at Cambridge and despite the similar institutional arrangements and obstacles at Edinburgh and Glasgow,[24] Tait and Thomson developed very different laboratories. This was due in part to Tait's lack of financial assistance; his salary of about two hundred pounds a year was modest, even when supplemented by a fee of three guineas for each student enrolled in his courses. The fees made it financially expedient to teach the elementary and nonmathematical courses that catered to larger audiences. But even with one hundred students in his elementary courses, Tait was unable to pay for laboratory assistants, which Thomson easily did with his income from patents. Whereas Thomson's assistants took some of the lecture load, Tait remained overburdened by elementary teaching all of his life. The difference between Thomson's and Tait's laboratories also was due in part to differences in personalities; unlike Thomson, Tait considered himself a pure mathematical physicist, uncontaminated with industrial interests. The appointment of Fleeming Jenkin to the engineering chair at Edinburgh in 1868 also may have influenced Tait's exclusive preoccupation with physics by making him wary of encroaching on his colleague's area of competence. Lacking connections with the electrical industry, Tait did not have access to employment opportunities for his students in the way that Thomson did. Nor did Tait's appointment allow him to give instruction that was specialized enough to train professional physicists, for his program in the advanced class was meant to satisfy the needs of students reading for honors in mathematics or for

[22] *Reports of the Royal Commission on Scientific Instruction and the Advancement of Science* (London, 1875), *2*, 12.

[23] Tait's testimony to the Royal Commission, appointed in 1870, to study British science education, *ibid.*

[24] For institutional arrangements and the educational philosophy in Scotland see George Elder Davie, *The Democratic Intellect: Scotland and Her Universities in the Nineteenth Century* (Edinburgh, 1961).

the diploma in engineering. Only after the engineering curriculum and an engineering laboratory had become part of Edinburgh University in 1890 could physics be taught as preparation for a career in its own right rather than as preparation for another.

In London, the development of educational institutions was more affected than it was in Edinburgh by the demand for trained electricians and telegraph engineers. The events of the Crimean War had revealed such incompetence and disorganization in the army and navy that the need for reform of the civil service became obvious. To eliminate patronage a government-appointed commission recommended the selection of candidates for the civil service by competitive examinations.[25] In most branches the suggested changes were introduced gradually, but in the Indian Telegraph Service the competitive examinations were introduced immediately. Thomson made up weekly practice problems for civil service candidates and participated in their examination. Similarly, in 1864–1865 James Clerk Maxwell, then professor of natural philosophy at King's College, London, offered an evening course on electricity and magnetism, culminating in the treatment of electric telegraphs, at least in part to prepare civil service candidates; for this purpose he used the college's shop facilities and a general laboratory that served all the needs of the institution. He retired that same year and only repeated the course once.[26] When, in 1866 and 1868, George C. Foster[27] and Wil-

[25] R. K. Webb, *Modern England from the Eighteenth Century to the Present* (New York, 1968), pp. 293–294.

[26] Sir John Randall, "Aspects of the Life and Work of James Clerk Maxwell," in C. Domb, ed., *Clerk Maxwell and Modern Science* (London, 1963), pp. 17–18. Among other things, Maxwell covered systems of measurement of electrical magnitudes, methods of finding the position of a fault in a cable, heat produced by electric currents, the electromagnetic field and its lines of force, effects of a disturbance of the electromagnetic field, and conservation of energy in electromagnetic phenomena.

[27] George Carey Foster (1835–1919) studied chemistry at University College, London. Upon graduation in 1855 he worked as an assistant to his former chemistry professor, leaving in 1858 for the University of Ghent to continue his studies under August Kekulé. He also studied physics under J. C. Jamin in Paris and G. H. Quincke in Heidelberg. In 1862 he accepted the professorship of natural philosophy at the Andersonian University in Glasgow and became acquainted with William Thomson. (*Transactions of the Chemical Society, 115* [1919], 412–427.)

liam G. Adams[28] set up physics laboratories at University College and King's College, respectively, they strove to meet the needs of civil service candidates. Foster, who had only recently moved to London from a professorship of natural philosophy at the Andersonian College in Glasgow, transferred his research interests from chemistry to electricity and became an expert on accurate electrical measurements.[29] Of his early laboratory students a significant portion were candidates training to take the Indian Telegraph Service or other civil service examinations. Their number declined once Adams opened his laboratory at King's and charged much lower laboratory fees, attracting many students there. Adams visited several Continental physics laboratories in the spring of 1868, and got the authorities of King's College to agree to build separate physics and chemistry laboratories, at a cost of two thousand pounds, to replace the one general purpose laboratory. The physics laboratory, opened in the fall, consisted of an apparatus room, a research room, a student laboratory, and a store and battery room, and there was also a lecture room for eighty students. Adams' laboratory quickly became famous for the great number of students from it that passed Indian Telegraph Service examinations; in 1870 and 1871, almost a third of the successful candidates were trained there. In addition some students were sent to him by the Indian government "to get up theoretical knowledge in telegraphy";[30] in their laboratory practice, they determined the electromotive forces and the internal resistances of batteries, measured the intensity of electric currents and the resistance of telegraph wires, tested telegraph cable, and learned the principles of telegraphy.[31]

Foster and Adams placed an even greater emphasis than Tait upon practical teaching in the laboratory, which meant that their own re-

[28]William G. Adams (1836–1915) was the younger brother of the famous Cambridge astronomer John Couch Adams, the co-discoverer of Neptune. He studied at Cambridge, graduating in 1859. In 1863 he became a lecturer at King's College, London, and in 1865 succeeded Maxwell in the natural philosophy chair. (*Nature, 95* [1915], 211–212.)

[29]George C. Foster, "On a Modified Form of Wheatstone's Bridge, and Methods of Measuring Small Resistances," *Journal of the Society of Telegraph Engineers, 1* (1872–1873), 196.

[30]*Reports of the Royal Commission on Scientific Instruction and the Advancement of Science* (London, 1875), *1,* 448 (6888).

[31]*Ibid.,* p. 443 (6887). The number of his students fluctuated around twenty.

search suffered. Foster did not at first have a private room for his own research and had to clear a table in the student laboratory for that purpose. Adams, who had better facilities, also complained that the excessive workload left no time for "original investigation."[32] Their teaching loads were excessive, their salaries low, and their laboratory help inadequate. Both were overburdened with elementary teaching, for the students who came to them were ill-prepared in mathematics.

When specialized private schools were established to offer training in telegraphy exclusively,[33] telegraphy began to decline as the main source of students in Foster's and Adams' laboratories. A new source was created by legislative fiat.[34] The rapid expansion of British education during the 1870's created many new positions for science teachers and thus new employment for laboratory students trained in physics. In addition to Foster's and Adams', a third teaching laboratory appeared in London to help supply the demand for science teachers. In connection with the transfer of the physical department of the Royal School of Mines to South Kensington in 1872, a new physical laboratory was established. Its director, Frederick Guthrie, was, like Foster, a chemist by training. Guthrie's laboratory courses on heat, electricity, and magnetism soon became the well-

[32]*Ibid.*, p. 447 (6822).

[33]One of the earliest and most successful schools of this kind was the School of Telegraphy established in London in 1863, which by 1888 had expanded into all areas of electricity; it graduated more than 3,300 students. Its total fees were higher than those paid in physical laboratories, but students were admitted and promoted by examinations and graduated after a year of concentrated practical and theoretical work. (A. D. Southham, *Electrical Engineering as a Profession and How to Enter It* [London, 1892], pp. 166–167, 170–171.) See also announcements in the *Electrician's Directory* for 1888 for information on other schools.

[34]British educational reforms came in the wake of the Reform Bill of 1867, which extended the right to vote to new groups, especially the artisans and workers of the northern industrial towns. William Edward Forster, a radical Quaker and woolen manufacturer, introduced a bill in 1870 that made primary education compulsory, and gave special boards in each district the power to tax to maintain schools. Between 1870 and 1876, a million and a half new school places were provided and filled. Beginning in 1880, there also was a corresponding expansion in higher education. (Howard C. Barnard, *A History of English Education from 1760* [London, 1963], pp. 115–119, and Charles Birchenough, *History of Elementary Education in England and Wales* [London, 1930], pp. 127–161.)

known and popular courses for "certificate science teachers." He taught his students both how to use physical instruments and how to make them; every teacher trained by him took back to his science classes a stock of apparatus that he had built himself.

Probably the most lavish teaching laboratory was the one established by Robert B. Clifton at the University of Oxford. A Cambridge graduate, Clifton was appointed first professor of natural philosophy at Owens College in 1860. Five years later he was elected to the Oxford chair. In 1866 he established a small private laboratory and gave a practical laboratory course to eight students. In 1868 he successfully campaigned for funds for an official laboratory.[35] He argued that Oxford needed a physical laboratory because students had to become acquainted "by actual experience with actual physical processes." He argued, too, that a physical laboratory would qualify graduates to become science teachers, who were in growing demand. Clifton pointed out that Oxford must recognize a need that already "has been recognized in other schools of science. For many years Physical Laboratories have been established in the University of Glasgow, and in the principal Universities of the Continent; and attempts are being actively made to organize similar institutions in London and Manchester."[36] Clifton went on to present a plan of the basic accommodations required of a laboratory. His plan for a separate large building with twenty-five rooms was approved at an estimated cost of over ten thousand pounds. Its construction began the same year, 1868, and was sufficiently advanced to be in partial use by 1870. To add to an already fine collection of instruments, Clifton received a lump sum of one thousand pounds for new precision equipment. He also was granted one hundred and fifty pounds a year for a demonstrator to help in the laboratory, and another one hundred and fifty pounds a year for the purchase of new equipment.[37]

[35] Robert B. Clifton, *The Offer of the Clarendon Trustees* (25 January 1868), p. 1. The money for the Clarendon trust came from the sale of Clarendon's *History of the Great Rebellion* and had been left to accumulate since 1751. The original goal of the trust had been to establish a riding academy.

[36] *Ibid.*, pp. 2–4.

[37] *Reports of the Royal Commission on Scientific Instruction and the Advancement of Science* (London, 1875), *1*, 186 (2984, 2986, 2990, 2991, 2993).

5. ELECTRICAL UNITS AND STANDARDS

Research on electrical standards was another stimulus for the development of physics laboratories in Britain. The research performed under the aegis of the electrical standards committee of the British Association for the Advancement of Science during the 1860's provided a training ground for several directors of physical laboratories. To understand the origins and work of the committee, I will sketch some of the scientific and technical issues involved in establishing electrical standards.

Work on the telegraph brought to the fore the need for a common standard of electrical resistance to put a stop to a chaotic proliferation of standards. Physicists introduced units of resistance that differed from those used by practical electricians. Each country had its own standards. Standards were made from a variety of materials, and some materials that were used exhibited marked secular changes. When a standard was reproduced, its resistance varied, so that data referred to copies of the same standard could not be compared. No one could measure the absolute value of a given resistance, and all that could be done was to compare resistances by means of a Wheatstone bridge.

During the 1840's the problem of experimentally investigating the interconversion and conservation of forces heightened the interest of physical scientists in the absolute value of electrical resistance. In Britain, James Prescott Joule sought an electrical means to check his extremely precise quantitative measurements of the mechanical equivalent of heat. In Germany, Wilhelm Weber related the unit of resistance to the units of mechanics and showed how absolute measurements of resistance could be made in electricity.[38] In 1851 Thomson, who was engaged in thermodynamic and electrical problems and was collaborating with Joule on experimental problems related to the mechanical equivalent of heat, was drawn to Weber's efforts to perfect instruments for measuring electrical magnitudes;[39]

[38]Wilhelm Eduard Weber (1804–1891) became professor of physics at the University of Göttingen and, except for a period in exile (1837–1848), remained there until his death. In a series of papers beginning in 1846 he showed how absolute measurements of resistance could be made by reducing all electrical magnitudes to measurements of mass, length, and time. (*Nature*, *44* [1891], 229–230.)

[39]Wilhelm Weber, "Messungen galvanischer Leitungswiderstände nach einem absoluten Maasse," *Ann. d. Phys.*, *82* (1851), 337–369.

he immediately recognized the fundamental character of Weber's work. Thomson applied Weber's approach to measure electromotive forces and electrical resistances in absolute units, correctly surmising that Weber's resistance unit, based on the centimeter, gram, and second, was much too small for laboratory use.[40] He therefore built a standard of resistance based on British units and multiplied by a factor of one hundred million. Although few other British scientists used absolute measurements of electrical magnitudes, the convenience of a system of electrical units related to mechanics appealed to many of them; in practice they still used arbitrary standards and deplored the lack of agreement on some common system.[41]

Practical electricians and telegraph engineers introduced their own convenient units because those used by scientists were much too small for their needs. Even Wheatstone did not use the standard he had suggested in 1843 once he himself had turned industrialist.[42] Electricians introduced units, too, to eliminate the dependence of resistance on temperature and on impurities. The German engineer-scientist Werner Siemens developed a standard based on mercury, which he was led to by the requirements of the Atlantic cable; his firm, Siemens and Halske, was associated with one of the cable makers, Newall and Company. Siemens appreciated the value of accurate standards of resistance in locating cable breaks and faults of insulation.[43] He hoped that his mercury standard, which was convenient, easy to reproduce, and invariable, would soon be used by all scientists and engineers. On the Continent, Siemens' standard met

[40] William Thomson, "Applications of the Principle of Mechanical Effect to the Measurement of Electromotive Force and of Galvanic Resistances, in Absolute Units," *Philosophical Magazine, 2* (1851), 551-562.

[41] A table with fourteen different units then in use is included in the report of the electrical standards committee for the year 1864. (*Reports of the Committee on Electrical Standards* [Cambridge, 1913], facing page 165. Hereafter, *Reports.*)

[42] Charles Wheatstone proposed as a standard of electrical resistance the resistance of a copper wire one foot long, weighing one hundred grains. ("An Account of Several New Instruments and Processes for Determining the Constants of a Voltaic Circuit," *Philosophical Transactions, 133* [1843], 312.)

[43] Werner Siemens, "Vorschlag eines reproducirbaren Widerstandsmaasses," *Ann. d. Phys., 110* (1860), 1-20. For the British public he published his standard in *The Electrician, 4* (1863), 63-64, and in *Philosophical Magazine, 21* (1861), 25-38. For the role of the Siemens brothers in the telegraph industry see Georg Siemens, *History of the House of Siemens* (Freiburg and Munich, 1957).

with great approval, but in Britain, owing to interfirm rivalries, electrical industrialists opposed it.

When the joint committee studying the failure of the Atlantic cable met late in 1859, Charles William Siemens advocated the use of his brother's standard for cable testing. Thomson, however, pleaded for testing electric equipment "entirely by comparison with absolute standards of resistance."[44] He found a firm supporter in Jenkin, then a young employee of Newall and Company: "There should be [Jenkin said] some well recognized standard of resistance, or resistance coil. I have myself found a great want of such a recognized method. I have now adopted Professor W. Thomson's units, which, I believe, are based upon very scientific data; but it would be extremely convenient if there were a generally recognized standard or unit which one could always mention . . . and I think there should be somewhere, in some public institution, some standard coil with which any resistance might be compared."[45] The idea of a public institution as a repository of standards was advocated a year later by Latimer Clark, a member of the joint committee, and Charles Bright, a well-known telegraph engineer. Clark and Bright, having formed a consulting engineering partnership in 1861, were competitors of Siemens and Halske and were not eager to support the standard of their commercial rivals. In a joint paper read at the Manchester meeting of the British Association in 1861, they suggested a coherent system of electrical units and requested the patronage of the Association. Their paper was well timed,[46] coming in the wake of the publication of the report of the joint committee. In the lively discussion that followed, Thomson moved that a committee be appointed to consider the question of electrical standards in detail.

Initially, six members were appointed to serve: Thomson, Wheatstone, Matthiessen, Jenkin, and two chemists, W. A. Williamson and W. H. Miller. All six were academically trained, and all except Jenkin, who received this distinction shortly thereafter, were fellows of the Royal Society. Practical electricians were not deliberately excluded from the committee; a majority of the members were deeply involved in problems of the electrical industry, though their ap-

[44] *Report of the Joint Committee,* p. 118.

[45] *Ibid.,* p. 139.

[46] Latimer Clark and Sir Charles Bright, "On the Formation of Electrical Standards of Quantity and Resistance," *British Association Reports* (1861), pp. 37–38.

proach was as scientists and not as engineers. The members decided, first, that the unit of resistance should have a definite relation to the other units used in electrical measurements and to that of work, "the great connecting link between all physical measurements."[47] They wanted a coherent system based on the French metrical system; Thomson again strongly advocated that they adopt the Weber absolute system of electrical units. Practical electricians failed to see the advantages of the absolute system and in principle opposed units based on the metric system. But in spite of their protests, the committee passed unanimously a resolution to adopt Weber's unit of resistance augmented by a factor of ten million. As suggested in Bright and Clark's paper, the unit received the name of "ohmad," but it soon became the now familiar "ohm." The chemists on the committee went to work to determine the best form and material for the standard representing the ohm.

In 1862 the committee reported on the decisions reached and the work done during the first year. They recommended that the committee be reappointed to allow the chemists additional time to complete their work. Seven new members joined: scientists Maxwell, Joule, Balfour Stewart, and Ernst Esselbach, and electricians Bright, Cromwell F. Varley, and Charles William Siemens. Joule was interested in the work of the committee because he wanted to use the new standard of resistance in an independent determination of the mechanical equivalent of heat. Stewart hoped to have the standard of resistance entrusted to the Kew Observatory, of which he was superintendent and which he wanted to convert into a laboratory of standards. Maxwell's interest was the result of his recent theoretical research in electromagnetism.

Owing to Maxwell, the committee extended its investigation from electrical resistance to electrical measurements in general, and at the same time its work became more mathematical. Taking over from Thomson, Maxwell supervised the design and construction of the equipment for the extremely accurate experimental determination of the value of the ohm. Then he, Jenkin, and Stewart carried out a laborious set of experiments at King's College, working intensively from April to August 1863. The results dispelled any reservations that members of the committee had had about the possibility of

[47]*Ibid.*, p. 2.

obtaining a sufficiently accurate value of the ohm.[48] The committee reported that they considered the system of absolute electrical units not only "the best yet proposed," but that it was "the only one consistent with our present knowledge both of the relations existing between the various electric phenomena and of the connection between these and the fundamental measurements of time, space and mass."[49]

With the theoretical issues resolved, the committee turned to the construction of standards, and its members began to work in small groups. The private London laboratory of Matthiessen emerged as one of the centers of research. There Matthiessen, who had earned a doctorate in chemistry at the University of Giessen and had spent three years working in Robert Kirchhoff's laboratory on the electrical conductivity of metals, together with Jenkin, Foster, and C. Hockin worked first on constructing an electrical resistance standard and then on constructing a standard of capacity. When Jenkin, the committee's secretary, left London for the new engineering chair at the University of Edinburgh, the work on the capacity standard languished. Once in Edinburgh, Jenkin gravitated toward Glasgow, where Thomson was directing work on constructing standards of electrical resistance and electromotive force.

The fragmentation of the committee into smaller groups continued until 1870, when its members did not ask to have the committee reappointed, but decided to continue its work in three subcommittees. A group composed of Clark, Varley, and Stewart took over the problem of a standard of capacity. In that same year Stewart was appointed to the chair of natural philosophy at Owens College, where he immediately established a physical laboratory; but due to injuries in a railway accident, his participation in standards research declined. For Clark and Varley, representing the electrical industry, the construction of a standard of capacity took priority over that of other standards. They could not do much by themselves until Alexander Muirhead took up the problem in his private laboratory during the mid-1870's.[50] A second group, composed of Hockin, Matthies-

[48] James Clerk Maxwell, Balfour Stewart, and Fleeming Jenkin, "Description of an Experimental Measurement of Electrical Resistance, Made at King's College," *British Association Reports* (1863), pp. 140–158.

[49] *Ibid.,* p. 64.

[50] Alexander Muirhead (1848–1920), son of an electrical industrialist, obtained a D.Sc. in electricity, having studied under Foster at University College, London. In 1875 his father became a partner of Clark, Warden, and Company,

sen, and Williamson accepted responsibility for continuing work on the development of a standard electrodynamometer and of methods to measure electrical currents. Like the first group, it failed to achieve its goals and disintegrated after a short time when Matthiessen died. His fine laboratory was lost to the members of the committee and found no successor for several years; his death marked the end of the standards group in London. The third subcommittee consisted of Maxwell, Thomson, and Jenkin, who were to work on measuring instruments of electromotive force and on the construction of an adequate electromotive standard. Although Maxwell's election to the physics chair at Cambridge within the year made it difficult to cooperate with Jenkin in Edinburgh and Thomson in Glasgow, he continued to work at Cambridge on problems, such as the accurate testing of Ohm's law, that were of interest to the former members of the electrical standards committee. Indeed, at Cambridge, in the new Cavendish Laboratory, Maxwell made electrical standards part of the research program; this research, which provided a connection between the Cavendish and the electrical industry, was severed only in 1900 when the National Physical Laboratory was established and put in charge of research on standards.

6. THE CAVENDISH LABORATORY

The Cavendish Laboratory, like other academic laboratories, was in part a response to industrialism and to the professionalization of physics. It was also part of the unique social and intellectual setting that was Cambridge. Outsiders considered Cambridge conservative and lacking in the necessary conditions for physics, or for science in general, to flourish. Joseph Norman Lockyer, a scathing critic of Cambridge, wrote in 1874 in an unsigned editorial in *Nature* that despite Maxwell it may take Cambridge thirty to forty years to reach the level of a second-rate German university in physical research.[51] The forecast was vitiated from the start by the fine performance of the Cavendish during the five years under Maxwell's direction.

---

and Muirhead established a private laboratory on the premises and became the firm's scientific adviser. (*Proceedings of the Royal Society, 100* [1921–1922], viii–ix, and *Journal of the Institute of Electrical Engineers, 59* [1921], 782–783.)

[51] Unsigned editorial, attributed to its editor Lockyer, *Nature, 9* (1874), 298.

Through a combination of leadership and talented graduate students, Maxwell established during his tenure a school of research in experimental physics. The high standards continued under Lord Rayleigh and Joseph John Thomson, when the Cavendish became the training ground for British Nobel prize winners in physics.

In founding the Cavendish Laboratory, Maxwell realized his ideas about the goals and structure of a physics laboratory that came from his unique background. Of independent means, he had established his own private laboratory in a garret at the age of fifteen after a visit to the workshops of two scientific instrument makers in Edinburgh.[52] Maxwell's remarkable experimental skills were developed under James David Forbes, professor of natural philosophy at the University of Edinburgh, who gave him the run of his personal laboratory. After three years at Edinburgh, Maxwell moved to the University of Cambridge, where, in January 1854, he graduated as second wrangler. He was one of the very few British physicists who combined the experimental and philosophical Scottish tradition with the mathematical training provided at Cambridge.

Maxwell remained nearly two years at Cambridge, reading and doing research, and in the fall of 1856 he went to Marischal College, Aberdeen, as professor of natural philosophy. In 1860 he came to King's College, London, resigning in 1865 to devote himself to reforming the mathematical tripos at Cambridge, and to writing textbooks to prepare students and tutors for questions on electricity, magnetism, and heat. In 1866 Maxwell served for the first time as examiner of the mathematical tripos, and the questions he set dealt with recent developments in physics and proved to be among the most challenging and interesting. His questions confirmed the general feeling that the time was ripe for reform, and he served on a committee that was appointed to suggest reforms.

The report of the committee in 1868 suggested, among other things, that a chair of experimental physics be created along with a physical laboratory. But Cambridge lacked funds until, in the fall of 1870, the Duke of Devonshire, who was Chancellor of the University of Cambridge and an industrialist who appreciated the practical value of physics, offered to endow a laboratory if the University would create a chair of experimental physics. The University did so, and Maxwell was elected, unopposed, in February 1871. He im-

[52]General information on Maxwell comes from Campbell and Garnett.

mediately began work on plans for the laboratory, adopting the good features of its predecessors and avoiding the bad. As at Oxford, the laboratory was constructed as a separate building; it was officially inaugurated in June 1874.[53]

A major advantage of the Cavendish was that, like Thomson's laboratory, it began as a research laboratory rather than as a teaching one. Students who had taken their mathematical tripos came to the Cavendish in search of experimental problems for their fellowship dissertations; there were always graduate students willing to stay and work at the Cavendish for several years. Instead of assigning minute problems that required only a couple of months to work out, Maxwell could assign students the much more laborious problems suggested by his own electromagnetic theory.

In the summer of 1879 Maxwell became so ill that work at the Cavendish came to a standstill. Under Rayleigh, who succeeded Maxwell as director of the Cavendish at the beginning of 1880, teaching in the laboratory was systematized and developed. He did not overburden his laboratory demonstrator; indeed, he divided the post of demonstrator into two posts, giving each demonstrator three-day teaching schedules and releasing the remaining days for their own research.[54] Each demonstrator had a fellowship from his respective college, so that there was no financial hardship.[55] Rayleigh was confronted immediately by a clamor about the inaccuracy of the British Association standard of electrical resistance. Many scientists and engineers were distressed that Maxwell's illness had prevented him from carrying out a new determination of the ohm; they thought it desirable to transfer back to London the standards equipment, which had been deposited at the Cavendish, and to begin a new program of research. Rayleigh's election to the Cavendish chair strengthened their intention, as Rayleigh was not known to have much interest in electricity or electrical standards. However, at the Cavendish, Rayleigh turned to the teaching of electricity, reorienting

[53] The laboratory building is described in *Nature, 10* (1874), 139-142.

[54] William Garnett, who had been Maxwell's demonstrator, expected to replace him at the Cavendish. But he was not even considered for the post, and Rayleigh was elected instead. Garnett resigned, and to replace him Rayleigh picked Richard T. Glazebrook, who later became the first director of the National Physical Laboratory, and William Napier Shaw.

[55] Fellowships provided a comfortable standard of living and added few duties, and thus demonstrators at the Cavendish had comparatively high incomes and light teaching loads.

his own research to the subject. Specifically, he decided to try Maxwell's idea of "experiments in concert."[56] In an attempt to identify the Cavendish Laboratory with a major area of research, he sought a problem important and broad enough to coordinate the efforts of all researchers in the laboratory. In June 1880 he forestalled the restless London standards group by announcing that he, in concert with his students, would redetermine all electrical standards, thus providing continuity with Maxwell's directorship and research. He established an apparatus fund that grew to about two thousand pounds and otherwise prepared the Cavendish for the project of the redetermination of electrical standards. At the British Association meeting of 1882, Rayleigh announced the establishment of a station at the Cavendish for electrical standards, where for a fee tests would be made and certificates issued of their correct value at specified temperatures. In this way Rayleigh secured this area of research for the Cavendish as well as a small but steady source of income.

7. THE SPREAD OF ACADEMIC PHYSICS LABORATORIES: PROFESSIONALIZATION AND SPECIALIZATION

In the 1870's a student could get academic training in engineering or in physics, even though there were few places that offered a degree in either.[57] Differentiation between physics and engineering was still incomplete. There were no engineering laboratories until 1878; engineering students received their practical training as apprentices in industry or by studying in physics laboratories in which applied electricity, the basis of electrical engineering, was a favored subject. When additional chairs of engineering were created and separate engineering laboratories were attached to the chairs, physics laboratories began to train fewer engineering students. The rise of electrical engineering laboratories, often staffed and directed by physicists, paralleled a slow realignment of interests and functions of

[56] Maxwell's inaugural lecture, in W. D. Niven, ed., *The Scientific Papers of James Clerk Maxwell* (New York, 1965; Dover edition), *2,* 245.

[57] The University of London began offering the B.Sc. and D.Sc. degrees in 1853, but it was only an examining body. The University of Glasgow first offered a B.Sc. degree in 1872, and the Victoria University in 1881. Before 1881 only classics and mathematics were avenues to a degree at the University of Cambridge. Then physics became one through the natural science tripos examination. In 1895 the University of Cambridge finally recognized degrees from universities and began to offer a graduate degree. Beginning with 1894 it also offered an engineering degree through the mechanism tripos examination.

physics laboratories. By 1895 it was clear that industrial needs no longer stimulated the growth and work of physics laboratories.

Between 1867 and 1885 new engineering chairs were established at no fewer than fifteen institutions of higher education.[58] Fifteen academic engineering laboratories were established as well between 1878 and 1900.[59] More specialized chairs of electrical engineering appeared at many institutions of higher learning between 1881 and 1898, together with electrical engineering laboratories.[60] Beginning in the 1890's, engineering, electrical engineering, physics, chemistry, and mathematics were separate avenues to a degree, and thus were recognized as distinct professions at the academic level.

Academic physics laboratories grew from ten in 1874 to twenty-four by 1885 (see Table 4). In the civic universities that proliferated after 1880, the directors of physics laboratories had big teaching loads, little time for their own research, poorly prepared students, little money to buy research equipment, and few auxiliary laboratory staff. The civic universities depended on private benefactors who, although prepared to endow chairs and construct laboratories for reasons of prestige, were seldom willing to finance the running costs of the laboratories and the salaries of assistants.

[58]The occupants of these new chairs often were distinguished scientists or engineers, among them Osborne Reynolds at Owens College (1867), Fleeming Jenkin at the University of Edinburgh (1863), James Stuart at the University of Cambridge (1875), and James A. Ewing at University College, Dundee.

[59]These engineering laboratories were established, in chronological order, at: University College, London (1878), Finsbury Technical College, London (1881), Mason Science College, Birmingham (1882), King's College, London (1882), Royal Indian Engineering College, Cooper's Hill (1883), University College, Bristol (1883), Central Institute (City and Guilds), London (1884), Firth College, Sheffield (1885), Yorkshire College, Leeds (1886), Owens College, Manchester (1886), University College, Liverpool (1887), University College, Dundee (1889), University of Edinburgh (1890), University of Cambridge (1893), University of Glasgow (1900).

[60]Electrical engineering chairs were established at Finsbury Technical College (1881), Central Technical College (1884), University College, Dundee (1884), Owens College, Manchester (1884), University College, London (1885, under John A. Fleming), University College, Liverpool (1889), King's College, London (1890, under John Hopkinson), Heriot-Watt College, Edinburgh (1890), Municipal Technical School, Manchester (1891), University of Cambridge (1891, courses within the engineering program), University College, Liverpool (1892), Yorkshire College, Leeds (1892), University of Glasgow (1898). A laboratory was generally attached to a chair.

TABLE 4
Later Physics Laboratories at British Institutions of Higher Learning

Institution	Year Founded	Founder-Director
Armstrong College of Science, Newcastle-on-Tyne	1875	Alexander S. Herschel
University College, Bristol	1876	Silvanus P. Thompson
City and Guilds Technical College, Finsbury	1879	William E. Ayrton
University of Aberdeen	1880	Charles Niven
Mason College of Science, Birmingham	1880	John H. Poynting
Queen's College, Galway	1880	Joseph Larmor
Trinity College, Dublin	1881	George F. Fitzgerald
University College, Liverpool	1881	Oliver Lodge
University College, Nottingham	1882	William Garnett
Firth College, Sheffield	1883	William H. Hicks
University College, Cardiff	1884	Ernest H. Griffiths
University College of North Wales, Bangor	1884	Andrew Gray
City and Guilds Central Technical College, London	1885	William E. Ayrton
Yorkshire College, Leeds	1885	William Stroud

The directors of the physics laboratories—both the earlier and the later ones—constituted a fairly homogeneous group. There was a predominance of Cambridge graduates who had taken the mathematical tripos; e.g., Thomson (1845), Tait (1852), Maxwell (1854), Adams (1859), Clifton (1859), Herschel (1859), Rayleigh (1865), Hopkinson (1871), and J. J. Thomson (1880). Others had taken the mathematical tripos and in addition had worked at the Cavendish Laboratory; e.g., Niven, Poynting, Larmor, Garnett, Hicks, Griffiths, and Stroud. Still others had received part of their training at the Cavendish; e.g., Arthur Schuster at Manchester and John A. Fleming at University College, London. Next to the Cavendish, William Thomson's Glasgow laboratory was the largest training ground for future directors of physics laboratories. Foster most likely did research there during his stay in Glasgow, and Everett, Ayrton, and Gray were Thomson's students. Among the remaining directors, Barrett had no university education, but he had been Tyndall's assistant at the Royal Institution for over five years; and Foster and Guthrie were chemists by training.

8. IN CONCLUSION: THE SPECIAL ROLE OF THE CAVENDISH LABORATORY

Beginning in 1891, the British government began a program of scholarships to encourage graduate education in science. Institutions could periodically select one or two candidates to receive scholarships; the scholars were free to choose any institution in which to carry out their studies and research. During the first twenty years, 1891–1910, 336 students received the so-called "1851 Exhibition Scholarships." Of these, 101 were physics students.[61] When the program began, the University of Cambridge did not accept graduates from other universities, since it did not then recognize any degrees but its own and Oxford's. But once Cambridge opened its

TABLE 5
Distribution of 1851 Exhibition Scholarships in Physics, 1891–1910

University of Cambridge		45
University of Glasgow		9
Owens College		7
University College, Bristol		4
City and Guilds Central Technical College, London		4
Durham College of Science		3
Others (British)		5
Others (Abroad):	Germany	25
	United States	7
	Switzerland	6
	Others	4
Total		119[62]

[61] *Particulars of the Science Research Scholarships Awarded by the Royal Commission for the Exhibition of 1851* (London, 1922), pp. 30–89.

[62] The total in Table 5 differs from the number of scholarships granted because some recipients chose more than one place to study.

doors to other graduates in 1896, the prestige of the Cavendish Laboratory and the opportunity of earning a Cambridge research degree proved a strong attraction. The Cavendish Laboratory emerged as the only powerful graduate research center in Britain, which is clear from the preferences of the 1851 Exhibition Scholars (see Table 5).

The preeminence of the Cavendish Laboratory was never really challenged. Despite the growth in the number of physics laboratories, the Cavendish retained its leadership as a research center, even strengthening its position after 1896. Its graduates began to dominate British physics very much as Cambridge wranglers traditionally dominated British mathematics. Indeed, fully half of the laboratory directors in 1896 were either Cambridge wranglers or students of the Cavendish.[63]

One of the fundamental differences that emerge from a comparison of the Cavendish with the other laboratories is that the Cavendish was a graduate student research laboratory and the only one of its kind in Britain. It remained so even when teaching and formal undergraduate laboratory training were gradually introduced after Maxwell's death. Beginning with Maxwell, the Cavendish had a sufficiently high number of full-time workers—twenty or so at any time—so that one person's research invariably stimulated another's. A similar, fruitful massing of researchers in one laboratory did not occur anywhere else in Britain, not even in Thomson's laboratory. Students stayed longer at the Cavendish than at other laboratories; often they had been awarded fellowships by their respective colleges, which removed financial pressures. A laboratory director could provide an entirely different research project if he knew that his student would stay two or three years instead of two or three months. Only at the Cavendish, for example, could Rayleigh have found the resources necessary to determine simultaneously by three different methods the value of the ohm. The influence of the productive environment is evident even on men who were already highly productive before coming to the Cavendish. It was even so with Rayleigh himself, whose peak of productivity coincided with his directorship of the Cavendish; despite a protracted illness and trips abroad, he wrote sixty-six papers, an average of one every month.

Rayleigh resigned after five years, but by then the Cavendish tradition largely had been formed. His successor, J. J. Thomson, who

[63]Among them Thomson (Glasgow), Schuster (Manchester), Tait (Edinburgh), Adams (King's College, London), Clifton (Oxford), Niven (Aberdeen), Poynting (Birmingham), Hicks (Sheffield), and Griffiths (Cardiff).

was elected at the comparatively young age of twenty-eight, had done research in Balfour Stewart's laboratory while still an undergraduate at Owens College. He had considered going into engineering, but a fellowship to Cambridge changed his plans. There he prepared for the mathematical tripos and while doing so missed an opportunity to work at the Cavendish under Maxwell; he began his research only after Rayleigh had become director. As director himself, Thomson kept the laboratory on the same course as his predecessors' for the next ten years. It was Maxwell's electromagnetic theory that provided continuity to the Cavendish research program during its first twenty years and across three directorships. Electrical standards, too, fitted into the research program of the Cavendish, but the research problems raised by the electromagnetic theory gradually forced industrial problems into the background. Many students at the Cavendish were ideally suited by their mathematical tripos training to handle the complex mathematics of electromagnetic theory and to carry out the difficult experiments suggested by the theory.

The Cavendish was a financially independent laboratory from the start.[64] By the donation of his own equipment, the transfer of the equipment of the electrical standards committee, and an arrangement made with Stuart's engineering workshop, Maxwell maintained its financial independence. As a result he was able to stress research as the main function of the laboratory, whereas directors of other physics laboratories were forced to stress elementary teaching. Two expansions of the laboratory—in 1894 and 1908—were paid for by student fees and by Rayleigh's donation of his Nobel Prize money, and the laboratory's independence was retained. The only other financially independent laboratory was that of Thomson, who covered most laboratory expenses from his own pocket.

The character of a British physics laboratory depended very much on the director and his scientific caliber. Even though at each institution the physics laboratory responded to different pressures and performed different functions, the director was decisive. It was he who made use of the opportunities that presented themselves and impressed on the laboratory its direction. William Thomson, Maxwell, Rayleigh, and J. J. Thomson were in a scientific class by themselves, and so were their laboratories.

[64] Romualdas Sviedrys, "The Rise of Physical Science at Victorian Cambridge," *Historical Studies in the Physical Sciences, 2* (1970), 127–145.

Most of the early laboratory directors enjoyed long tenures. Everett held his chair for thirty years, Foster his for thirty-three, Adams his for almost forty, and Tait his for forty. Clifton and William Thomson held theirs the longest—for fifty and fifty-three years, respectively. Only at the Cavendish was there a talented and young director who responded fully to the challenges of the discoveries of radioactivity, X rays, and the electron. But even J. J. Thomson, too, in time settled into a routine, and by 1910 the Cavendish saw competing research centers emerge. Arthur Schuster's University of Manchester laboratory became a competitor of the Cavendish only after 1907, when he offered his chair to the Cavendish graduate Ernest Rutherford.[65] A second research center of high caliber emerged at the University of Leeds when in 1909 William Stroud stepped down from his chair and the Cavendish graduate William Henry Bragg was elected to replace him.[66] In 1919 Rutherford succeeded J. J. Thomson at the Cavendish; the moment had come when practically every physicist of note in Britain was either a Cavendish graduate or had studied under one.

ACKNOWLEDGMENT

It is a pleasure to acknowledge the generous help and criticism I have received from Paul Forman, Russell McCormmach, and Arnold Thackray.

[65] J. B. Birks, ed., *Rutherford at Manchester* (London, 1962).

[66] E. N. da C. Andrade, "William Henry Bragg 1862-1942," *Obituary Notices of Fellows of the Royal Society of London, 4* (1943), 271-300.

The Establishment of the Royal College of Chemistry: An Investigation of the Social Context of Early-Victorian Chemistry

BY GERRYLYNN K. ROBERTS*

The Royal College of Chemistry opened in London in October 1845. During its first eight years when it relied exclusively on private support and consequently labored under severe financial constraints,[1] the College, under the academic direction of A. W. Hofmann, became one of England's most productive scientific centers in terms both of the individuals to whose training it contributed and, concomitantly, of the scientific work which it generated.[2] Retrospective accounts by contemporary witnesses of the establishment and development of the Royal College of Chemistry, as well as subsequent historical treatments, have tended to follow Hofmann's lead in evaluating the significance of the College in terms of these successes in teaching and research.[3] Hofmann evaluated the College against the standard of the science pursued in the German universities, and he tended to argue backwards from the achievements he valued to the goals of the English institution. The establish-

*Faculty of Arts, The Open University, Walton Hall, Milton Keynes, MK7 6AA, England.

[1]At its opening the Royal College of Chemistry was backed by pledges of about £3900 from some 760 individuals. By 1853, when the College was taken over by the government and incorporated into a general science school, the number of supporters had dwindled to a handful and student fees provided the bulk of its income. Total contributions, apart from fees, up to 1853 amounted to about £10,300. The names and addresses and the amounts pledged and given to the College up to 1852 are recorded in Royal College of Chemistry, "Alphabetical List of Members, 1844–1846," Imperial College London, College Archives, C/2; and Royal College of Chemistry, "Subscribers and Donors, 1846–1852," Imperial College London, College Archives, C2/2. Some summary figures are also in Royal College of Chemistry, *Report of the Royal College of Chemistry and Researches Conducted in the Laboratories in the Years 1848–49–50–51* (London, 1851); and in "Royal College of Chemistry," in *First Report of the Science and Art Department of the Committee of Council on Education* (London, 1854), p. 416.

[2]Between October 1845 and July 1853, some 356 students enrolled for varying periods of study. See Royal College of Chemistry, "Register of Students, 1845–1859," Imperial College London, College Archives, C/6/1.

ment of the College was, in Hofmann's view, a "natural" response in institutional terms to the claims of a new scientific discipline and a new pedagogy—analytical organic chemistry and systematic laboratory instruction in research methods—both of which had been developed abroad.[4] As Hofmann presented it, the College owed its establishment to the enthusiasm that his German teacher Justus Liebig generated in England during the early 1840's for the possibility of applying the fruits of chemical research.

Although Hofmann's view is not without foundation, it does not provide a completely satisfactory basis for evaluating the significance of the Royal College of Chemistry. Liebig's works of chemical propaganda were cited extensively in the prospectuses circulated to promote interest in establishing the College; a Liebig pupil was appointed its first director; and the academic objectives of the institution were derived from Liebig's laboratory program at the University of Giessen. However, it is unclear how much enthusiasm for chemistry Liebig would have generated had there been no previous interest in the subject in England. The College's propagandists cited Liebig's works to reinforce indigenous trends already underway before the 1840's, so Liebig received further publicity through

---

The College's productivity is investigated in Gerrylynn K. Roberts, *The Royal College of Chemistry (1845-1853): A Social History of Chemistry in Early-Victorian England* (Ph.D. diss., Johns Hopkins University, 1973), Chapter 8 and Appendix II; hereafter cited as Roberts, *RCC.*

[3] A. W. Hofmann, "A Page of Scientific History: Reminiscences of the Early Days of the Royal College of Chemistry," *Journal of Science, 8* (1871), 145-153; Sir F. A. Abel, "The History of the Royal College of Chemistry and Reminiscences of Hofmann's Professorship," *Journal of the Chemical Society, 69,* pt. 1 (1896), 580-596; Lyon Playfair, "Personal Reminiscences of Hofmann and of the Conditions which Led to the Establishment of the Royal College of Chemistry and the Appointment of its Professor," *ibid.,* pp. 575-579; Theodore G. Chambers, *Register of the Associates and Old Students of the Royal College of Chemistry, the Royal School of Mines and the Royal College of Science with Historical Introduction and Biographical Notices and Portraits of Past and Present Professors* (London, 1896); and more recently, John J. Beer, "A. W. Hofmann and the Founding of the Royal College of Chemistry," *Journal of Chemical Education, 37* (1960), 248-251; Sir Patrick Linstead, *The Prince Consort and the Founding of Imperial College* (London, 1961); see also *idem, Notes and Records of the Royal Society, 17* (1962), 15-32; Jonathan Bentley, "The Chemical Department of the Royal School of Mines: Its Origins and Development under A. W. Hofmann," *Ambix, 17* (1970), 153-181.

the movement to establish the College of Chemistry.[5] To focus on the German connections of the College is to stress unduly its academic and research functions and to neglect its other purposes. Hofmann's evaluation of the College from the German point of view obscures the fact that it was an English institution shaped by its early-Victorian context. The establishment of the Royal College of Chemistry must be investigated in the light of the contemporary needs and aspirations which gave it form.[6]

This study examines the issues that were of concern to the groups which received and responded to the appeal for support for an English chemical school during the early 1840's, and it shows how these issues came to be reflected in the institution itself. The founders of the College solicited support from medical men, chemists and druggists, agriculturists, and manufacturers, as well as from members of existing scientific societies. They also sought government funds. Few of the individuals most actively involved in the College's early history were widely influential scientific or public figures, and few were primarily interested in chemistry; the College was founded by people looking for solutions to what were, for the most part, particular chemical problems. Nonetheless, their efforts in founding the

---

[4]Hofmann, "Page of Scientific History," p. 145. It is not surprising that, writing from the vantage of his German chemical chair, Hofmann should have construed his English experience in terms of his subsequent German career. He noted with some irony that while preparing this idealized reminiscence of the College he was also preparing testimony on the teaching of chemistry in Germany for the Devonshire Commission, which was seeking educational models for application in Britain. If Hofmann's view of the College was accurate, the ideal model had already existed in England for twenty-five years (*ibid.*, p. 153).

[5]*Proposal for Establishing a College of Chemistry for Promoting the Science and its Application to Agriculture, Arts and Medicine* (London, 1844), British Museum, Additional MSS, No. 40553, ff. 21–29. See also, *To Agriculturists: Supplement to the Proposal for Establishing a College of Chemistry* (London, n.d.), and *Supplement to the Proposal for Establishing a College of Chemistry: To the Proprietors of Mines and Metallurgists* (London, n.d.). Both supplements can be dated, see Royal College of Chemistry, "Minutes of the Council of the College, 1845–1851," 5 March 1845 and 25 January 1845, Imperial College London, College Archives, C/3/3.

[6]J. B. Morrell, "Individualism and the Structure of British Science in 1830," *Historical Studies in the Physical Sciences, 3* (1971), 192–201, indicates how richly diversified the institutional context of English science was.

College reveal general early-Victorian attitudes towards science.[7] The Royal College of Chemistry was more a product of English perceptions of the functions of science than a response to continental scientific developments.

THE EARLY-VICTORIAN CHEMICAL CONSTITUENCY

Throughout the early nineteenth century, the study of chemistry was closely allied to the study of medicine in England. The educational demands of the medical profession had long influenced chemical curricula, and the medical schools remained important centers for the study of chemistry through the 1840's.[8] During the 1830's and 1840's the English medical profession underwent reorganization, and consequently education was a prominent issue within the medical community.[9] It was then that, for professional as well as scientific reasons, laboratory instruction in practical chemistry (other than by apprenticeship) became required of medical students. The traditional three-tiered division of medical practitioners into physicians, surgeons, and apothecaries still applied legally, but in practice the boundaries were becoming blurred. There was an increasing tendency for members of all divisions of the profession to assume the responsibilities of a new sort of medical man, the general practitioner, while still legally retaining their traditional functions and educational requirements as set out by the Royal College of Physicians, the Royal College of Surgeons, and the Worshipful Society of Apothecaries. Meanwhile, a newer division of medical practi-

[7]This approach to the study of institutions is discussed in Steven Shapin and Arnold Thackray, "Prosopography as a Research Tool in the History of Science: The British Scientific Community, 1700 to 1900," *History of Science, 12* (1974), 1-28.

[8]See the unpublished dissertation (University College London, 1969) by J. K. Crellin, "The Development of Chemistry in Britain through Medicine and Pharmacy: 1700-1850," pp. 14, 20.

[9]For a general view of the relationship between problems of professional reorganization and medical education, see S. W. F. Holloway, "Medical Education in England, 1830-1858: A Sociological Analysis," *History, 49* (1964), 299-324. Holloway evaluates changes in the structure of the medical profession as a response to long-term changes in English society brought about by the Industrial Revolution. See also, Charles Newman, *The Evolution of Medical Education in the Nineteenth Century* (London, 1957), and Sir George Clark, *A History of the Royal College of Physicians of London* (Oxford, 1966), *2,* 671-696.

tioners—the chemists and druggists—moved to consolidate its position.[10]

The emergence of the general practitioner was reflected in institutional developments. The Apothecaries' Act of 1815 recognized that the apothecary was by then essentially a general practitioner, and it granted the Society of Apothecaries the sole right to license general practitioners in England and Wales.[11] The license was awarded on the basis of examinations set by the Society, covering courses which candidates could, and generally had to, study at a number of recognized medical schools. A further institutional stimulus toward general practice was the founding in 1826 of the Medical Faculty of University College London.[12] Although aware that its medical graduates would probably have to qualify to practice through one of the English licensing authorities, University College modelled its medical curriculum mainly on that of the University of Edinburgh. It offered a four year course of systematic training in medicine and the ancillary sciences, providing a concerted program for general practitioners that could be followed within a single institution.

Keeping abreast of Scottish developments, University College offered practical chemistry to medical students in 1829,[13] the year that the Royal College of Surgeons of Edinburgh began to require of its candidates three-months training in practical chemistry. Senti-

[10] Traditionally, the functions of the three orders of medical practitioners were well defined: the sphere of the physician was internal medicine, and his function was to diagnose and write prescriptions; the sphere of the surgeon was external medicine, and his function was to perform operations; the original function of the apothecary was to compound medicines prescribed by physicians and required by surgeons. See Holloway, "Medical Education," pp. 304-311, and Newman, *Medical Education,* pp. 1-22. The emergence of chemists and druggists in the eighteenth century is discussed by Crellin, "Development of Chemistry," p. 234.

[11] S. W. F. Holloway, "The Apothecaries' Act of 1815: A Reinterpretation," *Medical History, 10* (1966), 107-129, 221-236. Holloway analyzes this Act in terms of the social history of the medical profession and its intraprofessional rivalries.

[12] For the sake of clarity, I will refer to this institution as University College, although from 1826 to 1836 it was known as the London University and sometimes called the University of London. See Hugh Hale Bellot, *University College London: 1826-1926* (London, 1929), pp. 143-168, 215-248; John Rose Bradford, "University College London and Medical Education," in *Centenary Addresses,* ed. R. W. Chambers (London, 1927), pp. 5-26.

[13] Bellot, *University College,* p. 125.

ment in favor of compulsory practical chemistry for medical students developed in Scotland during the 1820's among those who felt that practical training was essential for understanding chemical science. Before the Scottish Universities Commission formally recommended in 1831 that the subject be required in university medical curricula, practical chemistry had been taught at the University of Glasgow and, extramurally, in Edinburgh; but attendances at these voluntary classes were low and finances precarious.[14] The first Professor of Chemistry at University College, Edward Turner, had been an extramural teacher of practical chemistry in Edinburgh, which may have been the reason for his London appointment.[15] Scottish precedents may also have prompted the Society of Apothecaries to require in 1830 that its chemistry lecturers have a laboratory and "competent" apparatus, and in 1832 to recommend that they include training in chemical manipulation and analysis in their basic chemistry courses.[16] In October 1835, by which time the University College practical course had lapsed (except for an extramural class given by one of Turner's early students), the Society of Apothecaries began to require evidence of training in practical chemistry from candidates for its license.[17]

[14]J. B. Morrell, "Practical Chemistry in the University of Edinburgh," *Ambix, 16* (1969), 69–73. Also three months of practical chemistry were required for candidates of the Army Medical Department. See *Evidence, Oral and Documentary Taken Before the Commissioners for Visiting the Universities of Scotland with Appendix and Index–Edinburgh. Parliamentary Papers, XXXV* (1837), 533. Hereafter cited as *Edinburgh Evidence.*

[15]Bellot, *University College,* pp. 127–128.

[16]Society of Apothecaries, *Regulations to be Observed by Students Intending to Qualify Themselves for Practise as Apothecaries in England and Wales, 1832* (London, 1833), p. 4. There was some rivalry between the Royal College of Surgeons of Edinburgh and the Society of Apothecaries. The former, with some reason, considered the qualifications of the latter to be inferior to its own. Yet the Apothecaries' Act of 1815 had granted the Society of Apothecaries alone the right to license general practitioners in England ("Return by the Royal College of Surgeons, Edinburgh to the Commission," *Edinburgh Evidence,* Appendix, p. 210).

[17]Society of Apothecaries, *Regulations to be Observed by Students Intending to Qualify Themselves for Practise as Apothecaries in England and Wales, 1835* (London, 1835). Evidence for the extramural class is cited in Bellot, *University College,* p. 127. Ironically the practical chemistry course at University College probably lapsed due to low enrollment, since none of the English qualifying bodies required the subject; yet the Society of Apothecaries' more stringent 1835 requirements, including practical chemistry, were instituted as a result of the overall success of the University College medical curriculum (*ibid.,* pp. 215–248).

The Medical Faculty of University College, anxious to maintain its reputation, reacted to the Society's new requirement by suggesting that Turner institute a practical course. He prevaricated, arguing correctly that the requirement would not become operative for some time. The following year College authorities gave Turner a direct order at the behest of the Medical Faculty to organize a course of practical chemistry for the summer of 1837, but he died in February.[18] The ability to teach practical chemistry to medical students was undoubtedly a consideration in the choice of Turner's successor in the Chair of Chemistry, Thomas Graham.[19] Graham learned chemistry within the context of the Scottish practical tradition. He studied at the University of Glasgow under Thomas Thomson, and at the University of Edinburgh under T. C. Hope, whose failure to give practical instruction stimulated the establishment of extramural practical classes. In addition, Graham had some experience teaching practical chemistry as a private instructor and at the Andersonian Institution.[20]

During 1837-1838, Graham offered one term of practical chemistry, and during the following academic year, two terms. However, enrollment was not high, despite the Society of Apothecaries' requirement, the relatively low fee of £4 per term, and the minimal commitment of three hours per week.[21] The students were not alone in lacking enthusiasm for the new course; Graham himself was reluctant to teach it. To the College Secretary, he expressed a

18 University of London, "Senatus Minute Book," 9 January 1836, 19 March 1836, and 27 January 1837, University College London, Records Office. Hereafter cited as University College London, "Senate Minutes."

19 Graham realized that the appointment was essentially in the hands of the Medical Faculty (Thomas Graham, letter to James Graham, 29 April 1837, in R. Angus Smith, *The Life and Works of Thomas Graham* [Glasgow, 1884], p. 39). Certainly the other short-listed candidates, Richard Phillips and David Boswell Reid, were well qualified to teach practical chemistry. See Roberts, *RCC*, pp. 23-28.

20 Thomson's course is described in J. B. Morrell, "Thomas Thomson: Professor of Chemistry and University Reformer," *British Journal for the History of Science, 4* (1969), 245-265, and *idem,* "The Chemist Breeders: The Research Schools of Liebig and Thomas Thomson," *Ambix, 19* (1972), 1-46. On Hope's limitations, see *idem,* "Practical Chemistry." On Graham's experience, see Thomas Graham, letter to Mrs. Graham, 19 November 1827, in Smith, *Thomas Graham,* pp. 31-32.

21 University College London, *Annual Reports* (London, 1837, 1838, and 1839). Only about ten percent of those who should have taken it in the first

financial complaint that was to recur in all efforts to institute courses on practical chemistry in England: "The teaching of practical chemistry in London is likely to have the most beneficial influence on the progress of experimental science, otherwise I would not think of undertaking the extraordinary trouble and pecuniary loss which it will for some time involve."[22] Commenting that his first course had run at a deficit, Graham pointed out that the system was at fault. A professor's only funds were students' fees, from which he had to meet all course expenses (including outfitting a laboratory) as well as derive a personal income. High fees risked reduced enrollments; low fees forced the professor to take on other jobs to earn a living. If he took on other jobs, he had less time for teaching practical chemistry, which conflicted with one of the most important features of laboratory instruction, close personal supervision. For financial reasons, professors favored large lecture courses and viewed small laboratory courses with dismay.

At King's College London, where, as at University College, medical studies were important, the Society of Apothecaries' new requirement met resistance at first. J. F. Daniell, Professor of Chemistry, argued that there was more than one definition of the term "practical chemistry."

> It is probable that [the Society of Apothecaries] meant to require that students should be instructed in chemical manipulation—i.e.

---

year and about twenty percent of those who should have taken it in the second year enrolled. One possible reason was that Graham's course was not medically oriented: "The student will be exercised in conducting processes from all the departments of chemistry, & in the manipulations of testing and analysis. He will have an opportunity of becoming acquainted w. many mineral substances used in the arts particularly the metals and their ores, & receive instructions in assaying. A course of reading will also be pointed out to him from which he may derive explanations and additional information respecting the processes which engage his attention" (Thomas Graham, letter to C. C. Atkinson, late November 1838, University College Correspondence, No. 4416, University College London, College Archives).

[22] Thomas Graham, letter to C. C. Atkinson, 1 December 1838, University College Correspondence, No. 4440. Wanting to keep the course, the Medical Faculty forced the College to provide additional money (University of London, "Committee of Management Minutes," 14 March 1839 and 22 March 1839, University College London, Records Office; University College London, "Council Minutes," 7 August 1839, University College London, Records Office).

> that their hands as well as their heads should be instructed and that it should be certified that each had worked at chemical operations with his own hands, under competent superintendance.
>
> But practical chemistry has a much larger meaning than this of handicraft and the lectures which I am in the habit of delivering from this table are meant to convey in common with those of other schools, instruction in Practical as well as Theoretical chemistry. . . . [I] dwell upon the practical application of science to the Arts and illustrate the processes of the different manufactures, of Metallurgy, Domestic Economy, Pharmacy etc.—and all this might be, and frequently is, learnt without touching a retort, handling the crucible tongs, or soiling the fingers with aqua fortis.[23]

Daniell was content to sign the Society of Apothecaries' certificates of attendance in practical chemistry for students who attended his usual course, particularly in view of the financial problems of laboratory teaching. He had to alter his course, however, when the University of London issued the requirements for its new medical degrees in the spring of 1839. The curriculum included one term of practical chemistry, and no medical school was to be recognized unless it had a laboratory in which students gained practical experience in chemistry, pharmacy, and forensic chemistry. Although still opposed in principle to giving a laboratory course, partly because he felt that one term would allow only the most superficial training, Daniell capitulated. The first laboratory course in practical chemistry at King's College London began in the spring of 1840; it was taught by Daniell's assistant, whom he had to pay privately.[24]

London members of the medical profession had actively promoted the establishment of the new University of London, which was chartered in 1836 as a degree granting body with no teaching functions. The University was empowered to devise course requirements leading to degrees in the arts or medicine and to set examinations; success in these examinations was the student's route to a degree. It

[23] J. F. Daniell, "Inaugural Lecture for the Chemistry Course in 1840-1841," c. October 1840, copy in the Daniell Collection, King's College London, Library.

[24] *Ibid.* Daniell tried to prevent this requirement from entering the new university curriculum (J. F. Daniell, letter to Sir John Lubbock, 29 November 1838, Lubbock Papers, No. D-7, Royal Society Archives). The financing of the course is mentioned in King's College London, "Minutes of the Council," 13 March 1840, King's College London, Library.

was hoped that for intending medical men the new degree would constitute a license to practice, thus circumventing the authority of the three traditional English professional bodies.[25] Accordingly, the designers of the University's medical curriculum tried to make it as scientific and rigorous as possible. Based on the Scottish model, the University's course requirements led to a two-tiered division of medical practitioners. Every student received the same basic training; he qualified as a general practitioner by passing an examination for the M.B. degree after four years of course work. If he wished to specialize he took an examination for the M.D. degree two years later. Basic instruction in the sciences was fundamental to the M.B. course; modern, practical training helped define the special expertise of the University of London medical graduate. Practical training was emphasized to the extent that the examinations were to include laboratory operations.[26]

The University's practical chemistry syllabus was medically oriented: it included "Practical Exercises in conducting the more important processes of General and Pharmaceutical Chemistry; in applying tests for discovering the adulteration of articles of the materia medica, and the presence and nature of poisons; and in the examination of Mineral Waters, Animal Secretions, Urinary Deposits, Calculi, etc."[27] It is unlikely, however, that the practical chemistry course at King's College London followed this syllabus, since it was required by both the Medical Department and the Department of Civil Engineering and Science as Applied to the Arts and Manufactures. Furthermore, the University College London Medical Faculty, which favored a more general chemical training for its students, opposed the University's vocational emphasis in its practical chemistry syllabus.[28] By 1840, in addition to the courses at King's and University

[25] Bellot, *University College,* Chapter 7. University College London, "Council Minutes," *passim* and University College London, "Senate Minutes," *passim* are instructive on the expectations for the new medical degree in the early 1830's.

[26] University of London, "Minutes of the First Subcommittee of the Faculty of Medicine: Subcommittee to Consider the Course of Study Required of Candidates for Degrees in Medicine who shall hereafter Commence their Medical Studies, August 8, 1837–October 12, 1837," 22 August 1837. University of London, Senate House Library.

[27] University College London, *Annual Reports* (London, 1840), p. 19.

[28] King's College London, *Calendar for the Year 1840* (London, 1840), pp. 14, 34. For the University College view, see University of London, *Minutes of the Senate, Vol. 1: March 4th 1837 to June 21st 1843* (London, n.d.), 27 March 1839, University of London, Senate House Library. Also cf. Graham's syllabus, *supra,* n. 21.

Colleges, some training in practical chemistry was offered at four London medical schools and three provincial ones.[29]

From 1841, chemists and druggists also actively began to promote the study of practical chemistry to enhance their professional status. Practical chemistry was a major concern of the Pharmaceutical Society of Great Britain, which was established in 1841 with the immediate object of lobbying against a bill that would have effectively placed chemists and druggists under the jurisdiction of apothecaries.[30] The functions of these two medical groups overlapped and caused rivalry. Chemists and druggists were specifically exempted from the 1815 Apothecaries' Act, although their role was defined in a negative manner within it. Essentially, chemists and druggists were to be dispensing tradesmen; as such they duplicated in part the duties of apothecaries, who were entitled to dispense medicines as well as to act as general practitioners. Furthermore, as population grew, straining the resources of the medical profession, chemists and druggists became in many instances the poor man's medical practitioners, illegally prescribing over the counter. Indeed, in the country, chemists and druggists were often the only medical practitioners. Unlike apothecaries, who had to undergo formal training and whose premises were subject to periodic inspection, chemists and druggists had no educational requirements to fulfill before setting up shop (although apprenticeship was the normal route), and they were not subject to legal sanctions for fraudulent practice. Apothecaries resented the encroachment by "unqualified" chemists and druggists on their functions and income. In their defense, chemists and druggists argued that they performed a needed service and that apothecaries had abdicated their dispensing rights by becoming general practitioners. Chemists and druggists in rural areas faced additional competition from grocers and oilmen who undersold them on patent medicines.[31]

[29] *The Medical Almanack or Calendar of Medical Information, 1840* (London, 1840). London schools were St. Bartholomew's Hospital, Guy's Hospital, the Aldersgate Street Medical School, and Sydenham College. In the provinces, practical chemistry could be studied at medical schools in Birmingham, Bristol, and York.

[30] Jacob Bell and Theophilus Redwood, *Historical Sketch of the Progress of Pharmacy in Great Britain* (London, 1880), pp. 87-88.

[31] Holloway, "The Apothecaries' Act," p. 125; Bell and Redwood, *Historical Sketch,* pp. 80, 91; "Pharmaceutical Meetings," *Pharmaceutical Journal, 1* (1841), 6-7; Jacob Bell, "General Observations by the Editor," *Pharm. J., 1* (1841), 38-39; Holloway, "Medical Education," pp. 311-312.

Members of the Pharmaceutical Society objected to the threatened imposition of regulations by outside bodies. They admitted the need for professional control, but wanted to provide their own by means of pharmaceutical education. Their object was to create a new professional role for the English practitioner after the French and German model. By studying the sciences on which his art was based, particularly practical chemistry, the chemist and druggist would elevate himself from the role of tradesman to that of practicing scientist, the pharmaceutical chemist. Selling drugs was still to be the business of the new pharmaceutical chemist, but the addition of scientific research to his responsibilities would expand his role. With the cooperation of other medical practitioners, he would help to advance the theory and practice of medicine by studying the production and the mode of action of known medicines and by discovering new medicines. Furthermore, the pharmaceutical chemist's certified scientific knowledge would be seen by the consumer as a guarantee of high quality drugs, so that his business prospects as well as his professional status would be enhanced.

The Pharmaceutical Society was to have a central role in creating the new pharmaceutical chemist. It was to establish a school, serve as an examining board, become the legal licensing body for chemists and druggists, and operate as a learned society complete with a journal for pharmaceutical subjects. The members hoped that by raising their status by means of formal qualifications, they would eliminate their conflicts with other sectors of the medical profession.[32] The Pharmaceutical Society's school opened in the fall of 1842, but the promised facilities for a laboratory course in practical chemistry did not materialize until October 1844. When a laboratory accommodating only eight students finally opened, the first practical course disappointed the reformers, since it consisted of routine pharmaceutical operations rather than general practical chemistry. Late in May 1845, the Society's Council approved the expansion of the inadequate teaching laboratory to eighteen places, and it seems that these were filled sometime during the 1845-1846 academic year.[33]

[32] Jacob Bell, *Observations Addressed to the Chemists and Druggists of Great Britain on the Pharmaceutical Society* (London, 1841), pp. 6-8; "Pharmaceutical Meetings," pp. 4-5; Bell, "General Observations," pp. 42-43, 75-79.

[33] Bell, *Observations,* p. 10; T. E. Wallis, *History of the School of Pharmacy, University of London* (London, 1964), p. 3; Joseph Ince, "The History of the

Many chemists and druggists supported the objects of the Pharmaceutical Society, but not all were satisfied with its performance. The periodical *The Chemist or Reporter of Chemical Discoveries and Improvements and Protector of the Rights of the Chemist and Chemical Manufacturer* published vitriolic attacks on the Pharmaceutical Society and its founder, Jacob Bell. The editors pointed to the wildly fluctuating membership of the Society as a symptom of the profession's lack of confidence in it, and they claimed that members were paying high fees (two guineas per year) and receiving little in return. They particularly criticized the delay in providing a laboratory and the pedestrian practical course taught there. In their view, the Pharmaceutical Society had fulfilled few of its goals by the time its third anniversary arrived in March 1844, leaving the profession vulnerable in the face of the medical reform movement of 1844.[34]

During the 1830's and early 1840's, the tendency to promote practical chemistry as a medical subject was reinforced by developments internal to the subject matter of chemistry. Organic chemistry emerged as a new discipline from chemical studies of the constituent parts of plants and animals. Although there was great controversy about the fundamental theories of organic chemistry, the analytical methods developed for its investigation (primarily in the Giessen laboratory of Justus Liebig) were sound and accurate. In turn, the new methods began to be used to study more systematically the processes through which organic constituents were produced in both healthy and diseased living systems; organic chemistry came to be viewed as the key to the study of the physiology and pathology of both plants and animals.[35] Consequently, medical men, particularly

---

School of Pharmacy," *Pharm. J.*, pt. 1 (1903), p. 282. Ince suggests that Bell tried to push through plans for a laboratory in 1842 but was balked by the Council.

[34] "A Bill for the Better Regulation of Medical Practice throughout the United Kingdom," *The Chemist, 2* (1844), 418; see editorial comments under varying titles throughout (*ibid.*, pp. 126–127, 144, 178, 226–227, 323–324, 555). See also *Report from the Select Committee on the Pharmacy Bill: together with the Proceedings of the Committee, Minutes of Evidence and Index, Parliamentary Papers, XII 387* (1852), qq. 954–969, p. 56. See also, Roberts, *RCC*, p. 47.

[35] Justus Liebig, *Organic Chemistry in its Applications to Agriculture and Physiology*, ed. Lyon Playfair (London, 1840). Hereafter, cited as Liebig, *Agricultural Chemistry*. See also, Justus Liebig, *Animal Chemistry or Organic Chemistry in its Applications to Physiology and Pathology*, ed. William Gregory (London, 1842).

those involved with the University of London medical program, argued that organic chemistry should become part of the new curriculum.

The cause of analytical organic chemistry was ardently taken up by Thomas Wakley, radical politician, medical reformer, and editor of the polemical medical journal *The Lancet.* One of Liebig's earliest English promoters, Wakley linked the new discipline to the movement for medical reform. A strong supporter of the two-tiered division of medical practice as institutionalized in the University of London's medical curriculum, Wakley argued that the London M.D. should be a research degree and that the chemical investigation of physiological problems begged for the attention of researchers. He gave publicity to the system of chemical education evolved by Liebig, whose students learned how to do research while learning the subject of chemistry. Wakley argued that the techniques of practical chemistry, including elementary organic analysis, should be studied by candidates for the M.B., so that they would be prepared to do research at the M.D. level. Should the M.B. candidate not go on to the M.D., he would nonetheless find his knowledge of organic chemistry useful for making pathological tests and for understanding subsequent advances in treatment. The major obstacle that Wakley saw was the lack of suitable facilities for studying practical organic chemistry in England.[36]

Practical chemistry attracted the support of English medical men in the early 1840's because, during the preceding decade, it had become increasingly relevant to their changing professional and scientific interests, which in turn reflected broader changes in English society. Similarly, by the 1840's, practical chemistry was seen to be relevant to the concerns of the landed interest in a number of ways. The use of scientific knowledge to improve agriculture and the exploitation of mineral resources had long attracted the attention of

[36]See numerous articles in *The Lancet:* "Review of *Elementary Instruction in Chemical Analysis . . . ,*" 21 October 1843, pp. 101–102; "Attendance on Liebig's Class," 27 January 1844, p. 591; "The Laboratory at Giessen," 18 May 1844, p. 261; "Remarks on the New College of Chemistry," 7 September 1844, p. 736; "The Science of Chemistry," pp. 231–232. The science of organic chemistry was also considered to be important for the new professional pharmacist, since new medicines would depend on discoveries in this field and many known ones were of organic origin.

the landed interest in England.[37] During the decade preceding the establishment of the Royal College of Chemistry, the British Association for the Advancement of Science, the Royal Agricultural Society of England, and the government all began to promote the applications of chemistry, particularly of analytical chemistry, to the problems of landownership. Throughout the 1830's, landowners attributed perhaps more importance to geological than chemical knowledge. They came to recognize, however, that geology needed to be supplemented by chemistry; the former could describe subterranean structure, but only chemical analysis could reveal its constituents. Soil analysis became an accepted agent of agricultural improvement.[38]

During the late 1830's, the British Association took considerable interest in the application of science to agriculture. One of its implicit objects was to serve as a clearing house for the collection and distribution of information (provided by the study of local geology, natural history, and chemistry) that might be of interest to landowners.[39] To enable members to take advantage of the information, practicing scientists were asked to report on the latest developments in their various fields. Reports requested in the late 1830's included ones on organic chemistry, inorganic chemistry, the relationship between chemistry and agriculture, and the relationship between

[37]For an early-nineteenth-century example of the way this concern was expressed through scientific institutions, see Morris Berman, "The Early Years of the Royal Institution, 1799-1810: A Re-evaluation," *Science Studies, 2* (1972), 205-240.

[38]The view that geology without chemistry was insufficient was reinforced by a widely read monograph that suggested a classification of soils according to their underlying geological formation and that used data supplied by chemical analysis to correlate subsurface structure with soil content: John Morton, *On the Nature and Property of Soils, their Connection with the Geological Formation on which they Rest, and the Best Means of Permanently Increasing their Productiveness, and on the Rent and Profits of Agriculture* (London, 1838). See Sir E. John Russell, *A History of Agricultural Science in Great Britain, 1620-1954* (London, 1966), pp. 81-86.

[39]A. D. Orange, "The Origins of the British Association for the Advancement of Science," *Brit. J. Hist. Sci., 6* (1972), 152-176, and *idem,* "The British Association for the Advancement of Science: The Provincial Background," *Science Studies, 1* (1971), 315-329. In 1839, some members attempted unsuccessfully to establish a separate agricultural section; this was not accomplished until 1912 (O. J. R. Howarth, *The British Association for the Advancement of Science: A Retrospect, 1831-1931,* 2nd ed. [London, 1931], p. 85).

chemistry and geology.[40] At the 1838 meeting, Henry De la Beche, Director of H.M. Geological Survey, stressed the importance of the relationship between chemistry and geology for all landowners, agriculturists as well as mine owners. He successfully urged the government to establish a Mining Records Office as an adjunct to the Survey for use by landowners as a center for scientific information about land; he thus realized one of the British Association's objects.[41] In 1839, De la Beche successfully persuaded the government to spend £1500 on yet another extension to the Survey, an office of Curator to the Museum of Economic Geology. The Curator's duty was to serve as chemist to the Survey, analyzing and classifying samples sent in by field workers, solving chemical problems for other government departments, performing analyses for private individuals, and lecturing on analytical chemistry, agricultural chemistry, metallurgy, and mineralogy.[42] De la Beche's rhetoric in support of both these projects stressed the importance of soil analysis.

The Royal Agricultural Society of England, founded in 1838, was also concerned with the application of chemistry and geology to agriculture. By providing a means of communication among agriculturists, sponsoring experiments, and promoting improved agricultural education, the Society hoped to stimulate agricultural progress.[43] One of its activities was to offer prizes for work on topics of current interest. Among the prizes proposed in the first

[40] British Association, *Report for 1840,* p. xxiv.

[41] Margaret Reeks, *Register of the Associates and Old Students of the Royal School of Mines and the History of the Royal School of Mines* (London, 1920), pp. 13–14.

[42] Commissioners of Woods and Forests, letter to H. M. Treasury, Prospectus for the Museum of Economic Geology, 26 February 1839, Great Britain, Public Record Office, Treasury Papers, T. 1-3776. Compared to the government's previous financial record, £1500 was a large commitment. See Mr. Chawner, letter to the Commissioners of Woods and Forests, 20 September 1839, and T. C. Brooksbank and A. Van Spiegel, letter to the Treasury, 24 March 1840 (*ibid.*). It is difficult to determine how often the curator was consulted; during the final quarter of 1839, the income from analyses performed for the general public was only £18 (Henry De la Beche, letter to the Commissioners of Woods and Forests, 31 December 1839 [*ibid.*]). Reeks, *Royal School of Mines,* p. 15, suggests that the government used the office rather more than the public.

[43] J. A. Scott-Watson, *The History of the Royal Agricultural Society of England: 1839–1939* (London, 1939), pp. 18–19. The Society would not, however, directly finance experimental work (J. Hudson, letter to Sir Charles Gordon, 23 November 1840, English Agricultural Society, "Letter Book,"

year was one for a simple and cheap method for analyzing soils which could be used by practical farmers. Thus, the Society already recognized that analytical chemistry was an important subject for the agriculturist and even contemplated setting up its own laboratory with a chemist to perform soil analyses.[44] During 1841, Charles G. B. Daubeny, Professor of Chemistry, Botany, and Rural Economy at Oxford, urged the Society to take a more active interest in agricultural chemistry. Arguing that the future of agriculture rested in part on understanding the nature of soils and the mode of operation of chemical and other agents that altered soil conditions, Daubeny made a plea for systematic experimental research on this topic. Such research would require improved agricultural education, both for researchers and for ordinary agriculturists. He pointed out that extrinsic funding, either from agricultural societies or the government, would have to be found for such highly specialized training. To reinforce his plea, he pointed to continental institutions and suggested that English agriculture was not keeping pace with agricultural advances because it lacked similar institutions.[45]

Daubeny's somewhat belated report on chemistry and agriculture may well have been an attempt to capitalize on the interest aroused by the 1840 English edition of Justus Liebig's *Organic Chemistry in its Applications to Agriculture and Physiology.*[46] Liebig's intention

---

Royal Agricultural Society Archives). For the Society's view on financing educational institutions, see English Agricultural Society, "Minutes of the Committee of Management," 13 February 1839, Royal Agricultural Society Archives.

44 Royal Agricultural Society, "Committee of Management Minutes," 18 July 1838; Scott-Watson, *Royal Agricultural Society,* p. 19, indicates that the appointment of a chemist was discussed further in May 1839.

45 Charles B. Daubeny, "Lecture on the Application of Science to Agriculture, 9 December 1841," *Journal of the Royal Agricultural Society, 3* (1842), 136–157; *idem,* "Lecture on Institutions for the Improvement of Agriculture," 27 April 1842, Sherard MSS, No. 278, Oxford University, Bodleian Library; *idem,* "On Public Institutions for the Advancement of Agricultural Science which Exist in Other Countries, and on Plans which have been set on Foot by Individuals with a Similar Intent in Our Own," *J. Roy. Ag. Soc., 3* (1842), 386. Most of Daubeny's continental examples were French. He found his own research and teaching in Oxford restricted by lack of practical facilities (*idem,* "Diary: 1834–1867," 25 May 1846, 28 November 1846, Sherard MSS, No. 264).

46 This was Liebig's report on organic chemistry which the British Association assigned to him in 1838.

was to explain agriculture theoretically in terms of organic chemistry, justifying at the same time his chemical theories. The possibility that his theoretical approach might eventually lead to changes in agricultural practice was of course important; the book itself, however, was a program for research, rather than a guide to good practice.[47] Liebig began by discussing the chemical constituents of various parts of plants and their sources in nature. He then argued that the plant physiologist, the chemist, and the agriculturist together could determine what conditions (heat, light, components of the atmosphere, substances in the soil) influenced plant growth in general. Once they showed how all of these conditions could best be combined to promote the growth of specific plants, the practice of agriculture would become a simple matter of substituting analytical data into the following equation:

chemical composition of desired plant
(minus) ingredients available in soil and atmosphere
chemical composition of required fertilizer.

Liebig felt that much more research was needed before such a formula could be applied. For example, he suggested a nationwide series of plant-ash analyses that would look chemically at the same plants grown on different soils.[48] Essentially, Liebig suggested a program complementary to soil analysis which had already been accepted in England as a legitimate agricultural aid. He showed that plant-ash analysis and soil analysis should be pursued in parallel to aid agricultural practice most efficiently; fertilizers, natural or artificial, could only be understood as an intermediary between soil and plant. Liebig's ideas were favorably received in England partly because his research program complemented and even furthered ongoing ones.[49]

[47]Liebig feared, with some justification, that his English audience would misinterpret the book (T. Weymss Reid, *Memoirs and Correspondence of Lyon Playfair* [London, 1899], p. 47). Lyon Playfair constructed the second English edition in textbook fashion so that it could be appreciated by non-chemists (Justus Liebig, *Chemistry in its Applications to Agriculture and Physiology,* ed. Lyon Playfair, 2nd ed. [London, 1842], pp. vii–viii).

[48]Liebig, *Agricultural Chemistry,* pp. 138, 140.

[49]Certainly Liebig's contention that chemistry was totally ignored by English agriculturists was not justified (*ibid.,* p. 137). A particularly prominent supporter of Liebig's research program was Philip Pusey, an improving landowner and editor of *The Journal of the Royal Agricultural Society* (Philip Pusey, "On the Progress of Agricultural Knowledge during the Last Four Years," *J. Roy. Ag. Soc., 3* [1842], 169–215).

Partly because of Liebig's *Organic Chemistry,* Daubeny's efforts to institute training for agriculturists did not go unheeded. In late 1841 and early 1842, the government considered fostering scientific education for agriculturists. The Commissioners of Woods and Forests, who controlled the Mining Records Office and the Museum of Economic Geology, were particularly active in this regard, as was the Prime Minister, Sir Robert Peel. One plan was to add an agricultural department to an existing university, such as the University of Durham.[50] Another plan, proposed in the spring of 1842, was for the Museum of Economic Geology to undertake soil analysis on a nationwide scale. Such a plan would not offend agricultural conservatives, since its implementation had always been envisaged as the combined function of the Museum and the Records Office and since the Royal Agricultural Society had previously promoted a similar plan. At the same time, it was suggested that Liebig's former student Lyon Playfair should be employed by the government to carry out soil analyses at the Museum of Economic Geology. In reality, this plan was a thinly disguised attempt to establish a government-funded agricultural research institute and school.[51] Although the government project did not succeed, the attempt resulted in the creation by the Royal Agricultural Society of the (unsalaried) post of Analyst.[52]

Later in 1843, private individuals set about establishing a school at Cirencester for training agriculturists following a model suggested by Daubeny. In addition to a practical department, the school was to have an academic department. The latter would feature science courses ancillary to agriculture, which would include "manual participation" in teaching the techniques of soil analysis, "etc."[53] Daubeny ardently supported this proposed institution and suggested that the government finance it because such schemes as nationwide analyses were beyond the scope of individual efforts. He hoped that the school at Cirencester would be the first of a network of provincial agricultural schools deliberately sited on various geological

[50]Philip Pusey, letter to Sir Robert Peel, 16 January 1842, British Museum, Add. MSS, No. 40500, f. 166.

[51]Royal Agricultural Society, "Council Minutes," 2 March 1842, Royal Agricultural Society Archives. The disguise is mentioned in Reid, *Playfair,* p. 77. See Roberts, *RCC,* pp. 79-81.

[52]Royal Agricultural Society, "Council Minutes," 1 March 1843, 5 April 1843.

[53]"Agricultural College," *English Journal of Education, 1* (1843), 383-384; the "etc." is unfortunately not specified.

formations for undertaking national analytical projects.[54] By the end of 1844, the Agricultural College had received sufficient pledges of funds for its planners to apply for a charter; it opened in the fall of 1845. Meanwhile, a national plant-ash analysis project promoted by Liebig, Philip Pusey, and Daubeny was taken up at the British Association meeting in September 1844. Thomas Graham, Lyon Playfair, and Edward Solly were appointed to analyze the ashes of plants grown on different soils in the British Isles. The British Association voted £50 for the project with the proviso (ultimately fulfilled) that the Royal Agricultural Society also provide funds.[55] Liebig attended the 1844 British Association meeting which may have had something to do with the successful launching of this project. Commenting that the presence of country gentlemen at the sessions of the Chemical Section was a "new feature" at this meeting, Thomas Graham felt that they had been attracted by Liebig.[56]

Although agriculturists at the Chemical Section of the British Association may have been new in 1844, their acceptance of chemical analysis as a valid agricultural aid and their promotion of scientific education for agricultural ends were not. The possibility of the repeal of the Corn Laws increased their interest in scientific means for improving production; their increased interest was not overlooked by those seeking support for agricultural education.[57] Having weathered the severe depression of 1839–1842, English agriculturists did not welcome the threat of foreign competition. Whatever the out-

[54] Charles Daubeny, *A Lecture on Institutions for the Better Education of the Farming Classes, especially with reference to the Proposed Agricultural College near Cirencester, 14 May 1844* (Oxford, 1844), pp. 11, 27.

[55] British Association, *Report for 1844,* p. xxiii. The Royal Agricultural Society agreed and voted £350 for the project (Royal Agricultural Society, "Council Minutes," 5 February 1845). Solly was analytical chemist to the Royal Horticultural Society (Harold R. Fletcher, *The Story of the Royal Horticultural Society, 1804–1968* [London, 1969], p. 157).

[56] Thomas Graham, letter to Mrs. J. Reid, 29 September 1844, in Smith, *Thomas Graham,* p. 44.

[57] "As soon as the Corn-law question comes on for the discussion of parliament, the blow will be struck for the establishment of Agricultural Colleges, and your capital letter or rather the extract of it referring to the necessity of regenerating chemistry in England will privately be brought under the notice of our premier Sir Robert Peel by some of my friends at Court" (Lyon Playfair, letter to Justus Liebig, 27 December 1841, Liebigiana 58, Bayerische Staatsbibliothek, Munich). I am grateful to W. H. Brock of the University of Leicester for this reference.

come of the Corn Laws issue, it made sense to them to advocate science. If repeal succeeded, scientifically increased production would improve England's competitive position; if repeal failed, scientifically increased production would satisfy domestic demand. If the claims of scientific propaganda could be realized even partially, agriculturists' investment in science would be amply repaid.[58]

It was not only the possibility of applying chemical knowledge to agricultural problems that interested landowners in chemistry at this time. Philip Pusey pointed out that at the time that efforts were being made to install Lyon Playfair as organic chemist to the Museum of Economic Geology, landowners were more likely to consult a chemist about the mineral contents of their estates than about soils.[59] One reason was the rapid expansion of the railways which made demands on the nation's iron and coal resources. Moreover, in this era of large-scale land sales for railroad rights-of-way, it was prudent to avoid selling land that might prove valuable for other purposes. Landowners were also interested in the possibility of exploiting colonial resources.[60]

Practical chemistry was thought to bear on several contemporary interests of manufacturers and engineers no less than on interests of landowners and medical men. Indeed, some scientific propaganda of this period held up the use of chemistry by manufacturers as a

[58]Earl of Clarendon, "The Royal College of Chemistry," *London Medical Gazette, 2* (1846), 1011.

[59]Philip Pusey, letter to William Buckland, 27 October 1842, Papers and Correspondence of Lyon Playfair, No. 594, Imperial College London, College Archives.

[60]David Spring, "The English Landed Estate in the Age of Coal and Iron, 1830–1880," *Journal of Economic History, 11* (1951), 3–24, points out that landowners were seldom totally dependent on agriculture for their income. The profits they made from railways were more likely to result from land sales than from returns on capital investments. *The Mining Guide, Containing Particulars of Each Mine, British and Foreign, Its Situation and Produce* (London, 1853), pp. iii–iv, mentions that between 1843 and 1853, the number of English mining companies known in the market grew from under 50 to about 520. C. R. Fay has noted a growing "imperial" outlook during the 1830's and 1840's: "That a metropolitan school of science should start as a school of mines resting on geology was natural to a country which was itself rich in mineral resources and also had overseas possessions where new fields of study were constantly opening out; not as yet styled 'imperial', it was in principle imperial . . ." (*Palace of Industry: A Study of the Great Exhibition and its Fruits* [Cambridge, 1951], p. 112).

model for both agriculturists and chemists and druggists.[61] The manufacturing community had increasing opportunity to demonstrate its interest in science through metropolitan institutions such as the Chemical Society (established in 1841) and the Royal Society of Arts (revitalized in the mid-1840's). In addition, several new journals catering to the chemical interests of particular manufacturers were started during the early 1840's.[62] Furthermore, from the late 1830's a number of opportunities arose for manufacturers to take courses in applied science. Most of these courses belonged to programs for training engineers, all of which required chemistry.[63] Engineering was developing into a profession and the number of English engineers was growing rapidly. For engineering as for medicine, the promotion of strict academic entry requirements was a means of affirming professional status. The formal study of applied chemistry and other sciences was directly useful to prospective engineers and it also helped define their professional expertise.[64] The interest of engineers in chemistry was reinforced by the coincidence of new developments in engineering practice and in chemistry. Specialities—metallurgical, mining, and gas engineering—developed whose practitioners wanted to put their work on a more scientific basis.

[61] E.g., on agriculturists, see Daubeny, "Lecture on Institutions, 1844," p. 9; and William Stark, *A Letter to the Rt. Hon. Lord Wodehouse, President of the Norfolk Agricultural Association on the Use of Chemical Manures* (Norwich, 1844), p. 7. On chemists and druggists see, *The Chemist, 2* (1844), 2.

[62] Tom Sidney Moore and James Charles Phillips, *The Chemical Society, 1841–1941: A Historical Review* (London, 1947), p. 15. See also Derek Hudson and Kenneth V. Luckhurst, *The Royal Society of Arts, 1754–1954* (London, 1954). From the mid-1840's, the Society of Arts shifted from promoting specific scientific projects to fostering scientific education. The periodicals listing of the British Museum Catalogue is instructive on scientific periodicals, as is Samuel H. Scudder, *Catalogue of Scientific Serials of All Countries including the Transactions of Learned Societies in the Natural, Physical, and Mathematical Sciences, 1633–1876* (Cambridge, Mass., 1879).

[63] From 1838, there were formal engineering courses at King's College London and at the University of Durham; the Putney College for Civil Engineers opened in 1839. Chemical manipulation was taught in the engineering course at King's College and applied chemistry was taught at both of the other institutions, although it is not clear whether or not laboratory training was included (Roberts, *RCC*, pp. 91–99).

[64] The classic study of professionalization is A. M. Carr-Saunders and P. A. Wilson, *The Professions* (Oxford, 1933). The emergence of new professions as a response to (and agent of) the changing structure of nineteenth-century society is evaluated in Philip Elliot, *The Sociology of the Professions* (London,

Although one of the origins of chemistry was metallurgy, British metallurgical practice was for the most part carried on without resort to science. During the early 1840's, it was argued that unless production was improved through the use of science, Britain would lose its resource-based industrial lead and become dependent on foreign products; many of the problems requiring solution were chemical. It was also argued that science should be applied to a humanitarian problem made prominent by mining disasters in 1842 and 1843. Government reports on these disasters emphasized the necessity of scientific education for professional men connected with mining, and they compared Britain unfavorably with the continent in this regard.[65] The coal gas industry suddenly began to expand during the early 1840's; three new companies appeared in London alone in 1842 and 1843. With the additional competition, the industry, seeking ever larger markets and greater consumption, concentrated on making gas suitable for domestic lighting. The problems involved were mainly chemical; a thorough knowledge of organic chemistry, which was just beginning to sort out the products of the distillation of oils and coals, became increasingly important to the gas engineer. Indeed, an enquiry into faulty gas mains in the late 1840's resulted in the strong recommendation that every gas works employ a chemist, both for its own benefit and for the public's.[66]

1972). See also W. J. Reader, *Professional Men: The Rise of the Professional Classes in Nineteenth-Century England* (London, 1966). In 1845, the Institute of Civil Engineers had 513 members of various categories. The profession grew from 959 engineers in 1841 to 3,009 in 1851. See "Annual Report," *Minutes and Proceedings of the Institute of Civil Engineers, 4* (1845), 6. See also, *Census of Great Britain, 1841. Abstract of the Answers and Returns: Occupation Abstract, MDCCCXLI.* Part I, *Parliamentary Papers,* 1844 (587) xxvii–1. Part II, *Parliamentary Papers,* 1844 (588) xxvii–385; and *Census of Great Britain 1851,* Population Tables II. Ages, Civil Condition, Occupations and Birthplace of the People. *Parliamentary Papers,* 1852–1853, lxxxviii (I). Table 54, pp. cxxviii–xclix.

65 Royal College of Chemistry, *Supplement to the Proprietors of Mines,* pp. 4, 7; J. H. Morris and L. J. Williams, *The South Wales Coal Industry, 1841–1875* (Cardiff, 1958); O. O. G. M. MacDonagh, "Coal Mines Regulation: The First Decade, 1842–1852," in *Ideas and Institutions in Victorian Britain,* ed. Robert Robson (London, 1967), p. 58, discusses the motives of early mining regulations.

66 "City Commissioners of Sewers," *Patent Journal, 4* (1847–1848), 259–261. The state of the coal gas industry in the 1840's is discussed in Dean Chandler and A. Douglas Lacey, *The Rise of the Gas Industry in Britain* (London, 1949), pp. 73–76.

Thus, as even a cursory survey shows, in early-Victorian England the promotion of the study of practical chemistry was perceived to be relevant to the independent interests of a variety of groups. Indeed, given the diversity of chemical interests and the variety of existing opportunities to study chemistry, it is perhaps surprising that a scheme to establish a separate practical chemical school was suggested in 1843. However, in the first instance, this scheme was a personal project, the product of the efforts of individuals who saw the exploitation of the chemical issues of the period as a means of realizing their own ambitions. That the Royal College of Chemistry was eventually founded was due to their talent as entrepreneurs.

PROPOSALS FOR A SCHOOL OF PRACTICAL CHEMISTRY

The movement for establishing what eventually became the Royal College of Chemistry was initiated by two fairly insignificant early-Victorian figures. In November 1843, John Gardner and John Lloyd Bullock—by training an apothecary and a chemist and druggist, respectively—proposed to the Managers of the Royal Institution that a "Practical Chemical School" be appended to the existing Royal Institution.[67]

Typical entrepreneurs, Gardner and Bullock saw a possibility of great financial gain in the systematic application of science in their occupations. Gardner obtained the license of the Society of Apothecaries in 1829 and, from 1843, he suffixed "M.D." to his name. However, his degree was from the University of Giessen; his contemporary detractors suggested that it was one of that university's notorious "Doctors 'payable at sight'." During the early 1840's, Gardner translated from the German for *The Lancet,* and it may well have been in this capacity that he met Liebig. In any case, Gardner visited Giessen with Bullock during this period.[68] It was presumably

[67][John Gardner and John Lloyd Bullock], "For a Practical Chemical School," 1843, Royal Institution Archives. Hereafter cited as Gardner and Bullock, "RI, Proposal."

[68]The *DNB* lists Gardner as Licentiate of the Society of Apothecaries, but I could find no record of his application in the archives of the Society in The Guildhall Library, London. The first formal record I have found of Gardner's M.D. is in *The London and Provincial Medical Directory* (London, 1850). The unsavory reputation of the Giessen M.D. is discussed in *The Medical Directory of Great Britain and Ireland for 1845,* comp. and ed. by a "Country Surgeon

to further his pharmaceutical career that Bullock studied with Liebig during 1839; he also studied chemistry in Paris. Bullock was listed in contemporary postal directories as a "chemist and druggist" and "operative chemist." Those who styled themselves "operative chemists" generally took on one or more chemically related functions in addition to retailing; in short, they were the informal precursors of pharmaceutical chemists whose role the reformers of the 1840's hoped to institutionalize. Bullock, a member of the Chemical Society and the Pharmaceutical Society, was greatly concerned about the status of the pharmaceutical profession. During the early 1840's, he waged a vituperative battle with Jacob Bell, who opposed Bullock's idea of restricting Pharmaceutical Society membership to those with academic qualifications.[69]

Gardner and Bullock's proposal for establishing a practical chemical school was ambitious, and its adoption would have entailed considerable changes in the Royal Institution. The school was to have two departments, the first devoted to pure science, the second to applied chemistry. The former was to be administratively attached to the Royal Institution, sharing its physical plant; the latter was to be administered independently and located elsewhere, having only an informal "sister-school" relationship to the department of pure science. Like Liebig's university laboratory, Gardner and Bullock's pure science department was to be a center for training students in research methods through laboratory practice in chemical analysis. Ideally, such a program would equip students to follow any subse-

---

and General Practitioner" (London, 1845), pp. 665-666. The 1845 purchase price of a Giessen M.D. was £22. See also [G. L. M. Strauss], *Reminiscences of an Old Bohemian,* new ed. (London, 1883), p. 268. I thank W. H. Brock for bringing Strauss and his autobiography to my attention. Evidence for the Giessen visit is in B. Lespius, "Festschrift zur Feier des 50 jährigen Bestehens der Deutschen Chemischen Gesellschaft und des 100. Geburtstages ihres Begründers August Wilhelm Hofmann," *Berichte der Deutschen Chemischen Gesellschaft, 51,* pt 2 (1918), Sonderheft, photograph and caption opposite p. 8.

[69] Arnim Wankmüller, "Ausländische Studierende der Pharmazie und Chemie bei Liebig in Giessen," in *Tübinger Apothekensgeschichtliche Abhandlungen* (Stuttgart, 1967), *15,* 12. "The Late Mr. Bullock," *Chemist and Druggist, 66* (1905), 882-883, 886; J. Lloyd Bullock, "A Lecture on the State of Pharmacy in England and its Importance to the Public with Remarks on the Pharmaceutical Society," *The Chemist, 2* (1844), 276-282; Galen, "Passing Events," *Pharmaceutical Times,* 5 September 1846, pp. 17-18.

quent chemical career:

> Whether the object of the student be to qualify himself as a teacher of chemistry, to learn the bearing of that science on medicine and physiology, or to become a manufacturer, the same purely scientific education in the art of research is recommended to all. . . . [E]ven for directly practical purposes, the most purely scientific education is really the best, and is more certain to lead to improvements in practice than the most laborious experience in any one manufacture, gained as it generally is, at the expense of general principles.[70]

It was hoped that a suitably qualified (meaning German-trained) chemist would be appointed to a full-time salaried post as academic director of the department.

The financial issue was crucial; Gardner and Bullock insisted that funding, over and above students' fees, would be necessary for securing a full-time academic director. They argued that their proposed applied science department, which would be devoted both to research on applied problems and to vocational training, would help generate funds.

> In order to meet the especial exigencies of this country and at once to adopt the mature improvements of the best continental schools, we suggest that as an Appendage to the Scientific School, a prac-

[70] William Gregory, *Letter to the Rt. Hon. George, Earl of Aberdeen on the State of the Schools of Chemistry in the United Kingdom* (London, 1842), p. 24. The outline of the structure and activities of the pure science department drew heavily on this pamphlet by Gregory, one of Liebig's former British pupils and now Professor of Chemistry and Medicine at King's College Aberdeen. An appeal to the British government to subsidize laboratory instruction in practical chemistry in the Scottish universities, the pamphlet was perhaps the first description in English of the system of instruction developed and operated by Liebig in Giessen. Stressing the importance of chemical research to an industrial and commercial nation such as Britain, Gregory linked recent much-publicized German achievements in organic chemistry to the excellent facilities for studying the science at Giessen. He showed in some detail why laboratory instruction could not be self-financing, pointed to the government funds enjoyed by German universities, and urged the British government to follow the German example. It should be noted that, apart from the specific references to Giessen, the style of rhetoric and the arguments used by Gregory are similar to those used by D. B. Reid in *Remarks on the Present State of Practical Chemistry and Pharmacy with Suggestions as to the Importance of an Extended Practical Course* (Edinburgh, 1838).

> tical laboratory should be provided for the Application of Chemistry to Medicine, Arts and Agriculture.
>
> First because an adequate inducement should be held out for obtaining Subscribers and Students.
>
> Secondly, because it is desirable that the School should have a direct bearing upon the advancement of Pharmacy and Agricultural Chemistry. . . . In this department, the course of manipulation required by the Apothecaries' Company; the analysis of soils or commercial articles for subscribers; the preparation of all the articles in the Pharmacopoeia in a consecutive course; and afterwards the application of chemistry to the Arts, as Dyeing etc., might be taught if deemed expedient.[71]

Thus, the second department was to combine the functions of a department of technical chemistry (which Giessen had in the 1840's), a pharmaceutical institute (which Liebig ran at Giessen),[72] and an analytical practice. The proposal did not discuss the relationship between the teaching and research programs of the two departments except to emphasize that they should be kept separate to prevent the applied chemistry department from impeding the pure science department.

The emphasis on the separateness of the two departments is significant; the applied department would be in effect Gardner and Bullock's own school, the former serving as secretary and the latter as scientific director. That in 1843 they intended to use the applied department of the proposed Practical Chemical School for personal profit cannot be specifically documented, and, indeed, Bullock at least was genuinely concerned with providing scientific training for pharmacists. But however advantageous the school might have been

[71] Gardner and Bullock, "RI Proposal." Note the particular interests appealed to here: agriculture, possibly because of the Royal Institution's longstanding involvement with this subject (Berman, "Royal Institution," pp. 210–216), and pharmacy, Bullock's chief concern. A lecture course on chemistry, which was recognized by the Society of Apothecaries, had been given at the Royal Institution for some time. The proposed applied department would make it possible for students to fulfill the rest of the Society's requirement, and, conversely, the Society's practical chemistry requirement would attract students to the applied department.

[72] The importance of the pharmaceutical school which Liebig ran in parallel with his university course was pointed out to me by Bernard Gustin, Yale University, who is preparing a dissertation entitled "The German Chemical Profession, 1800–1867."

to the advancement of Gardner and Bullock's respective professions, their subsequent behavior indicates that they were well aware of its profit-making potential.[73] In charge of a department of applied chemistry affiliated, however informally, to the only English research school in chemistry, Gardner and Bullock were in an excellent position to take advantage of patentable discoveries. For example, in 1845, their entrepreneurial activity drew considerable criticism. In that year, Liebig discovered a process that converted a waste product into an alternative source of the expensive drug quinine. With Liebig's consent, the substance was secretly patented in England under Bullock's name because English patent laws prohibited foreign patentees. Meanwhile, the discovery was advertised as a scientific advance in *The Lancet,* while a consortium (including Gardner, Bullock, Liebig, and Hofmann) bought up the raw material. Later, according to the *Pharmaceutical Journal,* Gardner gave a speech at the College about the substance, inducing several manufacturers to try and make it; only then was it announced that the discovery had been patented previously. Contemporaries saw this attempted misuse of the College as a clear violation of the norms of science and business.[74]

In December 1843, the Managers of the Royal Institution rejected Gardner and Bullock's proposal after considerable discussion and internal political maneuvering.[75] Gardner and Bullock were undaunted, however; by the time the negative decision was announced, a provisional committee of their supporters, which included some

[73] In fact, Jacob Bell accused Gardner of mercenary motives for supporting the College ("Letter to the Editor: Vindication of Dr. Gardner," *Pharm. J., 6* [1846–1847], 150).

[74] "Amorphous Quinine," *Pharm. J., 6* (1846–1847), 160–172. For further references to this incident and their other entrepreneurial activities, see Roberts, *RCC,* pp. 143–150.

[75] Professors Faraday and Brande, "Report on the Proposed School of Practical Chemistry Delivered to the Managers of the Royal Institution, 19 December 1843," in Royal Institution, "Minutes of the Meetings of the Managers," 19 December 1843, Royal Institution Archives. The final decision is recorded in the "Minutes," 26 December 1843. See also "Correspondence on the Proposed Practical Chemical School," 1843, Royal Institution Archives. Two letters in this correspondence indicate that Gardner and Bullock's suspect personal motives may have influenced the outcome (Lord Prudhoe, letters to John Barlow, 13 December 1843 and 17 December 1843). See also Roberts, *RCC,* pp. 151–154.

members of the Royal Institution, had been assembled.[76] Up to the summer of 1844, Gardner and Bullock limited their activities to privately gathering support for a practical chemical school and to developing a more sophisticated proposal for making a public appeal. In June 1844, in an article attacking the Pharmaceutical Society, Bullock publicly announced plans to launch a school for the scientific training of prospective pharmaceutical chemists for improving the profession.[77] In July, Gardner and Bullock published their *Proposal for Establishing a College of Chemistry for Promoting the Science and its Applications to Agriculture, Arts, Manufactures, and Medicines.*[78]

The title, form, and contents of the new proposal showed that its authors had gained valuable promotional experience at the Royal Institution. The title immediately elevated the institution from a "school" to a "college," united what had been effectively two separate schools into a single institution, and clearly defined the groups from which support might be sought. Printed instead of hastily handwritten, the new proposal started with a list of thirty-five eminent public and scientific figures who were designated a provisional governing body;[79] a slate of officers, a London administrative office, and a London bank were listed as well. In an obvious effort to improve public relations, the proposal included not only specific plans for a college, but also propaganda for it derived mainly from Gregory's *State of the Schools of Chemistry in the United*

[76]Chambers, *Register of the Old Students of the RCC,* p. xlvii, suggests that Gardner and Bullock had been working to establish a practical chemical school since 1842 and that a provisional committee consisting of Sir Howard Elphinstone, the Earl of Essex, John Davy, Henry De la Beche, R. I. Murchison, Thomas Wyse, and the Professors Brande, Daubeny, and Gregory had been formed. There is confirming evidence that Gardner and Bullock had taken soundings on the idea by late 1842 (Royal College of Chemistry, *Supplement to the Proprietors of Mines,* p. 5); however, they definitely took the plan to the Royal Institution as their own with the help of Brande, Professor of Chemistry there. See also, Royal College of Chemistry, "Council," Preface.

[77]J. Lloyd Bullock, "Lecture on the State of Pharmacy."

[78]See above, n. 5; hereafter cited as Royal College of Chemistry, *Proposal.*

[79]Bullock's is the last name on the list of the provisional governing body; it is conspicuous as the only obscure name. Gardner appears on the title page as provisional Secretary. Evidently, those listed were not always aware of their membership; the merest expression of interest in the institution was sufficient for their names to go on the list ("Royal College of Chemistry," *Pharm. J., 6* [1846–1847], 157).

Kingdom of 1842 and Liebig's *Familiar Letters on Chemistry and its Relation to Commerce, Physiology, and Agriculture* of 1843.[80]

The proposal's propaganda began with the argument that chemistry was so central to all the pursuits of an integrated industrial and agricultural society that it was surprising to find Britain lagging Germany and France in the promotion of it. Emphasizing the theme of competition, the proposal attributed the German lead to the "establishment of schools, where not only practical and systematic instruction is given to students in qualitative and quantitative analysis, but where original researches are conducted, in concert by several individuals skilled in manipulation, and where the professors can work out their problems by the aid of many qualified hands."[81] In anticipation of the inevitable counter-suggestion that a new chemical institution would be superfluous given the variety of English opportunities to study chemistry (indeed, practical chemistry), the proposal argued that the College would have a unique function: it would make it possible for English students to prepare for a professional career in chemistry. Assuming that there were students who wanted to be professional chemists and, implicitly, employers who wanted their services, the proposal claimed that the only viable route to the profession was through systematic study in a research laboratory. To date, English students could learn to do research only by studying privately or by leaving England, both too expensive for the sort of student likely to be interested in chemistry. Arguing that student fees should be kept low at the College and citing Gregory on the probable cost of such an institution, the proposal appealed to both the general public and the government for extrinsic funding. In return, private individuals would receive some direct services from the new institution, and the nation would benefit generally. The national benefits were those that Liebig's school was said to have conferred on Germany; the proposal asserted that most of Liebig's students became manufacturers who were

[80]For Gregory, see above, n. 70. Gardner's edition of Liebig's *Familiar Letters* appeared just in time to support the Royal Institution proposal. Liebig's preface indicates that the purpose of the work was to arouse "the attention of governments and an enlightened public, to the necessity of establishing Schools of Chemistry" (Justus Liebig, *Familiar Letters on Chemistry and its Relation to Commerce, Physiology, and Agriculture,* ed. John Gardner [London, 1843], p. v).

[81]Royal College of Chemistry, *Proposal,* p. 3.

"silently but surely enabling the productions of Germany to compete with those of this country."[82] Furthermore, a body of science teachers and qualified pharmacists was spreading throughout Germany, serving as scientific advisors and accelerating economic progress.

Stressing current chemical preoccupations and issues, the proposal gave specific examples of the importance of chemistry to each of the groups from which it hoped to receive support. The proposal drew the attention of manufacturers to Liebig's discussion of the sulphuric acid industry in his *Familiar Letters.* There Liebig showed that sulphuric acid was a crucial industrial chemical and that the principal current method of manufacture had been discovered by scientific investigation. The proposal pointed out that there were still many problems within the industry, such as substituting indigenous sources of sulphur for foreign ones, that required chemical solutions. Several references to Liebig's work also appeared in the section of the proposal directed to agriculturists. Although it indicated that profit from the results of research in organic chemistry was definitely a long-term prospect, the proposal's rhetoric may have encouraged the agricultural reader to believe that an analyst would be immediately useful to him. Furthermore, the propaganda connected the repeal of the Corn Laws with the need for research and extended the theme of foreign competition to mineral resources, another of the British landowner's interests. The proposal suggested that chemical investigation of subsoil resources would reveal domestic sources of currently imported minerals. To tie in this aspect of landownership with the prosperity of manufactures and agriculture, the proposal pointed to Liebig's recent suggestion that phosphates, which were generally applied to crops in the form of expensive imported guano, were present and accessible in native fossil beds and could be exploited by chemical knowledge. Finally, the sections of the proposal that discussed the benefits of practical chemistry to medical men were cast in reformist terms. The propaganda hoped to attract the support of the new pharmaceutical chemist, the new broadly educated general practitioner (the M.B.), and the new M.D. by showing the relevance of analytical and organic chemistry to their current professional and scientific interests.

[82]*Ibid.,* p. 6. Proposed services to individual contributors included copies of College publications on applied science, cut-rate analyses, and the right to nominate scholarship students.

The proposal ended with a brief description of the intended divisions and functions of the College of Chemistry:

> 1st A Laboratory for original investigations, and for extending the boundaries of this most important national Science, on the model of the Giessen laboratory.
>
> 2nd A College for the instruction of students in analysis and scientific research, upon terms as to encourage young men of talent and scientific taste to apply themselves to Chemistry, and for qualifying public lecturers and teachers.
>
> 3rd Departments for the application of Chemistry to especial purposes, as Agriculture, Geology, Mineralogy, and Metallurgy, by the analysis of soils, rocks, etc.; to Medicine, Physiology, and the Arts.
>
> 4th The Employment of such means as may appear expedient to the Council for encouraging and facilitating the pursuit of Scientific Chemistry throughout the country, and for making it a branch of general education.[83]

To serve any one of these functions would have been a sufficient goal for a single institution; yet, to appeal to enough groups to gain adequate financial support as well as to achieve their personal aims, the initiators of the College of Chemistry proposed to institute them all.

THE CAMPAIGN FOR THE ESTABLISHMENT OF THE COLLEGE OF CHEMISTRY

Gardner and Bullock's *Proposal for Establishing a College of Chemistry* served as the basis of an unsuccessful appeal to the government for public funding in the autumn of 1844. Since the government had a fairly strong record of at least investigating the feasibility of scientific projects bearing on agriculture, and since the Prime Minister, Sir Robert Peel, was noted for his personal interest in scientific farming, Gardner persuaded landowning supporters of the College to approach the government informally by seeking Peel's approval. After reading the proposal and meeting with Liebig and other scientists, Peel decided against supporting a separate school devoted only to chemistry. He maintained that it would be

[83]*Ibid.*, p. 16.

preferable to "unite the study of chemistry with other branches of knowledge in the universities and existing public institutions formed for the Education of youth."[84] Peel's position was reasonable on both practical and academic grounds. Every facility suggested in the proposal was either planned or already available or could be made available with only slight modifications of existing institutions, some of which had government financing. Furthermore, to achieve the broad interscientific objectives of the proposed institution, the isolation of chemistry from other sciences would be counterproductive. Peel had a further objection; although he firmly believed in the importance of applied science, he believed equally firmly that it was the responsibility of individuals, not the government, to promote it. He argued that the successful establishment of the private Putney College for Civil Engineers proved that such institutions, if genuinely needed, could manage independently.[85]

After this informal appeal to the government had failed, nine of the thirty-five members of the provisional governing body met in January 1845 and decided to continue the project of establishing a college of practical chemistry as a completely private venture. Seven of those attending formed a Provisional Committee charged with transforming vague ideas and aspirations into a working institution.[86] At this stage the project passed out of the sole control of Gardner and Bullock. The former served as Secretary to the Provisional Committee and had considerable initiative and executive responsibilities; but policy decisions were generally made by Com-

[84] Sir Robert Peel, letter to Lord Ducie, 28 October 1844, British Museum, Add. MSS, No. 40553, f. 31. Peel felt so strongly on this point that he was barely persuaded to contribute to the College as a private individual (Sir James Clark, letter to Sir Robert Peel, 3 December 1845, British Museum, Add. MSS, No. 40580, ff. 186–190).

[85] Sir Robert Peel, letter to William Buckland, 19 November 1844, British Museum, Add. MSS, No. 40554, f. 233. Peel's attitude toward scientific farming is discussed in Norman Gash, *Sir Robert Peel: The Life of Sir Robert Peel after 1830* (London, 1972), pp. 678–682. Subsequently, in July 1845, by which time it was apparent that the College would in fact open, Thomas Wakley, with the support of his radical colleagues William Ewart and Thomas Wyse, attempted to persuade Parliament to fund the College of Chemistry (*Hansard, 82,* 18 July 1845, cols. 715–716).

[86] This committee met fifteen times and expanded its membership to fifteen over the next six months, until it was replaced by the permanent administrative structure which it organized. Membership and attendance are recorded in Royal College of Chemistry, "Council," *passim.*

mittee members. Initially, the Committee's principal task was to secure funds for the College by means of a propaganda campaign. Besides preparing advertisements for newspapers and specialist periodicals, it circulated a series of supplements to the initial proposal "addressed to such particular classes as may be supposed to be interested in the progress of Scientific Chemistry."[87] These supplements were far more specific than was the general proposal about the benefits that potential supporters might expect from the College. Individual supporters no longer needed to have faith that their investment in chemical education and research eventually would yield benefits to the community at large; the supplements promised them immediate personal chemical services.

The supplement directed to agriculturists reiterated Liebig's theme: it should be the ideal of the enlightened farmer to increase productivity, and the key to realizing this ideal was to chemically analyze the relationship between soil, plants, and fertilizers. It suggested that research be undertaken to resolve a current theoretical controversy (whether plants derived their nitrogen from the soil or from the air) and to solve a problem of immediate practical interest (whether or not potash, an essential ingredient in fertilizers, could be manufactured from sea water). Further, it promised a geochemical survey of Britain: "a part of the design of the College of Chemistry, and one which the Council trust that they shall be supplied with funds to carry out, is to devote a department to the analysis of soils etc.; so that, for example, a chemical knowledge of the soil on the surface, and of the strata immediately beneath it, may be obtained corresponding to the geological map of the country."[88] Thus, the whole of Liebig's analytical program would finally be available to English landowners; the British Association and the Royal Agricultural Society were already organizing plant-ash analyses, and the College of Chemistry would organize complementary soil analyses. Furthermore, in line with Daubeny's plan for a series of agricultural colleges, the supplement envisaged a

[87] Advertisements were sent to the *Mining Journal, The Lancet, Medical Review, Farmer's Journal, The West Briton, The Falmouth Packet, The Times* (London), *The Herald, The Morning Chronicle, The Medical Gazette, The Pharmaceutical Journal, The Economist, Gardners' Chronicle, British and Foreign Medical Review, The Chemist, The Athenaeum, The Medical and Chirurgical Review* (Royal College of Chemistry, "Council," *passim*).

[88] Royal College of Chemistry, *Supplement to Agriculturists,* p. 8.

nationwide network of analytical chemists coordinated through London. The establishment of such an analytical network to aid agriculture would also benefit provincial pharmaceutical chemists, who presumably would be called upon to perform the analyses; the College was to be the point of contact between agriculturists and pharmacists, which may well account for its appointment of twenty-nine Provincial Agents in agricultural (rather than industrial) centers throughout the country. At least twenty-six of the twenty-nine were chemists and druggists, and, judging from what is known of their practices, at least ten of these would have been designated "operative chemists" had they lived in the metropolis.[89]

The analytical network would also interest metallurgists and the proprietors of mines, to whom another of the supplements was directed. This appeal was cast in imperial, not merely national, terms: "an incalculable amount of mineral wealth exists in Great Britain and its Colonies, and also in India, concealed from its proprietors only for want of knowledge."[90] The supplement stressed that the application of chemistry to the problems of mining and metallurgy would forestall foreign competition. For example, it was within the province of chemists to develop extraction and refining processes for known, abundant, domestic sources of metallurgical raw materials that were previously imported. Furthermore, research might be done on possible uses for metals such as tungsten that were known to be in abundant supply, but which were not mined because no one knew what to do with them. Another advantage of the chemical investigation of metallurgical processes was that products valuable for other purposes such as agriculture might be discovered in waste materials. All of these examples were selected to direct attention to the importance of training research chemists as well as analysts. The supplement pointed out that although there might already be enough analysts for metallurgical purposes, there were not enough scientists: "An analysis is a totally

89Agents were also responsible for soliciting and collecting funds for the College. They were handpicked by Gardner and Bullock. Royal College of Chemistry, "Subscribers," lists agents in Swansea, Cambridge, Shrewsbury, Salisbury, Collumpton, Norwich, Dudley, Hastings, Llandidlo, Annan (Dumfries), Leicester, Stourbridge, Leominster, Hereford, Melton Mowbray, Blandford, Hexham, Penrith, Chelmsford, Oswystry, Handsworth, Mansfield, Newport (Mons.), Stockton-on-Tees, Richmond, Axminster, and Tamworth (the site of Peel's country seat, a cunning choice).

90Royal College of Chemistry, *Supplement to the Proprietors of Mines*, p. 1.

different thing to an investigation. When the proprietor of a mine sends a specimen of ore for assay, he is told pretty accurately the amount of marketable metals it contains, but no account is given him of the residue. It is precisely in this residue, and in the refuse matter of mines, that many valuable materials lurk, waiting to be discovered and applied by chemistry."[91] Thus, the need for the professional chemist was evident from yet another quarter.

Meanwhile, what amounted to a supplement directed to the medical profession appeared as a series of editorials in *The Lancet*. Its editor, Thomas Wakley, who was a vocal supporter of the movement for the reorganization of the medical profession, placed particular emphasis on the importance for medicine of the laboratory and the discipline of organic chemistry. He argued that although other schools provided satisfactory facilities for the study of chemistry for limited purposes, the College was needed as a general research institute.

> We announce with feelings of extreme satisfaction that a plan has been organised and measures have been taken to establish an institution for the promotion of the science of chemistry . . . a LABORATORY where young men have the opportunity of acquiring skill in all chemical operations upon such terms as will exclude no person of promising abilities and small means—a LABORATORY wherein researches may be pursued by medical men or chemists under an able superintendance, and in concert,—in a word, where the boundary of the science may be extended.[92]

To serve the interests of medical men and indeed of the other groups from whom support was sought, a network of Provincial Secretaries was distributed throughout the country as a complement to the network of Agents. The Secretaries were to keep local supporters in touch with scientific matters that arose at the College, whereas the Agents were to be more concerned with financial and commercial

[91] *Ibid.*, p. 4. For an assessment of the contemporary state of mining science, see Roy Porter, "The Industrial Revolution and the Rise of Geology," in *Changing Perspectives in the History of Science: Essays in Honour of Joseph Needham*, eds. Mikulas Teich and Robert Young (London, 1973), pp. 320–343.

[92] "Remarks on the New College of Chemistry," *The Lancet*, 7 September 1844, p. 736. See also "The Science of Chemistry," *ibid.*, 16 November 1844, pp. 231–232; "The College of Chemistry," *ibid.*, 11 October 1845, pp. 403–404.

matters. Thus, supporters of the College would be put in touch quickly with metropolitan developments, and the isolation, which many provincial members of the pharmaceutical and other metropolitan societies felt, would be eliminated. Conversely, the Secretaries could suggest projects and report on provincial scientific activities to the College, which in turn would act as a distribution center for scientific information.[93]

SUPPORT FOR THE COLLEGE OF CHEMISTRY

Although there are few documentable instances where supporters of the College of Chemistry explicitly stated the reasons for their support, there is ample documentation linking individual supporters with the issues, chemical or otherwise, contained in the College propaganda. My approach is to look at the general composition of some of the groups that supported the College, and to examine in more detail the interests of some of the key members of the Provisional Committee, the administrative body that realized the institution. I have classified according to occupation or principal source of income or sphere of interest approximately seventy-four percent of the 760 individuals who pledged financial support to the College of Chemistry.[94] Most highly represented were members of the various branches of the medical community and landowners; together they accounted for about half of the College's supporters and seventy percent of the donated funds. Landowners contributed

[93] Royal College of Chemistry, "Council," 10 July 1845 and 4 August 1845. Secretaries were located in Alfreton, Ashford, Beaconsfield, Bishop's Wearmouth, Brighton, Bromsgrove, Cockermouth, Coleford, Derby, Downpatrick, Dublin (3), Halifax, Hertford, Jamaica, Leeds (2), Liverpool, Sunderland, Torquay, Windsor, and York. Of those identified, nine were medical men, four were involved in the metallurgical industry, two were textile manufacturers, and two were Irish academic chemists. On the isolation of provincial medical men and the consequent importance of local scientific activities, see William H. McMenemy, *The Life and Times of Sir Charles Hastings, Founder of the British Medical Association* (London, 1959).

[94] Academics 31 (chemistry 9, geology 4, medicine 5, veterinary medicine 2, school science 5), architects 5, brewers 10, chemists and druggists 110, engineers 25 (gas engineers 8), financiers 4, journalists 2, landowners 153 (those with agricultural interests 135, those with metallurgical interests 18), lawyers and solicitors 20, manufacturers 70 (chemical 44, glass 4, paper 3, textiles 14), merchants 5, metallurgical producers (non-proprietary) 11, military men 15, physicians 65, surgeons 34 (Roberts, *RCC,* p. 194 and Appendix I).

twice as much money as medical men, but members of the latter group were most active in establishing the College and administering it as a private institution. Groups other than the landowners and medical men were important, too, for all contributions were vital to so precariously financed an institution. These positive features of the pattern of support for the College highlight an important negative feature: prominent academic chemists were conspicuously absent from the list of supporters.[95]

Medical men alone comprised some twenty-eight percent of the College's supporters. One third of them practiced in London, which partly explains their extensive committee work for the College. Almost two thirds of the physicians and surgeons had Scottish qualifications and several were involved in promoting the University of London medical curriculum or in chemically based research. About three fourths of the chemists and druggists who pledged subscriptions to the College were scattered throughout the provinces; their support can in some measure be attributed to their well publicized disappointment with the metropolitan Pharmaceutical Society. Many of the chemists and druggists who supported the College were not narrowly pharmaceutical in their interests; seventeen of the twenty-eight metropolitan practitioners were designated "operative chemists," and many of the provincial ones could have been. In addition, some of the provincial chemists and druggists were actively involved in local efforts to promote pharmaceutical reform through education.[96]

[95] Indeed, only a small proportion of the groups that might have been viewed as its potential constituency supported the College. For example, only 37 members of the Royal Society of Arts contributed ("List of Contributing members of the Society, corrected up to 31st December 1844," *Trans. Soc. Arts, 55* [1845], 1-xxiii). Only 86 of the approximately 1800 members of the British Association supported it ("A List of Members of the British Association for the Advancement of Science, March 1, 1845," appended to British Association, *Report,* 1844). Only 46 of the 196 members of the Chemical Society supported it ("A List of Officers and Members of the Chemical Society of London," in *Mem. & Proc. Chem. Soc., 3* [1845-1848]).

[96] Bristol Chemists' Association, "Letter to the Editor," *Pharm. Times,* 27 November 1847, p. 137. See also *Select Committee on the Pharmacy Bill of 1852, Parliamentary Papers,* 1852 (387)-xii, qq. 1286-1319; "The Leicestershire Association of Chemists and Druggists," *Pharm. J., 4* (1846), 389-390; J. K. Crellin, "Leicester and Nineteenth Century Provincial Pharmacy," *Pharm. J., 195* (1965), 417-420.

Turning to the landowning supporters, the proposed College of Chemistry offered them little that was not, at least in theory, available through other institutions in England. It would provide an English parallel to the Agricultural Chemistry Association of Scotland, which was established in 1843, and the College's activities could supplement the training in agricultural analysis at the Royal Agricultural College and the analytical service of the Royal Agricultural Society. The Royal Agricultural Society's official attitude toward agricultural experimentation was one of constructive conservatism; it preferred to communicate the results of successful field trials rather than to promote speculative experiments. Its view was that the scientific principles of agriculture could only be discovered through farming, not through laboratory research. In 1844, it was justifiably skeptical about the immediate application of organic chemistry to agriculture, maintaining that the agricultural results could only be as valid as the scientific principles applied, and that the principles of organic chemistry were less than well established.[97] By supporting the College of Chemistry, members of the Royal Agricultural Society could help bring about the future utility of organic chemistry without risking the Society's resources. At the same time they were furthering a project based on inorganic chemistry—the soil analysis scheme—to which the Society was already committed, and they would receive private benefits as well. Of the landowners and farmers who supported the College, 108 belonged to the Royal Agricultural Society and 32 of them were officers or members of its Council. Further, 33 of them also supported the Royal Agricultural College.[98]

[97] Royal Agricultural Society, "Minutes of General Meetings," 14 December 1844, Royal Agricultural Society Archives.

[98] "List of Governors and Members to December 1845," *J. Roy. Ag. Soc., 6* (1845), 3–72. Royal Agricultural Society, "Council Minutes," 23 May 1842 gives the Society's total membership as 5,834, more than double the 1840 figure (*ibid.,* 6 May 1840). "List of Shareholders," Cirencester, Royal Agricultural College, *Prospectus, 1846* (Cirencester, 1 May 1846), pp. 9–15. Another twelve non-agricultural supporters of the College of Chemistry also belonged to the Royal Agricultural Society: seven proprietors of mines or metallurgical works, one gas engineer, two operative chemists, and two physicians. The last four all performed commercial analyses and the gas engineer promoted the use of the by-products of gas manufacture as fertilizers.

Given the emphases in the propaganda for the College, the strength and nature of the medical and landowning support are not surprising. The support highlights the inadequacy of Hofmann's evaluation of the College of Chemistry solely as a teaching and research institution. Furthermore, had the College been, as he suggested, a "natural" response to the stimulus of continental scientific and pedagogic developments, prominent English academic chemists might have been expected to support it more enthusiastically. Some, of course, may have feared the effects of such an institution on their own enrollments. Or, like Daniell, Graham, and Faraday, others may have feared that competition from such an institution would have forced them to teach unremunerative and time-consuming practical courses themselves.[99] Chemists may have objected to the obvious medical ties of the College, jealous of the autonomy that chemistry had recently achieved as a discipline. They may have objected, too, to the blatant commercialism of Gardner and Bullock, even though it was considered proper, and indeed financially necessary, for academic chemists to do consulting work for extra funds. In any case, even Daubeny and W. T. Brande, Professor of Chemistry at the Royal Institution, who initially had been enthusiastic, quickly lost interest in the College.

The interests of the members of the Provisional Committee are particularly important, for this was the group that did the work of seeking support for the College between January and July 1845.[100] The three chemists on this committee were marginal figures in terms of their scientific status. At least two of them promoted the College for professional as well as scientific reasons. Robert Warington, a principal founder of the Chemical Society, first studied chemistry

[99]For Daniell and Graham, *supra,* pp. 443-445. See also Thomas Graham, letter to J. Graham, 29 October 1844, in Smith, *Thomas Graham,* p. 45. Had a practical school been appended to the Royal Institution, Faraday would have been forced to be its director against his will (Lord Prudhoe, letter to John Barlow, n.d., "Correspondence on the Proposed Practical Chemical School," 1843).

[100]The Provisional Committee consisted of two physicians (Sir James Clark and Thomas Bevan), one surgeon (Alexander Nasmyth), one pharmaceutical chemist (John Lloyd Bullock), five "improving" agriculturists (William Bingham Baring, James Smith of Deanston, Sir Howard Elphinstone, the Earl of Essex, and James Adam Gordon), one gas engineer (George Lowe), one linen manufacturer (J. Marshall), three chemists (Robert Warington, David Boswell Reid, and Robert Porrett), and one radical educationalist (William Ewart).

as an apprentice to a chemical lecturer and manufacturer of fine chemicals. In 1828 he became Edward Turner's assistant at University College and did some extramural practical teaching in association with Turner's course. In 1831 he became a brewer's chemist, and finally in 1842, after three years with no official post, he was appointed chemical operator to the Society of Apothecaries; in the latter post he was responsible for dispensing operations, managing at the same time to do some research on organic chemistry. Warington was, not surprisingly, disillusioned with the state of chemistry. Improving chemistry professionally, by bringing science and practice into closer communication, was certainly one of his aims in promoting the establishment of the Chemical Society in 1841. Warington may well have viewed the rising professional aspirations of chemists and druggists as likely to subsume the practice of chemistry under pharmacy. Disapproving of the training then available, he undertook the education of his sons himself, introducing them to the laboratory at a very early age. It may have been in deference to his views that the Chemical Society's original plans included the establishment of a research laboratory. Warington even advised the young F. A. Abel, subsequently one of the College of Chemistry's most illustrious alumni, not to study chemistry because the professional prospects were so poor.[101]

Less is known about Robert Porrett, a typical amateur chemist employed as a civil servant. Before 1820 he did some sound, but pedestrian, work on inorganic acids; he did no more research until the 1840's, when he began to work on guncotton. His reawakened interest in chemistry and his friendship with Warington as well as his professional isolation may partly explain his support for an institution promising a supervised research laboratory for members' use.[102]

David Boswell Reid, the third chemist, had an unfortunate professional history. Trained in Edinburgh and subsequently employed there to teach practical chemistry as Hope's assistant, Reid lobbied

[101] J. H. S. Green, "Robert Warington (1807-1867)," *Proc. Chem. Soc.* (September 1957), pp. 241-246; "Robert Warington," *Proc. Roy. Soc., 16* (1867-1868), 1; Sir F. A. Abel, "History of the RCC," p. 581. On Warington's role in the establishment of the Chemical Society and the Society's original objects, see R. Warington, Jr., "The Foundation of the Chemical Society," in *The Jubilee of the Chemical Society of London* (London, 1896), pp. 115-122.

[102] *DNB;* "Obituary: Robert Porrett," *J. Chem. Soc.,* 7 (1869), vii–x. Porrett was a founding member of the Chemical Society.

vociferously and unsuccessfully for the establishment of a separate chair of practical chemistry at the University. His goal, apart from securing his own professional future, was to free the discipline of chemistry from its smothering association with medicine. In 1837, he was an unsuccessful candidate for the chemical chair at University College London, and in 1838 he lobbied the Royal College of Surgeons of Edinburgh to make their practical chemistry requirement more extensive.[103] In the early 1840's, Reid worked on the chemistry of public health problems in London. Under the auspices of the Privy Council's Committee on Education, he gave a course of lectures in 1842 entitled "Chemistry of Daily Life," arguing that it was essential for the general public to have a knowledge of such topics as pollution and air purification so that they could improve their own living conditions. In 1843, he became president of the Teacher's Scientific Association and organized the curriculum for its London normal school, which promoted "the general introduction of the elements of the sciences in the schools, in particular those facts and principles which explain what science has done for the comfort and economy of daily life; and which form the foundation of all measures for the improvement of health."[104] The Association's plans included a school for instructing teachers in practical science and experimental demonstrations, an experimental laboratory for members' research, and a scientific library. The College of Chemistry would promote scientific education by similar means and thus would fulfill many of the goals Reid had sought over several years.

Bullock, the only pharmaceutical member of the Provisional Committee, was on the periphery of both the chemical and the medical groups and prominent in neither. The most important medical man on the Provisional Committee was the Queen's

[103] Morrell, "Practical Chemistry," pp. 72–78; D. B. Reid, *Remarks on Practical Chemistry.* After his fulminations about the subservience of chemistry to medicine, Reid's search for support for the teaching of practical chemistry from the medical community is ironic.

[104] "Teachers' Scientific Association," *Eng. J. Ed., 1* (1843), 158. Dr. David Boswell Reid, letter to Sir Robert Peel, 8 March 1842, British Museum, Add. MSS, No. 40503, ff. 358, 360–361. See also, David Layton, *Science for the People: The Origins of the School Science Curriculum in England* (London, 1973), pp. 52–53.

physician, Sir James Clark.[105] He was a noted medical reformer, credited with being the chief architect of the University of London medical curriculum. Clark's reformism may well have been motivated by his own tortuous route to prominence as a medical practitioner. He took an Edinburgh M.D. in 1817 after eight years as a naval surgeon. From 1819 to 1826, he practiced on the continent, owing his eventual status as a London physician and his royal appointment to contacts made abroad.[106] Disappointed with the progress of medical reform by 1842, Clark felt that the University of London was too inefficient to achieve its goal of becoming a licensing board to replace the three traditional bodies. In pressing his scheme of reform on the Home Secretary, he expressed concern for the pharmaceutical as well as the medical profession, arguing that if the activity of the general practitioner were narrowed to that of a doctor, then pharmaceutical activity could be relegated solely to trained chemists and suitable educational qualifications could be required of them. Clark felt that it was essential for the pre-medical student to master the basic principles of chemistry to comprehend "the more complicated processes of that vital chemistry which is continually in action in the living body";[107] after that, in his subsequent medical curriculum, the student would learn the applications of chemistry to pathology, physiology, and therapeutics. Furthermore, Clark believed that it was only through training in practical chemistry that the student could master the basic

[105] Little is known about the second physician, Thomas Bevan, beyond that he qualified in Edinburgh, was physician to the Islington Dispensary, and was a member of the Linnean and Hunterian societies.

[106] "Sir James Clark," *Proc. Roy. Soc., 19* (1870-1871), xiii-xix; *DNB;* William Munk, *The Roll of the Royal College of Physicians of London,* 2nd rev. and enl. ed. (London, 1878), *3,* 222-226; "Sir James Clark," in Harley Williams, *The Healing Touch* (London, 1949), pp. 50-92. On Clark's view that English collegiate restrictions hampered the careers of Scottish-trained physicians, see *Report from the Select Committee on Medical Education: I. The Royal College of Physicians. Parliamentary Papers,* 602-I (1834), qq. 3724-3732.

[107] Sir James Clark, *Remarks on Medical Reform in a Second Letter Addressed to the Rt. Hon. Sir James Graham* (London, 1843), pp. 15, 18-20. See also *Select Committee on Medical Education: I,* qq. 3692-3695, 3698, 3739; and Sir James Clark, *Remarks on Medical Reform in a Letter Addressed to the Rt. Hon. Sir James Graham* (London, 1842), pp. 17, 42.

principles.[108] The chemical program at the proposed College of Chemistry clearly fitted into Clark's program for medical reform: it would help train pharmaceutical chemists; it would provide pre-medical students with basic training in chemistry; it would provide specialized training in physiological chemistry; and it would help ease the London chemical professors' burden of practical teaching.[109]

Although five Provisional Committee members had agricultural interests, William Bingham Baring was the only one of them to participate actively in planning the College of Chemistry. His support for the College was probably motivated as much by his keen interest in educational projects as by his commitment to agricultural improvement. From 1828 to 1834, he was active in the Society for the Diffusion of Useful Knowledge, which provided cheap textbooks for the poor.[110] The Society gave considerable publicity to the work of two Swiss educational innovators, J. H. Pestalozzi and Emmanuel von Fellenberg. It was common for English gentlemen to visit particularly Fellenberg's institution as part of a postgraduate continental tour, and there is evidence that Baring did so. In the early 1850's, he became involved in the movement for the Teaching of Common Things, and his ideas on it, as presented in a pamphlet launching a prize scheme for teachers, are readily traceable to those

[108] Sir James Clark, letter to Sir John Lubbock, [1838], Lubbock Papers, LUB-c. 260 bis, Royal Society Archives.

[109] The surgeon on the Provisional Committee was Alexander Nasmyth, a friend and perhaps protege of Clark's who had received specialist training in dental surgery in Edinburgh. His research on the pathology of teeth drew extensively on chemistry. He was an inveterate joiner of scientific societies and he lived conveniently next door to the original premises of the College (Dr. K. P. H. Marx, *Erinnerungen an England* [Braunschweig, 1842], pp. 121, 203; *Plarr's Lives of the Fellows of the Royal College of Surgeons of England,* rev. ed. Sir D'Arcy Power [London, 1930], *2,* 88).

[110] William Bingham Baring, letter to Thomas Coates, 7 March 1833, Society for the Diffusion of Useful Knowledge, "Letter Book," University College London, College Archives. During his membership, Baring served on several committees: General Committee, Committee for Entertaining Knowledge, Committee for the *Journal of Education,* Committee for Statistics on Trade, and the committee sponsoring the "Farmer's Series." Committee members are listed in the unpublished dissertation (University College London, 1933) by Monica C. Groebel, "The Society for the Diffusion of Useful Knowledge," *4.*

of Fellenberg.[111] He advocated teaching science to the poor to promote a contented, less socially disruptive laboring class. He argued that the application of scientific knowledge would ease daily life for workers and that the process of acquiring it would develop their skill and ingenuity, while enabling them to understand God's design and their place in it. Moreover, Baring argued that it was not only the poor who needed scientific training: "If we wish to hold our rank among nations, if we intend to maintain that manufacturing ascendancy which is the chief source of our national strength, we must carry this study of common things not only to the schools of the poor, but to our colleges and universities."[112] In the movement for the Teaching of Common Things, chemistry, because of the direct bearing it had on rural and urban life, was the crucial science. An institution such as the College of Chemistry would not only train teachers for the poor, but it would also provide advanced instruction for those charged with maintaining England's industrial supremacy.[113]

During the 1840's, Baring was also interested in agricultural chemistry, particularly in the problems surrounding the use of a

[111] Lord Ashburton [William Bingham Baring], *Ashburton Prizes for the Teaching of 'Common Things;' An Account of the Proceedings of a Meeting Between Lord Ashburton and the Elementary Schoolmasters Assembled at Winchester, 16th December 1853* (London, 1854); his visit to Fellenberg's school is suggested in *ibid.*, p. 8. The influence of Pestalozzi and Fellenberg on English education is discussed in W. A. C. Stewart and W. P. McCann, *The Educational Innovators, 1750–1880* (London, 1967), pp. 136–143, 211–213.

[112] Ashburton, *Ashburton Prizes*, p. 12. Baring was also neighbor to Richard Dawes' King's Somborne School, a famous experiment in the Teaching of Common Things (David Layton, "Science in the Schools: The First Wave—A Study of the Influence of Richard Dawes," *Brit. J. Educational Studies, 20* [1972], 38–57; *idem, Science for the People*, pp. 49–53).

[113] William Ewart was another Provisional Committee member with interests in education. Although he is now best known for his campaign for public libraries, he was also concerned with professional education. He was an active member of the Central Society of Education, which was set up during the late 1830's to survey education in England and abroad; the role that science might play at all levels of education was one of the Society's principal concerns. See Central Society of Education, "Prospectus," in Central Society of Education, *Papers* (London, 1838). Biographical information is from *DNB*, and W. A. Munford, *William Ewart, M. P., 1798–1869: Portrait of a Radical* (London, 1960). See also questions posed by Ewart to witnesses in *Select Committee on the Pharmacy Bill of 1852, passim.*

fertilizer known as "sewer manure." The fertilizer was suggested by Edwin Chadwick in his 1842 *Report on the Sanitary Condition of the Labouring Population of Great Britain.* He supported his proposal by referring to Liebig's *Agricultural Chemistry,* arguing that sanitary reform could be virtually self-financing since profits from the sale of town sewage to agriculturists could fund the drainage systems he advocated. The goal of the Health of Towns Commission, whose reports in 1844 and 1845 were widely publicized, was to test the practicality of Chadwick's scheme. By 1844, Chadwick had become unwilling to wait for the public sector to take up his scheme; he began to organize a private company which would finance drainage systems and reap the eventual profits from the sale of sewer manure. The improvement of towns would be a by-product of a fertilizer business.[114] Baring and three other members of the Provisional Committee, James Smith of Deanston, Sir Howard Elphinstone, and the Earl of Essex, all backed either Chadwick's company or its principal rival.[115] The implementation of a sewer manure scheme clearly would have required an army of well-trained medical, scientific, and technical personnel, particularly analysts for testing the purity of water supplies as well as the quality of fertilizers

[114]Edwin Chadwick, *Report on the Sanitary Condition of the Labouring Population of Great Britain,* ed. M. W. Flinn (Edinburgh, 1965), pp. 55, 121–122; Liebig, *Agricultural Chemistry,* pp. 189–203; S. E. Finer, *The Life and Times of Edwin Chadwick* (London, 1952), pp. 29, 224, 233.

[115]For Baring's involvement, see *Report from the Select Committee on Metropolitan Sewage Manure; Together with the Minutes of Evidence, Appendix and Index. Parliamentary Papers,* 1846 (474)–x–535, p. iv. For Smith, see James Smith of Deanston, "General Observations of the Sanatory State of Large Towns and the Means of Improvement," in *Report on the State of York and Other Towns. Health of Towns Commission* (London, 1845), pp. 26–37; *idem,* "On the Application of Sewer Water to Purposes of Agriculture with a View to the Establishment of an Independent Income for the Improvement of Towns," *ibid.,* pp. 38–45. James Adam Gordon, the fifth "agricultural" member of the Committee was an associate of Smith's. On Elphinstone, see Lyon Playfair, letter to Edwin Chadwick, 29 April 1842, Chadwick Papers, University College London, College Archives. On Essex, see *First Report from the Select Committee on the Sewage of Towns, together with Minutes of Evidence and Appendix. Parliamentary Papers,* 1862 (160)–xiv, qq. 1–44. Essex also experimented with artificial fertilizers (A. A. Croll, *Important to Agriculturists: Croll's Pure Sulphate of Ammonia and Prepared Essence of Guano* [London, 1844], p. 9). Three Provisional Committee members, Sir James Clark, D. B. Reid, and James Smith, served on Chadwick's Health of Towns Commission.

produced. The proposed College of Chemistry would be important both as the focus of a national analytical network and as a training center for analysts.

The final two Provisional Committee members were a gas engineer and a linen manufacturer. George Lowe, head of the Chartered Gas Company in London, was interested in devising chemical means for purifying coal gas and improving its illuminating power. He actively supported his assistant's work on the use of the by-products of gas manufacture as fertilizer, and he was a regular participant in a number of metropolitan scientific and technical institutions. The linen manufacturer, J. Marshall of Leeds, did not attend any of the Provisional Committee's meetings, although he might have served the College in other ways. After the College opened, he sent in an applied research topic on the cultivation of the flax plant.[116]

CONCLUSION

The Provisional Committee was responsible for recruiting supporters for the College of Chemistry and, not surprisingly, the range of chemical interests of the supporters in general was as diverse as that of the Committee members. In terms of the total number of supporters and the amount of capital it attracted, the College of Chemistry was a small project.[117] However, its significance for the historian does not lie in its scale, but in the way that the project was realized. The varied motives of the founders of the College, and particularly the still key role of medical interests in the 1840's, illustrate the complexity of the social relations of early-Victorian chemistry. To some extent the orientation given to the College by its founders substantiates Liebig's contemporary evaluation of early-

[116]On Lowe, see Stirling Everard, *The History of the Gas Light and Coke Company. 1812–1949* (London, 1949), and Bennet Woodcroft, *Alphabetical Index of Patentees of Inventions, March 2, 1617 to October 1, 1852* (London, 1854), nos. 6179, 6276, 8298, 8883, 11228, 11405. On Marshall, see W. G. Rimmer, *Marshalls of Leeds, Flax Spinners, 1788–1886* (Cambridge, 1960); J. E. Mayer and J. S. Brazier, "Analysis of the Mineral Constituents of the Flax Plant and of the Soils on which the Plants had been Grown," *Quart. J. Chem. Soc., 2* (1848–1849), 78–90.

[117]The College did not attract support in the same degree as did, for example, the Royal Agricultural College at Cirencester, which amassed 500 pledges of £30 before opening (Cirencester, Royal Agricultural College, *Prospectus*, p. 8).

Victorian science: "What struck me most in England was the perception that only those works which have a practical tendency awake attention and command respect; while the purely scientific, which possess far greater merit, are almost unknown. . . ."[118] But the establishment of the College of Chemistry also shows that English perceptions of science were not so narrow as Liebig's derogatory comment implied, for it was not merely potentially remunerative "practical" results that attracted supporters. More important, chemistry was seen as directly relevant to fundamental concerns of early-Victorian society: it would assist traditional groups to maintain their status, while simultaneously providing the expertise and methodology by which new, professional groups could gain status.

That it was necessary for the College of Chemistry to serve so many purposes illustrates the diversity of the institutional expectations (and past disappointments) of the early-Victorian scientific constituency. The College began as a personal venture; but it owed its eventual establishment to the reluctance, often for sound reasons, of existing English scientific institutions to deal with early-Victorian chemical issues. The efforts to establish the College helped to articulate these issues and prompted some hesitant contemporary institutions to begin to deal with them.[119]

Furthermore, the terms in which the College of Chemistry was put over to the public indicate that a professional role for the chemist, distinguishing him from workers in fields that used chemistry, was emerging against a background of English social change. Once established, the College, through its efforts to serve the diverse aspirations of its supporters, extended the institutional

[118]Bence Jones, *The Life and Letters of Faraday* (London, 1870), *2,* 188–189. Liebig's comment was made in 1844.

[119]Both University College London and King's College London provided teaching laboratories and professorships of practical chemistry from 1846 as a direct response to the competition of the new College of Chemistry (University College London, "Council Minutes," 8 February 1845, and University College London, "Committee of Management Minutes," 24 September 1845; King's College London, "Minutes of Council," 1 August 1845). It is also tempting to view the belated establishment of a teaching laboratory by the Pharmaceutical Society in 1844 as a response to the College's propaganda (*supra,* pp. 489, 491, 500). Finally, the College of Chemistry itself was taken over by the government in 1853 and incorporated in its Metropolitan School of Science which included the Museum of Economic Geology.

framework through which a profession based on the science of chemistry evolved in England.[120]

ACKNOWLEDGMENTS

I am very grateful to the following institutions for granting me access to their manuscript materials and, where appropriate, for permitting me to quote from them: Bodleian Library, British Museum, Imperial College London, King's College London, Norwich City Library, Public Record Office, Royal Agricultural Society, Royal Institution, Royal Society, University College London, and University of London. I am personally indebted to the archivists and librarians of these institutions, particularly to Mrs. Jeanne Pingree, College Archivist, Imperial College London, for their patient and enthusiastic help.

[120] The Royal College of Chemistry itself and the extent to which it achieved its diverse objectives are investigated in Roberts, *RCC,* chapters 6-8. This subject will form the basis of another paper.

NOTE ON CONTRIBUTORS

JOAN BROMBERG is Visiting Lecturer in the Department of History and the Institute for Fluid Dynamics and Applied Mechanics at the University of Maryland. She is presently preparing a taxonomy of approaches to atomicity in twentieth century physics.

R. G. A. DOLBY is Senior Lecturer in the Unit for the History, Philosophy and Social Relations of Science at the University of Kent at Canterbury, England. His main research interests are in the philosophy of scientific change and historical sociology of science.

STANLEY GOLDBERG is Associate Professor for the History of Science at Hampshire College, Amherst, Massachusetts. His research interests center on the history of early twentieth century physics and the relationship between the physics community and other social institutions. He is currently preparing a biographical study of Wilhelm Wien. In 1975 he was a postdoctoral fellow at the Smithsonian Institution.

HENRY GUERLAC is Professor of the History of Science and Director of the Society for the Humanities at Cornell University. He is the author of *Lavoisier–The Crucial Year* and numerous articles on the Chemical Revolution and other topics in the modern physical sciences.

P. M. HEIMANN is Lecturer in the History of Science at the University of Lancaster. He is working on eighteenth and nineteenth century natural philosophy, with particular emphasis on concepts of matter, force, and energy.

LEWIS PYENSON teaches at the Institut d'Histoire et de Sociopolitique des Sciences of the University of Montreal. He is interested principally in the social history of contemporary physical sciences.

GERRYLYNN K. ROBERTS is doing postdoctoral research on the social history of science in the Faculty of Arts at the Open University. She is currently working on the professionalization of chemistry.

ROMUALDAS SVIEDRYS teaches history of science and technology at the Polytechnic Institute of New York. His research centers on the sociology of nineteenth century science.

IN MEMORIAM:

Tetu Hirosige (1928–1975)

The many friends, colleagues, and followers of Tetu Hirosige, a member of the editorial committee of this journal, were shocked and saddened at the news of his death on the evening of 7 January 1975.

Dr. Hirosige may be said to have dedicated his twenty-year career as a historian of science to the establishment of the history of science as a field of learning in Japan. His researches were of a breadth and depth rare in Japan and in the world at large. He approached the history of science from two facets—the internal history of physics and the institutional history of science in modern Japan.

In his study of the history of physics, he directed his attention to the revolution in physics in the nineteenth and early twentieth centuries, with a special emphasis on the historical development from nineteenth century electromagnetism to the special theory of relativity. (The lead article of this issue is the fruit of that study.) He also did painstaking research on the history of the theory of atomic structure in that same period. A series of his research papers on these subjects was published in *Kagakusi Kenkyu* (*The Journal of the History of Science Society of Japan*), which Dr. Hirosige edited for the past four years. The major papers of that series were translated into English (some of them in revised form), and all but two of the translations have been published in *Japanese Studies in the History of Science.* The remaining translations are his paper on the formation of Lorentz' theory of electrons, which appeared in Vol. 1 (1969) of this journal, and the lead article in this issue. Dr. Hirosige also leaves us *Butsurigakusi* (*History of Physics,* 1968), the most authoritative book on its subject in the Japanese language.

His work in the institutional history of science began with *Sengo Nihon no Kagaku Undo* (*The Postwar Movement of Scientists in Japan,* 1960)—which also marked the beginning of his research on the internal history of physics—and culminated in his *Kagaku no Shakaisi* (*A Social History of Science: The Social System of Science in Modern Japan,* 1973). He set forth unique interpretations of the institutionalization of science and the integration of science into the Japanese societal system, basing his work on abundant source materials and precise analysis, just as he did in his study of the internal history of physics.

For Dr. Hirosige, these researches were closely related. He sought to ascertain the relationship between changes in the societal form of science and changes in the ideas and substance of science. His search was also related to his thorough criticism of modern science. Moreover, he was well aware that the same criticism may be justly directed to the study of the history of science as a field of learning. That he especially chose the history of modern physics as his research target was not unrelated to this awareness. (See the concluding remarks of his paper in this journal.)

Dr. Hirosige felt that the study of the history of modern physics has recently flourished, and that if it can be said that "the history of modern physics today is, as an active field in the history of science, comparable with the study of the scientific revolution in the sixteenth and seventeenth centuries," then it is time to "establish key factors which we would all agree to regard as critical in a historical analysis of the genesis and development of modern physics." Shouldering the responsibilities of program director of the organizing committee for the XIVth International Congress of the History of Science held in Japan last summer, Dr. Hirosige devoted his last energies to organizing and making a success of the symposium entitled "Critical Problems in the History of Modern Physics."

Dr. Hirosige was highly respected by others in his field in Japan and throughout the world, and his views on science have influenced students, scientists, and—through his numerous newspaper and magazine essays and his book for the layman—general readers alike. These continuing influences assure that his name will be long remembered.

Sigeko Nisio
13 March 1975